T0334280

Invasion Dynamics

Cang Hui

Centre for Invasion Biology, Department of Mathematical Sciences, Stellenbosch University; African Institute for Mathematical Sciences

AND

David M. Richardson

Centre for Invasion Biology, Department of Botany and Zoology, Stellenbosch University

UNIVERSITY PRESS

UNIVERSITY PRESS

Great Clarendon Street, Oxford, OX2 6DP,
United Kingdom

Oxford University Press is a department of the University of Oxford.
It furthers the University's objective of excellence in research, scholarship,
and education by publishing worldwide. Oxford is a registered trade mark of
Oxford University Press in the UK and in certain other countries

First Edition published in 2017

Impression: 1

Published in the United States of America by Oxford University Press
198 Madison Avenue, New York, NY 10016, United States of America

British Library Cataloguing in Publication Data
Data available

Library of Congress Control Number: 2016946832

ISBN 978–0–19–874533–4 (hbk.)
ISBN 978–0–19–874534–1 (pbk.)

DOI 10.1093/acprof:oso/9780198745334.001.0001

Printed and bound by
CPI Litho (UK) Ltd, Croydon, CR0 4YY

Acknowledgements

The year 2008 marked the peak of the most recent global financial crisis. In November of the same year we, and 135 other scientists from around the world, gathered in Stellenbosch, South Africa, to discuss another man-made crisis at a symposium hosted by the DST-NRF Centre of Excellence for Invasion Biology (C·I·B). That gathering commemorated the 50th anniversary of the publication of the iconic book, *The Ecology of Invasions by Animals and Plants*, by Charles Elton and reflected on progress in the scientific study of biological invasions over the last half century. Earlier that year, one of us (CH) joined the staff of the C·I·B at Stellenbosch University, while the other (DMR) was already ensconced at the C·I·B. as Deputy-Director: Science Strategy. We have been working together since then and have collaborated on more than 20 papers on a wide range of topics relating to invasions.

Over the years we have seen a rapid growth of publications documenting invasion dynamics. Since at least 2010 we talked occasionally about collaborating on a book, often in the corridor between our offices in the Stellenbosch hub of the C·I·B. When CH moved to his new office in Stellenbosch University's Department of Mathematical Sciences in 2014 to take up a research chair in mathematical biology we saw new opportunities to merge insights from current approaches in invasion ecology with issues more frequently discussed among mathematicians. Hence our interest in aspects of ecological complexity, and the decision to write this book on invasion dynamics.

During the writing of the book we received extremely helpful comments and suggestions on chapters (or parts of chapters) from Brett Bennett, Jane Catford, Jon Chase, Jane Elith, Janet Franklin, Mirijam Gaertner, Laure Gallien, Ruben Heleno, Steve Higgins, Jonathan Jeschke, Sabrina Kumschick, Guillaume Latombe, Sandy Liebhold, Jaco Le Roux, Mark Lewis, John Measey, Petr Pyšek, Tony Ricciardi, Wolf Saul, Nanako Shigesada, Rick Shine, and John Wilson. We are extremely grateful for all the generous comments and suggestions, not all of which we could accommodate. Any errors that remain are entirely ours.

Many ideas addressed in the book arose in collaborations and discussions with a large number of colleagues, postdoctoral fellows, and graduate students, many of them part of the extended 'C·I·B family' including: Karen Alston, Cecile Berthouly-Salazar, Oonsie Biggs, Tim Blackburn, Matt Brooks, Jim Carlton, Steven Chown, Rob Colautti, Jane Catford, Susana Clusella-Trullas, Richard Cowling, Carla D'Antonio, Sarah Davies, Jaimie Dick, Ulf Dieckmann, Genevieve Diedericks, Jason Donaldson, Benis Egoh, Franz Essl, Gordon Fox, Llewellyn Foxcroft, Mirijam Gaertner, Piero Genovesi, Sjirk Geerts, Jessica Gurevitch, Richard Hobbs, Brett Hurley, Donald Iponga, Haylee Kaplan, Rainer Krug, Christoph Kueffer, Christian Kull, Pietro Landi, David Le Maitre, Sue Milton, Jaco Le Roux, Andy Lowe, Matthew McConnachie, Sandra MacFayden, Denise Mager, Anne Magurran, Melodie McGeoch, Clifton Meek, Ony Minoarivelo, Jane Molofsky, Desika Moodley, Joice Ndlovu, Ana Novoa, Savannah Nuwagaba, David Phair, Luke Potgieter, Peter Prentis, Petr Pyšek, Mihaja Ramanantoanina, Marcel Rejmánek, Mark Robertson, Mathieu Rouget, Núria Roura-Pascual, Helen Roy, Phil Rundel, Sheunesu Ruwanza, Nate Sanders, Ross Shackleton, Mariska te Beest, John Terblanche, Farai Tererai, Genevieve Thompson, Wilfried Thuiller, Anna Traveset, Marinel van Rensburg, Brian van Wilgen, Nicola van Wilgen,

Vernon Visser, Olaf Weyl, Darragh Woodford, Stephanie Yelenik, Rafael Zenni, Feng Zhang, and Thalita Zimmermann.

Over the years, both of us have received generous financial and institutional support from the National Research Foundation of South Africa (grants 69866, 76912, 81825, and 89967 to CH; grant 85417 to DMR) and the DST-NRF Centre of Excellence for Invasion Biology (C·I·B), and the Working for Water Programme through their collaborative research project with the C·I·B on 'Integrated management of invasive alien species in South Africa'. CH acknowledges additional support from the African Institute for Mathematical Sciences and the South African Research Chairs Initiative (SARChI Chair in Mathematical and Theoretical Physical Bio-Sciences), Group on Earth Observations Biodiversity Observation Network (GEO BON), National Geographic, National Science Foundation of China (31000192, 31100308, and 31360104) and Australian Research Council (Discover Project DP150103017)

as well as the Elsevier Young Scientist Award in 2011. DMR is extremely grateful for financial support from the Hans Sigrist Foundation (through the Hans Sigrist Prize in 2006) and the Oppenheimer Memorial Trust. Both of us also acknowledge support from the US National Socio-Environmental Synthesis Center (SESYNC).

Special thanks to Lucy Nash, Ian Sherman, Janet Walker, Narmatha Vaithiyanathan, and the dedicated personnel at Oxford University Press, as well as Genevieve Diedericks, Corneile Minnaar, Lorraine Cilliers, and Vanessa du Plessis at Stellenbosch University for tremendous help.

Extra special thanks are due to our families for their support and forbearance (especially during the dash for the finish line): Beverley, Keira, and Zachary (CH), and Corlia and Sean (DMR). Zachary was born during the final preparation of the book.

Cang Hui and David M. Richardson
Stellenbosch, June 2016

Contents

CHAPTER 1

Setting the scene

1.1 Background

The geographic area occupied by a species changes over time in response to many interacting factors that affect the survival and reproduction of the species at locations within its current range and at sites within the potential dispersal reach of the species. Such dynamics have driven the evolution of organisms and the composition of ecosystems and biogeographic regions since life began on Earth.

The role of humans in rearranging the world's biota has changed radically over time. Humans have purposefully and accidentally moved organisms around the world for millennia. Until fairly recently, however, the scale of these movements was too limited to cause substantial changes to native species or ecosystems at regional as well as global scales. Few species were moved around, and those that were moved were transported in small numbers, over limited distances, and quite slowly. Only with the dawn of the Age of Discovery, the era of European colonialism (and the technological innovations associated with these phases of history) and, more recently, with post-Second World War globalization have humans generated the widespread invasions that are now a defining feature of the Anthropocene. Such changes are reshaping the Earth's biota and are leading to radical alterations to the functioning of ecosystems worldwide and their capacity to deliver services. Besides this huge challenge, biological invasions also provide intriguing natural experiments that offer fundamental insights to the forces that shape the composition and functioning of biotic communities and social-ecological systems at different spatial and temporal scales.

This book deals with the human-mediated introductions of organisms, especially to areas well outside their potential range as defined by their natural dispersal mechanisms and biogeographic barriers. The field covers all aspects relating to the introduction of organisms, their capacity to naturalize and invade (i.e. to reproduce and spread) new regions, and the many types of interactions with resident biota. The term 'invasion science' was coined recently to describe the full spectrum of fields of enquiry that address alien species and biological invasions. It embraces invasion *ecology*, but increasingly involves non-biological lines of enquiry, including environmental ethics, political ecology, environmental history, ecological economics, and traditional ecological knowledge. The expanding scope of research of invasions is increasingly being framed in the context of social-ecological systems, and more emphasis is being placed on inter- and transdisciplinary studies (Richardson *et al.* 2011). It is important to understand the origins of invasion science because events and trajectories in its short history have shaped the boundaries of the field and its relationships with other disciplines. As will become clear in this book, the study of invasions has borrowed concepts and methods from all fields of biology and many other disciplines. It is also contributing much to the science of global change and to ecology in general.

This chapter traces some of the key developments in the scientific study of invasions and outlines the scope of the book.

1.2 A history of invasion science

1.2.1 Preparing the stage

The Age of Discovery is a loosely defined period from the fifteenth century to the eighteenth century during which European powers began extensive

Invasion Dynamics. Cang Hui and David M. Richardson, Oxford University Press (2017).
© Cang Hui and David M. Richardson. DOI 10.1093/acprof:oso/9780198745334.001.0001

exploration overseas. This era overlaps with the European colonial period which lasted from the sixteenth century until the mid-twentieth century when several European powers established colonies in Africa, the Americas, the Indian subcontinent, and Australasia. European colonization was partly facilitated by technological innovations that provided the means for the rapid transfer of people, goods, and many organisms on ships over long distances. These events were momentous drivers in the history of the human-mediated reshuffling of the world's biota (Crosby 1986) which is now recognized as one of the defining features of the Anthropocene.

Initial introductions of species to areas colonized by Europeans were generally haphazard. Although many species introduced intentionally or accidentally at the start of the European colonial period became established, and later became invasive, many failed. As human populations in colonized areas grew, the desire for more introduced species increased. Starting in the mid-nineteenth century, acclimatization societies were formed in many parts of the world by diverse groups of people (aristocrats, landowners, biologists, agriculturalists, sportsmen, among others) to guide and encourage the introduction of alien animals and plants to improve domestic stock, to supply food, to provide animals for fishing and hunting, to satisfy nostalgic yearnings, to control pests, and for many other reasons (Lever 2011). Governments in settler colonies actively imported and experimented with alien species, especially for agriculture and forestry, and states continued to import species well into the twentieth century. These initiatives played a major role in sowing the seeds of many invasions.

Rapid changes in the scale and extent of human activities in ensuing decades and centuries created dispersal pathways similar in some respects to natural dispersal pathways, but they also forged completely novel opportunities for rapid species exchanges between previously isolated regions (Wilson *et al.* 2009). For example, engineering feats such as the Panama and Suez canals facilitated biotic interchange in essentially the same way as natural corridors. Human activities also enhanced existing natural dispersal pathways, for example through dramatic increases in the amount of floating oceanic debris in a way similar to historical rafting. Many other types of human activities and innovations created entirely new types of species movements. Such dispersal pathways grew in prominence with the rapid increase in trade routes between settled agricultural communities. This led to the movement of species in an increasingly organized fashion. Colonial traders and explorers transported many organisms in large numbers over major biogeographic barriers which had previously prevented the dispersal of many, if not most, of the species that now feature on lists of invasive species around the world. There is a large literature dealing with the vectors and pathways that have permitted the rapid shuffling of the world's biota in recent centuries (Ruiz and Carlton 2003; Hulme *et al.* 2008; Wilson *et al.* 2009; Essl *et al.* 2015a). These pathways provided the conduits for species exchanges and are a crucial facet of invasion science. This book, however, focuses mainly on the dynamics of species once they have arrived in new environments. With the benefit of hindsight, it seems strange that people did not foresee that opening the sluice gates to permit the mixing of biotas would have profound implications for life on Earth, and yet, the phenomenon of invasions was slow to grab the attention of scientists.

Several nineteenth-century naturalists, notably Charles Darwin (1809–82), Augustin Pyramus de Candolle (1778–1841), his son, Alphonse De Candolle (1806–93), Joseph Dalton Hooker (1817–1911), and Charles Lyell (1797–1875), mentioned invasive species in their writings. For example, Darwin, in Volume 2 of his *Journal of Researches*, made interesting observations on the European thistle cardoon, *Cynara cardunculus*, in several parts of South America. In commenting on various aspects of 'the invasion of the cardoon', Darwin discussed its weediness compared to other alien plants, the role of humans and livestock in its proliferation, and its impact on the native biota. Naturalized and invasive species were, however, mostly regarded as curiosities at the time and there is little evidence to suggest that observers perceived such invasions as important disrupters of ecosystems or posing a major threat to biodiversity.

In the first half of the twentieth century, Joseph Grinnell (1877–1939), Albert Thellung (1881–1928),

Frank Egler (1911–96), Herbert Baker (1920–2001), Carl Huffaker (1914–95), among others, published important contributions on introduced species. Thellung's work is particularly noteworthy, though overlooked. He devised concepts to classify alien species according to their degree of naturalization, introduction pathways, and residence time, and applied such notions systematically for the regional flora of Montpellier in France. He also introduced a population-based definition of naturalization which incorporated environmental barriers (Kowarik and Pyšek 2012). Until the mid-1900s, however, invasions were mostly documented on a case-by-case basis. The work of European researchers such as Thellung failed to infiltrate the mainstream literature, and no theoretical frameworks or scientific foundations were widely adopted as the foundation for understanding the phenomenon or for evaluating particular invasions.

Widespread invasions started to be documented in many parts of the world in the last few decades of the nineteenth century and the first few decades of the twentieth. By 1850, more than a dozen non-native forest insect species were damaging trees in the United States (Liebhold and McCullough 2011). In South Africa, the first records of widespread invasions of woody plants (e.g. *Hakea sericea* and *Pinus pinaster*) began in the 1850s (Richardson *et al.* 1997). At the same time, cheatgrass (*Bromus tectorum*) was spreading across North America. The European green crab (*Carcinus maenus*) first became established on the east coast of North America in about 1900. At about the same time, the Chinese mitten crab (*Eriocheir sinensis*) arrived in Europe from China in ballast water (Carlton 2011). In the 1930s, *Plasmodium relictum*, which causes avian malaria, was first documented in Hawaiian forest birds (Loope 2011). This period was the start of what Charles Elton (1958) would later term 'ecological explosions'—a prelude to the widespread invasions that we see today.

The *spread* (but not other aspects of invasions) of many alien species, especially insects, birds, and mammals, was well studied in the first half of the twentieth century. Hengeveld (1989) provides an excellent summary of the large body of work that focused on documenting patterns of spread of many species of insects (notably the Japanese beetle,

Popillia japonica, and the gypsy moth, *Lymantria dispar*, in North America), birds (notably common starlings, *Sturnus vulgaris*, in North America in 1890, and several species in Europe), and mammals (e.g. grey squirrels, *Sciurus carolinensis*, in the British Isles, red deer, *Cervus elaphus*, and Himalayan thar, *Hemitragus jemlahicus*, in New Zealand, European rabbits, *Oryctolagus cuniculus*, in Australia, and muskrats, *Ondatra zibethicus*, in Central Europe).

1.2.2 Charles Elton to 1980

Charles S. Elton's (1900–91) *The Ecology of Invasions by Animals and Plants* (Elton 1958) is widely acknowledged as the starting point for *focused scientific attention* on biological invasions. This is not to dismiss the contributions of the many naturalists who preceded him, but Elton's monograph was the first attempt to construct a global synthesis and framework for considering the multiple dimensions of invasions and to articulate the global scale of the phenomenon and the looming implications for global biodiversity (Richardson and Pyšek 2007).

It is worth reflecting on the contents and nature of Elton's book which was written in accessible prose and was based on a series of radio talks on the BBC. This relatively short book starts with a brief overview entitled 'The invaders', describing 'one of the great historical convulsions in the world's fauna and flora'. Chapter 2 provides a global biogeographic perspective, detailing the evolution of the world's biota and the importance of isolation in generating and maintaining biodiversity: 'Wallace's realms: the archipelago of continents'. The next chapter discusses the breakdown of isolation through human-induced movement of organisms around the world. Elton drew upon numerous case studies covering a wide taxonomic array from all parts of the world to outline the trends and motives that act as drivers for the widespread transfer of organisms to areas well outside their natural ranges (91 per cent of references are from the period 1920 to 1957; Carlton 2011). This chapter, called 'The invasion of the continents', summarized Elton's view of the phenomenon of invasions (Fig. 1.1). Chapters 4 and 5 described features of invasions of islands and the oceans, respectively. In Chapter 6, Elton began exploring the mechanisms of invasions in

Fig. 1.1 Charles Elton (1958) likened the continents to great tanks of water, connected by narrow tubing and blocked by taps. Using this analogy, he conceptualized the processes of biological invasions: 'fill these tanks with different mixtures of a hundred thousand different chemical substances in solution . . . then turn on each tap for a minute each day . . . the substances would slowly diffuse from one tank to another. If the tubes were narrow and thousands of miles long, the process would be very slow. It might take quite a long time before the system came into final equilibrium, and when this happened a great many of the substances would have been recombined and, as specific compounds, disappeared from the mixture, with new ones from other tanks taking their places. The tanks are the continents, the tubes represent human transport along lines of commerce.' Figure copyright: DST-NRF Centre of Excellence for Invasion Biology. Reproduced with permission from David M. Richardson, Director of the C.I.B.

a short essay entitled 'The balance between populations'. The next chapter expands on this theme and describes the various ways in which invasive species profoundly alter food chains in ecosystems. The last two chapters consider the implications of invasions for society and nature (had the terms 'biodiversity' and 'ecosystem services' been invented, Elton would have used them here). Chapter 8 discusses the various reasons for conserving nature and how the invasions described in the book threaten humankind's ability to conserve biodiversity. The final chapter poses philosophical questions (profound for the time) relating to economical, ethical, and other reasons for 'the conservation of variety' in the face of looming invasions and biotic homogenization.

Despite Elton's eloquent sketching, in 1958, of a road map for an exciting new field, researchers were slow to take up the challenge. In the decade between the publication of his book and his retirement in 1967, not even Elton himself worked on elucidating, expanding, or testing the bold generalizations

he had proposed. He mentioned invasions in several other publications, but the 1958 book was his magnum opus on biological invasions. Although it raises many profound questions, particularly those relating to the links between diversity and stability and diversity and invasibility (Richardson and Pyšek 2008; Fridley 2011), Elton never explored questions dealing explicitly with invasions in his later publications.

The Ecology of Invasions by Animals and Plants was recognized as an important contribution soon after its publication (e.g. Saville 1960; Waloff 1966; Pianka 1967), but it did not immediately stimulate a significant rallying of research effort to address issues directly associated with invasions. Also, unlike another famous popular book on environmental issues, Aldo Leopold's *Sand County Almanac*, which appeared in 1949, Elton's work had negligible impact on public perceptions and launched no environmental movement (Hobbs and Richardson 2011). Several reasons have been proposed for the slow uptake of Elton's ideas by ecologists. One is that although there were many widespread invasions at the time, it was only decades later that the scale of the overall phenomenon escalated to the point that the public and policy makers in many parts of the world began to acknowledge invasions as a serious environmental problem (Richardson 2011). Importantly, the field of ecology at the time was preoccupied with 'natural' ecosystems; invasions were seen as 'background noise' to be avoided as far as possible in field studies (Richardson and Pyšek 2007). Related to this is the notion that there was no neat space within the ecological paradigms of the day in which to slot concepts and hypotheses relating to human-mediated introductions and invasions. Invasion ecology still grapples with this awkward fit within ecology (which is why we wrote this book). As noted by Kitching (2011), 'since the 1930s there had been, more or less, a split in ecology along taxonomic and thematic lines. Population ecology was regarded as the very stuff of animal ecology (at least until John Harper's seminal book in 1977) whereas community ecology was principally about associations of plants.' Elton's treatise called for the integration of population and community ecology which 'ran counter to the mainstream' (Kitching 2011). Similarly, the science of biogeography was largely focused on other issues in the early twentieth century (notably palaeontology and patterns of variation within single species; Brown and Gibson 1983).

By the 1950s, colonizing species had already long been of interest to evolutionary biologists. Largely independently of the ideas advanced by Elton, researchers in this field were also gathering ideas and formulating hypotheses that would later be crucial for invasion ecology. A conference in California in 1964 led to the publication of an edited volume, *The Genetics of Colonizing Species* (Baker and Stebbins 1965), which provided the first synthesis on the genetics and evolution of colonizers. Contributions to this volume were crucial in focusing research on aspects that are now central tenets of invasion ecology, including determinants of invasion success, life-history trade-offs, generalist versus specialist strategies, general-purpose genotypes, adaptive phenotypic plasticity, mating systems, and the influence of bottlenecks on genetic variation (Barrett 2015).

Elton's (1958) ideas on diversity–stability relationships influenced theoretical ecologists such as Robert MacArthur (1930–72), whose work helped to hone hypotheses and develop the ecological theories that were to consider the many ways that introduced species potentially influence community dynamics (Fridley 2011; Rejmánek *et al.* 2013; see discussion in Chapter 5). Elton's (1958) book also influenced the work of Robert May, who used mathematical models to compare ecological communities with few species with those with more species (May 1973). May found that 'there could be no such simple and general rule [as had been proposed by Elton]; all things being equal, complex systems are likely to be more dynamically fragile' (May 2001). He showed that there is no automatic connection between complexity and stability as suggested by Elton's arguments. May's work and other recent studies (e.g. Naeem *et al.* 2000) have ensured sustained research interest in this topic. In 1969, a symposium at Brookhaven National Laboratory explored the diversity–stability relationship in ecological systems (Woodwell and Smith 1969).

One of Elton's most fundamental contributions to ecology was his formulation of concepts pertaining to the ecological niche. Invasions clearly gave him fertile ground for deliberating on the role of niches in structuring communities. The 'Eltonian niche' (Beals 1972) finds expression in the 1958 volume and it still enjoys attention in the search for robust theoretical frameworks for the consideration of geographic distributions (Soberón 2007) and biological invasions (e.g. Shea and Chesson 2002). Elton's generalization that islands are more invaded than mainland areas has held up to scrutiny (see Pyšek and Richardson 2006 for a status report for research on plant invasions). Many researchers who are now prominent in the field of invasion research attest to being inspired by Elton's book (e.g. Mooney 1998; Vitousek 2001) and/or still consider it a keystone reference in the field (Richardson and Pyšek 2007).

The decade after the publication of *The Ecology of Invasions by Animals and Plants* saw a flood of work on the implications of human modification/domination of the planet (Carson 1962; Ehrlich 1968; Hardin 1968). It also saw the birth of the field of ecological genetics (Ford 1964) and the publication of

MacArthur and Wilson's *The Theory of Island Bioge-ography* (1967). In the 1970s, papers on invasive species began appearing regularly in the mainstream ecology literature. Seminal contributions in plant ecology with important implications for studying invasive species included Baker's paper on 'The evolution of weeds' (1974) and John Harper's textbook, *Population Biology of Plants*, (1977), which both stressed the need to merge perspectives from community and population ecology to understand plant dynamics. Connell's classic paper, 'Diversity in tropical rain forest and coral reefs' (1978), paved the way for research to determine the role of disturbance in structuring communities.

1.2.3 1980–2010

Issues relating to the conservation of biological diversity began to be discussed by ecologists in earnest in the 1970s. The term 'conservation biology' was first used in the title of a 1978 conference that addressed tropical deforestation, disappearing species, and eroding genetic diversity within species. Along with many other drivers of biodiversity loss, biological invasions were now clearly an issue that required attention from scientists. The widespread invasion of South African fynbos vegetation by alien trees and shrubs and the marked alteration of ecosystem functioning caused by these invaders captured the attention of ecologists from many parts of the world during the Third International Conference on Mediterranean-Type Ecosystems held in Stellenbosch in 1980. At that time, the prevailing view, influenced by Elton's seminal book and lingering Clementsian orthodoxy, was that human-mediated disturbance was a fundamental prerequisite for the invasion of ecosystems by alien organisms. The situation in fynbos appeared to refute this notion. This provided the incentive for launching the SCOPE Programme on the Ecology of Biological Invasions (Mooney 1998; Bennett 2014). The aims of this initiative (1982–86) were to revisit Elton's key assumptions and generalizations, to review the current status of invasions worldwide, and in so doing to address three main questions:

1. What factors determine whether a species will be an invader or not?

2. What site properties determine whether an ecological system will be relatively prone to, or resistant to, invasion?

3. How should management systems be developed using the knowledge gained from answering questions 1 and 2?

The SCOPE programme attracted some of the world's top ecologists to address issues relating to invasions. It comprised national, regional, and thematic groups that produced a wealth of new research on many aspects of invasions (Drake *et al.* 1989). By marshalling intellectual forces and demonstrating the profound societal issues, SCOPE succeeded in putting invasion ecology firmly on the map as an exciting and attractive research topic in urgent need of dedicated attention (Simberloff 2011).

Another event that had a major role in shaping the trajectory of research on invasive species was a large international conference in Trondheim, Norway, convened by the United Nations and the government of Norway in 1996. Participants at this meeting concluded that invasions had become one of the most significant threats to global biodiversity and called for a global strategy and mechanism to address the problem (Sandlund *et al.* 1998; Mooney 1999). This led to the launching of the Global Invasive Species Programme (GISP Phase 1) in 1997. This initiative was explicitly more inter/transdisciplinary than the SCOPE programme; it acknowledged the need for work on economic valuation, stakeholder participation, and pathway analysis and management (Mooney *et al.* 2005). It primarily sought practical solutions to problems associated with invasions, but also recognized the need for improved understanding of fundamental ecological aspects (Barnard and Waage 2004). A product of GISP Phase 1 (1997–2005) was the *Global Strategy on Invasive Alien Species* (McNeely *et al.* 2001). Other initiatives at this time that helped marshal research effort on invasions included the Convention on Biological Diversity (CBD 1993), Article 8(h) which calls on member governments to prevent the introduction of, or the control or eradication of, those alien species that threaten ecosystems, habitats, or species. In 2000, the IUCN published their *Guidelines for the prevention of biodiversity loss caused by alien invasive species*.

The period between 1996 and 2010 saw an explosion of research dealing with concepts, frameworks, and terminology relating to invasive species (Richardson *et al.* 2000a), as well as major advances in elucidating the determinants of invasiveness (Rejmánek and Richardson 1996) and invasibility (Lonsdale 1999; Davis *et al.* 2000; Chytrý *et al.* 2008). Research on biological invasions was firmly embedded in a community ecology framework (Shea and Chesson 2002). Also, invasions were increasingly recognized as a crucial ingredient of global change (Fig. 1.2). Several protocols were proposed to synthesize the current understanding of invasiveness and invasibility into simple flow-chart models for assessing the risk of newly introduced species becoming invasive. The Australian Weed Assessment scheme (originally proposed by Pheloung *et al.* 1999) has been the most widely applied of these protocols. Phase 2 of GISP (2006–10) set out to improve the scientific basis for decision-making to enhance the ability to manage invasive species, assess the impacts of invasions on major economic sectors, and create a supportive environment for improved management of invasions. This initiative stimulated many new directions in research to elucidate the multiple dimensions of the impacts of invasive species and to utilize existing knowledge and incorporate new ideas and methodologies to inform options for management (e.g. Clout and Williams 2009). Numerous books and review articles have summarized approaches and advances in the study of biological invasions in recent decades (Richardson 2011).

Some areas of research that enjoyed considerable attention and where key innovations have emerged during this period include the following:

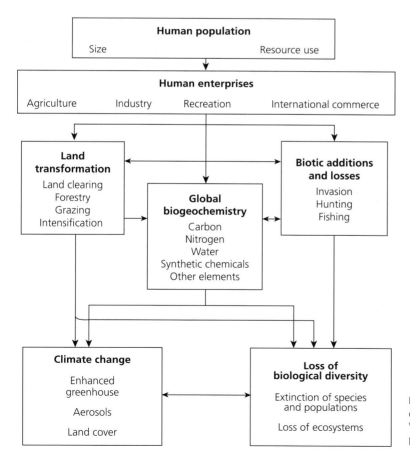

Fig. 1.2 Humanity's direct and indirect effects on the Earth system. Redrawn from Vitousek *et al.* (1997). Reprinted with permission from AAAS.

- Increasingly sophisticated experiments were devised to test theories and assumptions in invasion ecology, including work to quantify the importance of release from enemies (Keane and Crawley 2002), and the role of novel weapons (Calloway and Ridenour 2004) in mediating invasions.
- Much attention was given to the conceptualization of the role of positive interactions and facilitation (Bruno *et al.* 2003), in particular different types of mutualisms (Richardson *et al.* 2000b) and the role of soil biota (Callaway *et al.* 2004) in mediating plant invasions.
- Improved databases, protocols, and tools for analysing rates of naturalization and invasion for different taxa and regions shed new light on the tempo at which species progress along the introduction–naturalization–invasion continuum (e.g. Caley *et al.* 2008).
- The creation of large databases and improved conceptualization of invasion processes paved the way for the computation of better measures of propagule pressure and the evaluation of its role in invasions (Lockwood *et al.* 2005; Colautti *et al.* 2006; Simberloff 2009).
- Advances in molecular ecology provided new opportunities to gain new insights on many aspects of invasion ecology—for example, the elucidation of mechanisms involved in different stages of invasions (e.g. Dlugosch and Parker 2008)—as a tool for reconstructing introduction and invasion histories (e.g. Estoup and Guillemaud 2010), and to inform management (e.g. Le Roux and Wieczorek 2009).
- The development and refinement of techniques for modelling the distributions of invasive species led to the application of these to a wide range of ecosystems and taxa (e.g. Gallien *et al.* 2010).
- Many studies have sought general frameworks or approaches for measuring or evaluating impact (Parker *et al.* 1999). The notion of 'invasional meltdown' was proposed by Simberloff and von Holle (1999) and received much attention in a range of ecosystems (Ricciardi and MacIsaac 2000; O'Dowd *et al.* 2003; Grosholz 2005).
- The relative importance of 'applied' studies of invasions (those relating to impact, management, and risk assessment) increased over this period (Fig. 1.3). Broad areas that enjoyed substantial

research interest included risk assessment, pathway and vector management, early detection and rapid response, eradication, and mitigation and restoration (reviewed in Pyšek and Richardson 2010). Considerable work was also directed at biological control (e.g. McEvoy and Coombs 1999) and numerous approaches for optimizing or prioritizing management interventions (e.g. Zavaleta *et al.* 2001).

- The increasing occurrence, and often dominance, of invasive species in ecosystems across the globe, together with other forms of global change, led to new approaches for conceptualizing levels of degradation and/or deviation from pristine or reference ecosystems. One such concept, termed 'novel ecosystems', describes systems comprising species that occur in combinations and relative abundances that have never occurred previously in a given biome (Hobbs *et al.* 2006). Subsequent work elaborated on the forces that generate such non-historical configurations and expanded the concept to include both 'hybrid' systems that retain original and new elements, and 'novel' systems which comprise different species, interactions, and functions (Hobbs *et al.* 2009).

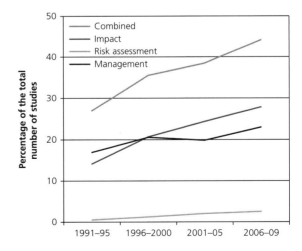

Fig. 1.3 Trends in studies on impact and management indicating an increase in research focus towards more practically oriented issues over the period 1991–2009. Values are percentages of the total number of studies that address impact and management, including risk assessment in five-year periods. Based on 8004 studies identified on the Web of Science by a search using the combination of terms alien, invasive, exotic, and naturalized with taxonomic affiliations. From Pyšek and Richardson (2010).

1.2.4 Since 2010

The three big questions from the SCOPE programme continue to underpin most work in invasion science, with substantial research effort in all three areas. In line with other problem-orientated fields, research on issues relating to biological invasions is increasingly clustered around three focal areas: systems knowledge (the analysis of casual relationships), target knowledge (clarification of conflicts of interest and values), and transformation knowledge (the quest for appropriate actions for management) (Kueffer and Hirsch Hadorn 2008), hence the increasing usage of the term 'invasion science' (see Chapter 11). Some of the key research directions since 2010 are discussed briefly in the next section.

Substantial progress was made towards deriving general models of invasion, such as Blackburn and colleagues' 'unified framework for biological invasions' (2011), which sought to merge insights from many previous attempts to conceptualize key aspects of invasion dynamics for all taxa (Fig. 1.4). This model reinforced the utility of conceptualizing the many processes implicated in phenomena of biological invasions using a series of barriers along an introduction–naturalization–invasion continuum and has been widely applied. It provides an objective framework for linking theoretical and applied aspects in invasion science, and a foundation for a standard lexicon of terms for the field of invasion science (Box 1.1).

There have been major advances in understanding the roles of traits in determining invasiveness, especially through the development of improved conceptual frameworks and the evaluation of global databases and improved methods of meta-analysis (van Kleunen *et al.* 2010a, 2010b). Increasing attention is being given to the detailed elucidation of the multiple dimensions of invasions through the multidisciplinary evaluation of natural experiments,

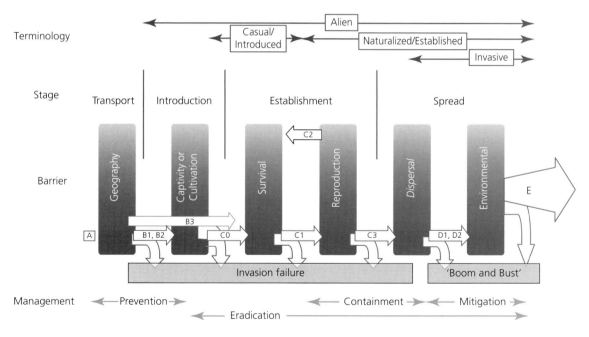

Fig. 1.4 The proposed unified framework for biological invasions. It recognizes that the invasion process can be divided into a series of stages, that in each stage there are barriers that need to be overcome for a species or population to pass on to the next stage, that species can be categorized according to their position on the 'introduction–naturalization–invasion continuum', and that different types of management interventions are required at different stages. Different parts of the framework emphasize views of invasions that focus on individuals, populations, processes, or species. The unfilled block arrows describe the movement of species along the invasion framework with respect to the barriers, and the alphanumeric codes associated with the arrows relate to the categorization of species with respect to the invasion pathway. From Blackburn *et al.* (2011). Reprinted with permission from Elsevier.

Box 1.1 Basic concepts in invasion science. Additional terms and concepts are elucidated in the Glossary.

Alien species (synonyms: adventive, exotic, foreign, introduced, non-indigenous, non-native)—Species whose presence in a region is attributable to human actions that enabled them to overcome fundamental biogeographical barriers (i.e. human-mediated extra-range dispersal). Some alien species (a small proportion) form self-replacing populations in the new region. Of these, a subset has the capacity to spread over substantial distances from introduction sites. Depending on their status within the naturalization–invasion continuum, alien species may be objectively classified as casual, naturalized, or invasive.

Biological invasions (synonyms: bioinvasions, biotic invasions, species invasions)—The phenomenon of, and suite of processes involved in determining: (1) the transport of organisms, through human activity (intentionally or accidentally) to areas outside the potential range of those organisms as defined by their natural dispersal mechanisms and biogeographical barriers; and (2) the fate of such organisms in their new ranges, including their ability to survive, establish, reproduce, disperse, spread, proliferate, interact with resident biota, and exert influence in many ways on and in invaded ecosystems. There is a school of thought that advocates that the concept of biological invasions should more broadly embrace both range expansions (involving no obvious human mediation), since the fundamental processes (except, critically, the means of negotiating a major biogeographic barrier) are the same (both involve the movement of individuals from a donor community into a recipient community).

Introduction—Movement of a species, intentionally or accidentally, due to human activity, from an area where it is native to a region outside that range ('introduced' is synonymous with alien). The act of an introduction (inoculation of propagules) may or may not lead to invasion.

Introduction–naturalization–invasion continuum—A conceptualization of the progression of stages and phases in the status of an alien organism in a new environment which posits the view that the organism must negotiate a series of barriers. The extent to which a species is able to negotiate sequential barriers (which is mediated by propagule pressure and residence time, and which frequently involves a lag phase) determines the organism's status as an alien: casual, naturalized, or invasive species.

Invasion—The multistage process whereby an alien organism negotiates a series of potential barriers in the naturalization–invasion continuum (the increasing complexity of questions in invasion science calls for a shift from the paradigm of invasions as such a linear continuum to a complex network).

Invasion ecology—The study of the causes and consequences of the introduction of organisms to areas outside their native range as governed by their dispersal mechanisms and biogeographical barriers. The field deals with all aspects relating to the introduction of organisms, their ability to establish, naturalize, and invade in the target region, their interactions with resident organisms in their new location, and the consideration of costs and benefits of their presence and abundance with reference to human value systems. This term is often used interchangeably with 'invasion biology' in the literature.

Invasion science (synonym: invasion research)—A term used to describe the full spectrum of fields of enquiry that address issues pertaining to alien species and biological invasions. The field embraces invasion ecology, but increasingly involves non-biological lines of enquiry, including economics, ethics, sociology, and inter- and transdisciplinary studies.

Invasive species—Alien species that sustain self-replacing populations over several life cycles produce reproductive offspring, often in very large numbers at considerable distances from the parent and/or site of introduction, and have the potential to spread over long distances. Invasive species are a subset of naturalized species; not all naturalized species become invasive. This definition explicitly excludes any connotation of impact, and is based exclusively on ecological and biogeographical criteria. It should be noted that the definition supported by the World Conservation Union (IUCN), the Convention on Biological Diversity (CBD), and the World Trade Organization (WTO) that explicitly assumes invasive species cause impacts on the economy, environment, or health. This important difference has implications for risk analyses of invasive species. Consequently, it is crucial for risk assessment protocols to assign dimensions of risk separately for elements of invasion and impact. Note: designation of a species as invasive should include a statement about the region under discussion; depending on the scale of observation.

Native species (synonym: indigenous species, sometimes referring to native species occurring over a broad range, and endemic species, sometimes used to refer to native species found nowhere else, over a much smaller range)—Species that have evolved in a given area or that arrived there by natural means (via range expansion), without the intentional or accidental intervention of humans from an area where they are native.

Naturalized species (synonym: established species)—Those alien species that sustain self-replacing populations for several life cycles or a given period of time (ten years is advocated for plants) without direct intervention by people, or despite human intervention. The term is currently mainly used with reference to terrestrial plants invasions, although it was previously widely used for mammals.

From Richardson *et al.* (2011b).

such as the global introductions, usage, and invasions of many plant and animal species (e.g. Richardson *et al.* 2011a). Improved global indicators of the extent of invasions were derived (e.g. McGeoch *et al.* 2010) and important new insights emerged about the extent and types of impacts of invasive species through advanced meta-analyses (Vilà *et al.* 2011; Pyšek *et al.* 2012). The availability of global databases is facilitating the elucidation of many macroecological patterns, such as the role of socio-economic factors in driving invasions and the conceptualization of the dimensions of invasion debt (Pyšek *et al.* 2010; Essl *et al.* 2011; Rouget *et al.* 2016). Progress has been made towards incorporating invasions into global frameworks for conceptualizing impacts of different facets of global change on biodiversity (Gilbert and Levine 2013; Essl *et al.* 2015b; Franklin *et al.* 2016; Blackburn *et al.* 2014). Detailed assessments of the relative contributions of invasions and other forms of global change to biodiversity declines were undertaken (Butchart *et al.* 2010; Bellard *et al.* 2016). Improved methods for modelling invasions are allowing researchers to move beyond correlative climate-envelope models to incorporate multiple drivers, thereby paving the way for an improved ability to explain existing invasion patterns and to predict the outcome of future introductions (e.g. Roura-Pascual *et al.* 2011; Gallien *et al.* 2012).

Greater public awareness of invasions and the growth of fields of study within the environmental humanities, and environmental history in

Table 1.1 A field guide to misleading criticisms of invasion science. From Richardson and Ricciardi (2013).

Criticism	Rebuttal
Modern invasions are nothing new. The magnitude and impacts of human-assisted invasions are similar to those in the fossil record (i.e. generally low) and thus do not merit major concern and concerted conservation action.	The current scale, impact, and evolutionary importance of invasions are unique. Under human influence, organisms are spreading faster, farther and in greater numbers than ever before. Human-mediated introductions create dispersal pathways that are fundamentally distinct from those possible for spread events not involving human actions. This facilitates colonization events that are inadequately explained by natural dispersal models.
Impacts of non-native species on biodiversity and ecosystems are exaggerated.	Global datasets clearly implicate invasions as a major and growing cause of population-level and species-level extinctions. Decades of experimental research have demonstrated the capacity for invasions to alter ecosystems. Impacts of invasions on plant extinction are frequently masked by the lengthy time lags inherent in plant extinctions: numerous species affected by invasions survive as 'the living dead'.
Increased species introductions raise biodiversity (e.g. by adding to regional species pools; generating new taxa through hybridization) and therefore do not merit concern.	Focusing on species richness counts ('the numbers game') is a misleading approach to quantifying impact, especially when the persistence of many species recorded over long time periods is not verified. Extinction may not be an appropriate measure of impact on ecosystem function. Assessment of the influence of invasions on the abundance and distribution of native species (and consequences of these changes on the functioning of ecosystems) is crucial. Hybridization has been shown to be a major contemporary extinction force, especially when accompanied by habitat homogenization, causing species declines through introgression, genetic swamping, and reproductive interference.
Positive (desirable) impacts of non-native species are understated and are at least as important as their negative (undesirable) impacts.	Non-native species are far more likely to cause substantial ecological and socioeconomic damage, such as ecosystem-level regime shifts, than are native species. Furthermore, many of the 'positive' impacts attributed to non-natives are likely to be transient, whereas 'negative' impacts are typically more permanent and often irreversible.
Invasions science is biased and xenophobic.	Xenophobes obsessed with eradicating all non-native organisms operate on the fringe of the conservation movement—as do those who link informed efforts to manage introduced species with xenophobia.
The biogeographic origin of a species has no bearing on its impact. The native/non-native dichotomy holds no value to science. Therefore, these factors should not guide management, and there is no rationale for invasion science.	Ignoring biogeographic origins as a mediator of impact ignores the importance of evolutionary context in species interactions. Non-native consumers inflict greater damage on native populations. The more 'alien' an established animal, plant or microbe is to its recipient community, the greater the likelihood that it will be ecologically disruptive.

particular, has fuelled new engagements between scholars in the humanities, social sciences, and natural sciences. A considerable variety of humanistic work is underway on introduced and invasive species, much of it focused on assessing changing cultural valuations of native and alien species (Kueffer 2013). One strand of historical research has described how popular and scientific attitudes towards invasive alien species turned increasingly negative in former settler societies during the second half of the twentieth century in response to concerns about the ecological and economic impacts of invasive species (Bennett 2014).

There has been a growing recognition that research on invasions is invaluable for understanding how ecosystems work. Studies of invasions have yielded novel insights about key ecological concepts, including, *inter alia*, the diversity–stability relationship, trophic cascades, keystone species, the role of disturbance in community assembly, ecological naiveté, ecological fitting, rapid evolution, island biogeography, ecosystem engineering, and niche construction. The field has also contributed concepts of its own (e.g. propagule pressure, biotic resistance, invasional meltdown, enemy release) that have stimulated productive research of both theoretical and applied importance. Despite the accumulation of rigorous evidence of its importance to science and society, invasion science has been the target of criticism from a relatively small but vocal number of scientists and academics who challenge various concepts, philosophical underpinnings, and methods of the field (Table 1.1).

1.3 The scope of the book

The field of invasion science has grown rapidly into a very large, multifaceted discipline involving thousands of researchers working on a great variety of topics related to alien species and invasions. Issues in invasion science can be subdivided in many ways. One approach is to separate pre-border, at-border, and post-border dimensions of invasions. Pre- and at-border topics deal with the many interacting issues that drive intentional and accidental introductions, legal and illegal translocations, survival during transfer and hitchhiking, quarantine and detection, risk assessment and the development

of international/regional watch lists, and cross-region coalition and conflicts to regulate introductions. Post-border topics deal with the spread, establishment, and impact of introduced species within the recipient social-ecological systems and the actions and responses from components of the system to mitigate and integrate the effect of invasions through natural and artificial means. This book focuses primarily on the post-border dimensions of biological invasions and touches only on pre- and at-border issues to the extent that aspects of introduction pathways have crucial implications for mediating many subsequent processes along the introduction–naturalization–invasion trajectory (Wilson *et al.* 2009). We seek to merge insights from ecology in general with those from the study of invasions to explore their intertwined paradigms and to find new directions that shed light on the workings of ecosystems and how to manage them in the face of rapid change.

Our focus is thus on the dynamics, behaviour, and impacts of invasive species on recipient systems and the responses of the system to such intrusions. We address invasion dynamics—the action of invasive species and reaction of recipient socio-ecological systems (Fig. 1.5), shifting the paradigm of a linear continuum of invasion processes to a complex network. We discuss key advances and challenges within the traditional domain of 'invasion ecology', but we have also tried to introduce methods and insights from other disciplines to propose new approaches to elucidate diverse issues in invasion dynamics. We thus cast our net wider, and in different waters, compared to most syntheses of the phenomenon of biological invasions.

One discipline where we have sought new insights that are relevant to invasions is complexity science (CS), a field that studies systems with many parts that interact to produce global behaviour (Bar-Yam 2002). Recent studies of CS, which is firmly rooted in mathematics and physics (Buldyrev *et al.* 2010; Parshani *et al.* 2010), have focused on issues pertaining to global trade, telecommunication, traffic, food security, health care, climate change, cloud computation, power grids, the emergence of life, artificial intelligence, robotics, brain science, intelligent machinery, neural networks, to list a few. Ecological systems are clearly an important model system

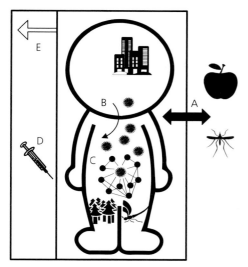

SOCIAL-ECOLOGICAL SYSTEM

Fig. 1.5 Invasion dynamics of alien species in a social-ecological system. Invasion dynamics is defined here as the action of invasive species and reaction of recipient social-ecological systems. Through industrialization and globalization, a regional social-ecological system has moved away from being autonomous to become open. (A) Being an open system, a social-ecological system needs to exchange goods and materials with other systems, in so doing inevitably exposing itself to cross-system benefits and risks, including biological invasions. This has become integral to the sustainability of modern social-ecological systems. (B) Invasive species can dissipate and spread within recipient systems in diverse ways. The recipient system can react and absorb the intrusion internally through biotic interaction and system processes (C), or through artificial means through diverse human interventions (D). Either way, the system boundaries and dynamics will shift (E), the structure and function of the system diverging from the one that existed when cross-system material exchange and levels of biotic disturbances were low. This book addresses issues within the extended box (B–E).

for CS, with their properties of self-organization, adaptation, and emergence (Levin 1999; Anand *et al.* 2010). They are the oldest truly complex adaptive systems and have contributed to the development of many concepts in CS, such as non-linearity, complexity, chaos, and catastrophe. Ecological systems can be organized at different scales, from local communities to the global biosphere, with intricate interactions between species and their environments (Solé and Goodwin 2000; Parrott and Meyer 2012). With the integration of advanced quantitative methods and the accumulation of large datasets on

biodiversity and environmental information, the traditional scope of ecology has expanded rapidly in recent decades to incorporate big data science and eco-informatics (Parrott 2010). Data and theory are transforming traditional ecology into a trans-disciplinary field which is merging with fields such as global change biology and invasion ecology, and increasingly with non-biological fields.

The study of invasions has much to gain from drawing insights from complexity science, but it is also a rich source of new principles and theories with much potential to advance complexity science. Biological invasions, besides posing great challenges to global biodiversity, provide important opportunities for us to delve into the workings of system assembly and functioning, and to gain new perspectives on system resilience to intrusions (Levine 2000). The science of invasion ecology developed largely from the traditional biological subfields of population ecology and community ecology (Richardson 2011). In the chapters that follow we have tried to synthesize knowledge on invasion dynamics, drawing mainly from the traditional domains of community and population ecology, but also considering perspectives from CS, using systems thinking and the complex adaptive system as a metaphor. We therefore seek to merge insights from ecological experiments, conceptualization, and simulation, but also to apply sophisticated mathematics and algorithms as tools to support our understanding of ecological systems and to aid management. As such, our chapters do not seek to present simple ecological stories but also set out to explore what the phenomenon of biological invasions means for complexity science.

We divide the book into three parts. The first, dealing with spread, details invasion dynamics from the point of view of the invasive species. Specifically, we systematically review the patterns of invasive spread (Chapter 2), discuss models for elucidating and predicting spread (Chapter 3), examine theories of boosted spread (Chapter 4), and drivers of population variability of non-equilibrium dynamics of invasion and tools for studying these (Chapter 5). The second part, focusing on impact, explores issues from the point of view of recipient social-ecological systems. Specifically, we examine different types of biotic interactions between invasive and resident species (Chapter 6), the potential of regime shifts in recipient ecosystems and opportunities for identifying warning signs of such shifts before they happen (Chapter 7), community assembly processes and how invasions interfere with these (Chapter 8), and theories on monitoring and management of invasive species (Chapter 9). In the last section, we merge insights to propose a framework of complex adaptive systems (Chapter 10), and envisage how knowledge gained could transform invasion ecology as a key agent in the paradigm shift of ecological and environmental sciences that are increasingly appreciating the role of change and the reality of a novel social-ecological system (Chapter 11).

References

Anand, M., Gonzalez, A., Guichard, F., et al. (2010) Ecological systems as complex systems: challenges for an emerging science. Diversity, 2, 395–410.

Baker, H.G. (1974) The evolution of weeds. Annual Review of Ecology and Systematics, 5, 1–24.

Baker, H.G. and Stebbins, G.L. (eds) (1965) The Genetics of Colonizing Species. New York, NY: Academic Press.

Barnard, P. and Waage, J.K. (2004) Tackling Species Invasions Around the World: Regional Responses to the Invasive Alien Species Threat. Cape Town: Global Invasive Species Programme.

Barrett, S.C.H. (2015) Foundations of invasion genetics: the Baker and Stebbins legacy. Molecular Ecology, 24, 1927–41.

Bar-Yam, Y. (2002) Complexity rising: from human beings to human civilization, a complexity profile. In: Kiel, D. (ed.) Knowledge Management, Organizational Intelligence and Learning, and Complexity, Encyclopedia of Life Support Systems (EOLSS), Vol II. Oxford: EOLSS Publishers/UNESCO, pp. 22–41.

Beals, E.W (1972) Fundamentals of ecology updated. Ecology, 53, 983–4.

Bellard, C., Cassey, P., and Blackburn, T.M. (2016) Alien species as a driver of recent extinctions. Biology Letters, 12, 20150623.

Bennett, B.M. (2014) Model invasions and the development of national concerns over invasive introduced trees: insights from South African history. Biological Invasions, 16, 499–512.

Blackburn, T.M., Essl, F., Evans, T., et al. (2014) Towards a unified classification of alien species based on the magnitude of their environmental impacts. PLoS Biology, 12, e1001850.

Blackburn, T.M., Pyšek, P., Bacher, S., et al. (2011) A proposed unified framework for biological invasions. Trends in Ecology and Evolution, 26, 333–9.

Brown, J.H. and Gibson, A.C. (1983) Biogeography. St Louis, MO: The C.V. Mosby Company.

Bruno, J.F., Stachowicz, J.J., and Bertness, M.D. (2003) Inclusion of facilitation into ecological theory. Trends in Ecology and Evolution, 18, 119–25.

Buldyrev, S.V., Parshani, R., Paul, G., et al. (2010) Catastrophic cascade of failures in interdependent networks. Nature, 464, 1025–8.

Butchart, S.H.M., Walpole, M., Collen, B., et al. (2010) Global biodiversity: indicators of recent declines. Science, 328, 1164–8.

Caley, P., Groves, R.H., and Barker, R. (2008) Estimating the invasion success of introduced plants. Diversity and Distributions, 14, 196–203.

Callaway, R.M. and Ridenour, W.M. (2004) Novel weapons: invasive success and the evolution of increased competitive ability. Frontiers in Ecology and the Environment, 2, 436–43.

Callaway, R.M., Thelen, G.C., Rodriguez, A., et al. (2004) Soil biota and exotic plant invasion. Nature, 427, 731–3.

Carlton, J.T. (2011) The inviolate sea? Charles Elton and biological invasions in the world's oceans. In: Richardson, D.M. (ed.) Fifty Years of Invasion Ecology. The Legacy of Charles Elton. Oxford: Wiley-Blackwell, pp. 25–34.

Carson, R. (1962) Silent Spring. New York, NY: Houghton Mifflin.

Chytrý, M., Jarošík, V., Pyšek, P., et al. (2008) Separating habitat invasibility by alien plants from the actual level of invasion. Ecology, 89, 1541–53.

Clout, M.N. and Williams, P.A. (eds) (2009) Invasive Species Management. A Handbook of Principles and Techniques. Oxford: Oxford University Press.

Colautti, R.I., Grigorovich, I.A., and MacIsaac, H.J. (2006) Propagule pressure: a null model for biological invasions. Biological Invasions, 8, 1023–37.

Connell, J.H. (1978) Diversity in tropical rain forest and coral reefs. Science, 199, 1302–10.

Crosby, A.W. (1986) Ecological Imperialism. The Biological Expansion of Europe, 900–1900. Cambridge: Cambridge University Press.

Davis, M.A., Grime, J.P., and Thompson, K. (2000) Fluctuating resources in plant communities: a general theory of invasibility. Journal of Ecology, 88, 528–34.

Dlugosch, K.M. and Parker, I.M. (2008) Founding events in species invasions: genetic variation, adaptive evolution, and the role of multiple introductions. Molecular Ecology, 17, 431–49.

Drake, J.A., Mooney, H.A., di Castri, F., et al. (eds) (1989) Biological Invasions: A Global Perspective. Chichester: John Wiley & Sons.

Ehrlich, P.R. (1968) *The Population Bomb*. New York, NY: Ballantine Books.

Elton, C.S. (1958) *The Ecology of Invasions by Animals and Plants*. London: Methuen.

Essl, F., Bacher, S., Blackburn, T.M., *et al.* (2015a) Crossing frontiers in tackling pathways of biological invasions. BioScience, 65, 769–82.

Essl, F., Dullinger, S., Rabitsch, R., *et al.* (2015b) Historical legacies accumulate to shape future biodiversity change. Diversity and Distributions, 21, 534–47.

Essl, F., Dullinger, S., Rabitsch, W., Hulme, P.E., *et al.* (2011) Socioeconomic legacy yields an invasion debt. Proceedings of the National Academy of Sciences of the USA, 108, 203–7.

Estoup, A. and Guillemaud, T. (2010) Reconstructing routes of invasion using genetic data: why, how and so what? Molecular Ecology, 19, 4113–30.

Ford, E.B. (1964) *Ecological Genetics*. London: Methuen.

Franklin, J., Serra-Diaz, J.M., Syphard, A.D., *et al.* (2016) Global change and terrestrial plant community dynamics. Proceedings of the National Academy of Sciences of the USA, 113, 3725–34.

Fridley, J.D. (2011) Biodiversity as bulwark against invasion: conceptual threads since Elton. In: Richardson, D.M. (ed.) *Fifty Years of Invasion Ecology. The Legacy of Charles Elton*. Oxford: Wiley-Blackwell, pp. 121–30.

Gallien, L., Douzet, R., Pratte, S., *et al.* (2012) Invasive species distribution models—how violating the equilibrium assumption can create new insights. Global Ecology and Biogeography, 11, 1126–36.

Gallien, L., Munkemuller, T., Albert, C.H., *et al.* (2010) Predicting potential distributions of invasive species: where to go from here? Diversity and Distributions, 16, 331–42.

Gilbert, B. and Levine, J.M. (2013) Plant invasions and extinction debts. Proceedings of the National Academy of Sciences of the USA, 110, 1744–9.

Grosholz, E.D. (2005) Recent biological invasion may hasten invasional meltdown by accelerating historical introductions. Proceedings of the National Academy of Sciences of the USA, 102, 1088–91.

Hardin, G. (1968) The tragedy of the commons. Science, 162, 1243–8.

Harper, J.L. (1977) *Population Biology of Plants*. London: Academic Press.

Hengeveld, R. (1989) *Dynamics of Biological Invasions*. London: Chapman and Hall.

Hobbs, R.J., Arico, S., Aronson, J., *et al.* (2006) Novel ecosystems: theoretical and management aspects of the new ecological world order. Global Ecology and Biogeography, 15, 1–7.

Hobbs, R.J., Higgs, E., and Harris, J.A. (2009) Novel ecosystems: implications for conservation and restoration. Trends in Ecology and Evolution, 24, 599–605.

Hobbs, R.J. and Richardson, D.M. (2011) Invasion ecology and restoration ecology: parallel evolution in two fields of endeavour. In: Richardson, D.M. (ed.) *Fifty Years of Invasion Ecology. The Legacy of Charles Elton*. Oxford: Wiley-Blackwell, pp. 61–9.

Hulme, P.E., Bacher, S., Kenis, M., *et al.* (2008) Grasping at the routes of biological invasions: a framework for integrating pathways into policy. Journal of Applied Ecology, 45, 403–14.

Keane, R.M. and Crawley, M.J. (2002) Exotic plant invasions and the enemy release hypothesis. Trends in Ecology and Evolution, 17, 164–70.

Kitching, R.L. (2011) A world of thought: 'The Ecology of Invasions by Animals and Plants' and Charles Elton's life work. In: Richardson, D.M. (ed.) *Fifty Years of Invasion Ecology. The Legacy of Charles Elton*. Oxford: Wiley-Blackwell, pp. 3–24.

Kowarik, I. and Pyšek, P. (2012) The first steps towards unifying concepts in invasion ecology were made one hundred years ago: revisiting the work of the Swiss botanist Albert Thellung. Diversity and Distributions, 18, 1243–52.

Kueffer, C. (2013) Integrating natural and social sciences for understanding and managing plant invasions. In: Larrue, S. (ed.) *Biodiversity and Society in the Pacific Islands*. Marseille: Presses Universitaires de Provence and Canberra: ANU Press, pp. 71–95.

Kueffer, C. and Hirsch Hadorn, G. (2008) How to achieve effectiveness in problem-oriented landscape research: the example of research on biotic invasions. Living Reviews in Landscape Research, 2, 2.

Le Roux, J. and Wieczorek, A.M. (2009) Molecular systematics and population genetics of biological invasions: towards a better understanding of invasive species management. Annals of Applied Biology, 154, 1–17.

Lever, C. (2011) Acclimatization Societies. In: Simberloff, D. and Rejmánek, M. (eds) *Encyclopedia of Biological Invasions*. Berkeley, CA: University of California Press, pp. 1–4.

Levin, S.A. (1999) *Fragile Dominion: Complexity and the Commons*. New York, NY: Basic Books.

Levine, J.M. (2000) Species diversity and biological invasions: relating local process to community pattern. Science, 288, 852–4.

Liebhold, A.M. and McCullough, D.G. (2011) Forest insects. In: Simberloff, D. and Rejmánek, M. (eds) *Encyclopedia of Biological Invasions*. Berkeley, CA: University of California Press, pp. 238–41.

Lockwood, J. L., Cassey, P., and Blackburn, T. (2005) The role of propagule pressure in explaining species invasions. Trends in Ecology and Evolution, 20, 223–8.

Lonsdale, W.M. (1999) Global patterns of plant invasions and the concept of invasibility. Ecology, 80, 1522–36.

Loope, L. (2011) Hawaiian Islands: invasions. In: Simberloff, D. and Rejmánek, M. (eds) *Encyclopedia of Biological Invasions*. Berkeley, CA: University of California Press, pp. 309–19.

MacArthur, R.H. and Wilson, E.O. (1967) *The Theory of Island Biogeography*. Princeton, NJ: Princeton University Press.

May, R.M. (1973) *Stability and Complexity in Model Ecosystems*. Princeton, NJ: Princeton University Press.

May, R.M. (2001) *Biological Diversity: Causes, Consequences, and Conservation. Blue Planet Prize 2001 Commemorative Lecture*. Yonbancho: The Asahi Glass Foundation.

McEvoy, P.B. and Coombs, E.M. (1999) Biological control of plant invaders: regional patterns, field experiments, and structured population models. Ecological Applications, 9, 387–401.

McGeoch, M.A., Butchart, S.H.M., Spear, D., et al. (2010) Global indicators of biological invasion: species numbers, biodiversity impact and policy responses. Diversity and Distributions, 16, 95–108.

McNeely, J.A., Mooney, H.A., Neville, L.E., et al. (2001) *Global Strategy on Invasive Alien Species*. Gland: IUCN.

Mooney, H.A. (1998) *The Globalization of Ecological Thought*. Oldendorf: Ecology Institute.

Mooney, H.A. (1999) A global strategy for dealing with alien invasive species. In: Sandlund, O.T., Schei, P.J., and Viken, A. (eds) *Invasive Species and Biodiversity Management*. Dordrecht: Kluwer Academic Publishers, pp. 407–18.

Mooney, H.A., Mack, R.M., McNeely, J.A., et al. (eds) (2005) *Invasive Alien Species: Searching for Solutions*. Washington, DC: Island Press.

Naeem, S., Knops, J.M.H., Tilman, D., et al. (2000) Plant diversity increases resistance to invasion in the absence of covarying extrinsic factors. Oikos, 91, 97–108.

O'Dowd, D.J., Green, P.T., and Lake, P.S. (2003) Invasional meltdown on an oceanic island. Ecology Letters, 6, 812–17.

Parker, I.M., Simberloff, D., Lonsdale, W.M., et al. (1999) Impact: toward a framework for understanding the ecological effect of invaders. Biological Invasions, 1, 3–19.

Parrott, L. (2010) Measuring ecological complexity. Ecological Indicators, 10, 1069–76.

Parrott, L. and Meyer, W. (2012) Future landscapes: managing within complexity. Frontiers in Ecology and the Environment, 10, 382–9.

Parshani, R., Buldyrev, S.V., and Havlin, S. (2010) Interdependent networks: reducing the coupling strength leads to a change from a first to second order percolation transition. Physical Review Letters, 105, 048701.

Pheloung, P.C., Williams. P.A., and Halloy, S.R. (1999) A weed risk assessment model for use as a biosecurity tool evaluating plant introductions. Journal of Environmental Management, 57, 239–51.

Pianka, E.R. (1967) On lizard species diversity—North American flatland deserts. Ecology, 48, 333–51.

Pyšek, P., Jarošík, V., Hulme, P.E., et al. (2010) Disentangling the role of environmental and human pressures on biological invasions across Europe. Proceedings of the National Academy of Sciences of the USA, 107, 12157–62.

Pyšek, P., Jarošík, V., Hulme, P.E., et al. (2012) A global assessment of invasive plant impacts on resident species, communities and ecosystems: the interaction of impact measures, invading species' traits and environment. Global Change Biology, 18, 1725–37.

Pyšek, P. and Richardson, D.M. (2006) The biogeography of naturalization in alien plants. Journal of Biogeography, 33, 2040–50.

Pyšek, P. and Richardson, D.M. (2010) Invasive species, environmental change and management, and ecosystem health. Annual Review of Environment and Resources, 35, 25–55.

Rejmánek, M. and Richardson, D.M. (1996) What attributes make some plant species more invasive? Ecology, 77, 1655–61.

Rejmánek, M., Richardson, D.M., and Pyšek, P. (2013) Plant invasions and invasibility of plant communities. In: van der Maarel, E. and Franklin, J. (eds) *Vegetation Ecology* (2nd edn). Oxford: Wiley-Blackwell, pp. 387–424.

Ricciardi, A. and MacIsaac, H.J. (2000) Recent mass invasion of the North American Great Lakes by Ponto-Caspian species. Trends in Ecology and Evolution, 15, 62–5.

Richardson, D.M. (ed.) (2011) *Fifty Years of Invasion Ecology. The Legacy of Charles Elton*. Oxford: Wiley-Blackwell.

Richardson, D.M., Allsopp, N., D'Antonio, C.M., et al. (2000b) Plant invasions: the role of mutualisms. Biological Reviews, 75, 65–93.

Richardson, D.M., Carruthers, J., Hui, C., et al. (2011a) Human-mediated introductions of Australian acacias—a global experiment in biogeography. Diversity and Distributions, 17, 771–87.

Richardson, D.M., Macdonald, I.A.W., Hoffmann, J.H., et al. (1997) Alien plant invasions. In: Cowling, R.M., Richardson, D.M., and Pierce, S.M. (eds) *Vegetation of Southern Africa*. Cambridge: Cambridge University Press, pp. 535–70.

Richardson, D.M. and Pyšek, P. (2007) Classics in physical geography revisited: Elton, C.S. 1958: The ecology of invasions by animals and plants. London: Methuen. Progress in Physical Geography, 31, 659–66.

Richardson, D.M. and Pyšek, P. (2008) Fifty years of invasion ecology—the legacy of Charles Elton. Diversity and Distributions, 14, 161–8.

Richardson, D.M., Pyšek, P., and Carlton, J.T. (2011b) A compendium of essential concepts and terminology in invasion ecology. In: Richardson, D.M. (ed.) *Fifty Years of Invasion Ecology. The Legacy of Charles Elton.* Oxford: Wiley-Blackwell, pp. 409–20.

Richardson, D.M., Pyšek, P., Rejmánek, M., *et al.* (2000a) Naturalization and invasion of alien plants—concepts and definitions. Diversity and Distributions, 6, 93–107.

Richardson, D.M. and Ricciardi, A. (2013) Misleading criticisms of invasion science: a field-guide. Diversity and Distributions, 19, 1461–7.

Rouget, M., Robertson, M.P., Wilson, J.R.U., *et al.* (2016) Invasion debt—quantifying future biological invasions. Diversity and Distributions, 22, 445–56.

Roura-Pascual, N., Hui, C., Ikeda, T., *et al.* (2011) Relative roles of climatic suitability and anthropogenic influence in determining the pattern of spread in a global invader. Proceedings of the National Academy of Sciences of the USA, 108, 220–5.

Ruiz G. and Carlton J.T. (eds) (2003) *Invasive Species: Vectors and Management Strategies.* Washington, DC: Island Press.

Sandlund, O.T., Schei, P.J., and Viken, A. (eds) (1998) *Invasive Species and Biodiversity Management.* Dordrecht: Kluwer Academic Publishers.

Saville, D.B.O. (1960) Limitations of the competitive exclusion principle. Science, 132, 1761.

Shea, K. and Chesson, P. (2002) Community ecology theory as a framework for biological invasions. Trends in Ecology and Evolution, 17, 170–6.

Simberloff, D. (2009) The role of propagule pressure in biological invasions. Annual Review of Ecology, Evolution, and Systematics, 40, 81–102.

Simberloff, D. (2011) Charles Elton—Neither founder nor siren, but prophet. In: Richardson, D.M. (ed.) *Fifty Years of Invasion Ecology: The Legacy of Charles Elton.* Oxford: Wiley-Blackwell, pp. 11–24.

Simberloff, D. and von Holle, B. (1999) Positive interaction of nonindigenous species: invasional meltdown? Biological Invasions, 1, 21–32.

Soberón, J. (2007) Grinnellian and Eltonian niches and geographic distributions of species. Ecology Letters, 10, 1115–23.

Solé, R. and Goodwin, B. (2000) *Signs of Life: How Complexity Pervades Biology.* New York, NY: Basic Books.

Vilà, M., Espinar, J.L., Hejda, M., *et al.* (2011) Ecological impacts of invasive alien plants: a meta-analysis of their effects on species, communities and ecosystems. Ecology Letters, 14, 702–8.

Vitousek, P.M. (2001) The ecology of invasions by animals and plants (book review). Biological Invasions, 3, 219.

Vitousek, P.M., Mooney, H.A., Lubchenco, J., *et al.* (1997) Human domination of Earth's ecosystems. Science, 277, 494–9.

Waloff, N. (1966) Scotch broom (*Sarothamnus scoparius* (L.) Wimmer) and its insect fauna introduced into Pacific Northwest of America. Journal of Applied Ecology, 3, 293–311.

Wilson, J.R.U., Dormontt, E.E., Prentis, P.J., *et al.* (2009) Something in the way you move: dispersal pathways affect invasion success. Trends in Ecology and Evolution, 24, 136–44.

Woodwell, G.M. and Smith, H.H. (eds) (1969) *Diversity and Stability in Ecological Systems. Brookhaven Symposium of Biology.* New York, NY: Brookhaven National Laboratory.

van Kleunen, M. Dawson, W., Schlaepfer, D., *et al.* (2010a) Are invaders different? A conceptual framework of comparative approaches for assessing determinants of invasiveness. Ecology Letters, 13, 947–58.

van Kleunen, M., Weber, E., and Fischer, M. (2010b) A meta-analysis of trait differences between invasive and non-invasive plant species. Ecology Letters, 13, 235–45.

Zavaleta, E., Hobbs, R.J., and Mooney, H.A. (2001) Viewing invasive species removal in a whole-ecosystem context. Trends in Ecology and Evolution, 16, 454–9.

PART I

Spread

CHAPTER 2

The dynamics of spread

2.1 Introduction

Once introduced and established, alien species have the potential to spread and expand their ranges in recipient ecosystems, resulting in different patterns of spread. Because spread is an emergent pattern of population dynamics, key components of population demography, such as fecundity, mortality, and dispersal, mediate the form and velocity of spread, as do habitat heterogeneity and dispersal vectors and barriers. Clarifying the mediators of spread is crucial for accurate modelling and efficient management of invasive species. Knowing the rate of spread of an invasive species in particular habitats can help to guide management activities and policy. Understanding the patterns of spread over space and time and how the dynamics of expansion change across landscapes paves the way for more efficient allocation of resources to different facets of management. In this chapter, we categorize typical spreading dynamics associated with the range expansion of invasive organisms.

Range expansion is the spatial manifestation of population growth. Population dynamics are largely regulated by two mechanisms: negative density dependence at high densities, and positive density dependence at low densities. In the absence of such regulating mechanisms, populations undergo exponential growth, as in a Malthusian population, and range expansion without any regulation should follow an exponential increase as long as habitat is available to accommodate such expansion. Similarly, once the niche-filling stage is almost complete, the exponential expansion will then change to a sigmoidal expansion and finally stop once all accessible habitat is occupied. Under a weak or thin-tailed dispersal strategy, exponential expansion often morphs into a linear expansion with a constant rate of spread as only limited numbers of individuals can be dispersed from the core to beyond the existing range into unoccupied area forming the new range front.

Most studies, especially those covering wide geographical ranges, document accelerating expansion that converges on a much higher asymptotic velocity until all available habitat is occupied. Accelerating range expansion can be achieved by reverting back to exponential expansion through fat-tailed dispersal, or continuously sorting those individuals to the range front with strong dispersal ability and capacity to handle difficulty when facing low propagule pressure at the range front. Accelerating range expansion can also comprise two phases: slow linear expansion, followed by a much faster expansion which may still be linear but at a higher rate or may be exponentially accelerating. The transition from slow to fast range expansion is often associated with changing habitat features and environmental drivers that could have altered the mechanism of demographic regulation. It can, of course, also be the result of stratified dispersal consisting of two dispersal rates (e.g. seed dispersal by both physical forcing, such as wind and ocean currents, and vectors, such as frugivores and humans).

In many cases, range expansion does not occur immediately after introduction. Rather, populations remain in a 'dormant' or 'lag' phase which may last for decades. The mediators of lag phases are not adequately understood for many invasions because introduction events are seldom documented in sufficient detail, but could be related to factors behind prolonged positive density dependence. Pre-adaption, propagule pressure, and suitability of habitat at sites of introduction all affect the presence and duration of lag phase.

Invasion Dynamics. Cang Hui and David M. Richardson, Oxford University Press (2017).

Rapid range expansion sometimes follows a boom-and-bust pattern. This is because the performance or the fitness of an alien species, normally expressed by its per-capita population growth rate, is frequency- or density-dependent. A successful invader possessing an 'r strategy' could be a superior colonizer when its population size is small, but could lose such advantage once negative density dependence kicks in. Most cases of boom-and-bust phenomena in invasion ecology are attributable to the success of control strategies and/or large shocks from strong perturbations or adaptive biotic interactions/responses, or simply a response to worsening environmental conditions.

To complicate matters, the same invasive species can show different forms and rates of spread in different regions, a completely contextually dependent phenomenon. Even under the same context, environmental and demographical stochasticity can also affect the population demography and produce diverse forms of realized spread dynamics.

Adding to the temporal complexity, spreading alien organisms can also show complex spatial patterns. For instance, the frontier of expansion is not always continuous. In most cases the leading front of an invasion is ragged, with small vanguard populations extending ahead of the front. The isolation of such populations may reflect habitat heterogeneity or fragmentation, but this is very often the result of stratified dispersal. Many theoretical studies have documented the possibility of highly synchronized spatial patterns (spirals and waves) from self-organized local interactions. For instance, Allee effects, trophic interactions, and density-dependent dispersal can form spatial chaos, leading to fascinating spatial patterns during range expansion.

Range expansion of invasive alien species, coupled with chance and adaptation, can last for decades or even centuries, continuously pushing range limits and negotiating novel habitat before all accessible habitats are occupied. Such a long time span complicates the reconstruction of range expansion, which typically involves assessing records from herbaria, museums, and even grey literature sources which have many biases and prejudices. Imperfect detection ability and low monitoring effort also hinder the documentation of newly arrived alien species, especially in the case of unintentional introductions. The public are often only interested in the fast-expanding phase and have little interest, for different reasons, in the two ends of expansion (the initial or final phases). Once the range of an introduced species covers large areas, even if the range remains much smaller than that defined by the extent of accessible and potentially suitable habitat, public interest typically wanes which results in a reduction in support of further monitoring.

The rate of range expansion is normally estimated by plotting the distance from the advancing front to the introduction point against time. For the spread over two-dimensional space, the rate of range expansion can be estimated along a transect, or replacing the distance by the square root of the area occupied (an indicator of the diameter of the range). The accumulative rate of detected records, after correcting for sampling biases, may also be used as a proxy for the spreading rate. Once the temporal dynamics of range expansion are reconstructed, linear, non-linear, and segmented regression applied to estimate the rate of spread over different periods. Methods of time-series analysis can also be used to detect trends and non-stationarity in the dynamics.

In the following sections we elaborate on the complexity of spreading dynamics by discussing seven aspects thereof, providing examples and expanding on potential mechanisms. In particular, we divide typical spreading dynamics into four patterns: (i) exponential and sigmoidal expansion; (ii) linear expansion with a constant rate of spread; (iii) biphasic expansion with a faster linear expansion following a slower linear expansion; and (iv) acceleration with a continuously increasing rate of spread. We also give attention to three particular phenomena associated with invasion dynamics: (v) lag phase, (vi) boom and bust, and (vii) contextual discordance of spread.

2.2 Exponential and sigmoidal expansion

Exponential range expansion and the subsequently sigmoidal phase are direct manifestations of the dynamics of population growth, in particular for species with good dispersal ability. Such patterns of spread have been reported for the range expansion of many invasive species. We discuss a few examples of such exponential and sigmoidal expansion.

The first example is the rapidly spreading pest, the hemlock woolly adelgid *Adelges tsugae*, in eastern North America, since its introduction in 1950s. The woolly adelgid feeds on the phloem sap of hemlock shoots and injects toxins while feeding; this causes trees to lose needles and stops new growth. The pest has been dispersed by both wild animals and humans (Turner *et al.* 2011) and has spread rapidly, killing large tracts of hemlock forests in the eastern United States (Orwig 2002). Many studies have attempted to predict the spread dynamics of hemlock woolly adelgid (Fitzpatrick *et al.* 2012; Ferrari *et al.* 2014). By resampling the data from Ferrari and his colleagues' work (2014), we see that its expansion across the eastern United States followed a typical exponential pattern:

$$A_t = A_0 \cdot \exp(r \cdot t) \qquad (2.1)$$

where A_t is the area occupied (or the number of counties occupied in this case) after t number of years elapsed since introduction, and A_0 and r are two model parameters. In particular, parameter A_0 represents the initial occupied area, and r the intrinsic population growth (expansion) rate. The invasive adelgid follows this exponential expansion, with $A_t = 0.86 \cdot \exp(0.099t)$ and $R^2 = 0.98$, suggesting a 10 per cent annual expansion rate. When estimating the rate of range expansion using the square root of occupied area, $A^{1/2}$, the exponential expansion often resembles a biphasic expansion (see section 2.4), as is the case of an apparent transition in 1985 in the spread dynamics of hemlock woolly adelgid in the eastern United States. Exponential expansion simply represents the spatial realization of Malthusian population growth with adequate dispersal to alleviate the crowding density at the core, whereas the biphasic expansion could, in contrast, point to either a change of internal demographic regulators or the effects of external drivers at the inflection point (section 2.4).

The second example is the extra-limital expansion of the Mexican grackle, *Quiscalus mexicanus*, in North America since 1880 (Wehtje 2003). Native to Central America, this blackbird species was still confined to breeding locations in southern Texas in 1900 but had bred in 20 southern states of the United States by 2001. It is sexually dimorphic and avoids forests and areas lacking surface water (Johnson and Peer 2001). The range expansion has been attributed to its large body size and its social, omnivorous, opportunistic, and ecologically flexible characteristics (Elton 1958; Mayr 1965). Wehtje (2003) reconstructed the range expansion of the species in the United States using specimens from museums and field surveys (Fig. 2.1). The spread exhibited a classic exponential expansion, where a small population expands its range slowly until population size has accumulated sufficiently to allow rapid expansion, with $A_t = 67662\exp(0.0323t)$ in km^2 and $R^2 = 0.985$, suggesting a 3.2 per cent annual increase rate. The asymptotic rate of northward expansion converges to 12.7 km/yr. The extra-limital range expansion of Mexican grackles shows two key features: a change in migratory behaviour in the novel range, and the rapid establishment of breeding populations from scarce vagrants, with the average interval between first breeding and first wintering being only 6.6 years (Wehtje 2003).

The third example is the sigmoidal expansion of the pine wilt disease caused by the pine wood nematode, *Bursaphelenchus xylophilus*, in the Korean Peninsula. Native to North America (Kiritani and Morimoto 2004), this tree-parasitic nematode is a serious pest of pine trees in many countries, including China (mainland and Taiwan), Japan, Portugal, and South Korea (Togashi and Shigesada 2006). Juveniles of the nematode enter the body of intermediate insect vectors through the tracheal system. In the Korean Peninsula, the Japanese pine sawyer, *Monochamus alternatus*, and long-horned beetle, *M. saltuarius*, are vectors of pine wood nematode, while the host trees include Japanese red pine (*Pinus densiflora*), Japanese black pine (*P. thunbergii*), and Korean white pine (*P. koraiensis*) (Shin 2008). Pine wilt disease was first reported in South Korea in the southern port of Busan in 1988, and three new occupied areas were reported in 1999—likely to have been the result of human-mediated dispersal (Shin 2008). The initial spread rate, from 1988 to 2000 in the Busan area, was around 1.1 km/yr, similar to the mean annual movement distance of the Japanese pine sawyer of 1.4 km/yr (Lee *et al.* 2007; Shin 2008); it later accelerated to 13.8 km/yr. A sigmoidal expansion follows the following dynamics of occupied area,

$$A_t = \frac{K}{1 + (1/A_0 - 1)\exp(-r \cdot t)} \qquad (2.2)$$

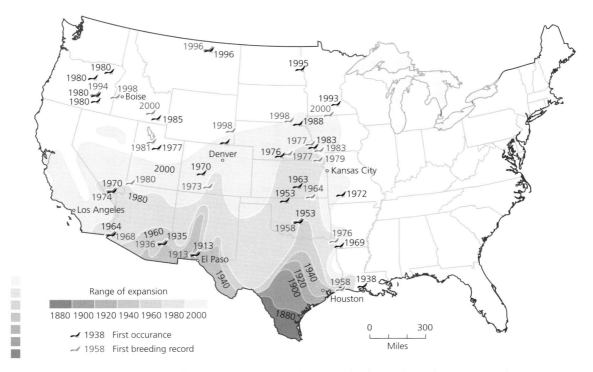

Fig. 2.1 Expansion of the breeding range of the Mexican grackle (*Quiscalus mexicanus*) in the United States between 1880 and 2000 (Wehtje 2003). Reproduced with permission from John Wiley & Sons.

where K is the maximum possible range of the species. The spread of pine wood nematode in the Korean Peninsula, after it was first detected in 1988, followed a sigmoidal pattern (Fig. 2.2), with $K = 73.3$ (number of areas damaged) and $r = 1/2.65 = 0.377$ (Choi and Park 2012), suggesting a nearly 40 per cent annual increase in the number of areas infested.

Sigmoidal expansion sometimes follows a seemingly three-segment mode of a slow initial phase, a fast middle phase around the inflection point of the dynamics, and a slow final phase when no further suitable habitat is available. For instance, the common waxbill *Estrilda astrild* was released in coastal Portugal in the early 1960s and subsequently spread across the Iberian Peninsula (Reino and Silva 1998). Silva and colleagues (2002) identified three phases in the range expansion of this species in Portugal

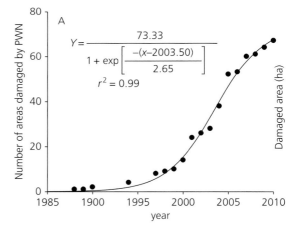

Fig. 2.2 Number of cities experiencing the disease caused by the pine wilt nematode (PWN, *Bursaphelenchus xylophilus*) along the year in the Korean Peninsula (Choi and Park 2012). Reproduced with permission from John Wiley & Sons.

between 1964 and 1999 (Fig. 2.3a): a slow phase between 1964 and 1978, when the species showed limited dispersal due to difficulties in finding mates and optimal habitat; a fast phase between 1978 and 1987, due to disperser build-up in high-quality habitats; and a slow rate after 1987 by which time most optimal habitats had been occupied.

A similar sigmoidal pattern was observed during the range expansion of the non-pest saproxylic beetle *Cis bilamellatus* in Britain (Orledge *et al.* 2010). This obligate fungivore which typically lives and breeds in the fruiting bodies of wood-rotting basidiomycetes may have entered Britain in fungal material sent to Kew Herbarium during the 1850s (Reibnitz 1999; Orledge and Reynolds 2005). Before 1930, the rate was limited to 1–1.6 km/yr; the rate jumped to 13.3 km/yr between 1931 and 1970 when *C. bilamellatus* spread across south-west England, north Wales, and the Midland Valley of Scotland; expansion then slowed to 2.9 km/yr after 1971 and stopped after 2000 after it had reached all accessible and suitable habitat (Fig. 2.3b).

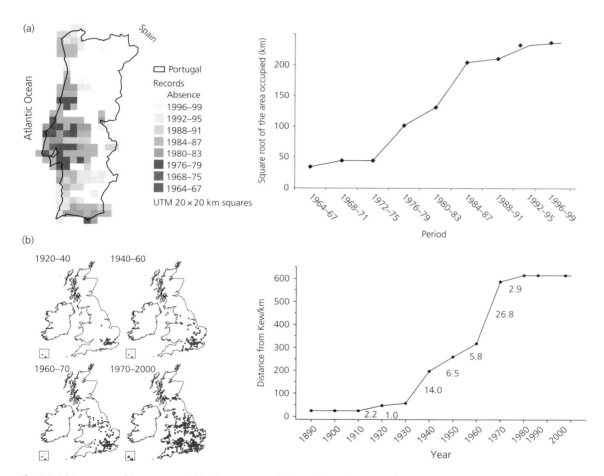

Fig. 2.3 (a) Square root of the area occupied by the common waxbill (*Estrilda astrild*) in Portugal, with a distribution map in 20 × 20 km grid (Silva *et al.* 2002). Reproduced with permission from John Wiley & Sons. (b) Cumulative recorded distance from Kew on mainland Britain for *Cis bilamellatus*, with annotations indicating mean spread rates in kilometres per year (Orledge *et al.* 2010). With permission of Springer.

2.3 Linear expansion

Linear expansion is the pattern of spread that was most widely studied in early works, especially following Skellam's (1951) classic work on estimating the constant rate of range expansion of muskrats (*Ondatra zibethica*) in Europe. A linear expansion can be depicted as

$$\sqrt{A_t} = c \cdot t \qquad (2.3)$$

where c is the rate or the velocity of spread. The square root of occupied area can also be replaced by the distance to the introduction location. The linear expansion is traditionally regarded as the null model of spread, due to Skellam's influence. However, it only applies to the expansion with limited or thin-tailed dispersal (e.g. the dispersal distance follows a negative exponential distribution). Here, we regard the exponential and sigmoidal expansion as the null, and linear expansion as a spreading mode constrained by limited dispersal. Although linear expansion has been thought to be common, recent examples from biological invasions of this type are rare, and most conform with linear expansion only over limited periods. We now discuss a few examples of linear expansion.

The first example is the Eurasian beaver, *Castor fiber*, a semi-aquatic herbivorous mammal and the largest European rodent. Beavers were widely hunted in Europe before the nineteenth century, almost leading to their extirpation in the twentieth century (Véron 1992). After a series of reintroductions, beavers have recolonized large parts of their former ranges (Halley and Rosell 2002). Barták and colleagues (2013) modelled the spatial spread of Eurasian beaver in the river networks of the Czech Republic. The expansion of three populations (Morava south, Morava central, and Berounka) was reconstructed from historical records and an ongoing long-term monitoring programme (Fig. 2.4). Although beavers can occasionally move across land, dispersal normally requires a river network (Novak 1987), usually for sub adults during the spring (Sun *et al.* 2000). Barták and colleagues (2013) reported an exponential increase of cumulative records and network range size (measured in this case as cumulative length of dispersal trajectories). The rate of expansion was 2.3, 3.4, and 1.5 km/yr for the Morava south, Morava central, and Berounka populations, respectively (Fig. 2.4). This suggests that an exponential expansion along river network (at a rate of 15–20 km/yr) can be equivalent to a linear expansion at a slower rate when measured as the diameter of the two-dimensional range. In this sense, linear expansion can be considered a crude realization on the two-dimensional space when the species actually follows a linear pathway with limited dispersal capacity.

The second example of a linear range expansion deals with the fall webworm, *Hyphantria cunea*, which defoliates deciduous trees. In its native North American range, the fall webworm usually targets

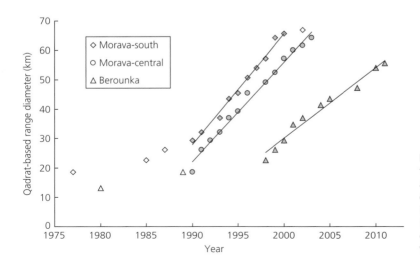

Fig. 2.4 Linear regression of the classical two-dimensional range diameter, computed as the diameter of a circle with area equalling the range size. The range size is estimated as the area of the occupied KFME (Kartierung der Flora Mitteleuropas) quadrats (Barták *et al.* 2013). Reproduced with permission from John Wiley & Sons.

trees in late summer when photosynthesis has nearly ceased for winter dormancy, and it therefore normally causes only limited damage to deciduous trees. It was confined to North America before the Second World War, but it dispersed to central Europe and eastern Asia during the 1940s probably with movement of larva in wooden artillery boxes (Gomi 2007). In East Asia, it was first reported in Japan in 1945 and in South Korea in 1958, as a result of the establishment of US Army bases (Chung *et al.* 1987), and then in China in 1979 due to the beginning of open trading in China (Yang and Zhang 2007). The dispersal capacity of a fall webworm is less than 300 m/day (Yamanaka *et al.* 2001), translating to a rate of 100 km/yr if establishment and dispersal seasonality are ignored. Although the spread of fall webworms has been categorized as a linear expansion in South Korea (Fig. 2.5) at an average speed of 31 km/yr (Choi and Park 2012), we can clearly see the sigmoidal shape of the expansion in the data; this suggests that many reported cases of linear expansion could actually be sigmoidal expansion. As the sigmoidal form includes three parameters, in contrast to the two-parameter linear form, we recommend the use of Akaike Information Criterion for selecting the best form when dealing with real data.

Other examples of linear expansion include the rapid spread of the quagga mussel, *Dreissena rostriformis*, in freshwater ecosystems. Dreissenids negatively impact freshwater ecosystems by altering macroinvertebrate composition, reducing plankton abundance, increasing macrophyte and benthic algal growth (Kelly *et al.* 2010), and causing the proliferation of invertebrate predators (Mörtl *et al.* 2010; Van Eerden and De Leeuw 2010). This species was first detected in North America in 1991, the result of ballast water discharge containing larvae (May and Marsden 1992). In Europe, the quagga mussel only occurred in southern river mouths of the Black Sea until the 1940s (Van der Velde and Platvoet 2007) when it entered the Caspian Sea via the Volga–Don Canal (Orlova *et al.* 2004; Son 2007; Zhulidov *et al.* 2005). It was detected in the river Danube in Romania in 2004 (Micu and Telembici 2004; Popa and Popa 2006) and in a dam of the Rhine River in The Netherlands in 2006 (Molloy *et al.* 2007).

Quagga mussels can move in two ways: the larvae can float and drift, and adult mussels can secrete proteinaceous byssal threads that attach to boats, thereby achieving long-distance and even upstream dispersal (Johnson and Carlton 1996). These two dispersal strategies led to large variation in the rates of spread, ranging between 23 and 383 km/yr (Matthews *et al.* 2014), with an average of 120 km/yr (Pollux *et al.* 2010; Therriault and Orlova 2010). Due to the large variation from multi-vector dispersal, linear expansion was assumed for the expansion of quagga mussels. Indeed, linear expansion has often been preferred due to its simplicity, especially when data quality is poor or when observed patterns encompass large variations. We could expect other forms of expansion to emerge with the accumulation of records.

Because linear expansion often occurs only over a limited period, due largely to limitations on dispersal, it can appear together with other forms of spread. For instance, plant invasions have transformed grasslands over large tracts of the western United States from native perennial grasslands to ecosystems dominated by alien annuals. The transformation of vegetation dominated by sagebrush to cheatgrass (*Bromus tectorum*) or red brome (*B. madritensis*) monocultures, both from the Mediterranean Basin, has initiated a fire regime that prevents the establishment of native perennial species (D'Antonio and Vitousek 1992). During El Niño Southern Oscillation (ENSO) events, which produce above-average winter precipitation in the south-western United States, red brome can become dominant, whereas its populations plummet during years of below-average winter precipitation (Hunter

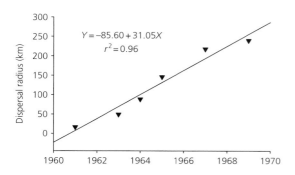

Fig. 2.5 Expansion of the fall webworm (*Hyphantria cunea*) 1958–70 in the Korean Peninsula (Choi and Park 2012). Reproduced with permission from John Wiley & Sons.

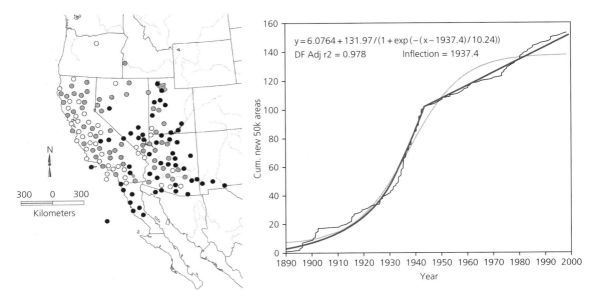

Fig. 2.6 The spread of red bromes (*Bromus madritensis*) in western North America depicted as area occupied (in 50 km units) (Salo 2005). Open circles: records 1880–1929; grey circles, 1930–1942; black circles, after 1942. Raw data: black line; fitted sigmoid function: grey line. With permission of Springer.

1991). Using herbarium records, Salo (2005) reconstructed the spread of red brome in western North America. A sigmoid curve was fitted to the cumulative number of new 50-kilometre areas occupied by red brome between 1890 and 1998 (Fig. 2.6). However, the records are better explained by an initial exponential expansion until 1940 followed by a linear expansion. Such a transition of spreading modes (from exponential to linear) coincided with the time when red brome had invaded the Mediterranean-climate region and started its eastward expansion into Arizona and south-western New Mexico.

2.4 Biphasic expansion

Biphasic range expansion is commonly observed in the spread of invasive species. The study of Okubo (1988) on the range expansion of common starlings, *Sturnus vulgaris*, in North America, and subsequent attempts to explain this phenomenon (e.g. Shigesada *et al.* 1995) are key examples. There are two primary drivers of such biphasic expansion: stratified dispersal and habitat transition. We can clearly see the two drivers in studies of the spread of common starlings in their invasive range in different parts of

the world. The species occurs naturally across Eurasia and is one of the most successful avian invaders (Blackburn *et al.* 2009). Approximately 80 per cent of introduced populations have become established (Sol *et al.* 2002), and the non-native range now includes North America, Australia, New Zealand, South Africa, and some Pacific islands (Long 1981). Okubo (1988) reported a biphasic expansion of common starlings in North America, after a ten-year lag phase, following the introduction of 160 birds to New York in 1880. Between 1900 and 1915 the spread was slow (11.2 km/yr) and linear. Between 1915 and 1950 the spread was fast (51.2 km/yr) and also linear. In Okubo's (1988) words, 'the faster rate of spread [in the second phase] is associated with invasion across the prairie, where fewer fruits and towns are available for the birds in comparison with the timbered east [in the first phase]'. This implies that the driver of phase change in spread was habitat transition. When Shigesada and colleagues (1995) re-examined this pattern, however, they designed a model that considers stratified diffusion of starlings, thus posing an alternative explanation for the biphasic expansion versus Okubo's (1988) original explanation by habitat transition.

Several recent studies have re-examined the relative importance of habitat transition and stratified diffusion in producing biphasic range expansions. Starlings were found to spread much slower following their introduction to southern hemisphere regions than was the case for North America. In South Africa, the species' range is still expanding following the release of 18 birds in Cape Town in 1897 (Harrison *et al.* 1997). A reconstruction of historical records also revealed a biphasic expansion for South Africa: a slower phase of linear expansion at the rate of 6.1 km/yr before 1940 and a faster phase of linear expansion at the rate of 25.7 km/yr after 1940 (Fig. 2.7; Hui *et al.* 2012)—half the speed of expansion recorded in North America. In Australia, 89 birds were introduced to Adelaide from the late 1850s through to 1870, and the initial spread proceeded at a similar pace to that observed during the fast phase in South Africa: 20.7 km/yr. Interestingly, there was no slow phase in Australia. The greater spread velocity in North America does not appear to be attributable to a difference in the diffusion rate. Rather, it appears to be driven by a higher intrinsic rate of increase (0.224 yr^{-1} in North America versus 0.115 yr^{-1} in South Africa) (Hui *et al.* 2012). This is perhaps a consequence of more second broods and larger clutch sizes in the northern hemisphere compared to equivalent southern latitudes with different climatic regimes (Lack 1968; Evans *et al.* 2005; Hui *et al.* 2012). Hui and colleagues (2012) reported an inverse power function for the dispersal kernel of starlings, frequency (d) $\sim d^{-3/2}$, and suggested that this fat-tailed long-distance dispersal strategy could well explain the two-phase range expansion of starlings. In contrast, spatial visualization and Bayesian assignment of genetic diversity revealed two subpopulations (Fig. 2.7) with a contact zone around 150–300 km east from the introduction site (Berthouly-Salazar *et al.* 2013). The contact zone represents an area where climatic conditions, especially winter precipitation, change rapidly. For this reason, both mechanisms (dispersal strategy and habitat transition) could be driving the biphasic range expansion.

Another classic example of biphasic range expansion is the case of the European map butterfly, *Araschnia levana*, which has a Palaearctic distribution across Eurasia and has been expanding westwards and northwards in the last century (Parmesan 2005). The recent expansion in Finland followed a typical biphasic mode. The species was first detected in eastern Finland 1973 and expanded its extra-limital range steadily until 1999 when the velocity of spread increased substantially (Mitikka *et al.* 2008). This species is a generalist in terms of habitat and host plants, and has been a model system for examining the effect of climate change on potential range shift in species (Parmesan and Yohe 2003; Mitikka *et al.* 2010). Its range has therefore been particularly well documented, for example through the efforts of the National Butterfly Recording Scheme in Finland. Mitikka and colleagues (2008) used four potential measures for quantifying the spread dynamics, including maximum annual dispersal distance, the annual average northwards shift in distribution, the number of newly occupied grid cells per annum, and the cumulative number of occupied grid cells. They examined three periods of the biphasic expansion (Fig. 2.8) and reported a 1.29–1.47 km/yr velocity before 1999 compared to 7.5 km/yr afterwards (Mitikka *et al.* 2008).

Map butterflies produce two broods in a season; first generations overwinter as pupae and develop second and third generations that are dependent on day length and temperature (Brakefield and Shreeve 1992). The second generation is normally bigger, has larger and less pointed wings which improves their long-distance flight capacity (Fric and Konvicka 2002), and normally occurs in summer and disperses actively (Virtanen and Neuvonen 1999). Mitikka and colleagues (2008) suggest that the acceleration from phase I (Fig. 2.8a and 2.8b) to phase II (Fig. 2.8c) could be due to higher average growing degrees and summer temperatures due to warmer climates since 1999 which have facilitated annual maximum dispersal distance and thus the rate of spread. Other studies on species with complex life cycles such as the cotton bollworm, *Helicoverpa armigera*, in East Asia (transition in 1992; Ouyang *et al.* 2014), and the rose chafer beetle, *Oxythyrea funesta*, in central Europe (transition in 1990; Horak *et al.* 2013) also suggest patterns of biphasic outbreak and expansion, largely due to climate change and human activities which weaken negative density dependence that is crucial in regulating population dynamics.

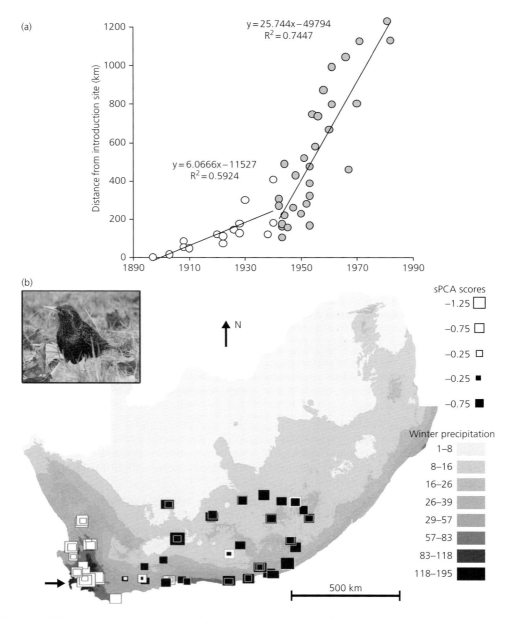

Fig. 2.7 Aspects of the range expansion of Common Starlings (*Sturnus vulgaris*) in South Africa. (a) Rates of expansion during the twentieth century (Hui *et al.* 2012). Reproduced with permission from John Wiley & Sons. (b) Spatial genetic patterns (Berthouly-Salazar *et al.* 2013). Shading indicates precipitation (in millimetres) during winter. The arrow indicates the site of introduction. Squares represent scores from the spatial principle component analysis of genetic diversity. Reproduced with permission from John Wiley & Sons.

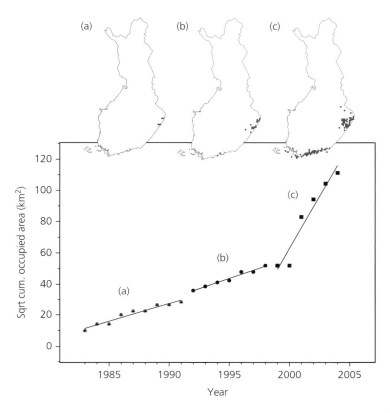

Fig. 2.8 Expansion of the map butterfly (*Araschnia levana*) in Finland between 1983 and 2004. The expansion is illustrated as the square root of the area occupied (km²) as a function of time (year) and is shown separately for three time periods: (a) the first years of expansion in eastern Finland, (b) expansion after 1992, when the first observation on the southern coast was made, and (c) expansion since the 1999 invasion of the southern coast (Mitikka *et al.* 2008). With permission of Springer.

A highly invasive pest that is currently experiencing biphasic expansion in North America is the emerald ash borer, *Agrilus planipennis*, a phloem feeder from Asia that arrived in Michigan in 2002 or earlier. It has caused massive damage to ash trees (*Fraxinus* spp.) in native forests and urban ecosystems (Cappaert *et al.* 2005; Poland and McCullough 2006). By 2014 it had been recorded in at least 22 states of the USA (Emerald Ash Borer Information 2014) and had become the most destructive and costly forest pest in North America (Aukema *et al.* 2011). Although it is only a secondary pest in its native range, where it kills only severely stressed ash trees (Liu *et al.* 2003), emerald ash borers attack both stressed and healthy trees in North America (McCullough *et al.* 2009; Siegert *et al.* 2010; Klooster *et al.* 2014; Herms and McCullough 2014).

To reconstruct the origin and progression of emerald ash borers, Siegert and colleagues (2014) collected increment cores from dead or declining ash trees in Michigan and traced the introduction back to 1997 in the small town of Canton in Ohio. A clear biphasic range expansion was reconstructed (Fig. 2.9), with a slow expansion of 3.84 km/yr from 1998 to 2001, and a more rapid expansion (12.97 km/yr) after 2001. The biphasic range expansion of emerald ash borer has been attributed to the coalescence of new satellite colonies and the primary population driven by natural dispersal and inadvertent human-assisted transport of infested ash material (Siegert *et al.* 2014).

Forest pests introduced from areas not far from the invaded region can also experience fast spread and biphasic expansion. For instance, the pine

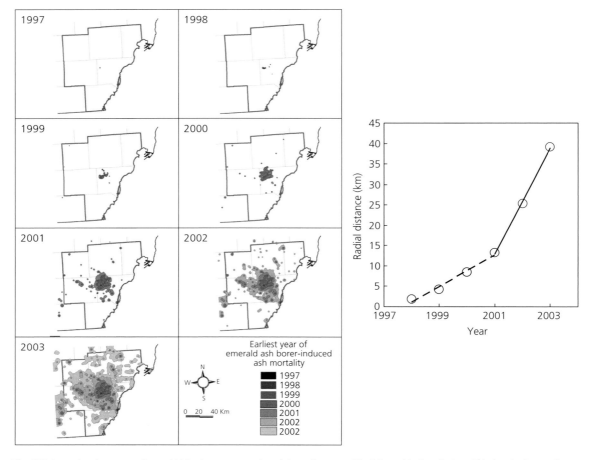

Fig. 2.9 Interpolated reconstruction and biphasic range expansion of the earliest emerald ash borer (*Agrilus planipennis*)-induced ash mortality across south-eastern Michigan, United States. The initial 2002–03 quarantine area is denoted by the thick outline (Siegert *et al.* 2014). Reproduced with permission from John Wiley & Sons.

needle gall midge, *Thecodiplosis japonensis*, is a major pest of pines (e.g. Park and Hyun 1983; Soné 1987). This midge was introduced to the southern Korean Peninsula during the early 1920s after Japan's annexation of Korea in 1910, probably due to inter-regional trade (Park and Hyun 1983). Infestations were first noted at the southern tip of Korea and then extended across the entire peninsula. Biphasic expansion predominated the spread of the midge, with a slow early phase of 2.2 km/yr and a high rate of expansion of 8.2 km/yr after the Korean War in 1954 (Lee *et al.* 2007). Although the introduction and transition of the biphasic expansion were clearly signalled by human activities in the

region, Allee effects and long-distance dispersal are also likely implicated in the biphasic expansion (Lee *et al.* 2007), but further work is required to confirm this.

Invasive alien plants also often show biphasic expansion. Many species of Asteraceae are highly invasive, including crofton weed, *Ageratina adenophorum*, a perennial semi-shrub native to Mexico that is poisonous to cattle. The seeds of crofton weed are small, 1.5–2.0 mm long, with parachute-like pappi that facilitate dispersal by wind. Seeds are also transported in hay as fodder or by sticking to machinery, vehicles, animals, and human clothes, which provides another mechanism for long-distance dispersal. A crucial factor in the invasiveness of this species is the large number of seeds produced (10 000–100 000 per mature plant per year in invaded areas). By forming dense populations and releasing allelopathic substances into the soil to reduce the growth of potential competitors (Song *et al.* 2000), crofton weed has successfully invaded many parts of the world (Auld 1970; Henderson 2001; Muniappan *et al.* 2009). First recorded in 1940 in South China, most likely from bridgehead populations in Burma and Vietnam (Ding *et al.* 2008), crofton weed has spread quickly northwards and eastwards since the 1960s, causing severe impacts on animal husbandry by poisoning livestock and reducing fodder quality and agricultural production (Auld 1970; Zhao and Ma 1989). Between the 1960s and the 1980s, a slow phase of 6 km/yr was observed, whereafter spread accelerated to 23 km/yr in the 1980s (Sang *et al.* 2010). Habitat complexity and multiple vectors of seed dispersal could have contributed to the biphasic expansion, with a spread rate through dispersal by wind of 3.24 km/yr along rivers by 270.4 km/yr and along roads by 5.3 km/yr in southern China (Horvitz *et al.* 2014).

2.5 Acceleration

The importance of accelerating range expansion was publicized by James S. Clark and colleagues (1998) in their classic paper on 'Reid's paradox' which showed that the reconstructed range expansion of trees, especially those with large seeds, from paleoecological records since last glaciation was too fast to have been driven by standard means of dispersal (see Chapter 4 for further discussion).

Acceleration can be largely explained by two alternative mechanisms. Firstly, spatial sorting of individuals with different dispersal capacities can accelerate the spread. Stronger dispersers are likely to win the race and locate themselves at the advancing range front and leave their progenies for further acceleration, thus pushing the rate of spread to increase gradually to a higher rate. Secondly, fat-tailed dispersal has been shown to be able to accelerate range expansion (Kot *et al.* 1996; Nathan *et al.* 2001). Here, we introduce a few cases of accelerating expansion in biological invasions.

Although biphasic expansion can be considered acceleration in a simplified form, many of the instances of biphasic expansion reported in the literature appear to have been triggered by habitat transition or altered disturbance regimes. As such, acceleration probably qualifies as a unique mode of spread.

Our first case focuses on the role of the dispersal kernel. The California sea otter, *Enhydra lutris nereis*, was hunted to near extinction before a small population was discovered and protected in 1914, allowing it to reinvade its previous range. Otters reproduce only once a year and inhabit near-shore habitats in social groups (Riedman and Estes 1990). Their movements include predominately localized plus rare long-distance dispersal events, with juvenile males being the strongest dispersers (Ralls *et al.* 1996). Krkošek and colleagues (2007) developed a stage-structured integrodifference equation (IDE) model (see Chapter 3) and simulated the range expansion of the species over the period 1914–86, at a rate of 2.63–4.35 km/yr. A key finding of this work was that the dispersal kernel (the probability of annual movement distance), $a \cdot \exp(-b \cdot |d|^{1/2})$, fitted the data best, while all tested dispersal kernels can nonetheless explain the accelerating range expansion rather well (Fig. 2.10). The authors emphasized the role of a high frequency of long-distance dispersers in driving the range acceleration. However, as dispersal is male-biased, they also suggested a typical pattern of spatial sorting; that is, a spatial transition of traits: males dominating the outer limits of the population, but females contributing to population expansion through reproduction. It worth noting that range acceleration in many cases resembles an exponential expansion or a biphasic expansion. For instance, an earlier study by Lubina and Levin (1988) explained the expansion of sea otters as biphasic and argued that the slow phase I spread was attributable to reduced dispersal in kelp forests and the fast phase II spread was due to high dispersal rates over sandy-bottom habitats, again highlighting the need to understand the relative roles of stratified dispersal and habitat transition in shaping biphasic expansion. Inferring processes from observed spread patterns is a key challenge in such studies.

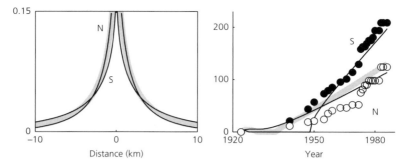

Fig. 2.10 The relationships between dispersal and range expansion for the California sea otter (*Enhydra lutris nereis*) Left: dispersal kernels fitted from dispersal data (thick grey lines), the southern expanding front (black lines; S), and the northern expanding front (black lines; N). Right: the range expansion of otters moving south (filled circles) and north (open circles) in kilometres. Thick grey lines show the predicted range expansion based on numerical simulations of the sea otter integrodifference model. Black lines are the least-square best fits of the model to the southern and northern fronts, obtained by leaving the coefficient a in the corresponding kernel as a free parameter (Krkošek *et al.* 2007). Reproduced with permission from Elsevier.

Our second case focuses on the role of spatial sorting of dispersal traits in explaining range acceleration. Cane toads, *Rhinella marina*, native to South America, were introduced from Hawai'i to Queensland in Australia, in an ill-advised attempt (and despite recommendations from scientists not to proceed) to control the grey-backed cane beetle (*Dermolepida albohirtum*), a pest of sugar cane (Seabrook 1991). The movement of cane toads between 1945 and 2005 was studied by Phillips and colleagues (2007), who reported an average progression of approximately 55 km/yr, which is five times the initial spread rate of about 11 km/yr (Fig. 2.11). Although the increase in velocity fits an exponential curve extremely well ($R^2 > 0.95$), this is different from the exponential expansion (comparing their vertical axis), but truly an acceleration of the expansion speed. This acceleration has been attributed to the spatial sorting of dispersal traits, in particular longer legs and faster straight-line movement for individuals at the vanguard of invasion (Phillips *et al.* 2006; Brown *et al.* 2014). We will revisit the debate between spatial sorting and fat-tail dispersal in driving range acceleration, and the debate around the relative importance of stratified dispersal and context-based dispersal in driving biphasic expansion, in Chapter 4.

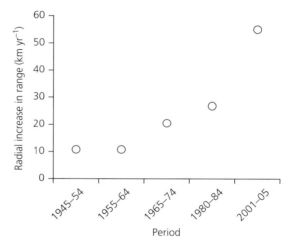

Fig. 2.11 The rate at which the invasion of cane toads (*Rhinella marina*) progressed through tropical Australia has increased substantially since toads arrived in 1935 (Phillips *et al.* 2006). Reprinted by permission from Macmillan Publishers Ltd.

2.6 Lag phase

The initiation of spread is often preceded by a phase of dormancy or inactivity. Understanding the phenomenon of lag phases is important for efficient management (Baker and Stebbins 1965; Hobbs and Humphries 1995; Kowarik 1995; Crooks and Soulé

1999; Binggeli 2001; Aagaard and Lockwood 2014). In biphasic expansion, we have already noted that initial spread is often extremely slow. In many cases, the expansion during phase I of the biphasic spread is undetectable due to the lack of records and/or poor detectability when an alien species first arrives. Crooks (2005) divided lags of invasion dynamics into three types: (i) a slower rate of expansion compared to the potential of the species to spread at a much faster rate; (ii) a delay of a normal spread after the species was introduced; (iii) the slow rate of expansion initially due to small population size or low density. Type (i) mainly deals with discordance and context-based spread and is discussed further in section 2.8. Type (iii) considers a lag phase as an artefact of exponential expansion, which starts extremely slowly. We focus only on the true type of lag in invasion (type ii). Mechanisms behind a true lag phase include: (a) genetic filtering or adaptation leading to more invasive genotypes; (b) the requirement for multiple introductions to overcome Allee effects or introduce new genotypes; (c) an abrupt change in biotic interactions (e.g. new association with a resident or introduced mutualist); and (d) environmental switches (e.g. major disturbance facilitating population growth).

An obvious reason for lag phases is the presence of Allee effects (Lewis and Kareiva 1993; Veit and Lewis 1996; Drake 2004; Leung *et al.* 2004; Parker 2004) which can lead to near-zero or negative population growth rates for small populations. Allee effects can be defined as a decline or low fitness at low population size or density, driven by a variety of mechanisms such as mating systems, predation, environmental modification, and social interactions (Courchamp *et al.* 2008). Pollen limitation in small, low-density patches can also cause prolonged lags in plants, such as that seen for the spread of cordgrass (*Spartina alterniflora*) (Davis *et al.* 2004; Parker 2004). Moreover, as the initial propagules often have low genetic diversity, organisms might need to adapt first to their new and often suboptimal environments (Baker and Stebbins 1965; Crooks and Soulé 1999; Petit 2004). This requires rapid adaptation (Reznick and Ghalambor 2001; Stockwell and Ashley 2004) due to strong directional selection under stress (Mooney and Cleland 2001; Lee 2002; Cox

2004) before suboptimal habitat and stochasticity cause extinction. An example is the velvetleaf plant (*Abutilon theophrasti*) which, although introduced to North America in the 1700s, only become a pest in the twentieth century; the species has evolved different life-history strategies to improve its competitive abilities depending on the species with which it is competing (Weinig 2000; see Chapter 6). Other factors, such as changing interactions between invaders and the biotic and abiotic environment can also cause population explosions of alien species (e.g., by changing carrying capacity) and will be important in affecting temporal dynamics of invader abundance and diversity (Crooks and Soulé 1999).

Detecting lag phases from real data is difficult as early records are often biased (not systematic). A lag phase can be defined as the time required for an invader to reach a point where a marked increase in records or range size occurs after arrival in an area (Pyšek and Prach 1993; Hobbs and Humphreys 1995; Daehler 2009). In general, this can be done in two ways using a piecewise regression for accumulative or spontaneous records. Aikio and colleagues (2010) used piecewise regression, a linear regression connected with a non-linear regression (e.g. sigmoid and exponential curves), on herbarium data for weed species in New Zealand. They show that a lag phase of 20–30 years was common for New Zealand weeds, with about 5 per cent of species having lag phases greater than 40 years, and about 9 per cent of species having no detectable lag phase. However, they caution that no detectable lag phase does not imply that their populations began to expand immediately after introduction, as species could have been introduced years before they were detected as the annual number of records collected can fluctuation dramatically (Fig. 2.12). The piecewise regression they proposed worked well in some cases. For instance, when conducting the piecewise regression for the cumulative records of Scotch thistle, *Cirsium vulgare*, in the South Island of New Zealand, they detected a 43-year lag (Fig. 2.12). Aikio and colleagues' (2010) exercise also clarified two interesting speculations. Firstly, their results support Sakai and colleagues' (2001) suggestion that species with large ranges spread faster. Secondly, species with large ranges normally have longer lags.

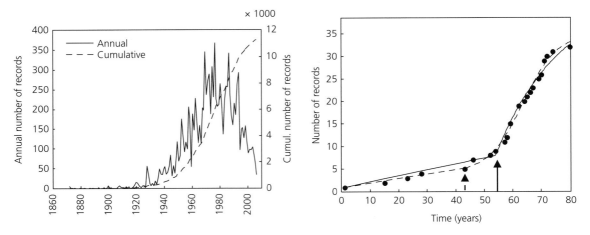

Fig. 2.12 The annual and cumulative total number of records of 400 weed species (left); example of the functions fitted to the cumulative records of Scotch thistle (*Cirsium vulgare*) in New Zealand (right). The curves represent the von Bertalanffy (solid line) and logistic (dotted line) functions fitted to the after-lag data. The arrows show the estimated lag (Aikio *et al.* 2010). Reproduced with permission from John Wiley & Sons.

Identifying the drivers behind lag phases is crucial for managing 'sleeper weeds' (*sensu* Groves 2006) during their 'sleeping' stage. Aikio and colleagues' (2010) findings on range size and lag are counter-intuitive and require further work. Larkin (2012) used Aikio and colleagues'(2010) piecewise regression and calculated lag lengths for alien plant species considered invasive or potentially invasive in parts of the upper Midwest of the USA. Of the 257 species with sufficient records for analysis, 197 cases showed lags, with a mean of 47.3 ± 34.6 years. Several factors were found to be significantly correlated with lag lengths. In particular, species with earlier first records appeared to have longer lags; however, this can be attributed to the fact that many recently introduced species are still 'sleeping' and the observed lag is only truncated from the future potential. Lag phases produce a backlog of invasion debt representing species that are already present but not yet spreading (Essl *et al.* 2011; Rouget *et al.* 2016). The long durations and low predictability of lags found in this study underscore the merit of a 'guilty until proven innocent' or 'white list' approach regarding species introductions (Mack *et al.* 2000; Simberloff 2006). This points to the need for a mechanistic understanding of what is limiting an invasion, for example, the lack of fire (Geerts *et al.* 2013). The weakness of the alternative

'innocent until proven guilty' or 'black list' strategy is illustrated by the fact that of the 139 species in this study for which the introduction pathway is known, 85 per cent were intentionally introduced. Further examination showed that several interacting factors influenced lag lengths but did not have high explanatory power (Fig. 2.13). In order of importance, significant variables were growth habit, breadth of invaded habitats, dispersal mode, and pollination mode (Larkin 2012).

The second approach that can detect lag phases also uses the piecewise regression but for annual records instead of cumulative records. Hyndman and colleagues (2015) advocated the use of annual collection rates and incorporated more realistic assumptions, especially regarding the fluctuation of collection effort. They assume that the number of specimens of the focal species collected in year t has a Poisson distribution with mean equal to $N_t \cdot \exp[f(t)]$, where N_t is the total number of records of alien species collected in year t (i.e. annual rate of collection effort) and $f(t)$ is a function of time allowing the number of specimens of the focal species in a particular year to change over time. They assume that a lag phase exists if $f(t)$ is constant until some year τ, and then takes some other higher values thereafter. When using this approach to examine the herbarium records of the European searocket, *Cakile*

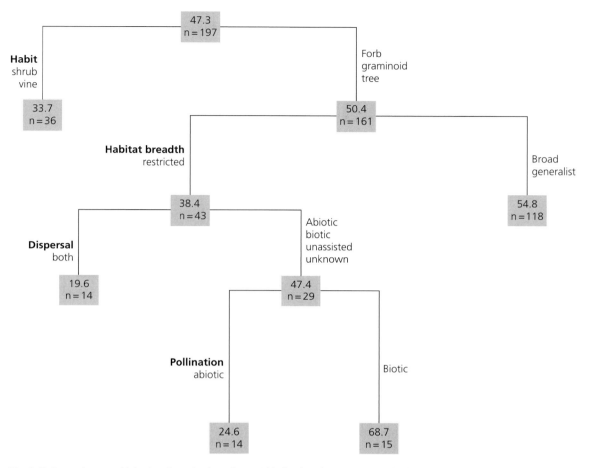

Fig. 2.13 Regression tree with lag lengths as the dependent variable for alien plant species considered invasive or potentially invasive in part of the upper Midwest of the United States (Larkin 2012). Boxes at nodes and leaves show mean lag and number of cases. With permission of Springer.

maritima, in Australia, Hyndman and colleagues (2015) detected a 52-year lag phase (Fig. 2.14). When re-analysing Aikio and team's (2010) data on New Zealand weeds, they detected only 28 per cent cases with detectable lags; when re-analysing Larkin's (2012) data from the Midwest, only 40 per cent species were shown to have exhibited lags. Compared to Aikio and his team (2010) and Larkin (2012), Hyndman and colleagues' (2015) approach of using annual rate detected far fewer and shorter lags. Either way, there could be a substantial portion of invasive species that undergo lag phases that last for decades, and the retrospective approaches to detect lags introduced high levels of uncertainty

due to the biases in herbarium data. A robust predictive model to forecast lags is yet to emerge and this presents a major challenge for invasion science (Wilson *et al.* 2017).

2.7 Boom and bust

Many invasive species experience seemingly spontaneous population crashes after they are considered established and already spreading (Simberloff and Gibbons 2004), with the invader populations often settling at a much lower density after the boom-and-bust spread (Williamson 1996). A trivial explanation would be that temporal proliferation

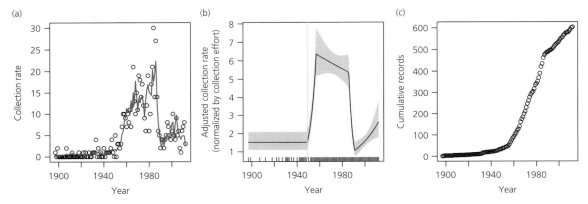

Fig. 2.14 Analysis of samples of the European searocket (*Cakile maritima*) collected in Australia (Hyndman *et al.* 2015). (a) Annual collection rates are shown as open symbols. (b) Collection rate adjusted by collection effort for all invasive species. This is equal to the estimated mean collection rate in each year normalized by the total number of alien species collected in each year. The estimated knots are breakpoints of the line (the first being the statistically significant lag phase ending in 1949), confidence intervals are shown using grey shading, and years in which collections were > 0 are shown as dashes on the horizontal axis. (c) Cumulative number of records, shown only in years when there were new collections. With permission of Springer.

of environmental conditions (e.g. favourable weather) could drive the boom phase and the invasion dynamics revert back to the bust phase when conditions return to normal levels that hinder establishment. One density-dependent explanation for such boom-and-bust patterns is that the fast population growth has overshot the carrying capacity and the population has to decline to return to its true equilibrium (Simberloff and Gibbons 2004). This is especially the case for species with complicated life cycles where the response of population growth to habitat quality and disturbance is slow and often delayed. Alternative explanations have linked this pattern to the mitigation of rescue effect in population dynamics (e.g. metapopulation or source-sink dynamics) from excessive emigration rate. Fahrig (2007) speculated that a high emigration probability does not always have a positive influence on the population dynamics. The rescue effect can sustain the range expansion only when the emigration probabilities from favourable and unfavourable habitats are balanced. When only a small proportion of the habitat is favourable for the growth, low emigration from favourable patches may not be enough to rescue populations in unfavourable patches from extinction. High emigration from favourable patches alone, on the other hand, can lead to the decline of rescuing individuals and yield a slower spread or even the extinction of the

population following a boom-and-bust pattern (Ramanantoanina and Hui 2016).

Another potential reason for boom-and-bust patterns in invasive species relates to adaptive biotic interactions. For instance, introduced plant species accumulate predators and pathogens over time (Hawkes 2007; Wingfield *et al.* 2011; Hurley *et al.* 2016), and native predators can adapt to non-native prey over time too, leading to increases in predator efficacy (Carlsson *et al.* 2009). An example of a boom-and-bust pattern driven by biotic interactions can be found in the invasion of the New Zealand mudsnail (*Potamopyrgus antipodarum*) (Moore *et al.* 2012). This herbivore is rapidly invading western North America and has the potential to change freshwater ecosystems radically. Hyperabundant mudsnails (500 000 individuals m^{-2}) can excrete enough ammonium to account for 66 per cent of the stream ammonium demand (Hall *et al.* 2003) and be responsible for the majority of invertebrate productivity (Hall *et al.* 2006). Moore and colleagues (2012) reported a ten-year boom-and-bust pattern in populations of the mudsnail (Fig. 2.15) and hypothesized an increased mortality of mudsnails due to predators which could have driven the bust phase. The abundance of predatory invertebrates increased sixfold during the increase of mudsnails, and predator populations remained elevated when New Zealand mudsnail populations crashed (Fig. 2.15).

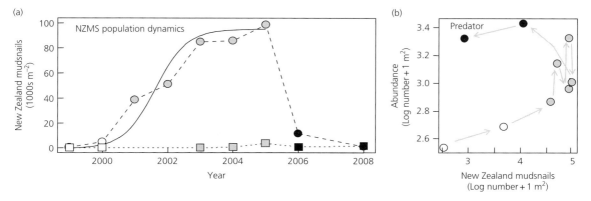

Fig. 2.15 (Left) Time series of average abundance of New Zealand mudsnail (*Potamopyrgus antipodarum*) in invaded (circles) and uninvaded sites (squares) in a Californian stream (Moore *et al.* 2012). Also shown is the best-fit logistic model for the invaded site, restricted to the before and during period. The periods of before, during, and decline are represented with white, grey, and black colours. (Right) The temporal dynamics of benthic invertebrates as a function of mudsnail abundance in the invaded site for predators. Each point denotes the mean annual abundance of native taxa and New Zealand mudsnail. Arrows show the temporal sequence. With permission of Springer.

For species interactions that are not highly specialized (for instance in food webs) predators select a diet comprising limited numbers of prey species from all accessible prey, to maximize the food intake rate (Stephen and Krebs 1986). Therefore, we could predict the diet composition and breadth based on the benefit and abundance of prey, following optimal foraging theory (Charnov and Orians 1973; Stephen and Krebs 1986). This will lead to the so-called zero-one rule where a forager will only consume resources from the optimal diet, completely ignoring other, less profitable, resources (Stephens and Krebs 1986). An effective forager should be able to detect the quality of resources in the environment and respond flexibly to environmental change (Freidin and Kacelnik 2011) according to the adaptive foraging strategy based on recent experience (Zhang and Hui 2014), leading to flexible interactions such as prey switching (see Chapter 6). As the population density is initially low for most invaders, it is inefficient for resident predators to target these novel resources. This could allow populations of prey species to grow until it becomes beneficial to the resident predators to include them in their diet, suppressing the invaders to a lower level (Fig. 2.16). Any trophic interactions related to efficient utilization or consumption of resources will lead to similar

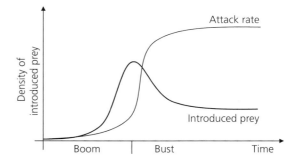

Fig. 2.16 A conceptual diagram of the boom-and-bust pattern in an introduced prey species. The phenomenon of prey switching in the foraging strategy of predators occurs once the density or abundance of the introduced prey has reached a tipping point, below which the predator simply ignores the prey, while above which the high encounter rate triggers the inclusion of the prey into the predator's optimal diet and the prey thus suffers from a much higher attack rate, suppressing the prey population to a low level.

'switching' behaviour so that an invader will face different selection regimes when its abundance is above a certain tipping point.

2.8 Concordance and discordance

Compared to the focus on temporal variation of spread in previous sections, the spatial variation in spread rates also provides a real challenge for deciphering the (inner and external) mechanisms that affect the spread patterns of species in different environments. To conclude this chapter, we bring more complicated, stochastic, and contextual realism into our consideration of the modes of spread. If we examine the invasion dynamics of similar species in different environments, a complicated picture emerges. Some species will show a conservative range, whereas others are expanding, retracting, or shifting their current ranges. As the geographic ranges of many species, native or introduced, are expected to change in response to ongoing global environmental changes (Thomas *et al.* 2004), invasive species will not only respond to these changes but also expand their ranges into suitable habitat as a process of niche filling. As such, the range dynamics of native species reflects the tracking of their suitable habitat. In contrast, the range dynamics of invasive species depicts both the spread into their potential suitable habitat and tracks the changes in suitable habitat. The range dynamics of natives and invasives, thus, differ fundamentally. For those invasive species that have filled their suitable habitat, we could expect to see concordance with native species in their range dynamics, both tending to track environmental changes in the region. Otherwise, discordance in range dynamics between native and invasive species can be expected. For instance, the two invasive sturnids in South Africa (common starling and Indian myna) are currently expanding their ranges northwards to occupy potentially suitable habitat, whereas the range of native sturnids mostly shifts towards south or west to track the shifting climatic niche (Simmons *et al.* 2004).

The con- and discordance of native and introduced species can be examined in two ways. Firstly, we need to determine whether the expanding invasives share common traits with expanding natives (trait concordance). Identifying such traits could be important for understanding how species, those with and without such traits, respond differently to the regional environmental changes. This could help conservation managers to identify those species that lack such traits and that are thus more vulnerable to environmental changes. Secondly, we need to know whether the ranges of native and invasive species are likely to expand or shift into the same areas (location concordance). Not only does this identify geographical hotspots that are important for conservation but also the environmental factors that characterize these areas. For instance, if climate change drives range shift, we would expect to see a clearer pattern of location concordance in the northern rather than in the southern hemisphere since no clear patterns of climate change is detected in the southern hemisphere (Friedman *et al.* 2013).

Invasive species often experience a large spatial variation in their spreading rates (spatial discordance). For instance, common starlings spread at half the speed in the southern hemisphere compared to in North America. It is important to determine whether differences in the environments in these locations are responsible for differences in the dispersal strategy and the population growth rate, and thus responsible for the spatial variation of spreading patterns. Other examples of spatial discordance include the speckled wood butterfly, *Pararge aegeria*, in Britain (Hill *et al.* 2001), with spread of 0.51 km/yr in England and 0.93 km/yr in Scotland attributable to the greater cover of woodland in Scotland. Lyons and Scheibling (2009) examined the spread dynamics of five prominent introduced algae (*Codium fragile, Caulerpa taxifolia, Grateloupia turuturu, S. muticum,* and *Undaria pinnatifida*) and four other native species with sufficient data (*Acrothamnion preissii* and *Caulerpa racemosa* in the Mediterranean, *Fucus serratus* in the north-west Atlantic, and *F. evanescens* in the north-east Atlantic). They detected a large variation in the rate of expansion for individual species between regions (Fig. 2.17), suggesting a complex interplay between algal traits, attributes of invaded regions and anthropogenic factors.

Such spatial variation in spreading rates could be explained by geographically specific abiotic and biotic conditions under which the newcomers must survive and reproduce (Carroll and Dingle 1996). Evolutionary response to these environmental challenges often leads to new life-history strategies in the invading population (Yoshida *et al.* 2007). Therefore, introduced populations of an invasive species may ultimately diverge in important life-history traits in response to selection pressures in the novel environment. This was shown for the Mediterranean fruit fly, *Ceratitis capitata* by comparing six populations originating from different regions (Diamantidis *et al.* 2009). Moreover, the discordance in the invasion dynamics can be caused by demographic or

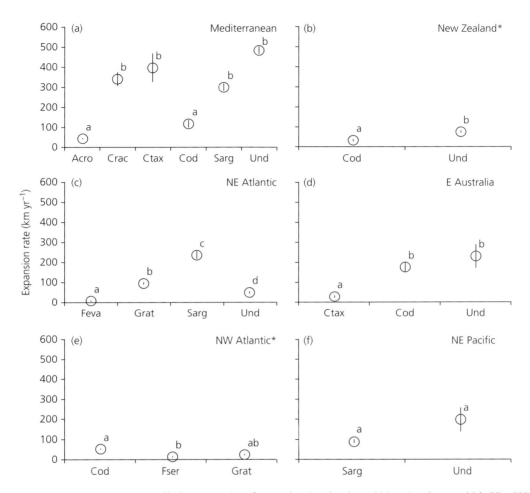

Fig. 2.17 Range expansion rates, as estimated by linear regression, of macro-algae introduced to multiple regions (Lyons and Scheibling 2009): north-west Atlantic, south-east Pacific, Mediterranean Sea, eastern Australia, New Zealand, north-east Atlantic, north-east Pacific, and south-west Atlantic. Reproduced with permission from John Wiley & Sons.

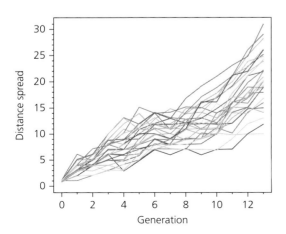

Fig. 2.18 Spreading dynamics of flour beetles in 30 replicate landscapes through 13 generations, all starting with 20 adult beetles in the leftmost patch of the landscape (Melbourne and Hastings 2009). Reprinted with permission from AAAS.

genetic stochasticity; such endogenously generated variance in spread rates can be remarkably high, as demonstrated in replicated spreading trials of the flour beetle *Tribolium castaneum* in laboratory microcosms (Fig. 2.18) (Melbourne and Hastings, 2009), which seriously compromises our ability to predicting spread dynamics. Overall, the patterns of spread follow major types and the rate of spread is highly context dependent. Subsequent chapters will examine some of these patterns in more detail, but much work remains to be done in this area.

References

Aagaard, K. and Lockwood, J. (2014) Exotic birds show lags in population growth. Diversity and Distributions, 20, 547–54.

Aikio, S., Duncan, R.P., and Hulme, P.E. (2010) Lag-phases in alien plant invasions: separating the facts from the artefacts. Oikos, 119, 370–8.

Aukema, J.E., Leung, B., Kovacs, K., *et al.* (2011) Economic impacts of non-native forest insects in the continental United States. PLoS One, 6, e24587.

Auld, B.A. (1970) Eupatorium weed species in Australia. Pest Articles and News Summaries, 16, 82–6.

Baker, H.G. and Stebbins, G.L. (eds) (1965) *The Genetics of Colonizing Species*. New York, NY: Academic Press.

Barták, V., Vorel, A., Šímová, P., *et al.* (2013) Spatial spread of Eurasian beavers in river networks: a comparison of range expansion rates. Journal of Animal Ecology, 82, 587–97.

Berthouly-Salazar, C., Hui, C., Blackburn, T.M., *et al.* (2013) Long-distance dispersal maximizes evolutionary potential during rapid geographic range expansion. Molecular Ecology, 22, 5793–804.

Binggeli, P. (2001) Time-lag between introduction, establishment and rapid spread of introduced environmental weeds. Proceedings of the III International Weed Science Congress, MS No. 8, International Weed Science Society. Oxford.

Blackburn, T.M., Lockwood, J.L., and Cassey, P. (2009) *Avian Invasion*. Oxford: Oxford University Press.

Brakefield, P.M. and Shreeve, T.G. (1992) Diversity within populations. In: Dennis, R.L.H. (ed.) *The Ecology of Butterflies in Britain*. Oxford: Oxford University Press, pp 178–216.

Brown, G.P., Phillips, B.L., and Shine, R. (2014) The straight and narrow path: the evolution of straight-line dispersal at a cane toad invasion front. Proceedings of the Royal Society B: Biological Sciences, 281, 20141385.

Cappaert, D.L., McCullough, D.G., Poland, T.M., *et al.* (2005) Emerald ash borer in North America: a research and regulatory challenge. American Entomologist, 51, 152–65.

Carlsson, N.O., Sarnelle, O., and Strayer, D.L. (2009) Native predators and exotic prey—an acquired taste? Frontiers in Ecology and the Environment, 7, 525–32.

Carroll, S.P. and Dingle, H. (1996) The biology of post-invasion events. Biological Conservation, 78, 207–14.

Charnov, E. and Orians, G.H. (1973) *Optimal Foraging: Some Theoretical Explorations*. Seattle, WA: University of Washington.

Choi, W.I. and Park, Y.S. (2012) Dispersal patterns of exotic forest pests in South Korea. Insect Science, 19, 535–48.

Chung, Y.J., Bae, W.I., and Yum, Y.C. (1987) Studies on the forecasting of major forest insect pests outbreak (II) prediction of the adult emergence of *Hyphantria cunea* Drury. Research Report of Forestry Research Institute in Korea, 34, 149–52.

Clark, J.S., Fastie, C., Hurtt, G., *et al.* (1998) Reid's Paradox of Rapid Plant Migration Dispersal theory and interpretation of paleoecological records. BioScience, 48, 13–24.

Courchamp, F., Berec, L., and Gascoigne, J. (2008) *Allee Effect in Ecology and Conservation*. Oxford: Oxford University Press.

Cox, G.W. (2004) *Alien Species and Evolution: The Evolutionary Ecology of Exotic Plants, Animals, Microbes, and Interacting Native Species*. Washington, DC: Island Press.

Crooks, J.A. (2005) Lag times and exotic species: the ecology and management of biological invasions in slow-motion. EcoScience, 12, 316–29.

Crooks, J.A. and Soulé, M.E. (1999) Lag times in population explosions of invasive species: causes and implications. In: Sandlund, O.T., Schei, P.J., and Viken, A. (eds) *Invasive Species and Biodiversity Management*. Dordrecht: Kluwer Academic Press, pp. 103–25.

Daehler, C.C. (2009) Short lag times for invasive tropical plants: evidence from experimental plantings in Hawai'i. PLoS One, 4, e4462.

D'Antonio, C.M. and Vitousek, P.M. (1992) Biological invasions by exotic grasses, the grass/fire cycle, and global change. Annual Review of Ecology and Systematics, 23, 63–87.

Davis, H.G., Taylor, C.M., Civille, J.C., *et al.* (2004) An Allee effect at the front of a plant invasion: *Spartina* in a Pacific estuary. Journal of Ecology, 92, 321–7.

Diamantidis, A.D., Papadopoulos, N.T., Nakas, C.T., *et al.* (2009) Life history evolution in a globally invading tephritid: patterns of survival and reproduction in medflies from six world regions. Biological Journal of the Linnean Society, 97, 106–17.

Ding, J., Mack, R.N., Lu, P., *et al.* (2008) China's booming economy is sparking and accelerating biological invasions. BioScience, 58, 317–24.

Drake, J.M. (2004) Allee effects and the risk of biological invasion. Risk Analysis, 24, 795–802.

Elton, C.S. (1958) *The Ecology of Invasions by Animals and Plants*. London: Methuen.

Emerald Ash Borer Information (2014) Emerald Ash Borer. Available at: <http://www.emeraldashborer.info/>.

Essl, F., Dullinger, S., Rabitsch, W., *et al.* (2011) Socioeconomic legacy yields an invasion debt. Proceedings of the National Academy of Sciences of the USA 108, 203–7.

Evans, K.L., Duncan, R.P., Blackburn, T.M., *et al.* (2005) Investigating geographic variation in clutch size using a natural experiment. Functional Ecology, 19, 616–24.

Fahrig, L. (2007) Non-optimal animal movement in human-altered landscapes. Functional Ecology, 21, 1003–15.

Ferrari, J.R., Preisser, E.L., and Fitzpatrick, M.C. (2014) Modeling the spread of invasive species using dynamic network models. Biological Invasions, 16, 949–60.

Fitzpatrick, M.C., Preisser, E.L., Porter, A., *et al.* (2012) Modeling range dynamics in heterogeneous landscapes: invasion of the hemlock woolly adelgid in eastern North America. Ecological Applications, 22, 472–86.

Freidin, E. and Kacelnik, A. (2011) Rational choice, context dependence, and the value of information in European starlings (*Sturnus vulgaris*). Science, 334, 1000–02.

Fric, Z. and Konvicka, M. (2002) Generations of the polyphenic butterfly *Araschnia levana* differ in body design. Evolutionary Ecology Research, 4, 1017–32.

Friedman, A.R., Hwang, Y.T., Chiang, J.C., *et al.* (2013) Interhemispheric temperature asymmetry over the twentieth century and in future projections. Journal of Climate, 26, 5419–33.

Geerts, S., Moodley, D., Gaertner, M., *et al.* (2013) The absence of fire can cause a lag phase—the invasion dynamics of *Banksia ericifolia* (Proteaceae). Austral Ecology, 38, 931–41.

Gomi, T. (2007) Seasonal adaptations of the fall webworm *Hyphantria cunea* (Drury) (Lepidoptera: Arctiidae) following its invasion of Japan. Ecological Research, 22, 855–61.

Groves, R.H. (2006) Are some weeds sleeping? Some concepts and reasons. Euphytica, 148, 111–20.

Hall Jr, R.O., Dybdahl, M.F., and VanderLoop, M.C. (2006) Extremely high secondary production of introduced snails in rivers. Ecological Applications, 16, 1121–31.

Hall Jr, R.O., Tank, J.L., and Dybdahl, M.F. (2003) Exotic snails dominate nitrogen and carbon cycling in a highly productive stream. Frontiers in Ecology and the Environment, 1, 407–11.

Halley, D.J. and Rosell, F. (2002) The beaver's reconquest of Eurasia: status, population development and management of a conservation success. Mammal Review, 32, 153–78.

Harrison, J.A., Allan, D.G., Underhill, L.H., *et al.* (1997) *The Atlas of Southern African Birds, including Botswana,* *Lesotho, Namibia, South Africa, Swaziland and Zimbabwe*. Johannesburg: BirdLife South Africa.

Hawkes, C.V. (2007) Are invaders moving targets? The generality and persistence of advantages in size, reproduction, and enemy release in invasive plant species with time since introduction. American Naturalist, 170, 832–43.

Henderson, L. (2001) *Alien Weeds and Invasive Plants. Plant Protection Research Institute Handbook No. 12*. Pretoria: Plant Protection Research Institute.

Herms, D.A. and McCullough, D.G. (2014) The emerald ash borer invasion of North America: history, biology, ecology, impacts and management. Annual Review of Entomology, 59, 13–30.

Hill, J.K., Collingham, Y.C., Thomas, C.D., *et al.* (2001) Impacts of landscape structure on butterfly range expansion. Ecology Letters, 4, 313–21.

Hobbs, R.J. and Humphries, S.E. (1995) An integrated approach to the ecology and management of plant invasions. Conservation Biology, 9, 761–70.

Horak, J., Hui, C., Roura-Pascual, N., *et al.* (2013) Changing roles of propagule, climate, and land use during extralimital colonization of a rose chafer beetle. Naturwissenschaften, 100, 327–36.

Horvitz, N., Wang, R., Zhu, M., *et al.* (2014) A simple modeling approach to elucidate the main transport processes and predict invasive spread: river-mediated invasion of *Ageratina adenophora* in China. Water Resources Research, 50, 9738–47.

Hui, C., Roura-Pascual, N., Brotons, L., *et al.* (2012) Flexible dispersal strategies in native and non-native ranges: environmental quality and the 'good-stay, bad-disperse' rule. Ecography, 35, 1024–32.

Hunter, R. (1991) Bromus invasions on the Nevada Test Site: present status of *Bromus rubens* and *Bromus tectorum* with notes on their relationship to disturbance and altitude. Great Basin Naturalist, 51, 176–82.

Hurley, B.P., Garnas, J., Wingfield, M.J., *et al.* (2016) Increasing numbers and intercontinental spread of invasive insects on eucalypts. Biological Invasions, 18, 921–33.

Hyndman, R.J., Mesgaran, M.B., and Cousens, R.D. (2015) Statistical issues with using herbarium data for the estimation of invasion lag-phases. Biological Invasions, 17, 3371–81.

Johnson, L.E. and Carlton, J.T. (1996) Post-establishment spread in large-scale invasions: dispersal mechanisms of the zebra mussel *Dreissena polymorpha*. Ecology, 77, 1686–90.

Johnson, K. and Peer, B.D. (2001) Great-tailed grackle (*Quiscalus mexicanus*). In: Poole, A. and Gill, F. (eds) *The Birds of North America*. Philadelphia, PA: The Academy of Natural Sciences, No. 576.

Kelly, D.W., Herborg, L.M., and MacIsaac, H.J. (2010) Ecosystem changes associated with Dreissena invasions: recent developments and emerging issues. In: Van der Velde, G., Rajagopal, S., and Bij de Vaate, A. (eds) *The Zebra Mussel in Europe*. Leiden: Backhuys Publishers, pp. 199–210.

Kiritani, K. and Morimoto, N. (2004) Invasive insect and nematode pests from North America. Global Environmental Research, 8, 75–88.

Klooster, W.S., Herms, D.A., Knight, K.S., *et al.* (2014) Ash (*Fraxinus* spp.) mortality, regeneration, and seed bank dynamics in mixed hardwood forests following invasion by emerald ash borer (*Agrilus planipennis*). Biological Invasions, 16, 859–73.

Kot, M., Lewis, M.A., and van den Driessche, P. (1996) Dispersal data and the spread of invading organisms. Ecology, 77, 2027–42.

Kowarik, I. (1995) Time lags in biological invasions with regard to the success and failure of alien species. In: Pyšek, P., Prach, K., Rejmánek, M., *et al.* (eds) *Plant Invasions: General Aspects and Special Problems*. Amsterdam: SPB Academic Publishing, pp. 15–38.

Krkošek, M., Lauzon-Guay, J.S., and Lewis, M.A. (2007) Relating dispersal and range expansion of California sea otters. Theoretical Population Biology, 71, 401–7.

Lack, D.L. (1968) *Ecological Adaptations for Breeding in Birds*. London: Methuen.

Larkin, D.J. (2012) Lengths and correlates of lag phases in upper-Midwest plant invasions. Biological Invasions, 14, 827–38.

Lee, C.E. (2002) Evolutionary genetics of invasive species. Trends in Ecology and Evolution, 17, 386–91.

Lee, S.D., Park, S., Park, Y.S., *et al.* (2007) Range expansion of forest pest populations by using the lattice model. Ecological Modelling, 203, 157–66.

Leung, B., Drake, J.M., and Lodge, D.M. (2004) Predicting invasions: propagule pressure and the gravity of Allee effects. Ecology, 85, 1651–60.

Lewis, M.A. and Kareiva, P. (1993) Allee dynamics and the spread of invading organisms. Theoretical Population Biology, 43, 141–58.

Liu, H., Bauer, L.S., Gao, R., *et al.* (2003) Exploratory survey for the emerald ash borer, *Agrilus planipennis* (Coleoptera: Buprestidae), and its natural enemies in China. Great Lakes Entomologist, 36, 191–204.

Long, J.L. (1981) *Introduced Birds of the World: The World Wide History, Distribution and Influence of Birds Introduced to New Environments*. London: Universe Books.

Lubina, J.A. and Levin, S.A. (1988) The spread of a reinvading species: range expansion in the California sea otter. American Naturalist, 131, 526–43.

Lyons, D.A. and Scheibling, R.E. (2009) Range expansion by invasive marine algae: rates and patterns of spread at a regional scale. Diversity and Distributions, 15, 762–75.

Mack, R.N., Simberloff, D., Lonsdale, W.M., *et al.* (2000) Biotic invasions: causes, epidemiology, global consequences, and control. Ecological Applications, 10, 689–710.

Matthews, J., Van der Velde, G., De Vaate, A.B., *et al.* (2014) Rapid range expansion of the invasive quagga mussel in relation to zebra mussel presence in The Netherlands and Western Europe. Biological Invasions, 16, 23–42.

May, B. and Marsden, J.E. (1992) Genetic identification and implications of another invasive species of dreissenid mussel in the Great Lakes. Canadian Journal of Fisheries and Aquatic Sciences, 49, 1501–6.

Mayr, E. (1965) Numerical phenetics and taxonomic theory. Systematic Zoology, 14, 73–97.

McCullough, D.G., Poland, T.M., Anulewicz, A.C., *et al.* (2009) Emerald ash borer (Coleoptera: Buprestidae) attraction to stressed or baited ash trees. Environmental Entomology, 38, 1668–79.

Melbourne, B.A. and Hastings, A. (2009) Highly variable spread rates in replicated biological invasions: fundamental limits to predictability. Science, 325, 1536–9.

Micu, D. and Telembici, A. (2004) First record of *Dreissena bugensis* (Andrusov 1897) from the Romanian stretch of River Danube. In: *Abstracts of the International Symposium of Malacology*. Sibiu, Romania.

Mitikka, V. and Hanski, I. (2010) Pgi genotype influences flight metabolism at the expanding range margin of the European map butterfly. Annales Zoologici Fennici, 47, 1–14.

Mitikka, V., Heikkinen, R.K., Luoto, M., *et al.* (2008) Predicting range expansion of the map butterfly in Northern Europe using bioclimatic models. Biodiversity and Conservation, 17, 623–41.

Molloy, D.P., Bij de Vaate, A., Wilke, T., *et al.* (2007) Discovery of *Dreissena rostriformis bugensis* (Andrusov 1897) in Western Europe. Biological Invasions, 9, 871–4.

Mooney, H.A. and Cleland, E.E. (2001) The evolutionary impact of invasive species. Proceedings of the National Academy of Sciences of the USA, 98, 5446–51.

Moore, J.W., Herbst, D.B., Heady, W.N., *et al.* (2012) Stream community and ecosystem responses to the boom and bust of an invading snail. Biological Invasions, 14, 2435–46.

Mörtl, M., Werner, S., and Rothhaupt, K.-O. (2010) Effects of predation by wintering water birds on zebra mussels and on associated macroinvertebrates. In: Van der Velde, G., Rajagopal, S., and Bij de Vaate, A. (eds) *The Zebra Mussel in Europe*. Leiden: Backhuys Publishers, pp. 239–50.

Muniappan, R., Raman, A., and Reddy, G.V.P. (2009) *Ageratina adenophora* (Sprengel) King and Robinson (Asteraceae). In: Muniappan, R., Reddy, G.V.P., and Raman, A. (eds) *Biological Control of Tropical Weeds using Arthropods*. Cambridge: Cambridge University Press, pp. 63–73.

Nathan, R., Safriel, U.N., and Noy-Meir, I. (2001) Field validation and sensitivity analysis of a mechanistic model for tree seed dispersal by wind. Ecology, 82, 374–88.

Novak, M. (1987) Beaver. In: Novak, M., Baker, J.A., Obbard, M.E., *et al.* (eds) *Wild Furbearer Management and Conservation in North America*. Toronto: Ontario Ministry of Natural Resources, pp. 283–312.

Okubo, A. (1988) Diffusion-type models for avian range expansion. In: *Acta XIX Congress International Ornithologici*, Vol. 1. Ottawa: University of Ottawa Press, pp. 1038–49.

Orledge, G.M. and Reynolds, S.E. (2005) Fungivore host-use groups from cluster analysis: patterns of utilisation of fungal fruiting bodies by ciid beetles. Ecological Entomology, 30, 620–41.

Orledge, G.M., Smith, P.A., and Reynolds, S.E. (2010) The non-pest Australasian fungivore *Cis bilamellatus* Wood (Coleoptera: Ciidae) in northern Europe: spread dynamics, invasion success and ecological impact. Biological Invasions, 12, 515–30.

Orlova, M.I., Golubkov, S., Kalinina, L., *et al.* (2004) *Dreissena polymorpha* (Bivalvia: Dreissenidae) in the Neva Estuary (eastern Gulf of Finland, Baltic Sea): is it a biofilter or source for pollution? Marine Pollution Bulletin, 49, 196–205.

Orwig, D.A., Foster, D.R., and Mausel, D.L. (2002) Landscape patterns of hemlock decline in New England due to the introduced hemlock woolly adelgid. Journal of Biogeography, 29, 1475–87.

Ouyang, F., Hui, C., Ge, S., *et al.* (2014) Weakening density dependence from climate change and agricultural intensification triggers pest outbreaks: a 37-year observation of cotton bollworms. Ecology and Evolution, 4, 3362–74.

Park, K.N. and Hyun, J.S. (1983) Studies on the effects of the pine needle gall midge, *Thecodiplosis japonensis* Uchida et Inouye, on the growth of the red pine, *Pinus densiflora* S. et Z.(II)-growth impact on red pine. Journal of Korean Forestry Society, 62, 87–95.

Parker, I.M. (2004) Mating patterns and rates of biological invasion. Proceedings of the National Academy of Sciences of the USA, 101, 13695–6.

Parmesan, C. (2005) Biotic response: range and abundance changes. In: Lovejoy, T.E. and Hannah, L. (eds) *Climate Change and Biodiversity*. New Haven, CT: Yale University Press, pp. 41–55.

Parmesan, C. and Yohe, G. (2003) A globally coherent fingerprint of climate change impacts across natural systems. Nature, 421, 37–42.

Petit, R. (2004) Biological invasions at the gene level. Diversity and Distributions, 10, 159–65.

Phillips, B.L., Brown, G.P., Greenlees, M., *et al.* (2007) Rapid expansion of the cane toad (*Bufo marinus*) invasion front in tropical Australia. Austral Ecology, 32, 169–76.

Phillips, B.L., Brown, G.P., Webb, J.K., *et al.* (2006) Invasion and the evolution of speed in toads. Nature, 439, 803.

Poland, T.M. and McCullough, D.G. (2006) Emerald ash borer: invasion of the urban forest and the threat to North America's ash resource. Journal of Forestry, 104, 118–24.

Pollux, B.J.A., Van der Velde, G., and Bij de Vaate, A. (2010) A perspective on global spread of *Dreissena polymorpha*: a review on possibilities and limitations. In: Van der Velde, G., Rajagopal, S., and Bij de Vaate, A. (eds) *The Zebra Mussel in Europe*. Leiden: Backhuys Publishers, pp. 45–58.

Popa, O.P. and Popa, L.O. (2006) The most westward European occurrence point for *Dreissena bugensis* (Andrusov 1897). Malacologica Bohemoslovaca, 5, 3–5.

Pyšek, P. and Prach, K. (1993) Plant invasions and the role of riparian habitats: a comparison of four species alien to central Europe. Journal of Biogeography, 20, 413–20.

Ralls, K., Eagle, T.C., and Siniff, D.B. (1996) Movement and spatial use patterns of California sea otters. Canadian Journal of Zoology, 74, 1841–9.

Ramanantoanina, A. and Hui, C. (2016) Formulating spread of species with habitat dependent growth and dispersal in heterogeneous landscapes. Mathematical Biosciences, 275, 51–6.

Reibnitz, V.J. (1999) Verbreitung und Lebensräume der Baumschwammfresser Südwestdeutschlands (Coleoptera: Cisidae). Entomologische Verein Stuttgart, 34, 1–76.

Reino, L.M. and Silva, T. (1998) The distribution and expansion of the Common Waxbill (*Estrilda astrild*) in the Iberian Peninsula. Biologia de Conservacion de la Fauna, 102, 163–7.

Reznick, D.N. and Ghalambor, C.K. (2001) The population ecology of contemporary adaptations: what empirical studies reveal about the conditions that promote adaptive evolution. Genetica, 112/113, 183–98.

Riedman, M. and Estes, J.A. (1990) *The Sea Otter (Enhydra lutris): Behavior, Ecology, and Natural History*. Washington, DC: US Department of the Interior, Fish and Wildlife Service. Biological Report, 90(14).

Rouget, M., Robertson, M.P., Wilson, J.R., *et al.* 2016. Invasion debt—quantifying future biological invasions. Diversity and Distributions, 22, 445–56.

Sakai, A.K., Allendorf, F.W., Holt, J.S., *et al.* (2001) The population biology of invasive species. Annual Review of Ecology and Systematics, 32, 305–32.

Salo, L.F. (2005) Red brome (*Bromus rubens* subsp. *madritensis*) in North America: possible modes for early introductions, subsequent spread. Biological Invasions, 7, 165–80.

Sang, W., Zhu, L., and Axmacher, J.C. (2010) Invasion pattern of Eupatorium adenophorum Spreng in southern China. Biological Invasions, 12, 1721–30.

Seabrook, W. (1991) Range expansion of the introduced cane toad *Bufo marinus* in New South Wales. Australian Zoologist, 27, 58–62.

Shigesada, N., Kawasaki, K., and Takeda, Y. (1995) Modeling stratified diffusion in biological invasions. American Naturalist, 146, 229–51.

Shin, S.C. (2008) Pine wilt disease in Korea. In: Zhao, B.G., Futai, K., Sutherland, J.R., *et al.* (eds) *Pine Wilt Disease*. Tokyo: Springer, pp. 26–32.

Siegert, N.W., McCullough, D.G., Liebhold, A.M., *et al.* (2014) Dendrochronological reconstruction of the epicentre and early spread of emerald ash borer in North America. Diversity and Distributions, 20, 847–58.

Siegert, N.W., McCullough, D.G., Williams, D.W., *et al.* (2010) Dispersal of *Agrilus planipennis* (Coleoptera: Buprestidae) from discrete epicenters in two outlier sites. Environmental Entomology, 39, 253–65.

Silva, T., Reino, L.M., and Borralho, R. (2002) A model for range expansion of an introduced species: the common waxbill *Estrilda astrild* in Portugal. Diversity and Distributions, 8, 319–26.

Simberloff, D. (2006) Risk assessments, blacklists, and white lists for introduced species: are predictions good enough to be useful? Agricultural and Resource Economics Review, 35, 1–10.

Simberloff, D. and Gibbons, L. (2004) Now you see them, now you don't!–population crashes of established introduced species. Biological Invasions, 6, 161–72.

Simmons, R.E., Barnard, P., Dean, W.R.J., *et al.* (2004) Climate change and birds: perspectives and prospects from southern Africa. Ostrich, 75, 295–308.

Skellam, J.G. (1951) Random dispersal in theoretical populations. Biometrika, 38, 196–218.

Sol, D., Timmermans, S., and Lefebvre, L. (2002) Behavioural flexibility and invasion success in birds. Animal Behaviour, 63, 495–502.

Son, M.O. (2007) Native range of the zebra mussel and quagga mussel and new data on their invasions within the Ponto Caspian Region. Aquatic Invasions, 2, 174–84.

Song, Q.S., Fu, Y., Tang, J.W., *et al.* (2000) Allelopathic potential of Eupatorium adenophorum. Acta Phytoecologia Sinica, 24, 362–5.

Soné, K. (1987) Population dynamics of the pine needle gall midge, *Thecodiplosis japonensis* Uchida et Inouye (Diptera, Cecidomyiidae). Journal of Applied Entomology, 103, 386–402.

Stephens, D.W. and Krebs, J.R. (1986) *Foraging Theory*. Princeton: Princeton University Press.

Stockwell, C.A. and Ashley, M.V. (2004) Rapid adaptation and conservation. Conservation Biology, 18, 272–3.

Sun, L., Müller-Schwarze, D., and Schulte, B.A. (2000) Dispersal pattern and effective population size of the beaver. Canadian Journal of Zoology, 78, 393–8.

Therriault, T.W. and Orlova, M.I. (2010) Invasion success within the family Dreissenidae: prerequisites, mechanisms and perspectives. In: Van der Velde, G., Rajagopal, S., and Bij de Vaate, A. (eds) *The Zebra Mussel in Europe*. Leiden: Backhuys Publishers, pp. 59–68.

Thomas, C.D., Cameron, A., Green, R.E., *et al.* (2004) Extinction risk from climate change. Nature, 427, 145–8.

Togashi, K. and Shigesada, N. (2006) Spread of the pinewood nematode vectored by the Japanese pine sawyer: modeling and analytical approaches. Population Ecology, 48, 271–83.

Turner, J.L., Fitzpatrick, M.C., and Preisser, E.L. (2011) Simulating the dispersal of hemlock woolly adelgid in the temperate forest understory. Entomologia Experimentalis et Applicata, 141, 216–23.

Van der Velde, G. and Platvoet, D. (2007) Quagga mussels *Dreissena rostriformis* bugensis (Andrusov, 1897) in the Main River (Germany). Aquatic Invasions, 2, 261–4.

Van Eerden, M.R. and De Leeuw, J.J. (2010) How Dreissena sets the winter scene for water birds: dynamic interactions between diving ducks and zebra mussels. In: Van der Velde, G., Rajagopal, S., and Bij de Vaate, A. (eds) *The Zebra Mussel in Europe*. Leiden: Backhuys Publishers, pp. 251–64.

Veit, R.R. and Lewis, M.A. (1996) Dispersal, population growth, and the Allee effect: dynamics of the house finch invasion of eastern North America. American Naturalist, 148, 255–74.

Véron, G. (1992) Histoire biogéographique du castor d'Europe, *Castor fiber* (Rodentia, Mammalia). Mammalia, 56, 87–108.

Virtanen, T. and Neuvonen, S. (1999) Climate change and macrolepidopteran diversity in Finland. Chemosphere: Global Change Science, 1, 439–48.

Wehtje, W. (2003) The range expansion of the great-tailed grackle (*Quiscalus mexicanus* Gmelin) in North America since 1880. Journal of Biogeography, 30, 1593–607.

Weinig, C. (2000) Plasticity versus canalization: population differences in the timing of shade-avoidance responses. Ecology Letters, 54, 441–51.

Williamson, M. (1996) *Biological Invasions*. New York, NY: Chapman and Hall.

Wilson, J.R., Panetta, F.D., and Lindgren, C. (2017) *Detecting and Responding to Alien Plant Incursions*. Cambridge: Cambridge University Press.

Wingfield, M.J., Roux, J., and Wingfield, B.D. (2011) Insect pests and pathogens of Australian acacias grown as nonnatives—an experiment in biogeography with far-reaching consequences. Diversity and Distributions, 17, 968–77.

Yamanaka, T., Tatsuki, S., and Shimada, M. (2001) Flight characteristics and dispersal patterns of fall webworm (Lepidoptera: Arctiidae) males. Environmental Entomology, 30, 1150–7.

Yang, Z.Q. and Zhang, Y.A. (2007) Researches on techniques for biocontrol of the fall webworm, *Hyphantria cunea*, a severe invasive insect pest to China. Chinese Bulletin of Entomology, 44, 465–71.

Yoshida, T., Ellner, S.P., Jones, L.E., *et al.* (2007) Cryptic population dynamics: rapid evolution masks trophic interactions. PLoS Biology, 5, e235.

Zhang, F. and Hui, C. (2014) Recent experience-driven behaviour optimizes foraging. Animal Behaviour, 88, 13–19.

Zhao, G.J. and Ma, Y.P. (1989) Investigation on the distribution and damage of *Eupatorium adenophorum* in Yunnan, China. Chinese Journal of Weed Sciences, 3, 37–40.

Zhulidov, A.V., Zhulidov, D.A., Pavlov, D.F., *et al.* (2005) Expansion of the invasive bivalve mollusk *Dreissena bugensis* (quagga mussel) in the Don and Volga River Basins: revisions based on archived specimens. Ecohydrology and Hydrobiology, 5, 127–33.

CHAPTER 3

Modelling spatial dynamics

3.1 Introduction

Understanding the spatio-temporal dynamics of populations, their expansion and retraction, has always been one of the main pursuits in ecology and biogeography. The spread of invasive species, while posing real and escalating threats to biodiversity conservation and ecosystem functioning, also provides superb natural experiments for unravelling the mechanisms and factors behind the dynamics of the geographical range of organisms. Studies of spread can be traced back to the dawn of mathematical ecology (Fisher 1937), and there have been steady advances in methods for analysing dispersal using partial differential equations and other spatial modelling techniques (Skellam 1951). Spatial modelling is currently a fundamental part of invasion science. A very wide range of methods, techniques, and philosophies underpin modern spatial modelling. This chapter provides a brief review of the current status of spatial modelling in invasion ecology, explores the emergence of different approaches, and discusses some key challenges.

Mathematical modelling has been extremely valuable for elucidating the dynamics of spread of introduced species (e.g. Okubo *et al.* 1989; van den Bosch *et al.* 1990; Shigesada and Kawasaki 1997; Caswell *et al.* 2003). However, most early mathematical ecologists lumped what we now call 'invasive' species (*sensu* Pyšek *et al.* 2004) with other spreading species; they were essentially interested in general dispersal and colonization dynamics of organisms. As defined by Richardson (2000) and Lockwood and their respective colleagues (2005), the invasion process also includes the phases of arrival, establishment, and naturalization of an alien species by overcoming geographic, environmental, and reproductive barriers. Breaching of dispersal barriers allows species to become 'invasive' (see also Blackburn *et al.* 2011). Although spatial dynamic models focus mainly on the last stage of the expansion of invasives and the integration by the recipient ecosystems, all parts of the invasion process must be considered as a cohesive continuum.

To thrive or to survive? Organisms must constantly be on the move—to alleviate intraspecific competition and inbreeding pressure and to exploit opportunities provided by disturbances. Such movements of individuals, either through random walk-like diffusion or via directed dispersal, lead to the collective phenomenon of advancing fronts in the geographic range of species—namely 'spread'. Classic examples of spreading organisms include the natural recolonization by trees of deglaciated landscapes in the northern hemisphere after the last ice age, the spread of common starlings (*Sturnus vulgaris*) across North America following the introduction of the species to the region by humans, and the expanding wave of the Canadian lynx (*Lynx canadensis*) in response to human-mediated environmental changes (see Hengeveld 1989). Although these examples of spread share some dynamic properties, this chapter focuses on models of spread for post-introduction alien species and extra-limital species.

The number and types of models used in invasion ecology have increased dramatically in the last few decades (Hastings *et al.* 2005). Models are now not only used to predict spread but also to explore options for intervention, for example by identifying the components of spread most amenable to alteration through management.

Invasion Dynamics. Cang Hui and David M. Richardson, Oxford University Press (2017).
© Cang Hui and David M. Richardson. DOI 10.1093/acprof:oso/9780198745334.001.0001

The essence of spread models has been broadened from a strictly mathematical orientation to a much more cohesive integration of data capture, mathematical analysis, and modelling realization. Following these developments, we suggest a conceptual framework representing an 'optimal model' for studying the spatial spread of invasive species (Higgins and Richardson 1996; Hui *et al.* 2011). It includes three components (Fig. 3.1): modelling core, context, and method. *Modelling core* refers to dynamic models used to describe the demography and spatial dynamics of species. *Modelling context* sets up the arena for the spatial dynamics and includes environmental factors that affect the spatial distribution and the rate of spread of the invasive species, including mainly biotic interactions, environmental/habitat heterogeneity, stochasticity, disturbance, and scales. *Modelling method* specifies how the modelling core is implemented within the framework of modelling context so as to produce the range dynamics of focal species.

A clear understanding of the assumptions, advantages, and drawbacks behind each modelling core, context, and method enables us to present a clear synthetic view of our current challenges and trends in modelling the spatial dynamics of invasive species, not only as a theoretical question but also as a fundamental requirement for efficient management of problematic invasive species. Here, we review

commonly used models of species distribution and spread within the proposed conceptual framework (Fig. 3.1) and introduce the basics of the modelling core, context, and methods in the next three sections. The rate of spread (i.e. the velocity of the travelling waves) and its commonly used metrics are presented analytically where possible. Building an ideal model for invasion dynamics requires us to first clarify these three components, and then to implement the modelling core correctively within the modelling context using the selected modelling method. We discuss modelling implementation in the last section. Overall, this chapter aims to provide the foundation to enable readers to build an appropriate spread model for exploring the spatial dynamics of invasive species. Models are flexible quantitative tools and can have different versions for specific purposes; some models (e.g. species distribution models) presented here are used in other chapters with different specifications and for different purposes.

3.2 Modelling core

The core of spread models can be divided into two groups according to the entity or unit of the model (Turchin 1998): those centred on moving individuals (Lagrangian framework), and those focusing on the density of populations in space (Eulerian framework). We introduce the Lagrangian framework in section 3.2.1 and the Eulerian framework in sections 3.2.2 and 3.2.3. More detailed descriptions of these models are provided by Lewis and colleagues in their 2016 work.

3.2.1 Movement and dispersal

Random walk models are the most basic yet important models under the Lagrangian framework. A random walk depicts the movement and the path $<X_t, Y_t>$ of an organism as a succession of random steps (Fig. 3.2). It can be largely depicted by two parameters: step length (L) and turning angle (θ). Step length is a random number describing the distance between the coordinates before and after a single step walk. Turning angle describes the difference in walking directions of two consecutive steps ($-\pi \leq \theta \leq \pi$). The term 'random walk' was coined

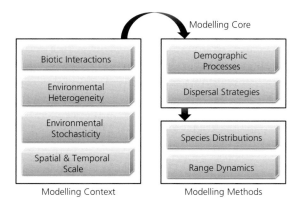

Fig. 3.1 A conceptual model for spread models for use in invasion science. For each modelling component a researcher must decide which sets of factors to include or remove, depending on the taxon, the spatial and temporal scale, and the research question. Arrows indicate the usual order in setting up a spatial model. Redrawn from Hui *et al.* (2011).

by the English mathematician and biostatistician Karl Pearson in 1905. Since then, the concept has been widely used in economics, physics, chemistry, and ecology. A basic formulation of random walks is as follows:

$$X_{t+1} = X_t + L \cdot \sin(\alpha - \theta)$$
$$Y_{t+1} = Y_t + L \cdot \cos(\alpha - \theta)$$

(3.1)

where $\alpha = \arctan\,((X_t - X_{t-1})/(Y_t - Y_{t-1}))$ is the direction of current move. A variety of random walk models exist; these can be categorized based on how the step length and turning angle are distributed and whether steps follow a constant duration. The rate of spread by a random walk can be measured by the number of distinct sites visited per unit time. This has been extensively studied in regular lattices and fractal landscapes, especially for studies on optimal foraging (Rammal and Toulouse 1983). It is often measured as the net squared displacement (NSD) which is the square of distance travelled after a certain number of steps or for a certain period of time. The mean (net) squared displacement (MSD) is the standard metric of dispersal capacity for a random walk and is closely related to diffusion rate D (Codling *et al.* 2008): MSD ~ $D \cdot t^a$. When $a = 1$, we have a normal diffusion such as the Gaussian random walk; $a > 1$ super-diffusive and $a < 1$ sub-diffusive motion.

The Gaussian random walk is the simplest model of this kind (Fig. 3.2), with the step length following the absolute value of a normal distribution, $N(0, \sigma)$, and no restrictions for turning angle and step duration (although they usually represent a regular pace walk). Due to the central limit theorem, a component of probability theory, random walks with the step length following the absolute value of any distributions of zero mean and a finite variance are statistically similar to a Gaussian random walk with an equal standard deviation, σ. The root mean square of travelled distance (MSD$^{1/2}$) after n steps of a Gaussian random walk is $\sigma \cdot n^{1/2}$, with 68 per cent chance that the individual travels less than this distance. Passive dispersal or diffusion-like movement, such as the Brownian motion of pollen grains in water, can be efficiently depicted as a Gaussian random walk. In contrast, a Lévy flight is a random walk where the step length follows a heavy-tailed distribution and an isotropic turning

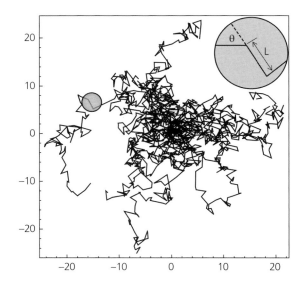

Fig. 3.2 Random walks in a homogeneous landscape. Fifteen individuals released at the origin randomly walked 100 steps, with no limit for the turning angle (θ) and step length (L) (illustrated in circles a few amplified steps) a normally distributed random number of zero mean and unit standard deviation. Random walks are normally implemented as the modelling core in individual-based models for simulating context-based spread dynamics of invasive species with known or assumed life cycles and management intervention.

angle. In ecology, the common heavy-tailed distributions include the Pareto (power law), log-normal, and Cauchy distributions (Hui *et al.* 2011). The movement of many animals follows a Lévy flight (Viswanathan *et al.* 1999; see also Chapter 4).

When turning angles follow a specific distribution other than isotropic, we are dealing with a correlated random walk (CRW). The turning angle is typically concentrated around zero (Fortin *et al.* 2005), indicating positive autocorrelation in consecutive steps of movement, or direction persistence. The net squared displacement after n moves of a CRW can be estimated as (Kareiva and Shigesada 1983): NSD $\approx n(m_2 + 2m_1\varphi/(1-\varphi))$, where m_1 is the mean step length, m_2 the mean squared step length, and φ the average cosine of turning angle. Directional persistence affects the rate of spread through the ratio $\varphi/(1-\varphi)$, with small turning angles producing larger ratios and thus larger displacement (faster spread). A reduced turning angle has been found in the vanguard population of invasive cane toads, *Rhinella marina*, in Australia (Brown *et al.* 2014).

As most data sources for parameterizing a random walk are from records with a regular frequency (e.g. telemetry data), the time interval between two consecutive steps is often considered constant. However, when using irregular records (e.g. recapture records), special consideration is needed for model parameterization. For instance, to generate potential movement paths from irregular records, we could interpolate the movement data using Brownian bridge or bivariate Gaussian bridge models (Horne *et al.* 2007; Sawyer *et al.* 2009; Kranstauber *et al.* 2012; Byrne *et al.* 2014), or other approaches such as the continuous correlated random walk model (Johnson *et al.* 2008).

A recent development is to explain or predict random walks by further considering environmental and habitat variables during the step selection phase. Specifically, using conditional logistic regression called Step Selection Function (SSF), we can compare the environmental conditions of each observed step with those of generated random steps of the same starting point but with different step length and/or direction (Fortin *et al.* 2005; Squires *et al.* 2013). SSF is originated from resource selection function (Manly *et al.* 2002) and follows the structure:

$$w(x) = \exp(\beta_1 x_1 + \beta_2 x_2 + \ldots + \beta_n x_n) \qquad (3.2)$$

where β_1 to β_n are regression coefficients and x_1 to x_n are environmental variables. The SSF, $w(x)$, is equal to the frequency of environmental attributes of observed steps divided by the frequency of available environmental attributes estimated from randomly generated steps (Thurfjell *et al.* 2014). The SSF can then be used in individual-based models to simulate animal movement in heterogeneous landscapes (e.g. Latombe *et al.* 2014). An alternative to the SSF setup is, instead of comparing the observed and available environmental attributes of each step, to compare the observed and available environmental gradient of each step (where gradient is defined as the difference of environmental attributes between the start and end points of a step).

3.2.2 Reaction and diffusion

Models with the goal of estimating the spread rate of invasive species focus on the dynamics of population density in space. Thus population-level models

follow the Eulerian modelling framework (Turchin 1998). This section introduces continuous models, while the next section (3.2.3) discusses discrete models. Continuous models are typically implemented using partial differential equations (PDEs), such as the reaction diffusion model, with the reaction term representing population growth and the diffusion term dispersal (e.g. Holmes *et al.* 1994; Turchin 1998; Petrovskii and Li 2006). They are also related to the random walk model as the diffusion part in the model can be directly translated into the accumulation of numerous individuals undergoing random walks. Although the following section handles models with one spatial dimension (i.e. those that consider spread along a linear habitat, such as coastline, rivers, and roads), models with two spatial dimensions can be implemented by adding an extra dimension to the diffusion term.

Two classical partial differential equations were first presented by Fisher (1937) and Skellam (1951) as the well-known reaction diffusion models (Fig. 3.3):

$$\frac{\partial n}{\partial t} = f(n) + D\frac{\partial^2 n}{\partial x^2} \qquad (3.3)$$

where $f(n)$, D, and n indicate the population growth rate, the diffusion rate, and the population density, respectively. For Skellam's (1951) model, the population follows the Malthusian growth $f(n) = r \cdot n$; for

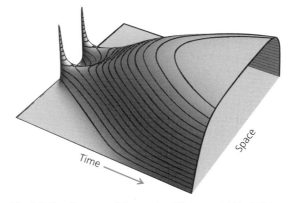

Fig. 3.3 Travelling wave of the reaction diffusion model. The initial propagules are introduced at two localities. Parameters: $r = K = 1$, $D = 0.05$, t from 0 to 8. Note, to use the PDE solver properly in any computational languages, both the initial condition ($n(x,0)$) and boundary condition ($n(0,t)$) need to be defined and compatibly at the intersection.

Fisher's reaction diffusion (1937) model, the population follows the logistic growth $f(n) = r \cdot n(1 - n/K)$, where r and K are the intrinsic growth rate and the carrying capacity, respectively. The model generates a travelling wave with an asymptotic velocity of $\hat{v} = 2\sqrt{f'(0)D}$. It is worth noting that this is the rate of spread caused by the Brownian-motion diffusion which may not be appropriate for representing realistic patterns of animal movement or the spread of propagules by vectors. Seed dispersal by different vectors defies random diffusion and is often examined by incorporating a dispersal kernel into the integrodifference equations (see section 3.2.3). Furthermore, when the reaction in the model is very fast, the asymptotic velocity can become arbitrarily large, and a solution to such an unrealistic property is to replace the above model by an integrodifferential equation (Branco et al. 2007).

It is clear that any effects causing changes to the linear approximation (first-order derivative) of population growth function at the initial point of invasion $f'(0)$ will affect the rate of spread. For instance, Lewis and Kareiva (1993) examined the rate of spread of the reaction diffusion model when the population dynamics is subjected to the Allee effect, $f(n) = r \cdot n(n - a)(1 - n)$, and found the asymptotic rate of spread to be sensitive to the intensity of the Allee effect: $\hat{v} = \sqrt{2rD}(0.5 - a)$ if $a < 1/2$ and $\hat{v} = 0$ if $a \geq 1/2$. Furthermore, in a two-dimensional landscape, the expanding wave can only start if the initial radius of the beachhead is greater than a threshold $\left(\sqrt{D/2r}(0.5 - a)\right)$, that is, there is a minimum initial population size for the range expansion (Lewis and Kareiva 1993).

Besides the random walk diffusion, if species dispersal is biased due to air and water currents, a drift (or convection) term can be added to the reaction diffusion model; it is then called an advection diffusion model:

$$\frac{\partial n}{\partial t} = f(n) - v_x \frac{\partial n}{\partial x} + D \frac{\partial^2 n}{\partial x^2} \qquad (3.4)$$

where v_x is the drift velocity along the x-axis. The travelling wave of the advection diffusion model is a simple overlap between the travelling wave of the diffusion model and the convection velocity (van den Bosch et al. 1990).

Another extension to the basic reaction diffusion model was made because of the realization that animals do not follow strictly random walks in their movements but tend to move with directional persistence. This correlated random walk can be depicted by the reaction telegraph equation (Holmes 1993):

$$\frac{\partial n}{\partial t} = f(n) + \frac{1}{2\lambda} \frac{\partial f}{\partial t} - \frac{1}{2\lambda} \frac{\partial^2 n}{\partial t^2} + \frac{\gamma^2 n}{2\lambda} \frac{\partial^2 n}{\partial x^2} \qquad (3.5)$$

where γ is the velocity of the individual and λ is the rate of changing direction. For logistic growth, $f(n) = r \cdot n(1 - n/K)$, Holmes (1993) found that the rate of spread (i.e. the speed of the travelling wave) is $v = \gamma \sqrt{8r\lambda}/(r + 2\lambda)$ if $0 < \sqrt{r/2\lambda} \leq 1$, and $v = \gamma$ if $\sqrt{r/2\lambda} \geq 1$. Furthermore, if we let $D = \gamma^2/2\lambda$, the comparison of the spreading rate between diffusion and telegraph models is made possible, which often predict relatively similar rates of spread in reality (Holmes 1993).

A general density-dependent diffusion model is presented by Okubo (1980) as the crowding-induced diffusion. Aronson (1980) investigated one such general reaction diffusion (crowding-induced) model:

$$\frac{\partial n}{\partial t} = f(n) + D \frac{\partial^2 n^m}{\partial x^2} \qquad (3.6)$$

where $f(n) = n(1 - n)$. This model has been rescaled so that the equation has only one parameter m. Obviously, if $m = 1$, this model is essentially the same as Fisher's reaction diffusion model. Individuals prefer to move towards crowding areas when $m > 1$, and the rate of spread thus slow down significantly. For instance, if $m = 2$, the rate of spread drops by a half, and the population density also becomes zero at a certain distance ahead of the wave. Individuals tend to avoid crowding areas when $m < 1$, suggesting a potentially high rate of spread. The crowding-induced diffusion comes from a biased random walk, instead of a pure or correlated random walk, and thus differs from the diffusion and the telegraph models (Aronson 1980; Turchin 1998). Individuals that move according to Lévy flights can be generally modelled by the Fokker–Planck equation with fractional derivatives of the diffusion part:

$$\frac{\partial n}{\partial t} = f(n) + D \frac{\partial^\alpha n}{\partial |x|^\alpha} \qquad (3.7)$$

with $\alpha = 2$ representing Brownian motion. Reaction diffusion models are particularly suitable for phrasing invasion dynamics in elegant mathematical terms; they are also sufficiently flexible to allow the incorporation of density-dependent recruitment and standard dispersal strategies. However, as partial differential equations seem to present a fairly formidable challenge to most ecologists, direct applications of reaction diffusion models in invasion ecology are largely in the domain of those with more theoretical interests in the field.

3.2.3 Integrodifference equations and dispersal kernels

The continuous Eulerian models discussed earlier can (although not easily) incorporate two components of realism: stage structure of the organism (e.g. age structure or egg–larvae–adult stages of insects) and different forms of dispersal kernel (i.e. the probability distribution of dispersal distance). The interplay of local adaptation and environmental heterogeneity further suggests the possibility of context-dependent dispersal (Matthysen 2005; Ramanantoanina *et al.* 2011; Bocedi *et al.* 2012; Pennekamp *et al.* 2014), with the dispersal strategy and spreading behaviour highly sensitive to the spatial and temporal variability of habitat quality, especially during the range shift and expansion (Dytham 2009; Hui *et al.* 2012; Henry *et al.* 2014; Kubisch *et al.* 2014).

To handle such realism, discrete Eulerian models enable us to incorporate complexities of both demography and dispersal. The most widely used of such models include the discrete-time dispersal kernel model, known as the *integrodifference equations*, and can be used to estimate the rate of spread based on any specific dispersal kernel (Weinberger 1982; Kot *et al.* 1996; Lewis *et al.* 2006). Moreover, integrodifference equations can be implemented with many modelling methods due to its flexibility to depict diverse demographic processes and dispersal strategies.

Let $k(x,y)$ denote the probability density function for the location x to which an individual at y disperses, we have the following integrodifference equation for calculating the population density in

locality x at time $t+1$ (Kot *et al.* 1996; Neubert and Caswell 2000):

$$n(x,t+1) = \int_{-\infty}^{\infty} k(x,y)b[n(y,t)]n(y,t)dy \quad (3.8)$$

where $b[n]n$ gives the per-capita population growth rate at locality y. The asymptotic rate of spread exists $\hat{v} = \min_{s>0} \ln[b(0)\Phi(s)]$ provided that $b[n]n$ increases monotonically with the population density n (also with no Allee effect) and that the dispersal kernel has a moment-generating function:

$$\Phi(s) = \int_{0}^{\infty} k(z)e^{sz}dz \quad (3.9)$$

where z indicates the distance between x and y. The integrodifference equation can also further incorporate the stage-structured population growth and dispersal, resulting in the stage-structured (matrix) model and enabling us to estimate the elasticity and sensitivity of different demographic stages (Neubert and Caswell 2000). A special focus of using integrodifference equations, which deserves further attention, is the estimation of rates of spread for fat-tailed long-distance dispersal (Higgins and Richardson 1999; Clark *et al.* 2003). To this end, Ramanantoanina and colleagues (2014) and Ramanantoanina and Hui (2016) have developed a model that uses integrodifference equations to incorporate individuals with different dispersal abilities in the initial propagules on both homogeneous and heterogeneous landscapes (see Chapter 4).

A closely related dispersal kernel model is the continuous-time kernel model (van den Bosch *et al.* 1992):

$$n(x,t) = \int_{-\infty}^{\infty}\int_{0}^{\infty} k(z,a)l(a)b(x-s,t-a)da \cdot ds \quad (3.10)$$

where $l(a)$ is the probability that an individual is still alive at age a. This leads to a rather similar rate of spread as the reaction diffusion model, suggesting that $\hat{v} = 2\sqrt{f'(0)D}$ is a robust measure of the rate of spread (van den Bosch *et al.* 1992; Grosholz 1996; Turchin 1998). Further work is needed to explore the relationship between the rate of spread and different forms of dispersal kernels (e.g. the general power law function for the long-distance jump dispersal) in different modelling contexts.

Integrodifference equations are extremely flexible tools for modelling invasion dynamics, especially because this modelling core can be easily used in lattice models (e.g. cellular automata) to handle real-world issues. However, modelling with a large number of sites or populations still requires strong computational power to implement.

3.3 Modelling context

Modelling context defines an arena for implementing spatial dynamics and also clarifies a few key factors that affect the features of invasion dynamics. Firstly, the spatial and temporal scales provide the physical bounds for the dynamics; the spatial extent and grain and the temporal resolution and duration are determined by the question being asked and the computational power at hand. Other contextual factors that must be specified before implementing the modelling core include biotic interactions, environmental heterogeneity, and stochasticity. We first address the implementation of environmental heterogeneity using basic species distributions models, and then discuss the implementation of biotic interactions, stochasticity, non-equilibrium, and assembly dynamics (also using the species distribution model in its broadest sense).

3.3.1 Environmental heterogeneity

Niche theory in ecology posits that the performance of organisms, such as their growth rate and mobility, changes along environmental gradients. Spatial spread of invasive species is also affected by environmental heterogeneity, and this can be incorporated in the modelling core using spatially dependent diffusion. Two frequently used methods include replacing the diffusion term in the model by the Fokker–Planck equation $\partial n / \partial t = \partial^2(\mu(x)n) / \partial x^2$ or the Fickian diffusion equation $\partial n / \partial t = \partial(D(x)\partial n / \partial x) / \partial x$, where $\mu(x)$ and $D(x)$ are motility and diffusivity at locality x that determine the rate of spread (Shigesada et al. 1986; Turchin 1998). A significant advance was made by Dewhirst and Lutscher (2009) who derived an approximate rate of spread in heterogeneous habitats using integrodifference equations (see also Higgins et al. 2003; Kawasaki and Shigesada 2007; Ramanantoanina and Hui 2016).

In theoretical studies, spatial heterogeneity is often generated by neutral landscape models (With and King 1997) which can be broadly categorized into random, hierarchical, and fractal types. Random and clumped neutral landscapes are first-generation models that seek the critical scales in detecting or comparing ecological processes based on the prediction of percolation theory (Gardner et al. 1987; Wiens et al. 1997; Hiebeler 2000). The second generation of neutral landscape models are hierarchical (Lavorel et al.1995). The most realistic neutral landscape models are those of fractal nature (Palmer 1992). Algorithms, such as the mid-point replacement and random Gaussian field, are often used to generate artificial landscapes for examining the distributional patterns.

For investigating the effect of realistic environmental heterogeneity on spread, species distribution models (SDMs) have become the standard approach for transforming environmental variables into suitability habitat maps on which the modelling core can be implemented. The recent development of SDMs has made it possible to project the potential equilibrium distribution of invasive species at regional and global scales (Guisan et al. 2006; Elith and Leathwick 2009; Franklin 2009; Elith 2016). Top-down correlative SDMs utilize the relationship between the occurrence of a species and predictor variables to define the potential distribution of that species (Fig. 3.4). Correlative SDMs have received much attention in the last two decades and major advances been made. Bottom-up mechanistic SDMs directly translate the effect of physical/environmental conditions to individual/population performance and have been used for mapping the fundamental niche (suitable habitat) in invaded regions (Morin and Lechowicz 2008; Kearney and Porter 2009). Knowledge gaps and uncertainties regarding these models, such as the debate on niche identified from SDMs, need further investigation (McInerny and Etienne 2012; Araújo and Guisan 2006; Elith 2016). Box 3.1 provides a simple guide to using SDMs.

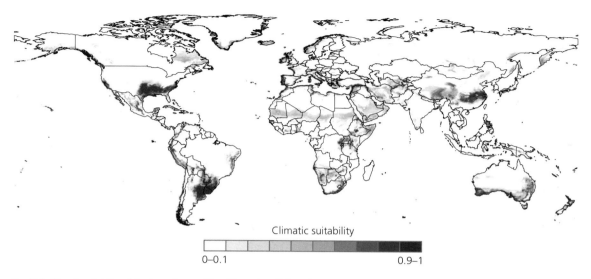

Fig. 3.4 Climatic suitability of the Argentine ant (*Linepithema humile*) based on boosted regression trees calibrated using global occurrence data from areas where the species is known to persist outside human buildings and five climate variables. Areas with a mean annual temperature below the lower thermal limits of the species (−10.4 °C) were not considered. From Roura-Pascual *et al.* (2011).

Box 3.1 Species distribution models

Correlative SDMs use a combination of environmental variables to explain the observed occurrence data and to project the potential range of a focal species. As with any multivariate approaches, the selection of these environmental predictors profoundly influences the prediction of species distribution (Guisan and Thuiller 2005). It is important to choose variables that are relatively independent of each other, and are relevant to the dynamics of the study species and to the resolution of the study (Pearson *et al.* 2004). Knowledge of the biology and demography of a species and its sensitivity to human disturbances are crucial for building a meaningful SDM. Variables that have a direct impact on the distribution should be used in preference to indirect variables. Spatial scales at which each process operates and the focal species distribute should be considered when choosing variables. It is essential to rely on *a priori* knowledge of which variables to include or exclude (Huntley *et al.* 2008; Elith and Leathwick 2009).

Collinearity between explanatory variables can be problematic for SDMs and should be handled with care (Dormann *et al.* 2013). For instance, the variance inflation factor (VIF), calculated for each continuous predictor as $1/(1-R^2)$ of the regression of predictor *i* on the remaining predictors, can be used to sequentially remove variables till VIFs for all remaining variables are less than 5 (Hugueny 1989; Guisan

and Zimmermann 2000; Obrien 2007). Stepwise model selection based on SDM performance can be used to remove redundant variables. A ten- or fivefold cross-validation with a tolerable fitting error can be used to decide how many and which variables to remove before the reduction in predictive performance exceeds a predefined threshold (Franklin 2009). The removal of one variable could reveal unexpected relationships among the remaining variables, possibly providing new insights on the species' habitat preferences.

Occurrence data used in an SDM can be presence only, presence/absence, or abundance data. Areas not currently occupied by a species are clearly not necessarily unsuitable habitat (Vaclavik and Meentemeyer 2012). To compensate for the lack of true absence data in most studies, background samples (Elith *et al.* 2011) and/or 'pseudo-absence' records (VanDerWal *et al.* 2009) are often used in SDMs (Elith *et al.* 2006). As a guideline, the background should remove areas where absence of the species is likely to be due to geographic barriers or insufficient time for spread or simply because the area has not been surveyed (Phillips *et al.* 2009; Elith *et al.* 2011). Presences of other species (Phillips *et al.* 2006) or features (Foxcroft *et al.* 2009) can potentially be used as absences; pseudo absences can also be created in a buffer around presence records, especially

(Continued)

Box 3.1 *Continued*

for locally restricted and patchy populations (Nakazato *et al.* 2010). Some of the early SDMs used only presence data, but as SDMs have developed, most methods now incorporate background or absence data as well, leading to an improvement in model accuracy. Because most presence records are not from systematic surveys, they are often spurious due to sampling bias and imperfect detection; recent developments in observation and hierarchical models have started to address these issues in SDMs (Dorazio 2014).

Choosing the correct spatial scale for species distribution modelling is crucial. Scale here refers to both extent and grain (grids) and will depend on the aim of the study, the available data, and the study system. Grid size needs to be appropriate for the properties of the data or analysis but also the characteristics of species records and the intended application (Elith and Leathwick 2009). It is important to find common ground where different environmental factors can be incorporated into a model. The key is to realize that species distributions are regulated by multiple biotic and abiotic processes working in concert but at different scales (Rouget and Richardson 2003; McGill 2010; Peterson *et al.* 2011). The scale of autocorrelation for significant variables often resonates with the scale of the study, a phenomenon we call scale resonance.

Regularization is an approach to smooth the predicted distribution through shrinking and penalizing model coefficients to allow for a less complex but more general prediction (Elith *et al.* 2011). A small regularization parameter (< 1) will overfit the model such that the predicted distribution primarily consists of those areas where the sample occurrences were found. The greater the regularization parameter, the larger the confidence interval and hence the smoother the response curves become, allowing for a more generalized distribution. Enforcing smoothness and a less complex model in this way has been shown to be misleading, especially when studying species distributions in novel climates (Elith *et al.* 2010). It could lead to unrealistic predictions of equal probabilities in divergent environments. It is therefore important to increase regularization when modelling invading species.

Most SDM methods are regression-like, and use additive combinations of predictors to model species' occurrence or abundance (Franklin 2009). Modelling methods range from straightforward environmental matching using generalized linear models (GLM) such as BIOCLIM and DOMAIN to increasingly complex models working with non-linear relationships such as generalized additive models (GAM) and maximum entropy models (MaxEnt) (Elith and Leathwick 2009). Machine learning and Bayesian methods are the most recent developments and have sophisticated model fitting

abilities, although at the cost of being computationally intensive. We see a surge of implementation of machine learning methods in SDMs, such as artificial neural networks (ANN), multivariate adaptive regression splines (MARS), regression trees (e.g. boosted regression trees, BRT), genetic algorithms, support vector machines, and MaxEnt (Franklin 2009). Elith and Graham (2009) reviewed five SDM methods, including GLM, Random forest (RF) and BRT for presence/absence data, and MaxEnt and genetic algorithm for rule set production (GARP) for presence only data. MaxEnt and BRT stand out as the most powerful and robust models among the five.

MaxEnt identifies species distribution as a probability distribution that maximizes the entropy under the constraints of known occurrence points (Phillips *et al.* 2006). When projecting into a novel climate it is likely that variable values not encountered during model training will be present. Extrapolation beyond the domain of model training needs to be undertaken with care. MaxEnt has a technique called 'clamping' for dealing with values that fall outside the training range. These values are treated as if they are the limit of the training range, remaining constant at those points outside the range. MaxEnt can further examine the extent of extrapolation by calculating multivariate environmental similarity surfaces (MESS) (Elith *et al.* 2010). It measures the similarity of the point to the distribution of reference points from the training range, with negative values given to dissimilar points.

Boosted regression trees (BRT) consist of two algorithms: boosting and regression trees. The former builds and combines models, while the latter forms part of the classification and regression tree-type models. Boosting is used to overcome inaccuracies inherent in a single tree model (Elith *et al.* 2006). Improvement on the performance of a single model is accomplished by fitting multiple models in a forward stepwise approach, gaining information from the errors at each step, making small modifications to achieve a better model fit, and finally combining them for predictions. Regression trees are good for selecting relevant variables and modelling interactions. Several studies have shown a preference for BRTs (Roura-Pascual *et al.* 2011; Fig.3.4) and have demonstrated their ability to outperform other methods (Elith *et al.* 2006; Bahn and McGill 2013). Presence data and absence (or pseudo-absence) data are required to implement a BRT model. BRT can to some degree distinguish between correlated variables, making it a superior method for identifying important variables while at the same time considering multicollinearity.

Mechanistic SDMs which translate the effect of environmental conditions to physiological performance (i.e. the reaction norm) have been used for mapping the fundamental niche (suitable habitat) in invaded regions (Kearney and Porter 2009; Morin and Lechowicz 2008). Besides physiological tolerance, mechanistic SDMs have used other bottom-up performance responses for projecting the suitable habitat; these include biophysical, life-history, phenological, and energetic factors (Chuine and Beaubien 2001; Morin and Thuiller 2009; Buckley *et al.* 2010). Population demography can be directly estimated based on these performance responses based on ambient environmental conditions (Higgins *et al.* 2012; Schurr *et al.* 2012). An important challenge for mechanistic SDMs is how best to estimate microclimatic factors (Storlie *et al.* 2014) that are most relevant to these performance responses. Such estimates are routinely derived from lab experiments. A recent SDM that calibrates a physiological model of plant growth to distribution data successfully predicted the invasive success of 749 Australian acacia and eucalypt tree species (Higgins and Richardson 2014).

Model verification assesses the ability of a model to fit the training data, whereas model validation assesses the ability of a model to predict events with independent test data (Araújo and Guisan 2006; Elith and Leathwick 2009). When simply trying to understand a system, model verification is appropriate and is used mainly to confirm the stability of selected variables. When evaluating a model's predictive capability one needs to focus on the model's ability to predict independent events. This becomes complicated when the aim of the modelling is to predict a species' distribution at a different scale, region, or time period, as is the case with invading species (Dormann *et al.* 2012). The most commonly used statistical measures are: area under the receiver operating curve (AUC), Cohen's Kappa statistic of similarity (κ) and correlation coefficients (Pearson *et al.* 2004; Elith *et al.* 2006). The validity and sensitivity of these tests to data characteristics continues to be debated.

AUC is probably the most widely used and unbiased measures of accuracy (Pearson *et al.* 2004), but there is ongoing discussion about its effectiveness (Lobo *et al.* 2008). The AUC is obtained from the receiver operating curve (ROC), which depicts the relationship between the proportion of true positives (sensitivity) and false positives (1–specificity) with varying probability thresholds. It thereby measures the ability of predictions to discriminate between presences and absences (Elith and Graham 2009). Good model performance is characterized by large areas under the ROC curves, hence a curve that maximizes sensitivity for low values of 1-specificity. AUC ranges from 0 to 1, 1 denoting a model that discriminates perfectly between presences and absences, and < 0.5 indicating that the model is no better than a random model. Moreover, 'ensembles' or combinations of different model predictions can be used and areas of uncertainty identified. This approach compensates for each model's specific strengths and weaknesses (e.g. using BIOMOD; Thuiller *et al.* 2009). Elith and colleagues (2010), however, do not recommend using ensemble methods for range-shifting species, as lack of congruence might not indicate model uncertainty but model error.

It is crucial to evaluate the model with independent data, because evaluation on training data will favour overfitted models. Overfitting leads to spurious conclusions regarding the role of the predictor variables and their relationships with the study species. There are various ways of obtaining independent data. These data could be collected independently, or could be temporally or spatially independent data. The original data can be split into a separate training and test set using random split, spatial split, or cross-validation resampling methods. If independent data are available these can be used for testing. A truly independent and spatially segregated dataset is necessary for testing if the model aims to make predictions in new areas/environments (Bahn and McGill 2013). Overall, the predictive performance of an SDM is influenced by the modelling method, the selection of predictor variables, the scale and the extent of extrapolation.

3.3.2 Other contextual factors

Other main factors that are part of setting up the modelling context include non-equilibrium dynamics and biotic interactions. The main challenge of using SDMs to project suitability maps or potential ranges for invasive species is that the expanding species is not at equilibrium with its novel environment and thus does not give a good indication of the conditions that the species can tolerate (Elith *et al.* 2010). To address this violation, various additions to SDMs have been proposed (Guisan *et al.* 2006; Franklin 2010; Higgins *et al.* 2012). Instead of using only records in the invaded range that are clearly transient, and thus not representative of the potential niche, attempts have been made to use records from native ranges or from combined records to assess whether invasive and native ranges can be reciprocally inferred (Randin *et al.* 2006; Hill *et al.* 2010). The practice of such reciprocal SDMs has currently focused on the potential of niche shift and rapid evolution during invasion (see Chapter 5).

Instead of focusing on the potential range, we can use similar statistical approaches to directly explain the colonization (or spread) events. A recent study has considered both the spatial autocorrelation of environmental variables and propagule pressure in explaining the extra-limital range expansion of a previously rare beetle in Europe. This requires a pre-transformation of both environmental variables and presence records to an environmental gradient and colonization events, followed by standard multivariate statistics (Horak *et al.* 2013). Results from such analyses can easily be applied using the modelling methods discussed in the next section. Overall, spatial autoregressive models (Kissling and Carl 2008; Schurr *et al.* 2012) can be used to address the issue of non-equilibrium dynamics in the modelling context (see Chapter 5).

Accounting for biotic interactions poses another challenge. The significant role that competition and other interactions play, over and above physiological constraints, in shaping species distributions might have profound consequences for species distributions, especially with changing climate and in novel environments. Models are now available for simultaneously considering biotic interactions between multiple species (Ovaskainen *et al.* 2010; Kissling *et al.* 2012; Linder *et al.* 2012; Wisz *et al.* 2013; Pollock *et al.* 2014; Svenning *et al.* 2014). Such models avoid using the distribution of other species as the modelling context of a focal species because their distributions will also be simultaneously affected by the presence of the focal species. We introduce these models and discuss the diverse effects of biotic interactions on invasion dynamics in Chapter 6.

When spread models are used to guide management of invasions, socio-economic factors (e.g. budget), risk and impact of invasion, monitoring and control strategies need to be considered as the modelling context. Some of these factors can already be implemented by SDMs. For instance, risk mapping combines typical SDMs with the effect of introduction and spread pathways (e.g. Kaplan *et al.* 2014). Other factors in invasion management are better implemented using specialized modelling techniques, such as bioeconomic models. We introduce invasion management as the modelling context in Chapter 9.

3.4 Modelling methods

The use of spatially explicit modelling methods in invasion ecology has advanced gradually. Such methods usually involve complex rule-based programming under the banner of agent-based models (ABMs), often with an intractable analytical solution. The system in an ABM comprises a collection of autonomous agents which make decisions based on a set of rules and their situation (Bonabeau 2002). Agents could be the populations of invasive and native species, conservation agencies, landowners, climate, or anything else for which rules of action can be defined. Repetitive interactions between agents are typical in ABMs which rely on computational power to explore potential system dynamics. Simple ABMs could also exhibit complex behaviour patterns due to non-linear feedbacks between agents, thereby providing valuable information about the dynamics of the real-world system that cannot be manipulated. Agents can evolve to have unanticipated behaviours, making ABMs appropriate for modelling complex adaptive systems (see Chapter 10). We introduce three varieties of ABMs, all of which have been widely used to model the spread and impact of invasive species.

3.4.1 Individual-based models

Methods for the spatial modelling of spread have been developed since the 1950s, with the advance of electronic computers. The earliest work on individual-based models (IBMs) can be traced back to the 1960s in a doctoral dissertation on the dynamics of Douglas fir, *Pseudotsuga menziesii* (Newnham 1964). One of earliest papers that applied IBMs in the ecological literature was by Michael Huston and his colleagues (1988). This visionary work separated IBMs from other types of models and called the development of IBMs 'a self-conscious discipline'. Several reviews have traced the history of IBMs for animals, plants, and in general (DeAngelis and Mooij 2005; Grimm and Railsback 2005). IBMs quantify differences between individuals both biologically and spatially (Huston *et al.* 1988), consistent with the reality in spread models. Given this advantage and increasing computational power, the use of IBMs has grown continuously (DeAngelis and Mooij 2005). Although IBMs are unlikely to lead to any general theories in ecology (DeAngelis and Mooij 2005), they are contributing substantially to predicting case-specific spread in invasion ecology.

Individual-based modelling is a bottom-up approach that focuses on individuals to understand properties of the system that emerges from these individuals and the interactions among them (Grimm 1999). An IBM can incorporate differences among individuals such as differences in experience and learning, genetic variability, phenotypic variability, behaviour, life cycles, movement, and local interactions (DeAngelis and Mooij 2005). These simulation techniques treat individuals as unique and discrete entities that often have at least one property that changes during the life cycle (Grimm and Railsback 2005). When modelling spread in a spatially explicit context, all individuals need to be assigned a location which may be static in the case of plants, or variable in the case of animals. The 'individual' in IBMs need not be a real individual but can also be a group of individuals that share a common entity (e.g. a colony of insects).

Evaluating IBMs is often challenging because of the nature of the modelling approach. We could use a similar approach to that used for evaluating SDMs, based on their goodness-of-fit using presence/absence as well as density data. Elith and Graham (2009) highlighted the importance of using more than one evaluation method for assessing model performance because of the different aspects of performance that each technique quantifies (see also Chatfield 1995; Rykiel 1996; Barry and Elith 2006; Franklin 2009). Statistical measures that modellers commonly use for evaluating presence/absence data include area under the receiver operating curve (AUC), Cohen's Kappa statistic of similarity (and other confusion matrix tests such as the true skill statistic), and correlation coefficients (Fielding and Bell 1997). The validity and sensitivity of these tests to data characteristics needs to be verified. As with SDMs, it is crucial to evaluate the model with independent data to avoid spurious conclusions.

When using a spatially explicit IBM as the modelling method, we still need to specify the modelling context and core. For illustration we show an IBM run on a spatially homogeneous and continuous landscape and with a simple life cycle of the individuals, representing the population dynamics of hermaphroditic perennial plants (Hui *et al.* 2010;

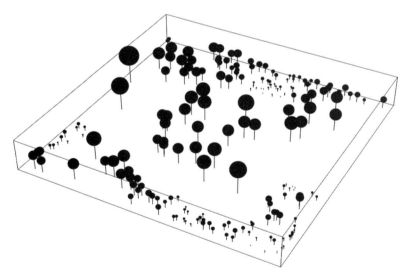

Fig. 3.5 An individual-based model of a box of plants. An individual in this simulation can be seen as a hermaphroditic perennial plant. A number of seeds are dispersed around the parent following an exponential dispersal kernel with a randomly chosen direction. The individuals that produce seeds are randomly chosen with a probability within the population of mature adults. Seeds only germinate if there are no other adult plants within a certain distance as a consequence of resource limitation and overcrowding. An individual also experiences a constant mortality. The simulation was performed in a two-dimensional homogeneous, square area with periodic boundaries (to exclude the edge effect). Rerun from Hui *et al.* (2010).

see online appendix for computer codes). During each time step (year), the plant faces the risk of death (*e*) and has a chance (*c*) to successfully reproduce *n* seeds once matured (age > 1). The seeds are dispersed isotropically around the plant according to the exponential distribution, $\mathrm{prob}(x) = \lambda e^{-\lambda x}$, where *x* is the seed dispersal distance and $1/\lambda$ the average seed dispersal distance. Those seeds that found no other adult plants nearby (minimum distance > *r*; over-crowding effect) can successfully germinate. The model generates a spatial map as the output (Fig. 3.5).

The above IBM is analogous to the forest growth simulators JABOWA, FORET, and SORTIE (see Levin *et al.* 1997), but differs from other lattice- or grid-based simulators in that the individuals are not constrained to grids. Overall, building an IBM requires detailed knowledge of the life cycle of the individual which needs to be translated into a set of context-dependent rules to allow each individual to automatically choose its fate (e.g. germinate, reproduce, disperse, etc.). Computation can then include loops over activities of all existing individuals. Desktop computers can normally handle < 2 million individuals for 50–100 generations within minutes. More demanding tasks require high-performance computer or lumping local individuals as one entity in the model. Combined with artificial or realistic heterogeneous landscapes and random walk models, IBMs can address many sophisticated questions, such as optimal resource utilization and adaptive life-history responses to environmental changes during the spread of invasive species (DeAngelis and Mooij 2005; see Chapter 9).

3.4.2 Cellular automata and lattice models

Cellular automata (CA) were first developed in the 1950s by the Hungarian-American mathematician, John von Neumannin; a failed attempt to prove the emergence of self-reproduction in life by solving partial differential equations discretely in lattices. Numerous attempts to find transition rules in CA to imitate life phenomena followed (e.g. 'the game of life'; Conway 1970). The development of CA was also accompanied by the design of parallel computers and digital image processing. In the 1960s, CA began to be used to explore neural networks,

reaction diffusion processes of active media as in heart and muscles, and the Ising model in physics (CA with randomness). In the late 1970s, when computers became widely available, research on CAs accelerated rapidly. In invasion ecology, the spread of invasive species is often simulated by CAs (also called lattice models) with continuous cell state across heterogeneous landscapes (e.g. Roura-Pascual *et al.* 2009).

CAs are an idealized 'system in which space and time are discrete' and where the cells normally have a finite number of possible states but can take continuous variables (Chopard and Droz 1998; Cannas *et al.* 2003; Lischke *et al.* 2006; Han and Hui 2014; Su *et al.* 2015). Classical CAs are also deterministic, and the state of a cell at the next time step is defined by the state of the cell and its neighbours in the previous time step (Conway 1970). As these conditions are quite limiting, the stochasticity condition can be relaxed and an infinite number of states for each cell allowed, resulting in the extended CA. The influence by neighbours can also be relaxed to allow states of all accessible cells to influence the state transition of a focal cell, after being weighted by the connectivity between cells, normally depicted by the dispersal kernel. CA modelling faces the modifiable areal unit problem (Openshaw 1984), given that the cell size is often artificially chosen. As a result, the linear dimension of the cell for any spread models should be less than the median of the dispersal kernel.

As an illustration, the spread of the invasive tree, *Acacia longifolia*, in the Western Cape of South Africa was modelled using an SDM–CA hybrid model (Donaldson *et al.* 2014). Suitable habitats were first identified from an SDM using MaxEnt. The demography (growth, survival, fecundity, and dispersal) and the suitable habitats were then used for building the CA in a lattice of 25 000 1 km^2 cells. In each time step, the tree population within a 1 km^2 cell first undergoes reproduction, then dispersal, and finally population growth up to a set carrying capacity. The potential number of new adult trees produced per cell is determined by the current population size and the number of reproductive trees that a single tree can produce during one time step. These potential individuals either remain in the same cell, move to adjacent cells (considering eight neighbouring

Bonabeau, E. (2002) Agent-based modelling: methods and techniques for simulating human systems. Proceedings of the National Academy of Sciences of the USA, 99, 7280–7.

Branco, J.R., Ferreira, J.A., and de Oliveira, P. (2007) Numerical methods for the generalized Fisher–Kolmogorov–Petrovskii–Piskunov equation. Applied Numerical Mathematics, 57, 89–102.

Brown, G.P., Phillips, B.L., and Shine, R. (2014) The straight and narrow path: the evolution of straight-line dispersal at a cane toad invasion front. Proceedings of the Royal Society B: Biological Sciences, 281, 20141385.

Buckley, L.B., Urban, M.C., Angilletta, M.J., et al. (2010) Can mechanism inform species' distribution models? Ecology Letters, 13, 1041–54.

Bullock, J.M. and Clarke, R.T. (2000) Long distance seed dispersal: measuring and modelling the tail of the curve. Oecologia, 124, 506–21.

Byrne, M.E., McCoy, J.C., Hinton, J.W., et al. (2014) Using dynamic Brownian bridge movement modelling to measure temporal patterns of habitat selection. Journal of Animal Ecology, 83, 1234–43.

Cannas, S.A., Marco, D.E., and Paez, S.A. (2003) Modelling biological invasions: species traits, species interactions, and habitat heterogeneity. Mathematical Biosciences, 183, 93–110.

Caplat, P., Hui, C., Maxwell, B.D., et al. (2014) Cross-scale management strategies for optimal control of trees invading from source plantations. Biological Invasions, 16, 677–90.

Caswell, H., Lensink, R., and Neubert, M.G. (2003) Demography and dispersal: life table response experiments for invasion speed. Ecology, 84, 1968–78.

Chatfield, C. (1995) Model uncertainty, data mining and statistical inference. Journal of the Royal Statistical Society, Series A, 158, 419–66.

Chopard, B. and Droz, M. (1998) Cellular Automata Modelling of Physical Systems. Cambridge: Cambridge University Press.

Chuine, I. and Beaubien, E.G. (2001) Phenology is a major determinant of tree species range. Ecology Letters, 4, 500–10.

Clark, J.S., Lewis, M.A., McLachlan, J.S., et al. (2003) Estimating population spread: what can we forecast and how well? Ecology, 84, 1979–88.

Clark, J.S., Silman, M., Kern, R., et al. (1999) Seed dispersal near and far: patterns across temperate and tropical forests. Ecology, 80, 1475–94.

Codling, E.A., Plank, M.J., and Benhamou, S. (2008) Random walk models in biology. Journal of the Royal Society Interface, 5, 813–34.

Conway, J. (1970) The game of life. Scientific American, 223, 120–3.

DeAngelis, D.L. and Mooij, W.M. (2005) Individual-based modeling of ecological and evolutionary processes. Annual Review of Ecology, Evolution, and Systematics, 36, 147–68.

Dewhirst, C. and Lutscher, F. (2009) Dispersal in heterogeneous habitats: thresholds, spatial scales, and approximate rates of spread. Ecology, 90, 1338–45.

Donaldson, J.E., Hui, C., Richardson, D.M., et al. (2014) Invasion trajectory of alien trees: the role of introduction pathway and planting history. Global Change Biology, 20, 1527–37.

Dorazio, R.M. (2014) Accounting for imperfect detection and survey bias in statistical analysis of presence-only data. Global Ecology and Biogeography, 23, 1472–84.

Dormann, C.F., Elith, J., Bacher, S., et al. (2013) Collinearity: a review of methods to deal with it and a simulation study evaluating their performance. Ecography, 36, 27–46.

Dormann, C.F., Schymanski, S.J., Cabral, J., et al. (2012) Correlation and process in species distribution models: bridging a dichotomy. Journal of Biogeography, 39, 2119–31.

Drake, J.M. and Lodge, D.M. (2004) Global hot spots of biological invasions: evaluating options for ballast-water management. Proceedings of the Royal Society B: Biological Sciences, 271, 575–80.

Dytham, C. (2009) Evolved dispersal strategies at range margins. Proceedings of the Royal Society B: Biological Sciences, 276, 1407–13.

Elith, J. (2016) Predicting distributions of invasive species. arXiv.org, no.1312.0851.

Elith, J., Ferrier, S., Guisan, A., et al. (2006) Novel methods improve prediction of species" distributions from occurrence data. Ecography, 29, 129–51.

Elith, J. and Graham, C.H. (2009) Do they? How do they? Why do they differ? On finding reasons for differing performances of species distribution models. Ecography, 32, 66–77.

Elith, J., Kearney, M., and Phillips, S. (2010) The art of modelling range-shifting species. Methods in Ecology and Evolution, 1, 330–42.

Elith, J. and Leathwick, J.R. (2009) Species distribution models: ecological explanation and prediction across space and time. Annual Review of Ecology, Evolution, and Systematics, 40, 677–97.

Elith, J., Phillips, S.J., Hastie, T., et al. (2011) A statistical explanation of MaxEnt for ecologists. Diversity and Distributions, 17, 43–57.

Epanchin-Niell, R.S. and Hastings, A. (2010) Controlling established invaders: integrating economics and spread dynamics to determine optimal management. Ecology Letters, 13, 528–41.

Fielding, A.H. and Bell, J.F. (1997) A review of methods for the assessment of prediction errors in conservation presence/absence models. Environmental Conservation, 24, 38–49.

Fisher, R.A. (1937) The wave of advance of advantageous genes. Annals of Eugenics, 7, 355–69.

Fortin, D., Beyer, H.L., Boyce, M.S., *et al.* (2005) Wolves influence elk movements: behavior shapes a trophic cascade in Yellowstone National Park. Ecology, 86, 1320–30.

Foxcroft, L.C., Richardson, D.M., Rouget, M., *et al.* (2009) Patterns of alien plant distribution at multiple spatial scales in a large national park: implications for ecology, management and monitoring. Diversity and Distributions, 15, 367–78.

Foxcroft, L.C., Rouget, M., and Richardson, D.M. (2007) Risk assessment of riparian plant invasions into protected areas. Conservation Biology, 21, 412–21.

Franklin, J. (2009) *Mapping Species Distributions*. Cambridge: Cambridge University Press.

Franklin, J. (2010) Moving beyond static species distribution models in support of conservation biogeography. Diversity and Distributions, 16, 321–30.

Gardner, R.H., Milne, B.T., Turner, M.G., *et al.* (1987) Neutral models for the analysis of broad-scale landscape pattern. Landscape Ecology, 1, 19–28.

Greene, D.F. and Calogeropoulos, C. (2001) Measuring and modelling seed dispersal of terrestrial plants. In: Bullock, J.M., Kenward, R.E., and Hails, R.S. (eds) *Dispersal Ecology*. Oxford: Blackwell, pp. 3–23.

Grimm, V. (1999) Ten years of individual-based modelling in ecology: what have we learned and what could we learn in the future? Ecological Modelling, 115, 129–48.

Grimm, V. and Railsback, S. (2005) *Individual-Based Modeling and Ecology*. Princeton, NJ: Princeton University Press.

Grosholz, E.D. (1996) Contrasting rates of spread for introduced species in terrestrial and marine systems. Ecology, 77, 1680–6.

Guisan, A., Broennimann, O., Engler, R., *et al.* (2006) Using niche-based models to improve the sampling of rare species. Conservation Biology, 20, 501–11.

Guisan, A. and Thuiller, W. (2005) Predicting species distribution: offering more than simple habitat models. Ecology Letters, 8, 993–1009.

Guisan, A. and Zimmermann, N.E. (2000) Predictive habitat distribution models in ecology. Ecological Modelling, 135, 147–86.

Han, X. and Hui, C. (2014) Niche construction on environmental gradients: the formation of fitness valley and stratified genotypic distributions. PLoS One, 9, e99775.

Hastings, A., Cuddington, K., Davies, K.F., *et al.* (2005) The spatial spread of invasions: new developments in theory and evidence. Ecology Letters, 8, 91–101.

Hengeveld, R. (1989) *Dynamics of Biological Invasions*. London: Chapman and Hall.

Henry, R.C., Bocedi, G., Dytham, C., *et al.* (2014) Interannual variability influences the eco-evolutionary dynamics of range-shifting. PeerJ, 2, e228.

Hiebeler, D. (2000) Populations on fragmented landscapes with spatially structured heterogeneities: landscape generation and local dispersal. Ecology, 81, 1629–41.

Higgins, S.I., Lavorel, S., and Revilla, E. (2003) Estimating plant migration rates under habitat loss and fragmentation. Oikos, 101, 354–66.

Higgins S.I., O'Hara R.B., and Roemermann, C. (2012) A niche for biology in species distribution models. Journal of Biogeography, 39, 2091–5.

Higgins, S.I. and Richardson, D.M. (1996) A review of models of alien plant spread. Ecological Modelling, 87, 249–65.

Higgins, S.I. and Richardson, D.M. (1999) Predicting plant migration rates in a changing world: the role of long-distance dispersal. American Naturalist, 153, 464–75.

Higgins, S.I. and Richardson, D.M. (2014) Invasive plants have broader physiological niches. Proceedings of the National Academy of Sciences of the USA, 111, 10610–14.

Hill, M.P., Hoffmann, A.A., MacFadyen, S., *et al.* (2010) Understanding niche shifts: using current and historical data to model the invasive redlegged earth mite, *Halotydeus destructor*. Diversity and Distributions, 18, 191–203.

Holmes, E.E. (1993) Are diffusion models too simple? A comparison with telegraph models of invasion. American Naturalist, 142, 779–95.

Holmes, E.E., Lewis, M.A., Banks, J.E., *et al.* (1994) Partial differential equations in ecology: spatial interactions and population dynamics. Ecology, 75, 17–29.

Horak, J., Hui, C., Roura-Pascual, N., *et al.* (2013) Changing roles of propagule, climate and land use during extralimital colonization of a rose chafer beetle. Naturwissenschaften, 100, 327–36.

Horne, J.S., Garton, E.O., Krone, S.M., *et al.* (2007) Analyzing animal movements using Brownian bridges. Ecology, 88, 2354–63.

Hugueny, B. (1989) West African rivers as biogeographic islands: species richness of fish communities. Oecologia, 79, 236–43.

Hui, C., Krug, R.M., and Richardson, D.M. (2011) Modelling spread in invasion ecology: a synthesis. In: Richardson, D.M. (ed.) *Fifty Years of Invasion Ecology: The Legacy of Charles Elton*. Oxford: Wiley-Blackwell, pp. 329–43.

Hui, C., Richardson, D.M., Pyšek, P., *et al.* (2013) Increasing functional modularity with residence time in the co-distribution of native and introduced vascular plants. Nature Communications, 4, 2454.

Hui, C., Roura-Pascual, N., Brotons, L., *et al.* (2012) Flexible dispersal strategies in native and non-native ranges: environmental quality and the 'good-stay, bad-disperse' rule. Ecography, 35, 1024–32.

Hui, C., Veldtman, R., and McGeoch, M.A. (2010) Measures, perceptions and scaling patterns of aggregated species distributions. Ecography, 33, 95–102.

Huntley, B., Collingham, Y.C., Willis, S.G., *et al.* (2008) Potential impacts of climatic change on European breeding birds. PLoS One, 3, e1439.

Huston, M., DeAngelis, D., and Post, W. (1988) New computer models unify ecological theory, BioScience, 38, 682–91.

Jerde, C.L. and Lewis, M.A. (2007) Waiting for invasions: a framework for the arrival of nonindigenous species. American Naturalist, 170, 1–9.

Johnson, D.S., London, J.M., Lea, M.-A., *et al.* (2008) Continuous-time correlated random walk model for animal telemetry data. Ecology, 89, 1208–15.

Jongejans, E., Skarpaas, O., and Shea, K. (2008) Dispersal, demography and spatial population models for conservation and control management. Perspectives in Plant Ecology, Evolution and Systematics, 9, 153–79.

Kaplan, H., van Niekerk, A., Le Roux, J.J., *et al.* (2014) Incorporating risk mapping at multiple spatial scales into eradication management plans. Biological Invasions, 16, 691–703.

Kareiva, P.M. and Shigesada, N. (1983) Analyzing insect movement as a correlated random walk. Oecologia, 56, 234–8.

Kawasaki, K. and Shigesada, N. (2007) An integrodifference model for biological invasions in a periodically fragmented environment. Japan Journal of Industrial and Applied Mathematics, 24, 3–15.

Kearney, M. and Porter, W. (2009) Mechanistic niche modelling: combining physiological and spatial data to predict species' ranges. Ecology Letters, 12, 334–50.

Kissling, W.D. and Carl, G. (2008) Spatial autocorrelation and the selection of simultaneous autoregressive models. Global Ecology and Biogeography, 17, 59–71.

Kissling, W.D., Dormann, C.F., Groeneveld, J., *et al.* (2012) Towards novel approaches to modelling biotic interactions in multispecies assemblages at large spatial extents. Journal of Biogeography, 39, 2163–78.

Kizuka, T., Akasaka, M., Kadoya, T., *et al.* (2014) Visibility from roads predict the distribution of invasive fishes in agricultural ponds. PLoS One, 9, e99709.

Kot, M., Lewis, M.A., and van den Driessche, P. (1996) Dispersal data and the spread of invading organisms. Ecology, 77, 2027–42.

Kranstauber, B., Kays, R., LaPoint, S.D., *et al.* (2012) A dynamic Brownian bridge movement model to estimate utilization distributions for heterogeneous animal movement. Journal of Animal Ecology, 81, 738–46.

Krug, R., Roura-Pascual, N., and Richardson, D.M. (2010) Clearing of invasive alien plants under different budget scenarios: using a simulation model to test efficiency. Biological Invasions, 12, 4099—112.

Kubisch, A., Holt, R.D., Poethke, H.-J., *et al.* (2014) Where am I and why? Synthesizing range biology and the eco-evolutionary dynamics of dispersal. Oikos, 123, 5–22.

Latombe, G., Parrott, L., Basille, M., *et al.* (2014) Uniting statistical and individual-based approaches for animal movement modelling. PLoS One, 9, e99938.

Lavorel, S., Gardner, R.H., and O'Neill, R.V. (1995) Dispersal of annual plants in hierarchically structured landscapes. Landscape Ecology, 10, 277–89.

Leung, B., Drake, J.M., and Lodge, D.M. (2004) Predicting invasions: propagule pressure and the gravity of Allee effects. Ecology, 85, 1651–60.

Leung, B. and Mandrak, N.E. (2007) The risk of establishment of aquatic invasive species: joining invasibility and propagule pressure. Proceedings of the Royal Society B: Biological Sciences, 274, 2603–9.

Levin, S.A., Grenfell, B., Hastings, A., *et al.* (1997) Mathematical and computational challenges in population biology and ecosystems science. Science, 275, 334–43.

Lewis, M.A. and Kareiva, P. (1993) Allee dynamics and the spread of invading organisms. Theoretical Population Biology, 43, 141–58.

Lewis, M.A., Neubert, M.G., Caswell, H., *et al.* (2006) A guide to calculating discrete-time invasion rates from data. In: Cadotte, M.A., McMahon, S.M., and Fukami, T. (eds) *Conceptual Ecology and Invasion Biology: Reciprocal Approaches to Nature*. Berlin: Springer, pp. 162–92.

Lewis, M.A., Petrovskii, S.V., and Potts, J.R. (2016) *The Mathematics Behind Biological Invasions*. Berlin: Springer.

Linder, H.P., Bykova, O., Dyke, J., *et al.* (2012) Biotic modifiers, environmental modulation and species distribution models. Journal of Biogeography, 39, 2179–90.

Lischke, H., Zimmermann, N.E., Bolliger, J., *et al.* (2006) TreeMig: a forest-landscape model for simulating spatio-temporal patterns from stand to landscape scale. Ecological Modelling, 199, 409–20.

Lobo, J.M., Jimenez-Valverde, A., and Real, R. (2008) AUC: a misleading measure of the performance of predictive distribution models. Global Ecology and Biogeography, 17, 145–51.

Lockwood, J.L., Cassey, P., and Blackburn, T. (2005) The role of propagule pressure in explaining species invasions. Trends in Ecology and Evolution, 20, 223–8.

Manly, B.F.J., McDonald, L.L., Thomas, D.L., *et al.* (2002) *Resource Selection by Animals: Statistical Design and Analysis for Field Studies*. Boston, MA: Kluwer Academic Publishers.

Matthysen, E. (2005) Density-dependent dispersal in birds and mammals. Ecography, 28, 403–16.

McGill, B.J. (2010) Matters of scale. Science, 328, 576.

McInerny, G.J. and Etienne, R.S (2012) Pitch the niche—taking responsibility for the concepts we use in ecology and species distribution modelling. Journal of Biogeography, 39, 2112–18.

Moilanen, A., Wilson, K.A., and Possingham, H. (2009) *Spatial Conservation Prioritization: Quantitative Methods and Computational Tools*. Oxford: Oxford University Press.

Moody, M.E. and Mack, R.N. (1988) Controlling the spread of plant invasions: the importance of nascent foci. Journal of Applied Ecology, 25, 1009–21.

Morin, X. and Lechowicz, M.J. (2008) Contemporary perspectives on the niche that can improve models of species range shifts under climate change. Biology Letters, 4, 573–6.

Morin, X. and Thuiller, W. (2009) Comparing niche-and process-based models to reduce prediction uncertainty in species range shifts under climate change. Ecology, 90, 1301–13.

Nakazato, T., Warren, D.L., and Moyle, L.C. (2010) Ecological and geographic modes of species divergence in wild tomatoes. American Journal of Botany, 97, 680–93.

Nathan, R. (2008) An emerging movement ecology paradigm. Proceedings of the National Academy of Sciences of the USA, 105, 19050–1.

Neubert, M.G. and Caswell, H. (2000) Demography and dispersal: calculation and sensitivity analysis of invasion speed for structured populations. Ecology, 81, 1613–28.

Newnham, R.M. (1964) The development of a stand model for Douglas-fir. PhD thesis, University of British Columbia, Vancouver.

Obrien, R.M. (2007) A caution regarding rules of thumb for variance inflation factors. Quality and Quantity, 41, 673–90.

Okubo, A. (1980) *Diffusion and Ecological Problems: Mathematical Models*. Berlin: Springer.

Okubo, A., Maini, P.K., Williamson, M.H., *et al.* (1989) On the spatial spread of the grey squirrel in Britain. Proceedings of the Royal Society B: Biological Sciences, 238, 113–25.

Openshaw, S. (1984) *The Modifiable Areal Unit Problem*. Norwich: GeoBooks.

Ovaskainen, O., Hottola, J., and Siitonen, J. (2010) Modeling species co-occurrence by multivariate logistic regression generates new hypotheses on fungal interactions. Ecology, 91, 2514–21.

Palmer, M.W. (1992) The coexistence of species in fractal landscapes. American Naturalist, 139, 375–97.

Pearson, R.G., Dawson, T.P., and Liu, C. (2004) Modelling species distributions in Britain: a hierarchical integration of climate and land-cover data. Ecography, 27, 285–98.

Pennekamp, F., Mitchell, K.A., Chaine, A., *et al.* (2014) Dispersal propensity in Tetrahymenathermophila ciliates—a reaction norm perspective. Evolution, 68, 2319–30.

Peterson, A.T., Soberón, J., Pearson, R.G., *et al.* (2011) *Ecological Niches and Geographic Distributions*. Princeton, NJ: Princeton University Press.

Petrovskii, S.V. and Li, B.L. (2006) *Exactly Solvable Models of Biological Invasion*. London: CRC Press.

Phillips, B.L., Brown, G.P., Travis, J.M.J., *et al.* (2008) Reid's paradox revisited: the evolution of dispersal kernels during range expansion. American Naturalist, 172, S34–8.

Phillips, S.J., Anderson, R.P., and Schapire, R.E. (2006) Maximum entropy modeling of species geographic distributions. Ecological Modelling, 190, 231–59.

Phillips, S.J., Dudik, M., Elith, J., *et al.* (2009) Sample selection bias and presence-only distribution models: implications for background and pseudo-absence data. Ecological Applications, 2009, 181–97.

Pollock, L.J., Tingley, R., Morris, W.K., *et al.* (2014) Understanding co-occurrence by modelling species simultaneously with a Joint Species Distribution Model (JSDM). Methods in Ecology and Evolution, 5, 397–406.

Porter, J.H. and Dooley Jr, J.L. (1993) Animal dispersal patterns: a reassessment of simple mathematical models. Ecology, 74, 2436–43.

Pyšek, P., Richardson, D.M., Rejmánek, M., *et al.* (2004) Alien plants in checklists and floras: towards better communication between taxonomists and ecologists. Taxon, 53, 131–43.

Ramanantoanina, A. and Hui, C. (2016) Spread in heterogeneous landscapes. Mathematical Biosciences, 275, 51–6.

Ramanantoanina, A., Hui, C., and Ouhinou, A. (2011) Effects of density-dependent dispersal behaviour on the speed and spatial patterns of range expansion in predator–prey metapopulations. Ecological Modelling, 222, 3524–30.

Ramanantoanina, A., Ouhinou, A., and Hui, C. (2014) Spatial assortment of mixed propagules explains the acceleration of range expansion. PLoS One, 9, e103409.

Rammal, R. and Toulouse, G. (1983) Random walks on fractal structures and percolation clusters. Journal of Physique Letters, 44, 13–22.

Randin, C.F., Dirnböck, T., Dullinger, S., *et al.* (2006) Are niche-based species distribution models transferable in space? Journal of Biogeography, 33, 1689–703.

Revilla, E., Wiegand, T., Palomares, F., *et al.* (2004) Effects of matrix heterogeneity on animal dispersal: from individual behavior to metapopulation-level parameters. American Naturalist, 164, E130–53.

Richardson, D.M., Iponga, D.M., Roura-Pascual, N., *et al.* (2010) Accommodating scenarios of climate change and management in modelling the distribution of the invasive tree *Schinus molle* in South Africa. Ecography, 33, 1049–61.

Richardson, D.M., Pyšek, P., Rejmánek, M., *et al.* (2000) Naturalization and invasion of alien plants: concepts and definitions. Diversity and Distributions, 6, 93–107.

Rouget, M. and Richardson, D.M. (2003) Understanding patterns of plant invasion at different spatial scales:

quantifying the roles of environment and propagule pressure. In: Child, L.E., Brock, J.H., Brundu, G., *et al.* (eds) *Plant Invasions: Ecological Threats and Management Solutions*. Leiden: Backhuys Publishers, pp. 3–15.

Roura-Pascual, N., Bas, J.M., Thuiller, W., *et al.* (2009) From introduction to equilibrium: reconstructing the invasive pathways of the Argentine ant in a Mediterranean region. Global Change Biology, 15, 2101–15.

Roura-Pascual, N., Hui, C., Ikeda, T., *et al.* (2011) The relative roles of climate suitability and anthropogenic influence in determining the pattern of spread in a global invader. Proceedings of the National Academy of Sciences of the USA, 108, 220–5.

Rykiel Jr, E.J. (1996) Testing ecological models: the meaning of validation. Ecological Modelling, 90, 229–44.

Sawyer, H., Kauffman, M.J., Nielson, R.M., *et al.* (2009) Identifying and prioritizing ungulate migration routes for landscape-level conservation. Ecological Applications, 19, 2016–25.

Schneider, D.W., Ellis, C.D., and Cummings, K.S. (1998) A transportation model assessment of the risk to native mussel communities from zebra mussel spread. Conservation Biology, 12, 788–800.

Schurr, F.M., Pagel, J., Cabral, J.S., *et al.* (2012) How to understand species' niches and range dynamics: a demographic research agenda for biogeography. Journal of Biogeography, 39, 2146–62.

Shigesada, N. and Kawasaki, K. (1997) *Biological Invasions: Theory and Practice*. Oxford: Oxford University Press.

Shigesada, N., Kawasaki, K., and Teramoto, E. (1986) Traveling periodic waves in heterogeneous environments. Theoretical Population Biology, 30, 143–60.

Skellam, J.G. (1951) Random dispersal in theoretical populations. Biometrika, 38, 196–218.

Squires, J.R., DeCesare, N.J., Olson, L.E., *et al.* (2013) Combining resource selection and movement behavior to predict corridors for Canada lynx at their southern range periphery. Biological Conservation, 157, 187–95.

Storlie, C., Merino-Viteri, A., Phillips, B., *et al.* (2014) Stepping inside the niche: microclimate data are critical for accurate assessment of species' vulnerability to climate change. Biology Letters, 10, 20140576.

Su, M., Hui, C., and Lin, Z. (2015) Effects of the transmissibility and virulence of pathogens on intraguild predation in fragmented landscapes. BioSystems, 129, 44–9.

Svenning, J.C., Gravel, D., Holt, R.D., *et al.* (2014) The influence of interspecific interactions on species range expansion rates. Ecography, 37, 1198–209.

Thuiller, W., Lafourcade, B., Engler, R., *et al.* (2009) BIOMOD—a platform for ensemble forecasting of species distributions. Ecography, 32, 369–73.

Thurfjell, H., Ciuti, S., and Boyce, M.S. (2014) Applications of step-selection functions in ecology and conservation. Movement Ecology, 2, 4.

Truscott, J. and Ferguson, N.M. (2012) Evaluating the adequacy of gravity models as a description of human mobility for epidemic modelling. PLoS Computational Biology, 8, e1002699.

Tsoar, A., Shohami, D., and Nathan, R. (2011) A movement ecology approach to study seed dispersal and plant invasion: an overview and application of seed dispersal by fruit bats. In: Richardson, D.M. (ed.) *Fifty Years of Invasion Ecology: The Legacy of Charles Elton*. Oxford: Wiley-Blackwell, pp. 103–19.

Turchin, P. (1998) *Quantitative Analysis of Movement: Measuring and Modeling Population Redistribution in Animals and Plants*. Sunderland, MA: Sinauer Associates.

Vaclavik, T. and Meentemeyer, R.K. (2012) Equilibrium or not? Modelling potential distribution of invasive species in different stages of invasion. Diversity and Distributions, 18, 73–83

van den Bosch, F., Hengeveld, R., and Metz, J.A.J. (1992) Analyzing the velocity of animal range expansion. Journal of Biogeography, 19, 135–50.

van den Bosch, F., Metz, J.A.F., and Diekmann, O. (1990) The velocity of spatial population expansion. Journal of Mathematical Biology, 28, 529–65.

Vanderwal, J., Shoo, L.P., Graham, C., *et al.* (2009) Selecting pseudo-absence data for presence-only distribution modeling: how far should you stray from what you know? Ecological Modelling, 220, 589–94.

Viswanathan, G.M., Buldyrev, S.V., Havlin, S., *et al.* (1999) Optimizing the success of random searches. Nature, 401, 911–14.

Weinberger, H.F. 1982. Long-term behavior of a class of biological models. SIAM Journal on Mathematical Analysis, 13, 353–96.

Wiens, J.A., Schooley, R.L., and Weeks, R.D. (1997) Patchy landscapes and animal movements: do beetles percolate? Oikos, 78, 257–64.

Williamson, M. (2009) Variation in the rate and pattern of spread in introduced species and its implications. In: Perrings, C., Mooney, H., and Williamson, M. (eds) *Bioinvasions and Globalization: Ecology, Economics, Management, and Policy*. Oxford: Oxford University Press, pp. 56–65.

Wisz, M.S., Pottier, J., Kissling, W.D., *et al.* (2013) The role of biotic interactions in shaping distributions and realised assemblages of species: implications for species distribution modelling. Biological Reviews, 88, 15–30.

With, K.A. and King, A.W. (1997) The use and misuse of neutral landscape models in ecology. Oikos, 79, 219–29.

Woodford, D.J., Hui, C., Richardson, D.M., *et al.* (2013) Propagule pressure drives establishment of introduced freshwater fish: quantitative evidence from an irrigation network. Ecological Applications, 23, 1926–37.

Zipf, G.K. (1946) The P_1P_2/D hypothesis: on the intercity movement of persons. American Sociological Review, 11, 677–86.

CHAPTER 4

From dispersal to boosted range expansion

4.1 Boosted range expansion

Early theories suggested that the velocity at which a species expands its range depends on two demographic factors: population growth and dispersal rates (Skellam 1951). Models based on partial differential equations, specifically the reaction diffusion model (see Chapter 3), assume a normally distributed dispersal distances and yield the most widely used formula in movement ecology that depicts a constant rate of population spread under these two demographic factors, $c = 2(rD)^{1/2}$, where r and D denote the intrinsic population growth and diffusion rates, respectively (Fisher 1937; Skellam 1951; Van den Bosch *et al.* 1992). However, a growing body of evidence suggests that many species expand their range much faster than would be expected from this estimate. This is especially true for invasive species that are still in the phase of rapid expansion (Cohen and Carlton 1998). Such unexpectedly rapid spread often results in delayed or even failed management. Elucidating the mechanisms behind such boosted range expansion (and associated augmented invasion performance) is one of the hottest research topics in invasion ecology.

Range expansion is a phenomenon involving populations that continuously move and establish in new ranges; the rate of expansion is thus ultimately determined by two factors: population growth and dispersal at the invasion front (Fisher 1937). Consequently, mechanisms that have been proposed to explain boosted range expansion are largely related to factors that can boost population growth and dispersal. For instance, purely ecological mechanisms such as long-distance dispersal, density-dependent dispersal, Allee effects, landscape connectivity, and disturbance (Kot *et al.* 1996; Veit and Lewis 1996; Shigesada and Kawasaki 1997; Ellner and Schreiber 2012; Perkins *et al.* 2013), and evolutionary mechanisms that affect the adaptation of growth- and dispersal-related traits (Holt *et al.* 2005; Perkins *et al.* 2013) are all well covered in the recent literature. In particular, introduced species are released from co-evolved biotic interactions and face a suite of new associates in the novel environment. Such altered biotic interactions will inevitably affect the invasion performance and the rate of range expansion; we deal with such altered biotic interactions facing introduced species in Chapter 6. A variety of stabilizing and destabilizing forces (e.g. Allee effects, negative density dependence, and disturbance) control the non-equilibrium invasion dynamics. Factors that perturb such regulating forces of population dynamics could also undoubtedly change the spreading dynamics; we deal with the changing forces of the non-equilibrium invasion dynamics in Chapter 5.

Altered biotic interactions are primarily used to discuss invasion performance, whereas non-equilibrium dynamics focuses on factors responsible for the variability of population size. To address boosted range expansion and enhanced rate of spread of many invasive species directly, we need to clarify which factors control the rate and distance of dispersal. Before we get to non-equilibrium dynamics (Chapter 5) and biotic interactions (Chapter 6), we need to deal with dispersal-related mechanisms that are implicated in boosted range expansion.

Invasion Dynamics. Cang Hui and David M. Richardson, Oxford University Press (2017).
© Cang Hui and David M. Richardson. DOI 10.1093/acprof:oso/9780198745334.001.0001

In this section we briefly review the evidence for boosted range expansion in invasion ecology and discuss the phenomenon known as Reid's paradox. We then introduce dispersal syndromes and discuss the effects of dispersal heterogeneity and spatial selection/sorting on the rapid evolution of dispersal-related traits at the vanguard of range expansion. In section 4.3 we introduce the components of dispersal and discuss augmented dispersal through long-distance dispersal, multiple dispersal vectors, human-mediated dispersal and models of fat-tailed dispersal kernels. Finally, we introduce an integrated model that combines mechanisms of augmented dispersal and spatial sorting, together with propagule pressure, to update Skellam's estimator of spreading velocity.

4.1.1 Boosted invasion dynamics

Although many forms of range expansion have been documented for invasion dynamics (exponential, linear, biphasic, acceleration, boom and bust, lag phase, and context dependence; see Chapter 2),

the most widely discussed and probably the most interesting phenomenon is boosted range expansion where the rate of spread accelerates continuously or in a stepwise fashion (biphasic) (e.g. Veit and Lewis 1996; Crooks and Soulé 1999; Urban *et al.* 2008). Several examples of boosted range expansion were discussed in Chapter 2. Here, we provide one additional example that is related to mechanisms later discussed. Further examples to illustrate each mechanism are included in the relevant sections.

The harlequin ladybird, *Harmonia axyridis*, is native to Asia but was intentionally introduced to many countries as a biological control agent of pest insects. The dramatic spread of this generalist top predator in many countries has been met with considerable trepidation (Roy *et al.* 2016). Lombaert and colleagues (2014) documented its accelerating range expansion in Europe and found a marked increase in the flight speed of the insects from the core to the front of the invasion range in two independent sampling transects (Fig. 4.1). The mean distance travelled by individuals during a one-hour

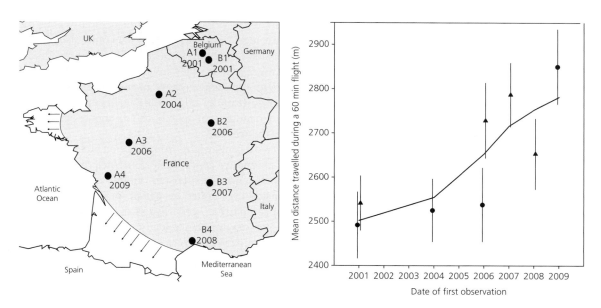

Fig. 4.1 (Left) Geographical origins of samples of harlequin ladybird (*Harmonia axyridis*) collected along two transects in Western Europe. The core area of the invasive outbreak is located in Belgium. The grey area roughly corresponds to the distribution of the species in 2010. Arrows indicate the direction of expansion. Black dots indicate sampled populations. The years indicated below the dots indicate when the species was first observed at that site. (Right) Mean distance travelled in 1 hectare (h) by ladybirds in flight mills, as a function of dates of first observation of the species during range expansion. From Lombaert *et al.* (2014). Reproduced with permission from John Wiley & Sons.

flight sampled from populations with different residence time increased from the core to the margin by 20 per cent. However, two other traits associated with dispersal (endurance and motivation to fly off) showed no signs of change, suggesting the lack of trade-offs between these dispersal traits. This chapter addresses the potential reasons behind such boosted range expansion and adaptation in particular dispersal-related traits.

4.1.2 Reid's paradox

The dilemma of boosted range expansion pre-dated the advent of invasion ecology and was discussed in the early interpretation of the spread of plant populations from fossil remains (Clark *et al.* 1998). Reid (1899) discussed the paradox of the rapid spread of oak trees which, despite having heavy seeds that are seldom dispersed beyond the canopy, expanded 1000 km northwards to north Britain between the recession of last Ice Age and the Roman occupation of Britain. Such colonization required a continuous expansion for 10^4 years at a rate of 100 m/yr. Similar evidence from paleoecological records, including plants largely from the northern hemisphere, has increased the interest in what is termed Reid's paradox (e.g. Davis 1981; Huntley and Birks 1983; Wright *et al.* 1993). These paleoecological records all suggest a much faster spread of trees after the Ice Age than expected from dispersal by wind or gravity, such as is embodied in Skellam's (1951) formula based on population demography and restricted dispersal, which predicts a much slower constant rate of range expansion (Fig. 4.2).

Evidently, the dispersal of large seeds is an obvious impediment to the rapid range expansion of trees after glacial periods. One way to resolve Reid's paradox is to argue that the estimates of population growth and diffusion should match the observed velocity of range expansion according to Skellam's (1951) formula, suggesting a much higher intrinsic growth rate (in particular, fecundity) or a much higher rate of diffusion. However, although this could allow the prediction to match the observed rate of migration for trees and herbaceous species, it often requires biologically unreasonable parameterisation of reaction diffusion models (Higgins and Richardson 1999). Relevant to management

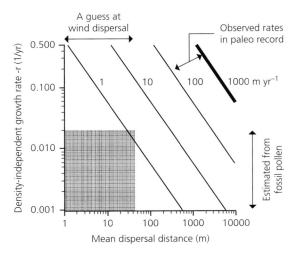

Fig. 4.2 Rates of spread predicted by a diffusion model for mean dispersal distances with a Gaussian dispersal kernel and density-independent growth rates (Skellam's formula). The shaded region indicates parameter space conforming to estimated dispersal distances from field data and growth rates typically reported from rates of increase in fossil pollen data. From Clark (1998). Reproduced with permission from University of Chicago Press.

for climate change adaptation and biological invasion, if tree populations migrated rapidly after the most recent glacial period, they could also exhibit boosted range shift from current and future climate change, or when they invade a novel environment where apparently empty niches exist for them to occupy (Rundel *et al.* 2014). A better mechanistic understanding of such boosted range expansion is clearly needed.

Two recent insights from invasion biology suggest an alternative concept that could help in explaining boosted range expansion. Firstly, instead of only considering limited dispersal, models incorporating leptokurtic seed dispersal can yield a much higher asymptotic velocity of spread (Mollison 1977; van den Bosch *et al.* 1990; Kot *et al.* 1996). Some rare long-distance dispersal events or fat-tailed dispersal kernels can have an overwhelming effect on rates of range expansion, accelerating the spreading speed to become a polynomial or even exponential function of time (Kot *et al.* 1996). Secondly, dispersal strategy is a density- and context-dependent adaptive trait. Many cases support positive density-dependent dispersal (Matthysen 2005); this, in the

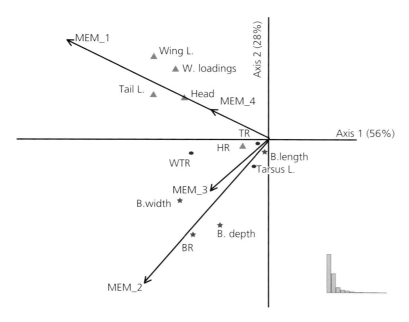

Fig. 4.5 Results from the environmental and morphological analysis using the multiscale pattern analysis for female Indian mynas (*Acridotheres tristis*) in South Africa. Triangles indicate traits related to flight, circles indicate traits related with tarsus, and stars indicate traits related with the bill. MEM_1 (Moran's eigenvector map) corresponds to the distance to the introduction locality; MEM_2 corresponds to environmental complexity. L: length; W: wing; TR: tarsus-to-body ratio; HR: head-to-body ratio; WTR: wing-tail ratio; B: bill; BR: bill ratio. From Berthouly-Salazar *et al.* (2012).

4.3 Augmented dispersal

4.3.1 Dispersal and its kernels

Dispersal involves the movement of individuals following diverse behaviours and modes, and leads to gene flow in space (Clobert *et al.* 2012) and can be divided into a three-stage movement: departure, transience, and settlement (Fig. 4.6). It is energetically costly and different risks exist at each movement stage (Matthysen 2012). However, dispersal provides a means for the organism to escape unfavourable conditions and it increases the fitness variance by distributing offspring over different conditions (i.e. a bet-hedging strategy), thus often exposing dispersal traits and strategies to strong selection. A disperser can reduce its inbreeding and kin competition but increase the likelihood for outbreeding and losing cooperation among kin. A decision to disperse is conditionally dependent on an individual's condition (e.g. biomass and nutritional status) or on the environmental context (e.g. density and harshness of the physical condition of the habitat).

A dispersal vector is important for the movement of hitchhikers and passive dispersers; for example, endozoochorous seed dispersal that depends strongly on the behaviour of the dispersing vectors

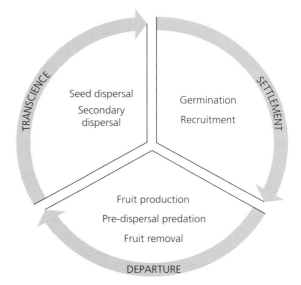

Fig. 4.6 Schematic representation of the life cycle of fleshy-fruited plant species, showing the stages of the dispersal process. Redrawn based on Lehouck *et al.* (2012).

(Lehouck *et al.* 2012). For instance, frugivorous seed dispersal comprises a multistage process that is critically dependent on the interactions between plants, their dispersers and the biotic/abiotic environment (Fig. 4.6). The dispersal of many introduced plants in their invaded ranges depends on animal vectors (Richardson *et al.* 2000) and their movement through heterogeneous landscapes (Theoharides and Duke 2007). For instance, the spread of invasive black cherry, *Prunus serotina*, in Europe depends on the movement patterns of pollinators, seed dispersal agents, and landscape structures (e.g. the presence of roosting trees for birds to perch and defecate seeds) (Deckers *et al.* 2005).

The Euclidian distance from the departure to the settlement of a propagule is called the net displacement, and the square of this value is the net squared displacement (NSD; Turchin 1998); the expected value of the NSD over time is called the mean net squared displacement (MSD). For a Brownian particle, we have MSD = $4D \cdot t$, where D is the diffusion rate and t time (multiplier 4 here for movement on two-dimensional space and needs to be changed to 2 for linear movements). Movement modes can be classified according to how MSD scales with time; in general, MSD = $D \cdot t^a$. The motion is called super-diffusive if $a > 1$ and sub-diffusive if $a < 1$ (Borger and Fryxell 2012). The statistical distribution of displacement distance in a population is termed the dispersal kernel, which can be thin-tailed (e.g. Gaussian and exponential) or fat-tailed (e.g. power law and log-normal) (Nathan *et al.* 2012). Spread models that incorporate both dispersal and population growth are promising tools for assessing the spread capacity of alien species in novel and future environments (Jeltsch *et al.* 2008; Thuiller *et al.* 2008; Jongejans *et al.* 2008; Soons and Bullock 2008).

Based on good data on dispersal kernels, Bullock (2012) estimated the wave speed under moderate population growth rate ($\lambda = 2$) for 32 plant species with different life forms and dispersal modes; this ranged from 0.3 to 706 m/yr. Some examples of invasive plants are the wind-dispersed dog-strangling vine, *Vincetoxicum rossicum* (4.4 m/yr; Cappuccino *et al.* 2002), the bird-dispersed shrub Darwin's barberry, *Berberis darwinii* (49.8 m/yr), and the tree of heaven, *Ailanthus altissima* (97 m/yr), which has seeds that are mainly dispersed in water. Although the velocities of species dispersed by vertebrates are an order of magnitude higher than those dispersed by wind and ants, these estimates are still low if we consider the observed rapid

spread of these species, due to human-mediated dispersal (see section 4.3.3).

Fat-tailed dispersal kernels (i.e. movements with a substantial portion of long-distance dispersal) have been shown to be capable of boosting range expansion (Kot *et al.* 1996). In particular, Kot and colleagues (1996) discussed three types of dispersal kernel using the integrodifference equation (see Chapter 3): (i) kernels possessing a moment-generating function have exponentially bounded thin tails and lead to a constant rate of range expansion; (ii) kernel with finite moments of all orders but without a moment-generating function, leading to accelerating range expansion with the rate of spread increasing linearly with time; (iii) kernels with infinite moments that could push range expansion faster than geometrically accelerating over time. A problem associated with the fat-tailed dispersal kernels (scenarios ii and iii) in explaining boosted range expansion is that the rate of spread keeps increasing without an upper bound (Phillips *et al.* 2008) which is clearly unrealistic.

4.3.2 Long-distance dispersal

Rates of local spread depend largely on both population growth and dispersal mode, as predicted from Skellam's (1951) formula, with an average rate ranging from 2 to 370 m/yr for plants (Pyšek and Hulme 2005). Importantly, however, range expansion at regional scales is driven not only by slow and steady local spread but more importantly by rare long-distance dispersal (LDD) events (Higgins and Richardson 1999; Theoharides and Duke 2007). While local distance-limited dispersal often results in range expansion with a constant slow rate from the introduction location, LDD adds non-linear and often accelerating rates of expansion (Lewis and Kareiva 1993; Kot *et al.* 1996; Neubert and Caswell 2000; Hastings *et al.* 2005). Rare LDD events can have a major influence on the spread dynamics. For instance, the average spread rate of *Opuntia stricta* in Australia was 370 m/yr, but satellite populations were able to establish up to 14 km away from the introduction site in the first 2 years of the invasion (Theoharides and Duke 2007). In their review of the spatio-temporal dynamics of plant invasions, Pyšek and Hulme (2005) showed that spread rates for over 100 taxa ranged between 3 and 500 km^2/yr; for many taxa this is orders of magnitude faster than could be realized through local distance-limited dispersal. Spreading dynamics are thus notably decoupled from the typical dispersal barriers evident at landscape scales (With 2004).

Theoretical studies show that rapid tree migration is possible with a fat-tailed dispersal kernel. However, the mechanisms for boosting such a fat tail in the dispersal kernel are less known, given the rare occurrence of such cases. For plant dispersal, LDD of seeds by vectors such as birds transporting fleshy fruits more than 1 km is observable (Clark *et al.* 1998). Blue jays, *Cyanocitta cristata*, can cache acorns and beech nuts 4 km away from parent trees (Johnson and Adkisson 1985), an observation anticipated by Reid (1899). Trapping data show that frugivorous mammals, such as foxes and bears, can travel more than 10 km per day (Storm and Montgomery 1975; Willson 1993), making them extremely efficient vectors for long-distance seed dispersal. Due to the rarity of such LDD events, their role has been largely neglected in theories on the global distribution of flowering plants.

Renewed interest in extreme LDD has resurfaced recently, as new tools and evidence have become available to infer, sometimes indirectly, rare LDD events (De Queiroz 2014). Phylogenetic and population-genetic approaches are shedding light on the puzzling relationship between widely separated but closely related taxa. For instance, Gildenhuys *et al.* (2015) showed that the desert-adapted balloon vine, *Cardiospermum pechuelii*, in southern Africa most likely evolved from *C. corindum* following transoceanic dispersal from South America 5.9 million years ago on ocean currents. Seeds of these species can float in seawater and remain viable for months. The freshwater amphibian *Ptychadena newtoni* on São Tomé island in the Gulf of Guinea (West Africa) most likely evolved from *P. mascareniensis* after the latter taxon reached the islands via long-distance rafting in freshwater plumes in the ocean following catastrophic climatic events between 5.6 and 18.6 Ma (Measey *et al.* 2007). These cases of LDD events happened long before the existence of modern humans.

The longest single natural LDD event documented to-date involves *Acacia heterophylla* and *A. koa*, until recently considered two endemic species on their respective island homes: La Réunion Island in the southern Indian Ocean and the Hawai'in

Islands in the northern Pacific Ocean, respectively. Using a molecular phylogeny in conjunction with the known ages of the Hawai'in Islands, Le Roux and colleagues (2014) showed that the mysterious distribution of these taxa is due to natural LDD from Australia to the Hawai'in Islands, followed by an even longer leap between Hawai'i and Réunion, more than 18 000 km apart, in the last 1.4 million years, again long before the existence of modern humans (thus refuting hypotheses linking the distribution of these taxa to human-mediated transfer). Because *Acacia* seeds are not adapted to survive in seawater and since both taxa are mid- to high-elevation trees that do not grow anywhere near the shore, such LDD cannot be attributed to dispersal in ocean currents. The only possible explanation is that seabirds, in particular petrels such as the endemic Barau's Petrel, *Pterodroma baraui*, from Réunion Island, transported the seeds. In Réunion, petrels dig burrows at elevations that coincide with the distribution of the acacias. Ingested seeds can be retained in the stomachs of petrels for weeks and even months—enough time for wind-swept petrels from Hawai'i to deliver seeds that would later become *A. heterophylla* on La Réunion.

Due to the rarity of LDD events and the multiple mechanisms involved, dispersal with a mixed dispersal kernel (Clark *et al.* 1998) should be considered when modelling the spread of alien species. A mixed-dispersal kernel that has two components k_1 and k_2 with weight p and $(1-p)$, respectively, can be expressed as $k(r) = p k_1(r) + (1-p)k_2(r)$. In particular, stratified diffusion has been studied through such mixed dispersal, where the population expands its range by both neighbourhood diffusion and LDD (Shigesada *et al.* 1995). While LDD allows the formation of new colonies ahead of the advancing front (at a rate $\lambda(r)$ from a colony of radius r), the diffusive dispersal drives the steady expansion from individual colonies (at a rate c). Different colonies can coalesce and form a large block of invaded region (Shigesada *et al.* 1995). The model yields an initially slow range expansion, which is determined by the diffusion of the initial population, followed by an accelerating range expansion which depends on the colonization rate $\lambda(r)$.

LDD events are rare but play a crucial role in boosting invasive spread (Higgins and Richardson 1999; Nathan *et al.* 2008). This means that explanations for, and predictions of, large-scale range expansions and range shifts demand elucidation of the mechanisms behind LDD, rather than local dispersal (Trakhtenbrot *et al.* 2005). Plant dispersal can be augmented through the presence of morphological features facilitating dispersal via natural means (e.g. wings or pappi for wind dispersal) or via vectors (e.g. fleshy fruits to attract animals, seeds covered in hooks or spines to attach to animals). Dispersal by multiple vectors (polychory) is common in plants (e.g. Berg 1983; Calvino-Cancela *et al.* 2006; Russo *et al.* 2006; Levey *et al.* 2008). Many plants can use the same vector. For instance, white-tailed deer, *Odocoileus virginianus*, occur in forests and suburbs in the eastern United States and serve as an important vector for the LDD of many invasive plants (Vellend 2002; Williams and Ward 2006). Extreme LDD events are rare because they rely on either irregular behaviour of common vectors or through encounters of unusual vectors—both make the prediction of extreme LDD events difficult. Fat-tailed dispersal kernels depend largely on rare LDD events from non-standard vectors (Fig. 4.7).

Many invasive species often arrive in a new region with low genetic diversity, and the fact that they can overcome the founding effect from low genetic diversity and rapidly adapt to their new environment poses an apparent paradox (Gross-niklaus *et al.* 2013; Rollins *et al.* 2015). Two mechanisms can come to the aid of such invaders. Firstly, LDD events redistribute genetic diversity across the expanding range and can therefore transport beneficial mutations to peripheral populations from anywhere across the species' range (Fayard *et al.* 2009), thereby increasing evolutionary potential (Buckley *et al.* 2012). LDD would be selected for in populations facing intermediate local-extinction pressure (Bohrer *et al.* 2005). Dispersal strategy, and specifically enhanced dispersal, can therefore be regarded as an evolutionary response to overcome the loss of genetic diversity during rapid range expansions (Berthouly-Salazar *et al.* 2013). In a recent landscape genetics study on the expansion of common starlings, *Sturnus vulgaris*, in South Africa, Berthouly-Salazar and colleagues (2013) showed that besides accelerating in its expansion (see Chapter 2), invasive starlings have succeeded in preserving their genetic diversity (Fig. 4.8). This lends support to the hypothesis that relatively frequent LDD events mitigated the effects of sequential founding events

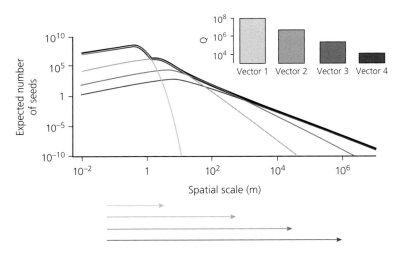

Fig. 4.7 The importance of four different dispersal vectors (different shadings) for different parts of the total dispersal kernel (thick black line) of a hypothetical plant. Vector 1 is a standard vector generating a thin-tailed dispersal kernel (e.g. wind). Vectors 2–4 represent three non-standard vectors (e.g. animals) that generate dispersal kernels with increasingly fatter tails, indicating LDD mechanisms that operate over larger spatial scales. Although vector 4 transports 104 times fewer seeds than vector 1 (see inset for the distribution of vector seed loads (Q)), the range of dispersal distances (horizontal arrows at the bottom) is far greater for vector 4 than for vector 1. Thus, non-standard dispersal vectors can dominate LDD, even if only a very small fraction of all seeds are thus dispersed. From Nathan *et al.* (2008). With permission from Elsevier.

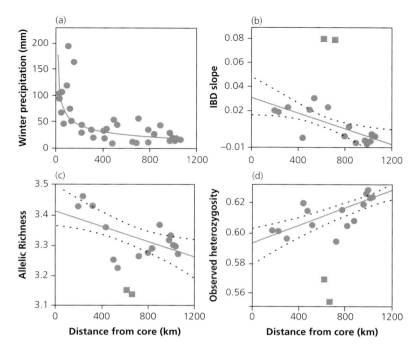

Fig. 4.8 Relationships between the geographic distance from the core distribution and: winter precipitation (A), isolation-by-distance (IBD) slopes (B), allelic richness (C), and observed heterozygosity (D) for the invasive common starling (*Sturnus vulgaris*) in South Africa. Open circles indicate outliers and were removed from regression analyses. From Berthouly-Salazar *et al.* (2013). Reproduced with permission from John Wiley & Sons.

during the range expansion of this species. However, it should be noted that spatial sorting and LDD are most likely to be opposing factors behind the fast expansion of species. This is because frequent LDD could override any signals from spatial sorting: a fat-tailed dispersal kernel results in panmictic populations which may inhibit the effects of spatial sorting (Klein *et al.* 2006). Secondly, heritable epigenetic variation may also allow invasive species to adapt to novel environments in situations where genetic variation is low (Allendorf and Lundquist 2003; Pérez *et al.* 2006; Rollins *et al.* 2013). There is increasing evidence from experiments to support this hypothesis (Perez *et al.* 2006; Angers *et al.* 2010; Kilvitis *et al.* 2014; Van Pategem *et al.* 2015).

4.3.3 Human-mediated dispersal

Different causal factors influence the three key steps of movement (Fig. 4.9). Human-aided dispersal is an important vector for moving many species around. Intentional human-facilitated dispersal can significantly boost all three stages of dispersal, while unintentional human facilitation occurs especially at the departure and transience stages, but it could also facilitate the settlement of aliens through

anthropogenic disturbance (e.g. the germination of alien grasses along road verges; Trombulak and Frissell 2000). Any introduced species have already undergone extra-range dispersal initially—the movement of propagules from its native range to a new area, through a variety of pathways: diffusion, corridor, jump, and long-distance dispersal, mass dispersal (from multiple sources), and cultivation (Wilson *et al.* 2009). Human activities can greatly facilitate post-introduction spread, for example through human-mediated jump and mass dispersal and cultivation of invasive species at multiple loci. Consequently, the dispersal rate and subsequently the rate of spread for most invasive species are much higher than expected from natural dispersal (Hastings *et al.* 2005; Meyerson and Mooney 2007).

Humans are clearly the most important dispersal vector behind modern biological invasions (Fig. 4.10). Humans can fragment landscapes into habitat patches that promote the establishment of alien species within small patches with long edges (Timmens and Williams 1991; Gelbard and Harrison 2003; Ohlemuller *et al.* 2006) and enhance connectivity (Damschen *et al.* 2006; Murphy *et al.* 2006). Disturbance corridors such as roads and trails facilitate augmented dispersal of alien plants by providing

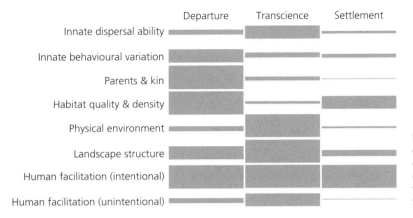

Fig. 4.9 The relative contribution of different factors to the three key stages of dispersal. The first six factors are based on Matthysen (2012). The importance of the factor increases with the thickness of the bar.

pathways for dispersal vectors (e.g. humans and vehicles; Timmens and Williams 1991; Lonsdale and Lane 1994; Campbell and Gibson 2001; deer and other small mammals; Meyers *et al.* 2004; Williams and Ward 2006), and reduce the size of native seed banks in disturbed soil, thereby making sites more amenable to the establishment of alien plant species (D'Antonio *et al.* 2000; Trombulak and Frissell 2000). Human-mediated dispersal mechanisms are crucial for boosting the speed of spread (Hodkinson and Thompson 1997; Higgins *et al.* 2003).

Human-mediated dispersal kernels can be developed using easily accessible data, experiments, modelling, and observations (Buckley and Catford 2016). For example, with a maximum wind dispersal distance of 70 m and assuming an annual doubling of population size for the wild cabbage, *Brassica oleracea* ssp. *oleracea*, the rate of spread is only estimated at 5.5 m/yr. Wichmann and colleagues (2009) used a semi-mechanistic model to show that adding human-mediated dispersal involving hikers carrying just 1 per cent of seeds boosted

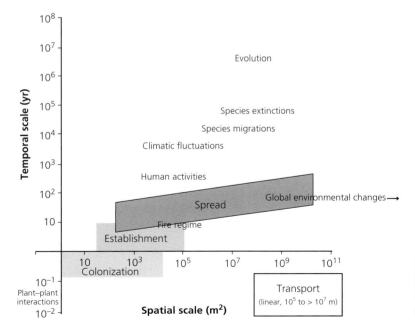

Fig. 4.10 The spatial and temporal scale of the four stages of invasion relative to other key biological processes. From Theoharides and Dukes (2007). Reproduced with permission from John Wiley & Sons.

the spread rate to 38.8 m/yr. Several studies have shown that human-mediated dispersal of livestock results in extremely fast spread rates. For example, Manzano and Malo (2006) showed that large numbers of seeds with different morphologies are frequently dispersed over hundreds of kilometres while attached to the fleece of sheep (both wild and domestic). Such results attest to the overwhelming importance of human-mediated dispersal in biological invasions.

Disturbance regimes strongly affect dispersal, especially the departure stage, by influencing the presence and persistence of different species (Mouquet *et al.* 2003; Theoharides and Duke 2007). Disturbance is crucial for enabling propagules to make contact with their dispersal vectors (D'Antonio *et al.* 2000), and the interface of suburban and natural landscapes is a hotspot for the departure stage of dispersal (Alston and Richardson 2006; Williams and Ward 2006). Natural disturbance regimes are often linked to physical site characteristics, extrinsic factors (e.g. weather and fire), and the biotic community. Anthropogenic disturbance tends to differ from natural disturbance, and may alter the regional disturbance regime (D'Antonio *et al.* 2000). Alterations of the natural disturbance regime can boost range expansion of invasive plants by creating opportunities for colonization and establishment (e.g. Hobbs and Huenneke 1992; Burke and Grime 1996; D'Antonio *et al.* 2000; Davis *et al.* 2000). For instance, the spread of yellow starthistle, *Centaurea solstitialis*, under uniform disturbance treatments was faster in invaded ranges than in its native range, due to escape from soil microbes that suppress its establishment in its native range (Hierro *et al.* 2006). Natural disturbance regimes create windows of opportunity for the establishment of alien plants, whereas human-mediated disturbance regimes can greatly promote invasion success (Pyle 1995; Kim 2005). Invasion success could further transform landscapes and alter natural disturbance regimes, creating a reinforcing feedback in recipient ecosystems (Hobbs 2000; see also Chapter 7).

To understand human-mediated dispersal of many invasive species during introduction and spread, we need a good picture of how humans move across local and regional landscapes, which could be depicted by the dispersal kernel. A dispersal kernel can be defined as the probability, P(r), of an individual having a displacement of distance r during a specific time period or a specific phase of life stage. Diffusion-based models (see Chapter 3) are only applicable when the variance of P(r) is finite (e.g. a Gaussian dispersal kernel). When P(r) lacks a finite variance (i.e. lacking a characteristic length scale), for instance when $P(r) \sim r^{-(1+\beta)}$ with $\beta < 2$, the estimate of diffusion rate from the dispersal kernel becomes infinite. A random movement following such a power-law dispersal kernel with an infinite variance is known as Lévy flights (Kot *et al.* 1996; Brockmann *et al.* 2006). Recently, the related notion of long-distance dispersal (LDD) has become well established in dispersal ecology (Nathan 2001); this takes into account the observation that dispersal kernels of many species show power law tails owing to long-range movements (Viswanathan *et al.* 1996; Nathan *et al.* 2002; Levin *et al.* 2003; Hui *et al.* 2012).

Humans move faster and over longer distances than most other animal vectors. Not only have humans and their vehicles already served as an important vector for the trans-regional and intercontinental voyages of many invasive species, they also play a huge role in boosting intraregional dispersal (Theoharides and Duke 2007). Although it is often unclear whether boosted intraregional dispersal by humans is due to more propagules being translocated or propagules moving over greater distances (Pyšek and Hulme 2005), humans are responsible for the increasing incursions of alien species in many areas, as indicated by the strong correlation between the number of visitors to national parks in North America and South Africa with the number of alien species in the park (MacDonald *et al.* 1989; Lonsdale 1999).

Humans travel at many spatial scales, ranging from a few to thousands of kilometres over short periods of time. Brockmann and colleagues (2006) analysed the dispersal of human movement using the proxy of more than a million records on the movement of bank notes in the United States and found a power law dispersal kernel for distance r > 10 km (Fig. 4.11), with an exponent $\beta = 0.59 \pm 0.02$. Surprisingly, P(r) increases linearly with r for r < 10 km, suggesting a uniform distribution with

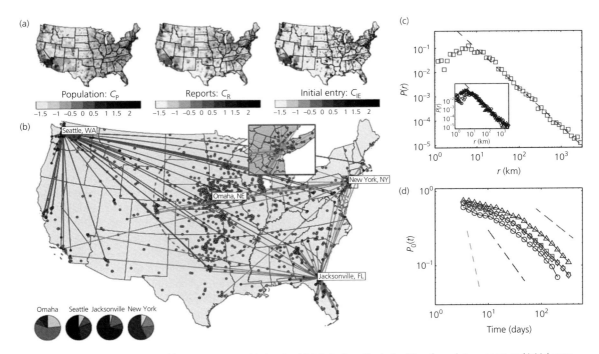

Fig. 4.11 Dispersal of bank notes and humans on geographical scales. (a) Relative logarithmic densities of population, report, and initial entry as functions of geographical coordinates. (b) Trajectories of bank notes originating from four different places. City names indicate initial location, symbols secondary report locations. Lines represent short-time trajectories with travelling time $T < 14$ days. The inset depicts a close-up view of the New York area. Pie charts indicate the relative number of secondary reports sorted by distance. The fractions of secondary reports that occurred at the initial entry location (dark), at short ($0 < r < 50$ km), intermediate ($50 < r < 800$ km) and long ($r > 800$ km) distances are ordered by increasing brightness. (c) The short-time dispersal kernel. The measured probability density function P(r) of traversing a distance r in less than $T = 4$ days is depicted in square symbols. The dashed black line indicates a power law P(r) ~ $r^{-(1+\beta)}$ with an exponent of $\beta = 0.59$. The inset shows P(r) for three classes of initial entry locations (triangles for metropolitan areas, diamonds for cities of intermediate size, circles for small towns). Their decay is consistent with the measured exponent $\beta = 0.59$. (d) The relative proportion $P_0(t)$ of secondary reports within a short radius ($r < 20$ km) of the initial entry location as a function of time. Squares show $P_0(t)$ averaged over 25 375 initial entry locations. Triangles, diamonds, and circles show $P_0(t)$ for the same classes as (c). All curves decrease asymptotically as t-η with an exponent $\eta = 0.60 \pm 0.03$ indicated by the top right dashed line. Ordinary diffusion in two dimensions predicts an exponent $\eta = 1.0$ (dashed line in the middle). Lévy flight dispersal with an exponent $\eta = 0.6$ as suggested by (b) predicts an even steeper decrease, $\eta = 3.33$ (dashed line on the left). From Brockmann et al. (2006). Reprinted by permission from Macmillan Publishers Ltd.

the 10 km circle. Dispersal from three different initial entry locations shows similar power laws (Fig. 4.11c). However, simple Lévy flights do not fit human movement because the proportion of displacement distance as a function of time declines much slower than expected from Lévy flights (Fig. 4.6d). After solving a bifractional diffusion equation for continuous-time random walks, Brockmann and colleagues (2006) derived the probability density of being at distance r at time t being, $W_r(r, t) = t^{-\alpha/\beta} L_{\alpha,\beta}(r / t^{\alpha/\beta})$, where $L_{\alpha,\beta}$ is a universal scaling function representing the characteristics of the process, and the typical distance travelled

scales according to $r(t) \sim t^{1/\mu}$, with $\mu = \beta/\alpha$. Fitting the data suggests $\alpha = 0.60 \pm 0.03$ which indicates a super-diffusive pattern of human movements with scale-free jumps and long waiting times between displacements.

By contrast, González and colleagues (2008) traced the movement of mobile phone users (Fig. 4.12) and found that the dispersal kernel fits a truncated power law: P(r) = $(r + r_0)^{-\beta} \exp(-r / k)$, with $r_0 = 1.5$ km and $\beta = 1.75 \pm 0.15$ and cut-off values $k = 400$ km for the 6-month data and $k = 80$ km for the week-long data; that is, mobile phone users follow a truncated Lévy flight. This observed

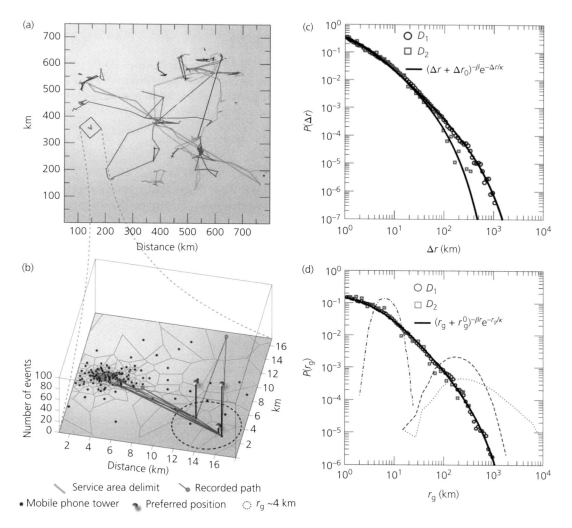

Fig. 4.12 Basic human mobility patterns. (a) Week-long trajectory of 40 mobile phone users indicates that most individuals travel only over short distances, but a few regularly move over hundreds of kilometres. (b) The detailed trajectory of a single user. The different phone towers are shown as dots, and the Voronoi lattice in grey marks the approximate reception area of each tower. The circle represents the radius of gyration centred in the trajectory's centre of mass. (c) Probability density function $P(r)$ of travel distances obtained for the two studied datasets, D1 and D2. The solid line indicates a truncated power law. (d) The distribution $P(r_g)$ of the radius of gyration measured for the users, where $r_g(T)$ was measured after $T = 6$ months of observation. The solid line represents a similar truncated power law fit. The dotted, dashed, and dot-dashed curves show $P(r_g)$ obtained from the standard null models (random walk, Lévy flight, and truncated Lévy flight, respectively). From González *et al.* (2008). Reprinted by permission from Macmillan Publishers Ltd.

pattern portrays a population-based heterogeneity coexisting with individual Lévy trajectories. There is a strong tendency of humans to return to places they have visited before; this describes the recurrence and temporal periodicity inherent in human mobility. There is a high degree of regularity in the daily travel patterns of individuals which is captured by the high return probabilities to a few highly frequented localities. Individuals live and travel in different regions, yet each user can be assigned to a well-defined anisotropic area. In contrast with the random trajectories predicted by the prevailing Lévy flight and random walk models, human trajectories show a high degree of temporal and spatial regularity, each individual being characterized by a time-independent characteristic travel distance and a significant probability of returning to a few highly frequented locations. After correcting for differences in travel distances and the inherent anisotropy of each trajectory, the individual travel patterns collapse into a single spatial probability distribution. This indicates that, despite considerable diversity in the travel history of humans, they follow simple, reproducible patterns. This inherent similarity in travel patterns could impact human-mediated dispersal in invasive species.

4.4 Skellam's formula revised

Multiple ecological and evolutionary processes contribute to invasion performance and dynamics (Fig. 4.13). In their synthetic invasion meta-framework (SIM), Gurevitch and colleagues

(2011) attempted to incorporate major conceptual frameworks, including human impacts. Concerning the boosted range expansion of many invasive species, this chapter has focused on two factors that can strongly affect the process of dispersal. Firstly, multiple dispersal vectors, especially with humans directly dispersing propagules or indirectly enhancing dispersal by disrupting natural vectors. Invasive species thus experience frequent LDD events which could substantially fatten the dispersal kernel. Such augmented fat-tailed dispersal kernel is a key cause of boosted range expansion (Fig. 4.13: human impacts → dispersal processes → range expansion). Secondly, strong evidence, primarily from core-edge trait comparisons, suggests that the dispersal-related traits of invasive species are under constant selection at the advancing range front, through spatial sorting (and [enhanced] spatial selection). Such assortment of ever stronger dispersers at the

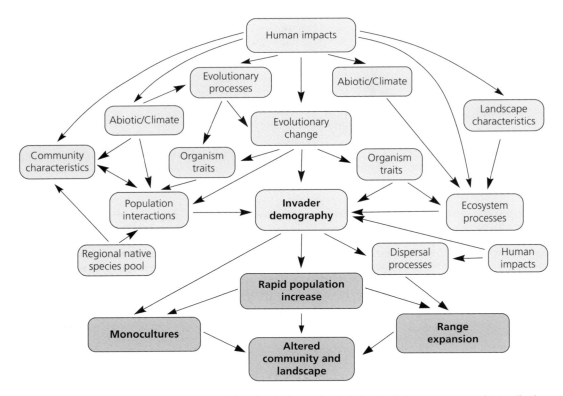

Fig. 4.13 A conceptual synthetic invasion meta-framework based on fundamental ecological and evolutionary processes and states. The three different characteristics of invasions and their effect on altering communities and landscapes are in bold capital letters. Transitions between the processes and states are indicated by arrows. Components found in more than one position affect or are affected by more than one set of other processes. From Gurevitch *et al.* (2011). Reproduced with permission from John Wiley & Sons.

vanguard is another important driver of boosted range expansion (Fig. 4.13: organism traits → invader demography → dispersal processes (and rapid population increase) → range expansion).

As mentioned in section 4.1.2, understanding plant spread and predicting the ability of species to track projected climate change remains a formidable challenge. Nathan and colleagues (2011) modelled the spread of North American wind-dispersed trees under current and projected future conditions, accounting for variation in five dispersal parameters and five demographic (population growth) parameters (Fig. 4.14). A stepwise rank-regression analysis revealed that the natural inter-specific variation in maturation time has the strongest impact on the wind-driven spread of trees, followed by other factors, such as post-dispersal survival, seed terminal velocity, fecundity, mean horizontal wind speed, and tree height. Dispersal factors (black boxes) were outweighed by the importance of demographic factors (grey boxes), with the total absolute coefficients 1.09 < 1.54. The estimated rate of range shift falls short of the temperature shifts

projected for temperate conifer forests (100–200 m/yr; Loarie *et al.* 2009), suggesting a typical wind-dispersed tree could only reach such a speed under the particular circumstance of high post-dispersal survival and seed abscission biased to strong winds. It is likely that two missing factors, human-mediated dispersal and spatial sorting, could boost the range expansion of conifers to catch up with temperature-driven shifts of potential habitat under future climates.

Given the two key insights from invasion biology that we have considered in this chapter (human-mediated fattened dispersal and spatial sorting), the reaction diffusion model of population spread needs to be revisited and Skellam's (1951) formula $c = 2(rD)^{1/2}$ revised to allow for the accurate prediction of the velocity of boosted range expansion. Using integrodifference equations, Ramanantoanina and colleagues (2014) considered both factors and derived a new formula for inheritable dispersal trait in an asexual population. The model describes the spreading dynamics of an invasive species with mixed dispersal-related phenotypes in a one-dimensional habitat. Both the instantaneous and

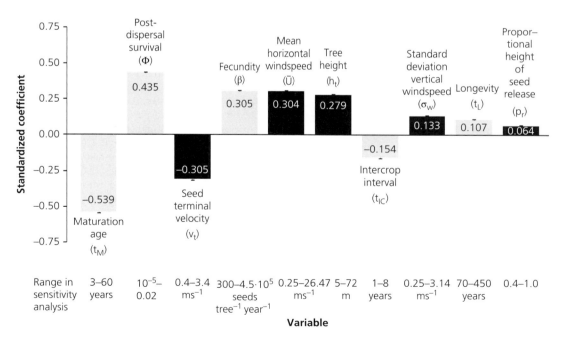

Fig. 4.14 Results of a stepwise rank regression for five dispersal (black boxes) and five demographic (grey boxes) parameters that determine the spread rate of North American wind-dispersed tree species. The parameters are arranged according to the absolute value of the regression standardized coefficients. The range of values for each parameter is given at the bottom line. From Nathan *et al.* (2011). Reproduced with permission from John Wiley & Sons.

average rate of spread approach the following asymptotic rate of spread:

$$c \approx \sqrt{2r\sigma_n^2}\left(1 + \frac{r}{12}\gamma\right) \qquad (4.1)$$

where r is the intrinsic population growth rate, and σ_n^2 and γ are the variance and kurtosis of dispersal kernel for the strongest disperser phenotype. The variance of one phenotype's dispersal kernel reflects its dispersal ability. Evidently, for a pureline population with a Gaussian kernel ($\gamma = 0$ and $\sigma_n^2 = \bar{\sigma}^2$, where $\bar{\sigma}^2$ is the dispersal kernel for the population), the above formula reverts back to Skellam's formula for one-dimensional spread. More scenarios were also included in the study, such as stratified dispersal, instantaneous speed, and propagule pressure (Ramanantoanina et al. 2014). In particular, assuming that individual dispersal ability follows a log-normal distribution in the population, $\ln N(\mu,\delta)$, propagule pressure (the initial number of introduced individuals, N_0) then comes into the equation by affecting σ_n^2:

$$\sigma_n^2 = \exp\left[\mu + \left(2\delta^2\right)^{\frac{1}{2}} / \mathrm{erf}\left(2^{1-\frac{1}{N_0}} - 1\right)\right] \qquad (4.2)$$

where erf($\cdot$) is the Gaussian error function. This new formula suggests that range expansion can be boosted by both demographic factors that enhance the intrinsic population growth rate (increasing r) and dispersal factors that fatten the dispersal kernel (increasing γ) and sort the fastest phenotypes to the range front ($\bar{\sigma}^2$ eventually replaced by σ_n^2). Moreover, high propagule pressure (large N_0) and high compositional diversity in propagules (large δ) will result in a faster expansion.

Boosted range expansion means that dispersal kernels built from samples of individuals across the invaded range or based on records from native ranges cannot produce good estimates of the actual speed of range expansion in invaded ranges. The spreading dynamics of invasion is largely driven by factors associated with boosting the dispersal of frontal populations. Such non-equilibrium dynamics of invasion demand the application of new sets of analytic tools to examine the stabilizing and destabilizing forces that influence only dispersal

but also population demography (Chapter 5), such as human-mediated disturbance. Contrasting traits between core and edge populations indicate that many hypotheses linking biotic interactions to invasion performance could be more relevant for edge populations, as core populations could have started to be absorbed into the resident communities (Chapter 6). Such boosted range expansion could alter community assemblages and their assembly processes (Chapter 8), potentially leading to regime shifts in recipient ecosystems (Chapter 7). To contain the invaded range or to slow down the range expansion, managers should give priority to the advancing range front, rather than core populations (Chapter 9). When building trait-based frameworks for predicting invasiveness and for assessing invasion risks, more attention needs to be given to traits sampled from the advancing edge and the trait difference between core and edge populations.

References

Allendorf, F.W. and Lundquist, L.L. (2003) Introduction: population biology, evolution, and control of invasive species. Conservation Biology, 17, 24–30.

Alston, K.P. and Richardson, D.M. (2006) The roles of habitat features, disturbance, and distance from putative source populations in structuring alien plant invasions at the urban/wildland interface on the Cape Peninsula, South Africa. Biological Conservation, 132, 183–98.

Altwegg, R., Collingham, Y.C., Erni, B., et al. (2013) Density-dependent dispersal and the speed of range expansions. Diversity and Distributions, 19, 60–8.

Amiel, J.J., Tingley, R., and Shine, R. (2011) Smart moves: effects of relative brain size on establishment success of invasive amphibians and reptiles. PLoS One, 6, e18277.

Angers, B., Castonguay, E., and Massicotte, R. (2010) Environmentally induced phenotypes and DNA methylation: how to deal with unpredictable conditions until the next generation and after. Molecular Ecology, 19, 1283–95.

Araújo, M.B. and Pearson, R.G. (2005) Equilibrium of species' distributions with climate. Ecography, 28, 693–5.

Baldwin, M.W., Winkler, H., Organ, C.L., et al. (2010) Wing pointedness associated with migratory distance in common-garden and comparative studies of stonechats (Saxicola torquata). Journal of Evolutionary Biology, 23, 1050–63.

Berg, R.Y. (1983) Plant distribution as seen from plant dispersal: general principles and basic modes of plant dispersal. In: Kubitzki, K. (ed.) *Dispersal and Distribution*. Hamburg: Paul Parey, pp. 13–36.

Berggren, H., Tinnert, J., and Forsman, A. (2012) Spatial sorting may explain evolutionary dynamics of wing polymorphism in pygmy grasshoppers. Journal of Evolutionary Biology, 25, 2126–38.

Berthouly-Salazar, C., van Rensburg, B.J., Le Roux, J.J., *et al.* (2012) Spatial sorting drives morphological variation in the invasive bird, *Acridotheris tristis*. PLoS One, 7, e38145.

Berthouly-Salazar, C., Hui, C., Blackburn, T.M., *et al.* (2013) Long-distance dispersal maximizes evolutionary potential during rapid geographic range expansion. Molecular Ecology, 22, 5793–804.

Blackburn, T.M., Prowse, T.A.A., Lockwood, J.L., *et al.* (2013) Propagule pressure as a driver of establishment success in deliberately introduced exotic species: fact or artefact? Biological Invasions, 15, 1459–69.

Bocedi, G., Heinonen, J., and Travis, J.M.J. (2012) Uncertainty and the role of information acquisition in the evolution of context-dependent emigration. American Naturalist, 179, 606–20.

Bohrer, G.I.L., Nathan, R.A.N., and Volis, S. (2005) Effects of long-distance dispersal for metapopulation survival and genetic structure at ecological time and spatial scales. Journal of Ecology, 93, 1029–40.

Börger, L. and Fryxell, J. (2012) Quantifying individual differences in dispersal using net squared displacement. In: Clobert, J. Baguette, M., Benton, T.G., *et al.* (eds) *Dispersal Ecology and Evolution*. Oxford: Oxford University Press, pp. 222–30.

Braendle, C., Davis, G.K., Brisson, J.A., *et al.* (2006) Wing dimorphism in aphids. Heredity, 97, 192–9.

Brockmann, D., Hufnagel, L., and Geisel, T. (2006) The scaling laws of human travel. Nature, 439, 462–5.

Brown, G.P., Phillips, B.L., Dubey, S., *et al.* (2015) Invader immunology: invasion history alters immune system function in cane toads (*Rhinella marina*) in tropical Australia. Ecology Letters, 18, 57–65.

Brown, G.P., Ujvari, B., Madsen, T., *et al.* (2013) Invader impact clarifies the roles of top-down and bottom-up effects on tropical snake populations. Functional Ecology, 27, 351–61.

Buckley, Y.M. and Catford, J. (2016) Does the biogeographic origin of species matter? Ecological effects of native and non-native species and the use of origin to guide management. Journal of Ecology, 104, 4–17.

Buckley, L.B., Hurlbert, A.H., and Jetz, W. (2012) Broad-scale ecological implications of ectothermy and endothermy in changing environments. Global Ecology and Biogeography, 21, 873–85.

Bullock, J.M. (2012) Plant dispersal and the velocity of climate change. In: Clobert, J. Baguette, M., Benton,

T.G., *et al.* (eds) *Dispersal Ecology and Evolution*. Oxford: Oxford University Press, pp. 366–77.

Burke, M.J. and Grime, J.P. (1996) An experimental study of plant community invasibility. Ecology, 77, 776–90.

Cadotte, M.W., Mai, D.V., Jantz, S., *et al.* (2006) On testing the competition-colonization trade-off in a multispecies assemblage. American Naturalist, 168, 704–9.

Calviño-Cancela, M., Dunn, R.R., Van Etten, E.J.B., *et al.* (2006) Emus as non-standard seed dispersers and their potential for long-distance dispersal. Ecography, 29, 632–40.

Campbell, J.E. and Gibson, D.J. (2001) The effect of seeds of exotic species transported via horse dung on vegetation along trail corridors. Plant Ecology, 157, 23–35,

Cappuccino, N., Mackay, R., and Eisner, C. (2002) Spread of the invasive alien vine *Vincetoxicum rossicum*: trade-offs between seed dispersability and seed quality. American Midland Naturalist, 148, 263–70.

Cheplick, G.P. and Quinn, J.A. (1982) Amphicarpum purshii and the 'pessimistic strategy' in amphicarpic annuals with subterranean fruit. Oecologia, 52, 327–32.

Cheptou, P.O., Carrue, O., Rouifed, S. *et al.* (2008) Rapid evolution of seed dispersal in an urban environment in the weed *Crepis sancta*. Proceedings of the National Academy of Sciences of the USA, 105, 3796–9.

Cheptou, P.O. and Massol, F. (2009) Pollination fluctuations drive evolutionary syndromes linking dispersal and mating system. American Naturalist, 174, 46–55.

Chesson, P. (2000) Mechanisms of maintenance of species diversity. Annual Review of Ecology and Systematics, 31, 343–66.

Chevin, L.M., Lande, R., and Mace, G.M. (2010) Adaptation, plasticity, and extinction in a changing environment: towards a predictive theory. PLoS Biology, 8, e1000357.

Clark, J.S. (1998) Why trees migrate so fast: confronting theory with dispersal biology and the paleorecord. American Naturalist, 152, 204–24.

Clark, J.S., Fastie, C., Hurtt, G., *et al.* (1998) Reid's Paradox of Rapid Plant Migration Dispersal theory and interpretation of paleoecological records. BioScience, 48, 13–24.

Clobert, J., Baguette, M., Benton, T.G., *et al.* (eds) (2012) *Dispersal Ecology and Evolution*. Oxford: Oxford University Press.

Clobert, J., Galliard, L., Cote, J., *et al.* (2009) Informed dispersal, heterogeneity in animal dispersal syndromes and the dynamics of spatially structured populations. Ecology Letters, 12, 197–209.

Cohen, A.N. and Carlton, J.T. (1998) Accelerating invasion rate in a highly invaded estuary. Science, 279, 555–8.

Colautti, R.I., Grigorovich, I.A., and MacIsaac, H.J. (2006) Propagule pressure: a null model for biological invasions. Biological Invasions, 8, 1023–37.

Cote, J. and Clobert, J. (2010) Risky dispersal: avoiding kin competition despite uncertainty. Ecology, 91, 1485–93.

Cote, J., Fogarty, S., Weinersmith, K., et al. (2010) Personality traits and dispersal tendency in the invasive mosquitofish (Gambusia affinis). Proceedings of the Royal Society B: Biological Sciences, 277, 1571–9.

Crooks, J.A. and Soule, M.E. (1999) Lag times in population explosions of invasive species: causes and implications. In: Sandlund, O.T., Schei, P.J., and Viken, A. (eds) Invasive Species and Biodiversity Management. Amsterdam: Kluwer Academic Publishers.

Cwynar, L.C. and MacDonald, G.M. (1987) Geographical variation of lodgepole pine in relation to population history. American Naturalist, 129, 463–9.

Damschen, E.I., Haddad, N.M., Orrock, J.L., et al. (2006) Corridors increase plant species richness at large scales. Science, 313, 1284–6.

D'Antonio, C.M., Dudley, T.L., and Mack, M. (2000) Disturbance and biological invasions: direct effects and feedbacks. In: Walker L.R., (ed.) Ecosystems of Disturbed Ground, Vol. 16. New York, NY: Elsevier Science, pp. 429–68.

Davis, M.A., Grime, J.P., and Thompson, K. (2000) Fluctuating resources in plant communities: a general theory of invasibility. Journal of Ecology, 88, 528–34.

Davis, M.B. (1981) Quaternary history and the stability of forest communities. In: West, D.C., Shugart, H.H., and Botkin, D.B. (eds) Forest Succession: Concepts and Application. New York, NY: Springer-Verlag, pp. 132–53.

Deckers, B., Verheyen, K., Hermy, M., et al. (2005) Effects of landscape structure on the invasive spread of black cherry Prunus serotina in an agricultural landscape in Flanders, Belgium. Ecography, 28, 99–109.

De Fraipont, M., Clobert, J., John, H., et al. (2000) Increased pre-natal maternal corticosterone promotes philopatry of offspring in common lizards Lacerta vivipara. Journal of Animal Ecology, 69, 404–13.

De Jong, Y.A., Butynski, T.M., Isbell, L.A., et al. (2009) Decline in the geographical range of the southern patas monkey Erythrocebus patas baumstarki in Tanzania. Oryx, 43, 267–74.

Denno, R.F., Roderick, G.K., Peterson, M.A., et al. (1996) Habitat persistence underlies intraspecific variation in the dispersal strategies of planthoppers. Ecological Monographs, 66, 389–408.

De Queiroz, A. (2014) The Monkey's Voyage: How Improbable Journeys Shaped the History of Life. London: Basic Books.

Dietz, H. and Edwards, P.J. (2006) Recognition that causal processes change during plant invasion helps explain conflicts in evidence. Ecology, 87, 1359–67.

Dornier, A., Pons, V., and Cheptou, P.O. (2011) Colonization and extinction dynamics of an annual plant metapopulation in an urban environment. Oikos, 120, 1240–6.

Ducatez, S., Crossland, M., and Shine, R. (2016) Differences in developmental strategies between long-settled and invasion front populations of the cane toad in Australia. Journal of Evolutionary Biology, 29, 335–43.

Duckworth, R.A. (2008) Adaptive dispersal strategies and the dynamics of a range expansion. American Naturalist, 172, S4–17.

Duckworth, R.A. (2012) Evolution of genetically integrated dispersal strategies. In: Clobert, J., Baguette, M., Benton, T.G., et al. (eds) Dispersal Ecology and Evolution. Oxford: Oxford University Press, pp. 83–94.

Dytham, C. (2009) Evolved dispersal strategies at range margins. Proceedings of the Royal Society B: Biological Sciences, 276, 1407–13.

Dytham, C. and Travis, J.M.J. (2006) Evolving dispersal and age at death. Oikos, 113, 530–8.

Ellner, S.P. and Schreiber, S.J. (2012) Temporally variable dispersal and demography can accelerate the spread of invading species. Theoretical Population Biology, 82, 283–98.

Fayard, J., Klein, E.K., and Lefèvre, F. (2009) Long distance dispersal and the fate of a gene from the colonization front. Journal of Evolutionary Biology, 22, 2171–82.

Fisher, R.A. (1937) The wave advance of advantageous genes. Annals of Eugenics, 7, 355–69.

Forsman, A., Merilä, J., and Ebenhard, T. (2011) Phenotypic evolution of dispersal-enhancing traits in insular voles. Proceedings of the Royal Society B: Biological Sciences, 278, 225–32.

García-Ramos, G. and Rodríguez, D. (2002) Evolutionary speed of species invasions. Evolution, 56, 661–8.

Gelbard, J.L. and Harrison, S. (2003) Roadless habitats as refuges for native grasslands: interactions with soil, aspect, and grazing. Ecological Applications, 13, 404–15.

Gertzen, E.L., Leung, B., and Yan, N.D. (2011) Propagule pressure, Allee effects and the probability of establishment of an invasive species (Bythotrephes longimanus). Ecosphere, 2, 1–17.

Giladi, I. (2006) Choosing benefits or partners: a review of the evidence for the evolution of myrmecochory. Oikos, 3, 481–92.

Gildenhuys, E., Ellis, A.G., Carroll, S.P., et al. (2015) Combining natal range distributions and phylogeny to resolve biogeographic uncertainties in balloon vines (Cardiospermum, Sapindaceae). Diversity and Distributions, 21, 163–74.

Gonzalez, M.C., Hidalgo, C.A., and Barabasi, A.L. (2008) Understanding individual human mobility patterns. Nature, 453, 779–82.

Grossniklaus, U., Kelly, W.G., Ferguson-Smith, A.C., et al. (2013) Transgenerational epigenetic inheritance: how important is it? Nature Reviews Genetics, 14, 228–35.

Gurevitch, J., Fox, G.A., Wardle, G.M., *et al.* (2011) Emergent insights from the synthesis of conceptual frameworks for biological invasions. Ecology Letters, 14, 407–18.

Gyllenberg, M., Kisdi, É., and Utz, M. (2008) Evolution of condition-dependent dispersal under kin competition. Journal of Mathematical Biology, 57, 285–307.

Hamilton, M.A., Murray, B.R., Cadotte, M.W., *et al.* (2005) Life-history correlates of plant invasiveness at regional and continental scales. Ecology Letters, 8, 1066–74.

Hanski, I. and Saccheri, I. (2006) Molecular-level variation affects population growth in a butterfly metapopulation. PLoS Biology, 4, e129.

Harrison, R.G. (1980) Dispersal polymorphisms in insects. Annual Review of Ecology and Systematics, 11, 95–118.

Hastings, A., Cuddington, K., Davies, K.F., *et al.* (2005) The spatial spread of invasions: new developments in theory and evidence. Ecology Letters, 8, 91–101.

Henry, R.C., Bocedi, G., and Travis, J.M.J. (2013) Eco-evolutionary dynamics of range shifts: elastic margins and critical thresholds. Journal of Theoretical Biology, 321, 1–7.

Herrera, C.M. (1987) Components of pollinator 'quality': comparative analysis of a diverse insect assemblage. Oikos, 50, 79–90.

Hierro, J.L., Villarreal, D., Eren, Ö., *et al.* (2006) Disturbance facilitates invasion: the effects are stronger abroad than at home. American Naturalist, 168, 144–56.

Higgins, S.I., Nathan, R., and Cain, M.L. (2003) Are long-distance dispersal events in plants usually caused by nonstandard means of dispersal? Ecology, 84, 1945–56.

Higgins, S.I. and Richardson, D.M. (1999) Predicting plant migration rates in a changing world: the role of long-distance dispersal. American Naturalist, 153, 464–75.

Hobbs, R.J. (2000) Land-use changes and invasions. In: Mooney, H.A. and Hobbs, R.J. (eds) *Invasive Species in a Changing World*. Washington, DC: Island Press, 55–64.

Hobbs, R.J. and Huenneke, L.F. (1992) Disturbance, diversity, and invasion: implications for conservation. Conservation Biology, 6, 324–37.

Hodkinson, D.J. and Thompson, K. (1997) Plant dispersal: the role of man. Journal of Applied Ecology, 34, 1484–96.

Holt, R.D., Keitt, T.H., Lewis, M.A., *et al.* (2005) Theoretical models of species' borders: single species approaches. Oikos, 108, 18–27.

Howe, H.F. and Smallwood, J. (1982) Ecology of seed dispersal. Annual Review of Ecology and Systematics, 13, 201–28.

Hudson, C.M., Phillips, B.L., Brown, G.P., *et al.* (2015) Virgins in the vanguard: low reproductive frequency in invasion-front cane toads. Biological Journal of the Linnean Society, 116, 743–7.

Hughes, C.L., Dytham, C., and Hill, J.K. (2007) Modelling and analysing evolution of dispersal in populations at expanding range boundaries. Ecological Entomology, 32, 437–45.

Hui, C., Roura-Pascual, N., Brotons, L., *et al.* (2012) Flexible dispersal strategies in native and non-native ranges: environmental quality and the 'good-stay, bad-disperse' rule. Ecography, 35, 1024–32.

Huntley, B. and Birks, H.J.B. (1983) *An Atlas of Past and Present Pollen Maps for Europe 0–13,000 Years Ago*. Cambridge: Cambridge University Press.

Jeltsch, F., Moloney, K.A., Schurr, F.M., *et al.* (2008) The state of plant population modelling in light of environmental change. Perspectives in Plant Ecology, Evolution and Systematics, 9, 171–89.

Johnson, W.C. and Adkisson, C.S. (1985) Dispersal of beech nuts by blue jays in fragmented landscapes. American Midland Naturalist, 113, 319–24.

Jongejans, E., Skarpaas, O., and Shea, K. (2008) Dispersal, demography and spatial population models for conservation and control management. Perspectives in Plant Ecology, Evolution and Systematics, 9, 153–70.

Kelehear, C., Brown, G.P., and Shine, R. (2012) Rapid evolution of parasite life history traits on an expanding range-edge. Ecology Letters, 15, 329–37.

Kilvitis, H.J., Alvarez, M., Foust, C.M., *et al.* (2014) Ecological epigenetics. Advances in Experimental Medical Biology, 781, 191–210.

Kim, K.D. (2005) Invasive plants on disturbed Korean sand dunes. Estuarine, Coastal and Shelf Science, 62, 353–64.

Kisdi, É. (2002) Dispersal: risk spreading versus local adaptation. American Naturalist, 159, 579–96.

Kisdi, É. (2004) Conditional dispersal under kin competition: extension of the Hamilton–May model to brood size-dependent dispersal. Theoretical Population Biology, 66, 369–80.

Kisdi, É. (2012) Year-class coexistence in biennial plants. Theoretical Population Biology, 82, 18–21.

Klein, E.K., Lavigne, C., and Gouyon, P.H. (2006) Mixing of propagules from discrete sources at long distance: comparing a dispersal tail to an exponential. BMC Ecology, 6, 1.

Kolar, C.S. and Lodge, D.M. (2001) Progress in invasion biology: predicting invaders. Trends in Ecology and Evolution, 16, 199–204.

Korsu, K. and Huusko, A. (2009) Propagule pressure and initial dispersal as determinants of establishment success of brook trout (*Salvelinus fontinalis* Mitchill 1814). Aquatic Invasions, 4, 619–26.

Kot, M., Lewis, M., and van den Driessche, P. (1996) Dispersal data and the spread of invading organisms. Ecology, 77, 2027–42.

Kubisch, A., Holt, R.D., Poethke, H.-J., *et al.* (2014) Where am I and why? Synthesizing range biology and the eco-evolutionary dynamics of dispersal. Oikos, 123, 5–22.

Lehouck, V., Bonte, D., Spanhove, T., *et al.* (2012) Integrating context- and stage-dependent effects in studies of frugivorous seed dispersal: an example from south-east Kenya. In: Clobert, J., Baguette, M., Benton, T.G., *et al.* (eds) *Dispersal Ecology and Evolution.* Oxford: Oxford University Press, pp. 50–59.

Léotard, G., Debout, G., Dalecky, A., *et al.* (2009) Range expansion drives dispersal evolution in an equatorial three-species symbiosis. PLoS One, 4, e5377.

Le Roux, J.J., Strasberg, D., Rouget, M., *et al.* (2014) Relatedness defies biogeography: the tale of two island endemics (*Acacia heterophylla* and *A. koa*). New Phytologist, 204, 230–42.

Levey, D.J., Tewksbury, J.J., and Bolker, B.M. (2008) Modelling long-distance seed dispersal in heterogeneous landscapes. Journal of Ecology, 96, 599–608.

Levin, S.A., Muller-Landau, H.C., Nathan, R., *et al.* (2003) The ecology and evolution of seed dispersal: a theoretical perspective. Annual Review of Ecology, Evolution, and Systematics, 34, 575–604.

Lewis, M.A. and Kareiva, P. (1993) Allee dynamics and the spread of invading organisms. Theoretical Population Biology, 43, 141–58.

Lindström, T., Brown, G.P., Sisson, S.A., *et al.* (2013) Rapid shifts in dispersal behavior on an expanding range edge. Proceedings of the National Academy of Sciences of the USA, 110, 13452–6.

Llewelyn, J., Phillips, B.L., Alford, R.A., *et al.* (2010) Locomotor performance in an invasive species: cane toads from the invasion front have greater endurance, but not speed, compared to conspecifics from a long-colonised area. Oecologia, 162, 343–8.

Lloret, F., Médail, F., Brundu, G., *et al.* (2005) Species attributes and invasion success by alien plants on Mediterranean islands. Journal of Ecology, 93, 512–20.

Loarie, S.R., Duffy, P.B., Hamilton, H., *et al.* (2009) The velocity of climate change. Nature, 462, 1052–5.

Lockwood, J.L., Cassey, P., and Blackburn, T. (2005) The role of propagule pressure in explaining species invasions. Trends in Ecology and Evolution, 20, 223–8.

Lombaert, E., Estoup, A., Facon, B., *et al.* (2014) Rapid increase in dispersal during range expansion in the invasive ladybird *Harmonia axyridis*. Journal of Evolutionary Biology, 27, 508–17.

Lomolino, M.V. (1984) Immigrant selection, predation, and the distributions of *Microtus pennsylvanicus* and *Blarina brevicauda* on islands. American Naturalist, 123, 468–83.

Lonsdale, W.M. (1999) Global patterns of plant invasions and the concept of invasibility. Ecology, 80, 1522–36.

Lonsdale, W.M. and Lane, A.M. (1994) Tourist vehicles as vectors of weed seeds in Kakadu National Park, Northern Australia. Biological Conservation, 69, 277–83.

Lopez, D.P., Jungman, A.A., and Rehage, J.S. (2012) Nonnative African jewelfish are more fit but not bolder at the invasion front: a trait comparison across an Everglades range expansion. Biological Invasions, 14, 2159–74.

MacDonald, I.A., Loope, L.L., Usher, M.B., *et al.* (1989) Wildlife conservation and the invasion of nature reserves by introduced species: a global perspective. In: Drake, J., Di Castri, F., Groves, R., *et al.* (eds) *Biological Invasions: A Global Perspective.* Chichester: Wiley, pp. 215–55.

Malcolm, J.R., Liu, C., Neilson, R.P., *et al.* (2006) Global warming and extinctions of endemic species from biodiversity hotspots. Conservation Biology, 20, 538–48.

Manzano, P. and Malo, J.E. (2006) Extreme long-distance seed dispersal via sheep. Frontiers in Ecology and the Environment, 4, 244–8.

Matthysen, E. (2005) Density-dependent dispersal in birds and mammals. Ecography, 28, 403–16.

Matthysen, E. (2012) Multicausality of dispersal: a review. In: Clobert, J., Baguette, M., Benton, T.G., *et al.* (eds) *Dispersal Ecology and Evolution.* Oxford: Oxford University Press, pp. 3–18.

Measey, J.G., Vences, M., Drewes, R.C., *et al.* (2007) Freshwater paths across the ocean: molecular phylogeny of the frog *Ptychadena newtoni* gives insights into amphibian colonization of oceanic islands. Journal of Biogeography, 34, 7–20.

Meyerson, L.A. and Mooney, H.A. (2007) Invasive alien species in an era of globalization. Frontiers in Ecology and the Environment, 5, 199–208.

Mikheyev, A.S., Tchingnoumba, L., Henderson, A., *et al.* (2008) Effect of propagule pressure on the establishment and spread of the little fire ant *Wasmannia auropunctatain* a Gabonese oilfield. Diversity and Distributions, 14, 301–6.

Møller, A.P. (2010) Brain size head size and behaviour of a passerine bird. Journal of Evolutionary Biology, 23, 625–35.

Mollison, D. (1977) Spatial contact models for ecological and epidemic spread. Journal of the Royal Statistical Society Series B, 39, 283–326.

Morse, D.H. and Schmitt, J. (1985) Propagule size, dispersal ability, and seedling performance in *Asclepias syriaca*. Oecologia, 67, 372–9.

Mouquet, N., Munguia, P., Kneitel, J.M. , *et al.* (2003) Community assembly time and the relationship between local and regional species richness. Oikos, 103, 618–26.

Murphy, H.T., Van Der Wal, J., Lovett-Doust, L., *et al.* (2006) Invasiveness in exotic plants: immigration in an

of metapopulations cannot be ensured if all sub-populations are undergoing truly random walks, as the metapopulation as a whole must also randomly walk to eventual extinction (see section 5.5; Chesson 1981; Cappucino 1995). Chesson and Case (1986) argued against the classical equilibrium assumptions and further provided four reasons as to why most observed populations are exhibiting non-equilibrium dynamics: (i) populations are not at a point equilibrium but competition still occurs continuously and is important, permitting more species to coexist than the diversity of resources; (ii) when fluctuations in population density or environmental variables are dominant, population dynamics may be density-independent; (iii) means and variances of environmental fluctuations are not constant over time; (iv) population size undergoes random walk, but time to extinction is so long that species is considered persistent over a long time (i.e. slow competitive displacement).

Charles Elton was highly sceptical of the existence of a 'balance in nature' (Simberloff 2014), and he dedicated much effort to studying the phenomenon and the reasons for population oscillations (Elton 1924); he argued strongly for a non-equilibrium view of population dynamics (Elton and Nicholson 1942). Elton's views are supported by early theoretical proof: Volterra's (1926) mathematical article on population oscillations showed that trophic interactions between predators and prey can lead to non-equilibrium dynamics of population fluctuations and oscillations. Volterra's paper was brought to Elton's attention by his Oxford tutor, Julian Huxley (Rohde 2006). Even without cross-trophic interactions, non-linear population growth could also lead to non-equilibrium dynamics (May 1975). The debate around the presence or absence of population equilibria evolved into discussions on the role of density dependence in regulating population dynamics in the 1950s. During the 1957 Cold Spring Harbor Symposium, Nicholson's (1957) theory of population regulation through density dependence was challenged by Andrewartha (1957) as lacking a 'substantial body of empirical facts'. Without density dependence, emphasizing the role of competition becomes misleading as natural population densities cannot be high enough to trigger noteworthy resource competition (Andrewartha and Birch

1954). To reconcile matters, without excluding the possibility of density dependence, Strong (1986) introduced the term density vagueness to describe parameters of birth and death rates that are only weakly explained by density.

The premise that there is a 'balance of nature' has permeated efforts to model ecological systems mathematically (DeAngelis and Waterhouse 1987), which would otherwise require only means of statistics and probability. Models are abstractions of how we think nature operates. Mathematical models have usually been structured on the premise of an equilibrium state, while admitting that species can deviate from equilibrium and that the equilibrium state may be unstable in many cases, giving rise to population oscillations. Criticisms of this premise not only hinge on the mere question of system stability, but on whether it is valid to define the existence of an equilibrium state at all, as the existence of such a population equilibrium state requires density-dependent population growth, which is often contestable. Moreover, the practical difficulty with models based on the assumed existence of an equilibrium state is that they cannot easily be extrapolated to fine spatial scales. It seems obvious that at sufficiently fine spatial scales the dynamic behaviour is short-lived or transient, and thus that the system will eventually lose balance when examined at increasingly finer scales. These issues raise crucial questions regarding temporal and spatial scales that are appropriate for discussing the long-term survival of species (and the dynamics of invasion).

The debate on the balance of nature and the role of density dependence divides ecology into two paradigms (Hengeveld and Walter 1999): demographic versus autecological. The former emphasizes the importance of competition and the potential of co-evolving of species towards optimization. The autecological paradigm discards the idea of optimization due to the high variability of the environment in space and time. If this is the case, what could be the sources of observed population variability? Perhaps the forces of stability and change must be first carefully studied and then applied to the subject of the balance of nature. To this end, DeAngelis and Waterhouse (1987) conceptualized a number of possible explanatory hypotheses of population dynamics and its variability (Fig. 5.1).

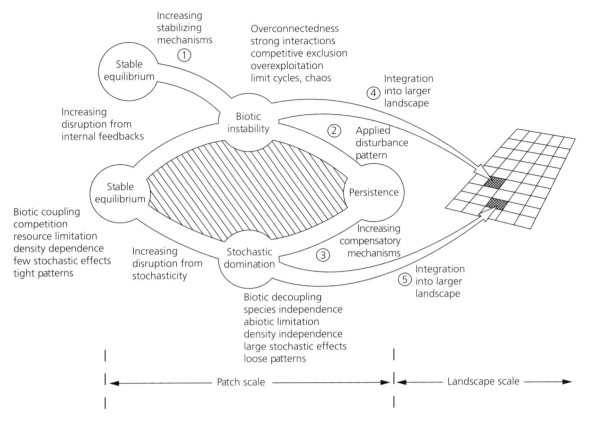

Fig. 5.1 A schematic representation of five general types of hypotheses to explain why ecological systems tend to be stable despite the prevalence of biotic instabilities and environmental stochasticity. From DeAngelis and Waterhouse (1987). Reproduced with permission from John Wiley & Sons.

Firstly, functional relationships between the species and community structure as a whole may counteract biotically induced instabilities. In particular, conditions for feasibility (positive population size) often also favour stability (stable population size). This means that existing systems and populations, through a long process of natural selection at the system level, could indirectly improve the stability of persisting members. Secondly, environmental disturbances may lessen the effect of instabilities. Disturbance could interrupt the process of competitive displacement, thereby facilitating persistence and coexistence of species. Thirdly, compensatory mechanisms acting at low population density levels may help to prevent extinctions caused by environmental stochasticity. Modification in life-history strategies and the behaviour of prey switching in predators could buffer the pressure on rare species, thus slowing down or averting local extinctions or

extirpations of populations. These three mechanisms could, in principle, act to maintain the persistence of populations and ecological communities even on the small spatial scale (Rohde 2006). Fourthly, the quasi-isolation of subpopulations in weakly connected subregions or cells could contribute to persistence. Weak covariance between local populations could reduce spatial synchrony and enhance the persistence at the level of metapopulations (section 5.5). Finally, spatial heterogeneity could create disparity in population dynamics and buffer large variations in population dynamics through a spatial storage effect (see Chapter 8), where favourable habitat and refuges persist even when populations in poor habitats fluctuate drastically due to environmental disturbance (DeAngelis and Waterhouse 1987).

Theorists argue that the mechanistic basis for population ecology, such as the non-equilibrium dynamics of population size, can only be provided by entities from lower hierarchical levels, such as by behaviour, physiology, and traits of individuals (Metz and Diekmann 1986; Caswell *et al.* 1997). Quantitative ecologists generally agree with this argument but suggest that it is rather unfeasible in practice because population dynamics is often recorded as time series, without comprehensive knowledge of the biology of the species. Early research directed at detecting density dependence primarily used life-table data to establish potential relations between vital rates (e.g. relative growth rate and mortality) and observed population density, with correlations (or measured using other related indices; e.g. Hassell 1975) signalling the existence of density dependence (Stearns 1976; Hassell *et al.* 1989; Solow and Steele 1990). In this regard, time-series analysis of population variability is extremely helpful and consistent with these early works for detecting density dependence when underlying mechanisms of population regulation are unclear (Box 5.1).

We use invasion dynamics as a model system to explore the concepts of the balance of nature and to explore the tools and factors that are regulating invasion dynamics at both local and regional scales. Invasion dynamics are fundamentally non-equilibrium in nature, and yet the dynamics and variation within the dynamics are bounded due to habitat suitability and physiological limitations. This means that invasive species in novel environments often experience the quasi-equilibrium of halted expansion during a lag phase or fast expansion to fill the opportunity niche (see Chapter 2 for a systematic review), during which they face many forces that act to stabilize or destabilize their population dynamics. Firstly, as the initial propagule pressure of invasive species is often lower than that required to exceed the carrying capacity of the recipient ecosystem, strong demographic stochasticity would be experienced initially. Because of the small size of founding populations, most invasive species experience positive density dependence known as the Allee effect which influences both population variability and the spatial dynamics of the invasive species (section 5.2). Secondly, once the population of the invading species has grown to reach carrying capacity, negative density dependence will begin regulating population dynamics within the bound

---

Box 5.1 Time-series analysis of population variability

Sources of the temporal variability in population size, depicted by a time series, are often grouped into three categories (Morris and Doak 2002): lags in response, environmental condition, and demographic stochasticity (Turchin 2003). Firstly, cross-trophic interactions can foster feedbacks and thus delays between the action of disturbance and the response experienced by its actor, thus destabilizing population dynamics. Besides cross-trophic interaction, time lags in model structures could also bring instability. Adding a time lag to continuous logistic equations (Hutchinson 1948) could change the stable equilibrium of population size into unstable fluctuations (May 1981), and discrete-time models developed from continuous models bring a one-generation lag in the system, triggering limit cycles and even chaos (May 1974; Gurney and Nisbet 1998). Age- and stage-structured models can obviously also lead to oscillations due to lags in development (Caswell *et al.* 1997).

Secondly, many exogenous factors, such as environmental stochasticity, seasonality, and disturbance, can lead to population instability and drive population oscillations.

Environmental stochasticity describes temporal variation in vital rates driven by changes in the environment that are inherently erratic or unpredictable (Morris and Doak 2002). This does not include consistent trends in the environment that cause parallel trends in vital rates (e.g. seasonality). The values of different vital rates are usually inter-correlated. The net effect is to increase the variability in overall population growth rates, which will decrease population viability. Temporal variability is also strongly affected by, often infrequent, catastrophes (extreme bad years with very low survival or reproduction) and bonanzas (extremely good years). Both environmental stochasticity and catastrophes/bonanzas are due to alterations in environmental conditions, independent of population size. In particular, environmental stochasticity often leads to a log-normally distributed population growth rate, making the best predictor of whether a population N_t will increase or decrease over the long term being the geometric mean of the annual population growth rates. Given a specific population $N_{t+1} = \lambda_t N_t$, where λ_t is either 0.86 or 1.16 with an equal chance, using the arithmetic mean of

(Continued)

Box 5.1 *Continued*

λ_t would ignore the stochasticity and lead to the misleading estimation of population growth at a rate of $\lambda = 1.01$, whereas using the geometric mean will reveal the actual population dynamics, showing an actual population decline due to stochasticity (Morris and Doak 2002).

Thirdly, demographic stochasticity is the temporal variation in population growth driven by chance in the actual fates of different individuals within a year, and its magnitude is strongly dependent on population size. Demographic stochasticity is essentially the same as the randomness that causes variation in the numbers of heads and tails when repeatedly flipping a coin (Morris and Doak 2002). It is a powerful force to consider when introducing new species to novel environments. As above, the intrinsic population growth rate, μ, and the variance of population growth σ^2 are defined as the average and variance of log(λ) in observed years. The probability density of the time t (first-passage time) when the population size reaches the quasi-extinction threshold for the first time is given by the inverse Gaussian distribution (Lande and Orzack 1988):

$$prob(t) = \frac{d}{\sqrt{2\pi\sigma^2 t^3}} \exp\left(\frac{-(d - |\mu| t)^2}{2\sigma^2 t}\right) \qquad (5.1)$$

where $d = \ln(N_c / N_x)$ is the difference between the log of the current population size N_c and the log of the extinction threshold N_x, note $N_c > N_x$. The above formula is typically used for the case of population extinction, with $\mu < 0$ and $d > 0$ (i.e. $N_c > N_x$). For the case of invasion dynamics, we could redefine the meaning of N_c and N_x as the establishment threshold and initial propagule pressure, and equation (5.1) will then indicate the probability of an invasive species growing from its initial propagule size N_x to a certain threshold N_c that would be considered a successful invasion. A more complicated formulation is needed if we add more realism to the invasion dynamics. For instance, individual variability is likely to reduce the importance of demographic stochasticity in small populations due to the buffer effect (Fox and Kendall 2002).

Time-series analysis can provide the means for quantifying population fluctuation (mean, variance and autocorrelation) or even the structure of density dependence (process order, shape between population change rate and lagged population densities, trajectory stability, and signal/noise ratio), especially when the underlying drivers of population regulation are unclear (Turchin 2003). Density dependence depicts a non-constant functional relationship between the per-capita rate of population change and population density, possibly involving lags (Murdoch 1994), that is,

$$\ln(n_t / n_{t-1}) = f(n_{t-1}, n_{t-2}, \ldots, \varepsilon_t), \qquad (5.2)$$

with the last parameter ε_t represents the action of exogenous factors. In short, density dependence occurs when the population change rate responds to previous population sizes (Hixon and Johnson 2009). Time-series analysis aims to decompose the variance of a series into trend, seasonal variation, other cyclic oscillations, and the remaining irregular fluctuations (Chatfield 1989).

Trends are often long term, and are exogenously driven, systematic changes in the environment that bring non-stationarity into the dynamics. Periodic changes in the environment (e.g. seasonality) are another source of non-stationarity. Before formal analyses, missing data can be imputed, and non-stationarity addressed through detrending and splitting. The most useful diagnostic tool for time-series analysis is the autocorrelation function (ACF), where the correlation coefficient between pairs of log-transformed population densities $n_{t-\tau}$ and n_t are plotted as a function of lag τ in a correlogram. The dominant period, T, is the lag when ACF reaches maximum. If the ACF correlation at lag T, ACF(T), is greater than, $2/\sqrt{n}$ (where n is the number of data points), it is taken as strong evidence of statistical periodicity. If not, we need to check whether the half period is the dominant period. If ACF($T/2$) is less than $-2/\sqrt{n}$, we have weak evidence of statistical periodicity (Turchin 2003). An alternative approach for detecting periodicity is to use spectral analysis in a periodogram. The lags of population fluctuation, known as the order of the time-series process, can be decided using partial rate correlation functions (PRCF; Berryman and Turchin 2001). It is worth noting that shorter lags can create longer periods in population dynamics; a model with two lags, n_{t-1} and n_{t-2}, can generate limit cycles with period of ten years (Turchin 2003).

---

of resilience (section 5.3). Many species which experience extra-limital expansion were initially limited by negative density dependence, but they begin expanding once the strength of density dependence weakens. Many invasive alien species are released from their co-evolved natural enemies in their novel range, and thus experience less regulated population dynamics, while many introduced species in their lag phase are waiting for novel associations with mutualists to overcome the Allee effect (see Chapter 6). Thirdly, many invasive species could experience strong niche shifts as a result of altered regulating/selection forces (section 5.4). Fourthly, because invasions are essentially a spatial phenomenon, we need to understand how to quantify the dynamics of a collection of local populations (spatial synchrony; section 5.5), how to connect local temporal dynamics with regional spatial dynamics (space-for-time substitution; section 5.6), and how to explore demographic or environmental drivers behind the nonequilibrial invasion dynamics statistically (spatial autoregressive models; section 5.7).

5.2 Positive density dependence: the Allee effect

The Allee effect is defined as a positive relationship between any components of fitness of a species and either numbers or density of conspecifics (Stephens *et al*. 1999). An individual that is subject to an Allee effect will suffer a decrease in some aspect of its fitness when conspecific density is low. Without Allee effects, demographic and environmental stochasticity could still reduce the likelihood of establishment for invasive species with small initial propagule size (Lande 1998). However, this is not a genuine Allee effect as stochasticity should not be a covariate of fitness (Stephens *et al*. 1999). Stephens and colleagues (1999) distinguish between component and demographic Allee effects. When a population experiences a component Allee effect, some component of individual fitness has a positive relationship with density, that is, it is reduced at low density. A component Allee effect could lead to a demographic Allee effect where the overall fitness has a positive relationship with density that results in growth rate per capita of the species being reduced at low density.

A demographic Allee effect can be either weak (non-critical) or strong (critical) (Wang and Kot 2001; Deredec and Courchamp 2003). Populations subject to a strong Allee effect experience negative growth rates per capita when density falls below a critical threshold. Under deterministic dynamics, a population that does not exceed this threshold will become extinct. The critical threshold in Allee effects does not need to be initial propagule size but can also be the initial area occupied; this is a special case of a hypersurface threshold in a spatially structured population (Schreiber 2004). Unlike the effect of demographic stochasticity, a strong Allee effect can produce an inflection point in the first passage

probability as a function of population size at the point of critical density (Dennis 2002). Interestingly, with critical Allee effects, both demographic and environmental stochasticity can increase the probability of establishment for populations even smaller than the critical threshold (Taylor and Hastings 2005) but reduce the probability of establishment for populations larger than the threshold (Dennis 1989, 2002; Grevstad 1999; Liebhold and Bascompte 2003). This confounding effect of stochasticity and Allee effects further intensifies as the degree of stochasticity increases (Liebhold and Bascompte 2003).

Although many studies consider only strong (critical) Allee effects, the examples in Warder Allee's original work clearly show that the definition should also include weak (non-critical) effects (Allee 1958; Fowler and Baker 1991; Stephens *et al*. 1999; Wang and Kot 2001; Wang *et al*. 2002). Populations with weak Allee dynamics experience lower growth rates per capita at low densities, but never experience negative growth rates per capita and therefore have no critical threshold to exceed (Chen and Hui 2009). Weak Allee effects can promote coexistence. Two competitors in a single patch can coexist at different densities if both experience Allee effects (Gyllenberg *et al*. 1999). In two patches with two competitors, Allee effects can promote spatial segregation and asymmetric competition where the population density in one patch is much higher than the other population (Gruntfest *et al*. 1997; Amarasekare 1998; Gyllenberg *et al*. 1999; Ferdy and Molofsky 2002), thus promoting coexistence (Keeling *et al*. 2003). However, with increasing strength of the Allee effect, coexistence becomes less likely and eventually impossible (Ferdy and Molofsky 2002).

Allee effects can be driven by many mechanisms, the most prevalent being mate shortage in sexually reproducing species (Dennis 1989; Boukal and Berec 2002). Lacking co-operation, or simply asociality in sexually reproducing species, could give rise to Allee effects (Odum and Allee 1954; Philip 1957; Bradford and Philip 1970), while other factors that affect reproduction success, such as sex ratio changes, asynchrony in reproductive timing of potential mates, sexual selection, and density-dependent performance, such as more effective predator avoidance in large groups and obligate co-operation, also produce Allee effects. Allee effects are common in plants (Groom 1998; Levin *et al.* 2009) and in all major groups of animals (Allee 1958; Fowler and Baker 1991; Myers *et al.* 1995; Liermann and Hilborn 1997; Gascoigne and Lipcius 2004). As most invasions start from small populations, it is highly likely that populations of most invasive species will go through the phase of Allee effects (Taylor and Hastings 2005). This has major implications for understanding invasions and for managing risks associated with alien species (Drake and Lodge 2006; Tobin *et al.* 2011).

Allee effects can cause different spatial structures and dynamics of spread, slowing or even halting range expansion, thereby providing windows of opportunity for implementing optimal control strategies. Numerous theoretical studies have incorporated Allee effects into spatial population models; there is a substantial body of theory relating to the consequences of Allee effects in invasion dynamics (Table 5.1). In particular, high rates of dispersal could destabilize spreading dynamics (Ferdy and Molofsky 2002), while Allee effects can stabilize population dynamics (Scheuring 1999; Fowler and Ruxton 2002; Hui and Li 2003). Allee effects can explain the lack of range expansion even when suitable habitat is accessible (Keitt *et al.* 2001); this known as range pinning or propagation failure (Fath 1998; Keitt *et al.* 2001; Hadjiavgousti and Ichtiaroglou

2004). Allee effects can slow the spatial colonization in metapopulations and pin down the range dynamics even in homogenous landscapes (Taylor and Hastings 2005), with the final population size critically dependent on the initial propagule size (Hui and Li 2004). Final spatial distributions resulting from dynamics that include Allee effects can be spatially heterogeneous, without the necessity of underlying habitat heterogeneity, forming irregular spatial clusters (Soboleva *et al.* 2003) where aggregations of individuals are linked by local interactions that form static borders (Fig. 5.2). Consequently, the shape of the invasion front can affect the rate of spread—invasions with more ragged fronts spread faster than those with planar boundaries (Lewis and Kareiva 1993).

As a classic example, the slow rate of spread of the gypsy moth, *Lymantria dispar*, in North America was traditionally attributed to the lack of mobility because females in invading populations are flightless. There is, however, considerable variability in female flight capabilities across the native range of Eurasia, from the flightless European populations to Asian populations with good flight capability (Mikkola 1971; Schaefer *et al.* 1984; Baranchikov 1989; Ponomarev 1994; Koshio 1996; Keena *et al.* 2008; Liebhold *et al.* 2008). However, it is possible that

Table 5.1: Ecological consequences of Allee effects on dynamics of invasive species (Taylor and Hastings 2005).

Consequence	Model type	Type of Allee effect
Non-spatial consequences. Species must be introduced at higher than critical threshold for invasion to succeed	Deterministic	Critical
Probability of establishment declines sharply at critical density	Stochastic	Critical
No establishment possible if Allee effect is too strong	Deterministic	Critical
Spatial consequences. Rate of spread slower	Deterministic or stochastic	Critical or non-critical
Range pinning	Deterministic or stochastic with discrete (patchy) space	Critical
Accelerating invasions converted to finite speed invasions	Deterministic, integrodifference with fat-tailed dispersal kernels	Critical or non-critical
Different levels of occupancy of patches and segregation of competing species	Discrete patches	Critical
Patchy invasion in continuous landscape	Stochastic models with continuous space or lattice models	Critical
Initial population must occupy area larger than critical spatial threshold	Deterministic	Critical

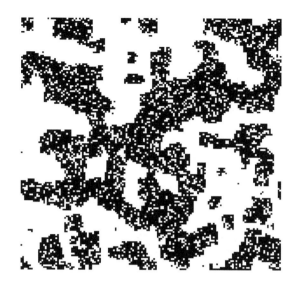

Fig. 5.2 The spatial distribution of metapopulations in a lattice of 150 × 150 dimension. Rerun of the model of Hui and Li (2004), with the colonization rate c = 0.5, local extinction rate e = 0.1, and Allee effect threshold a = 0.3. Dispersal happens among the four adjacent neighbouring cells. The distribution is a snapshot at t = 200, with 6000 randomly selected cells occupied initially.

populations with flight-capable females could have been accidentally introduced to North America from the 1990s (Bogdanowicz *et al.* 1993), thereby removing the lack of mobility as an explanation for slow spread. Analyses of time series from invading populations provide good evidence of strong demographic Allee effects that caused the extinction of low-density isolated populations (Liebhold and Bascompte 2003; Whitmire and Tobin 2006; Tobin *et al.* 2007). Failure to locate a mate produces a component Allee effect which ultimately contributes to a demographic Allee effect (Sharov *et al.* 1995; Robinet *et al.* 2007, 2008; Tobin *et al.* 2009). Temporal asynchrony in male and female sexual maturation can further intensify Allee effects due to mate location failure, which consequently leads to the extinction and slow spread of many invading populations (Calabrese and Fagan 2004; Robinet *et al.* 2007).

To illustrate the latter explanation, Robinet and Liebhold (2009) explored how female dispersal interacts with the probability of finding a mate to foster a demographic Allee effect and slowing the spread of the gypsy moth. They found that dispersal could aggravate Allee effects by reducing the

probability of finding mates (Hopper and Roush 1993; South and Kenward 2001; Jonsen *et al.* 2007). Populations with flight-capable females may actually be easier to eradicate because of the stronger Allee effect than populations with more limited dispersal capabilities (Hulme 2006). Dispersal as a trait could, therefore, undermine the invasiveness of a species (Fig. 5.3); this contradicts the overall view in the literature that dispersal promotes invasion success (Ehrlich 1989; Lodge 1993; Schöpf and Sih 2004). With the interaction of dispersal and Allee effects, spread into adjoining areas can only proceed when the number of migrants is sufficient to exceed the Allee threshold (Hui and Li 2004). The pulse invasion waves of the gypsy moth observed during the spread in North America (Johnson *et al.* 2006) can thus be explained by the critical demographic Allee effect.

Low density at the advancing edge or low initial propagule size provides opportunities for Allee effects to have a substantial effect on invasion dynamics. Allee effects in an invader should result in longer lag times, slower spread, and a decreased probability of establishment (Kot *et al.* 1996; Lewis and Kareiva 1993; Hastings 1996). Consequently, rapid evolution in successful invaders or at the advancing range front should favour life-history

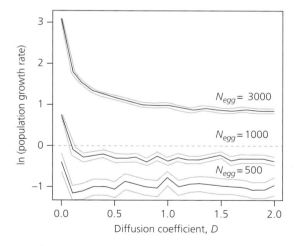

Fig. 5.3 Log-transformed growth rate of the gypsy moth (*Lymantria dispar*) as a function of the diffusion coefficient for adult female dispersal (D; unit: km²/generation) N_{egg} represents the number of eggs initially introduced. From Robinet and Liebhold (2009). Reproduced with permission from John Wiley & Sons.

strategies that can efficiently overcome Allee effects (e.g. self-fertilization facilitates the invasion of the Formosa lily, *Lilium formosanum*; Rodger *et al.* 2013). Importantly, invasion management could aim to exploit the Allee effect by reducing the propagule size during establishment or spread to below a critical threshold, using five management strategies (Tobin *et al.* 2011). Specifically, culling (e.g. pesticides and trapping) can decrease the population density, modify the existing Allee effect, and even create a new Allee effect (e.g. shooting 50 000 birds per year eradicated the great cormorant, *Phalacrocorax carbo sinensis*, within a few decades; Frederiksen *et al.* 2001). Disruption of successful mating could trigger Allee effects in invertebrates, reptiles, and mammals through the deployment of synthesized pheromones (Yamanaka 2007) or by releasing sterile males (Boukal and Berec 2009). Augmentation of generalist natural enemies could be used to trigger a component Allee effect in fish, invertebrates, and mammals (Gascoigne and Lipcius 2004); the collapse of the brown-tail moth, *Euproctis chrysorrhoea*, following the release of the generalist insect parasitoid *Compsilura concinnata* in the eastern United States (Elkinton *et al.* 2006) is a good example. Chemical enhancement of host plant defences can target invertebrate herbivores (Gatehouse 2002). Spatial tactics such as selective and spatially limited habitat modification can be used to eradicate terrestrial and aquatic animals (Jonsson *et al.* 2010; Suckling and Brockerhoff 2010; Hopkins *et al.* 2011).

5.3 Negative density dependence

Negative density dependence happens when vital rates (e.g. of survival and fecundity) decrease as population density increases (Lorenzen and Enberg 2002; Nicoll *et al.* 2003). It is therefore an important stabilizer (Weisberg and Reisman 2008) and regulator for the dynamics of animals (Sibly *et al.* 2005; Brook and Bradshaw 2006; Betini *et al.* 2013) and plants (Johnson *et al.* 2012). For instance, Johnson and colleagues (2012) found strong evidence of conspecific negative density dependence in forests worldwide but little effect of heterospecific density; this provides support for the Janzen–Connell hypothesis that proximity to adults of the same species reduces seedling survival through attack by host-specific enemies (Janzen 1970; Connell 1971). Both endogenous density-dependent factors (e.g. conspecific resource competition, interspecific interaction such as prey switching where predators tend to target disproportionally abundant prey but ignore rare ones) and exogenous density-independent factors (e.g. climate and human activity) (Sibly *et al.* 2005; Brook and Bradshaw 2006; Russell *et al.* 2011) drive population dynamics. Efficient management of biological invasions requires a fundamental understanding of how these factors interact to affect the recruitment and dynamics of the invasive species (Liebhold and Tobin 2008).

The existence of negative density dependence implies that the population is regulated and we therefore should be able to identify the long-term stationary probability distribution of population densities (Turchin 1995). To this end, time-series analysis of population density is widely used to detect negative density dependence and associated covariates (Box 5.1). For instance, the population dynamics of ungulates are influenced by the combined effects of density-dependence and stochastic factors (Grenfell *et al.* 1998; Solberg *et al.* 1999). Aanes and colleagues (2000) investigated the trigger for the polar expansion of Svalbard reindeer, *Rangifer tarandus platyrhynchus*, which live further north than any other cervid population. These animals experience negligible predation, almost no hunting, and are unaffected by interspecific interactions with other large herbivores. Time series for this species thus provide a unique opportunity to analyse the relative importance of density dependence and environmental stochasticity for the dynamics of an ungulate species unaffected by species interactions at the same or higher trophic level. Aanes and colleagues (2000) modelled the fluctuations in the population growth rate $R_t = \ln(N_{t+1}/N_t)$ using an autoregressive model with d-orders (Box and Jenkins 1970) plus an environmental variable U_t:

$$R_t = \beta_0 + \beta_1 R_{t-1} + \ldots + \beta_d R_{t-d} \\ + \omega_1 U_{t-1} + \ldots + \omega_k U_{t-k} + \varepsilon_t \quad (5.3)$$

No significant effects, however, of either direct or delayed density dependence were found to mediate the population dynamics of the Svalbard reindeer. Annual variation in population growth rate was strongly negatively related to the amount of precipitation during winter (i.e. high growth rates occurred when winters were dry), with the effect of climate stronger at high densities. These results support the view that population fluctuations of arctic ungulates are strongly influenced by stochastic variation in climate (Aanes *et al.* 2000).

There are many lessons to be learnt from long-term monitoring and management of pests, and from native species undergoing range expansion. For instance, the cotton bollworm, *Helicoverpa armigera*, which is characterized by its polyphagy, high mobility, high fecundity, and facultative diapause, is one of the most damaging crop pests in the tropics and subtropics (Drake and Gatehouse 1996; Scott *et al.* 2005). This species has moved into cooler regions such as central Europe where it previously faced overwintering problems (Farrow and Daly 1987). Unexpectedly large numbers of the bollworm have been observed in Hungary since 1993 (Keszthelyi *et al.* 2013). Using a 37-year time series of light-trapping records, Ouyang and colleagues (2014) provided quantitative evidence of how negative density dependence regulates the population dynamics of the species. Using the first-order time-series analysis of the R-function (Berryman and Turchin 2001), $R_t = \ln(N_t / N_{t-1})$, as dependent variable in a generalized additive model for the autoregression, Ouyang and colleagues (2014) identified strong negative density dependence in the overwintering, second and third generations, but none in the first generation. Strong density dependence is reflected in the behaviour of cannibalism and interactions with natural enemies (Kakimoto *et al.* 2003; Tschinkel 1981), while long-distance migratory movement releases the first generation population from density dependence.

Although the negative density dependence can effectively regulate the population change rate of *H. armigera* to fluctuate around zero at stable equilibrium levels before and after the East Asia outbreak in 1992, the population equilibrium jumped to a higher density level with apparently larger amplitudes after the outbreak (Ouyang *et al.* 2014), potentially suggesting a regime shift (see Chapter 7). Therefore, exogenous factors (warming climate and agricultural activity) seem to have loosened the regulatory mechanism (here, negative density dependence), allowing the population size to increase

to a higher level. This could be the reason for the bursts of many pests and extra-limital invasive species when regulatory mechanisms are disrupted by environmental changes and disturbances. This poses considerable risks to the provision of agro-ecosystem services and regional food security.

It is often stated that many invasive species experience positive density dependence (see section 5.2), but this is not to say that invasive species do not experience negative density dependence. Fauvergue and colleagues (2007) set out to explore the role of the Allee effect of propagule size on the establishment of invasive parasitoid wasp, *Neodryinus typhlocybae*, which was initially introduced to France as a biocontrol agent for the phytophagous flatid planthopper, *Metcalfa pruinosa*. Instead, they found clear negative density dependence but no signs of an Allee effect. This contrasts with many studies that have demonstrated the importance of the Allee effect in biological control (Hopper and Roush 1993; Grevstad 1999; Shea and Possingham 2000; Fagan *et al.* 2002; Memmott *et al.* 2005) and biological invasions (see section 5.2). Fauvergue and colleagues (2007) proposed six possible reasons for the absence of an Allee effect and the presence of strong negative density dependence in parasitoid species, which can be considered good indicators of density-dependence switching signs, from positive to negative.

Firstly, parasitoids generally experience intense conspecific competition for hosts due to interference and superparasitism (Godfray 1994); this prevents the possibility of a demographic Allee effect. Secondly, parasitoids are normally extremely efficient in locating hosts and mates through pheromones (Fauvergue *et al.* 1999); this prevents component Allee effects such as difficulty in finding mates when density is low. Thirdly, when arriving in isolated patches with low density, individuals could shift their behaviour by reproducing and foraging more intensively (Boivin *et al.* 2004; Thiel and Hoffmeister 2004; Tentelier *et al.* 2006) thereby countering the Allee effect. Fourthly, parasitoids belonging to the order Hymenoptera are haplodiploids (males develop from unfertilized eggs and are haploid, and females develop from fertilized eggs and are diploid) which alleviates the difficulty of finding mates (Godfray 1990; Hopper and Roush 1993). Fifthly, haplodiploids remove genetic load through haploid males, and thus experience less severe inbreeding depression at low population size than diploids (Henter 2003). With one exception, homozygosity at the sex locus in species with single-locus complementary sex determination results in sterile males and thus a strong Allee effect (Zayed and Packer 2005; Hedrick *et al.* 2006). This means that the genetics of sex determination could be an indicator of the type of density dependence. Last, the dispersal of individuals can rescue small populations from extinction (Hanski 1998) thereby releasing small sink populations from experiencing a critical Allee effect when nearing a source population. Fauvergue and colleagues' (2007) six clues provide criteria which can be used to anticipate whether density dependence will be positive or negative.

Because of the obvious implications of Allee effects for managing invasive species (Tobin *et al.* 2011), managers could also exploit negative density dependence, especially to contain invasive populations. Efficient and timely management of biological invasions in the Anthropocene will increasingly need to focus on strengthening and revitalizing weakening negative density dependence caused by other global change factors such as climate change and human activities. As negative density dependence largely reflects the effects of resource competition and antagonistic interactions with natural enemies, management aiming at manipulating density dependence should focus on reducing resource availability and boosting natural enemies. There are other successful lessons to be learnt from integrated pest management (Kogan 1998) and ecologically based pest management (Landis *et al.* 2000; Altieri and Nicholls 2003). For instance, management of agricultural habitats can be achieved by reducing fertilizer inputs, through crop rotation, by establishing semi-natural habitats in agricultural landscapes and through manipulating landscape design, and by formulating policies that promote co-ordinated management (Cumming *et al.* 2014). Overall, management of invasion dynamics by manipulating density dependence should boost positive density dependence when the population size is small, but shifts to enhancing negative density dependence when the population size is large.

5.4 Niche shifts

Besides being regulated by positive and negative density dependence, and other density-independent factors, invasive species often experience shifts in their realized niches, exhibiting a non-equilibrium niche shift. Based on Petitpierre and colleagues' (2012) initial proposal, Guisan and colleagues (2014) further unified climatic niche shifts during biological invasions into a framework (Fig. 5.4). They suggest that the niche space of an alien species can be classified into niche unfilling (C), stability (D), expansion (E), and centroid shift (IR arrow in Fig. 5.4). Many examples of evidence for (Broennimann *et al.* 2007; Fitzpatrick *et al.* 2007; Rodder and Lotters 2009; Medley 2010; Lauzeral *et al.* 2011) and against (e.g. Peterson 2011; Petitpierre *et al.* 2012; Strubbe *et al.* 2013) niche shift during invasions have been published recently.

Niche shift is normally quantified using two approaches (Fig. 5.5): by comparing observations using ordination (e.g. principle component analysis), or by comparing predictions from reciprocal species distribution models (see Box 3.1 in Chapter 3). To standardize the detection and analysis of niche shift during invasions, Guisan and colleagues (2014) provided five recommendations: ordination should be preferred to species distribution models (Broennimann *et al.* 2012); the same sets of environmental variables should be used for comparison; all four measures of niche shift (centroid shift, niche unfilling, expansion and overlap) should be estimated; attention must be given to data quality and sampling bias; the effect of climate outliers and resembling versus contrasting environments on niche shift metrics should be assessed.

Observed niche dynamics during biological invasion can be attributable to either rapid adaptive evolution or rapid spread under relaxed geographical constraints (Petitpierre *et al.* 2012; Broennimann *et al.* 2014). Rapid adaptive evolution of fundamental niches can happen during biological invasions (Prentis *et al.* 2008) and is probably facilitated by high standing genetic variation and recombination (Hermisson and Pennings 2005). Such rapid adaptation is especially effective with the hybridization of different genotypes that are spatially separated in the native range but mixed in the novel range

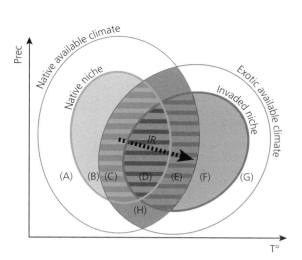

Fig. 5.4 Schematic representation of the indices of niche change (unfilling, stability, and expansion). The thin outer circles show the density of available environments in the native range and in the invaded range. The grey area shows the most frequent environments common to both ranges. The thick inner circles show the native and the invaded niches, respectively. The change of niche centroid is measured by inertia ratio (IR), shown with a thick broken arrow. Upper-case letters represent different components of niche shift: (A) available conditions in the native range, outside the native niche and novel to the invaded range. (B) Conditions inside the native niche but novel to the invaded range. (C) Unfilling; that is, conditions inside the native niche but outside the invaded niche, possibly due to recent introduction combined with ongoing dispersal of the alien species, which should at term fill these conditions. (D) Niche stability; that is, conditions filled in both native and invaded range. (E) Niche expansion; that is, conditions inside the invaded niche but outside the native one, due to ecological or evolutionary change in the invaded range. (F) Conditions inside the invasive niche but novel to the native range. (G) Available conditions in the invaded range but outside the invasive niche and novel to the native range. (H) Analog conditions between the native and invaded ranges. From Guisan *et al.* (2014). With permission from Elsevier.

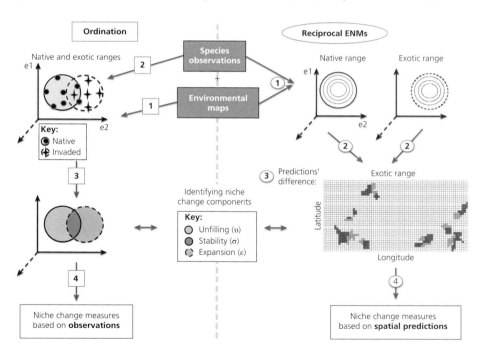

Fig. 5.5 The two approaches commonly used to quantify niche changes between ranges. Ordination is based on the observations directly, whereas ecological niche models (ENM; also known as correlative species distribution models) are based on predictions. Steps for ordination are (numbers in squares): 1. Definition of the reduced multidimensional environmental space. 2. Plot of the observations from each range in this space. 3. Comparison of the niche defined from observations in each range. 4. Calculation of the niche change metrics. Steps for ENMs are: 1. Fit of ENMs by relating field observations to environmental variables. 2. Projections of the ENMs in geographic space. 3. Compute difference in the projections. 4. Calculation of the niche change metrics. From Guisan *et al.* (2014). With permission from Elsevier.

(Novak and Mack 2005). By contrast, niche dynamics could instead result from changes in the realized niche following the release from biotic constraints in native ranges, such as predators, pathogens, or competitors, or simply from breaking constraints of physical barriers and a lack of suitable habitat in native ranges, without the need for any adaptive change to the fundamental niche (Tingley *et al.* 2014).

As one example, Dellinger and colleagues (2016) hypothesized that if niche dynamics in alien ranges are driven by rapid adaptation, apomictic species should be particularly conservative. On the other hand, if niche dynamics in alien ranges are triggered by rapid spatial spread following the relaxation of non-climatic constraints, apomixis should be advantageous. Six scenarios of niche shift during invasions were outlined (Fig. 5.6). Higher rates of expansion into novel niche space and niche broadening can be expected in sexually reproducing species (scenarios C and E in Fig. 5.6). In contrast, alien apomicts will not be able to broaden their niches (scenarios A and D in Fig. 5.6).

Dellinger and colleagues (2016) detected both shifts in niche optimum and changes in niche breadth for two-thirds of the 26 flowering plant species they examined (Fig. 5.7), lending support to scenarios A and C in Figure 5.6 (Glennon *et al.* 2014). Cases of niche contractions could be explained by residence times too short to allow for the introduced species to reach all accessible sites (Williamson *et al.* 2009). As no evidence of different niche shifts between sexual and apomictic species was found, rapid evolutionary adaptation is unlikely to be the main driver of niche dynamics, even though there might be some connection between genetic diversity and niche dynamics in individual cases. In other words, release from non-climatic restrictions may allow species to realize their already existing climatic potential more completely in the alien than in the native range (Early and Sax 2014). For example, niche shifts in introduced European plants that are naturalized in North America can also be attributed to the geographic range restriction in Europe but released and filled up after rapid spread in North America, which can be detected by changes in niche breadth (Early and Sax 2014).

Range expansion of currently invasive species is generally underpinned by a set of functional

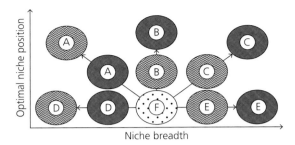

Fig. 5.6 Simplified scenarios of niche change in plant invasions; the native niche (bulb with a dotted pattern) occupies a specific position and breadth (F). During an invasion, niches may remain stable (F) or changes may occur in optimal niche position (A, B, and C) and/or niche breadth (A, C, D, and E). Shifts in optimal niche position should occur both in apomictic species (bulbs with a zig-zag pattern) and in sexual species (bulbs with a checked pattern), but sexual species should show more pronounced niche broadening. Types of patterns: (A) change in position, niche contraction; unfilling of native niche space and/or expansion into a narrower, new niche space; (B) change in niche position, breadth remains the same; expansion and unfilling are balanced; (C) change in niche position, niche broadens; expansion more pronounced than unfilling; (D) niche position remains the same, niche contraction; unfilling; (E) niche position remains the same, niche broadens at margins; expansion; (F) native niche position and breadth remain the same during naturalization; stability. From Dellinger *et al.* (2016).

traits related to their colonization and/or establishment capacities (Theoharides and Dukes 2007). Following this line, Gallien and colleagues (2016) proposed that if the traits driving invasion success are heritable across generations, then currently invasive species may belong to lineages that were also particularly successful at colonizing new regions in the past. After reconstructing the historical biogeography of the genus *Pinus* (pine trees), Gallien and colleagues (2016) confirmed this hypothesis and showed that currently invasive species indeed belong to lineages that were particularly good colonizers in the past (Fig. 5.8). Interestingly, these fast colonizing lineages of pines also showed significantly higher rates of climatic niche evolution, which suggests that their colonization success may be linked to a good capacity to adapt to new environmental conditions (see also Salamin *et al.* 2010; Lavergne *et al.* 2013; Quintero and Wiens 2013). However, in this study past rates of niche evolution could not explain the difference increase in climatic tolerances between species' native and invaded ranges, and thus

on spatial optimization in invasion management (Chapter 9).

5.6 Space-for-time substitution

The rate of expansion and retraction (i.e. the distributional trend) encompasses crucial information for population viability (Beissinger and Westphal 1998). To forecast future distributional trends, we need to collect long-term time-series data (see Box 5.1) which is often costly and time-consuming. Efficient invasion management cannot afford such a wait-and-see approach. Invasion managers would be better equipped if it was possible to convert the spatial data of a species' distribution into knowledge about temporal trends. In invasion ecology, it would be useful to convert insights of local population dynamics into knowledge about spreading dynamics. This is possible as changes of population sizes across time and space are entangled. For instance, space-for-time substitution has been applied in studies on plant succession after disturbance (Cooper 1923; Crocker and Major 1955) and the impact of biological invasions on native ecosystems (Fukami and Wardle 2005). Both have been criticized because spatial distribution and temporal dynamics of species are often governed by different ecological processes (Tokarska-Guzik *et al.* 2008). However, there is potential correlation between changes in population size (or range size) across time and space. As one example, by comparing the atlases of British butterflies calibrated from different periods (1970–82 and 1995–99), Wilson and colleagues (2004) discovered a strong correlation between the temporal rate of range expansion and the spatial structure of only one snapshot of species distributions. Measuring the scaling pattern of species distribution provides a short-cut way of assessing and forecasting aspects of the population dynamics of species (Wilson *et al.* 2004).

Ecologists need not wait to see the trend of a focal species—current species distributions and the scaling pattern of occupancy per se provide important clues on likely future dynamics. This allows us to include a metric of population trend as an indicator for invasion status of a species. For instance, historic records of well-known invasive species are readily available, but are rare for endangered species. Management plans for invasive species should be assessed *a priori* to ensure a negative population trend for mitigating its potential impact on the recipient environment. Based on a combination of models from the fields of range dynamics, occupancy scaling, and spatial autocorrelation, Hui (2011) derived a model for forecasting the population trend solely from its current spatial distribution. Applying this model in invasion management could provide a swift risk assessment for invasive species. For example, most invasive Australian eucalypts have lower power law exponents of their native geographic ranges than non-invasive congeners (Fig. 5.10). Moreover, the percolation exponent for eucalypts declines towards the end of the introduction–naturalization–invasion continuum. This is in contrast to the results for Australian acacias (Hui *et al.* 2011); naturalized and invasive *Acacia* species having a lower exponent than eucalypts, suggesting that this space-for-time exponent could be indicative of the discrepancy in invasiveness between the two taxa (Hui *et al.* 2014). This provides a tool for discriminating between invasive and non-invasive introduced species.

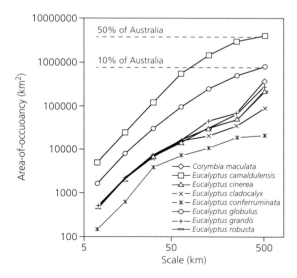

Fig. 5.10 Scaling patterns of the native range size, measured by the area of occupancy (AOO, km²), for eight invasive Australian eucalypts. From Hui *et al.* (2014). With permission from Springer.

Population dynamics and range dynamics are largely synchronized. When the population grows, the range expands. When species distributions are randomly distributed, the number of individuals in samples can be described by a Poisson distribution (Wright 1991). This could lead to the following relationship between range dynamics $\Delta p/\Delta t$ and population dynamics $\Delta\mu/\Delta t$ (Hui *et al.* 2012):

$$\frac{\Delta\mu}{\Delta t} = \frac{1}{1-p}\frac{\Delta p}{\Delta t} \tag{5.5}$$

Because the proportion of occupied sites p is less than one, it is obvious that the changes in abundance and occupancy for randomly distributed species are synchronized. Under reasonable restrictions, we can further conclude that, regardless of whether a species distribution is random or aggregated, the change in abundance is positively correlated (i.e. synchronized) with the change in occupancy. This means that spatial structure (e.g. aggregation, occupancy, and extent) of species distributions contains key information on the temporal trends of population dynamics and spread, and thus invasiveness. Moreover, Kuehn (2013) suggests that invasion monitoring at regional scales for assessing the range dynamics of invasive species could be simplified to only monitoring the temporal dynamics of one local invasive population. This suggests that for large-scale monitoring of biological invasions we could potentially reduce costs by focusing on selected sites for long-term monitoring. The range dynamics and the effectiveness of invasion management can be inferred from the temporal dynamics of their local populations in these selected sites (see Chapter 9). Space-for-time substitution provides us with flexible methods for risk assessment and invasion monitoring: spatial structure of geographical ranges could be used to infer invasiveness, and local temporal records can be used to infer spatial range dynamics.

5.7 Spatial autoregressive models

Any successful conservation effort requires an understanding of the factors that mediate range dynamics (i.e. expansion or retraction, Lockwood *et al.* 2005; Soberón and Peterson 2005). The spatial dynamics of invasive species are hardly ever at equilibrium. The presence and absence of a focal species at a particular site depends not only on the explanatory variables (e.g. environmental characteristics) of the site, but also the spatial autocorrelation of the dependent variables (i.e. the distribution of the focal species nearby). An empty site can be colonized by a species if the environmental characteristics fit with the species' niche requirement, or simply by the propagules from neighbouring populations (as in the case of source-sink dynamics). This process can be examined using the simultaneous autoregressive linear model (Kissling and Carl 2008):

$$y_i = \lambda\sum_{j=1}^{n} w_{ij}y_j + \sum_{k=1}^{s} \beta_k x_{ik} + \varepsilon_i \tag{5.6}$$

where the presence or absence of the species at site i, y_i ($=1$ or 0; can also be extended to represent population density), amongst the n sites surveyed can be explained by local propagule pressure, calculated by multiplying the possibility of a propagule from site j arriving and establishing at site i, w_{ij} (often take the form of a dispersal kernel), plus the environmental characteristics, depicted by s number of explanatory variables with the reading of explanatory variable k at site i, x_{ik}. Let $Y = \left(y_1, y_2, \ldots, y_n\right)^T$, $W = \left\{w_{ij}\right\}_{n\times n}$, $\beta = \left(\beta_1, \beta_2, \ldots, \beta_s\right)^T$, $X = \left\{x_{ik}\right\}_{n\times s}$, and $\varepsilon = \left(\varepsilon_1, \varepsilon_2, \ldots, \varepsilon_n\right)^T$, equation 5.6 can be written as $Y = \lambda WY + X\beta + \varepsilon$. The second part can also be set up as generalized additive models, called a simultaneous autoregressive additive model. Statistical approaches for fitting generalized linear and additive models can be used to fit the simultaneous autoregressive model. When cross-sectional data are available, we could explore the possible spatial dynamics using spatial autoregressive models. That is, y_i on the left-hand side is from observation at time $t+1$, whereas on the right-hand side, observation at time t. For example, Gallien and colleagues (2015) applied autoregressive linear models to explain the distribution of invasion plants in alpine communities.

When the spatial connectivity is ignored ($w_{ij} = 0$), this model essentially becomes a species distribution model (SDM; see Chapter 3) in the format of a generalized linear model, $y_i = \sum_{k=1}^{s} \beta_k x_{ik} + \varepsilon_i$, or an additive model (normally with a binomial error and a logit link function for presence absence data). Using SDMs to predict the range dynamics of an invasive species is problematic for two key reasons. Firstly, recent research has shown that niche shifts and rapid evolution are important in novel ranges. SDMs based on records from native ranges or other invaded areas therefore cannot capture the full context of the focal area. Secondly, the records in the invaded area are far from equilibrium (see discussion in Rouget *et al.* 2004), and thus cannot reflect the true relationship between the species and environmental characteristics in the invasive range. Hybrid models can overcome one fundamental limitation of SDMs (that species are not at equilibrium within their habitat), which makes them appropriate for modelling non-equilibrium invasion dynamics (Franklin 2010; Gallien *et al.* 2010). Consequently, hybrid models have been proposed to build the spatial autocorrelation component (i.e. colonization by neighbouring populations) on top of the suitability map created from SDMs (Roura-Pascual *et al.* 2009; Smolik *et al.* 2010). Such hybrid models are essentially set up as

$$y_{t+1,i} = \left(\lambda \sum_{j=1}^{n} w_{ij} y_{t,j} \right) \left(\sum_{k=1}^{s} \beta_k x_{ik} \right) + \varepsilon_i \qquad (5.7)$$

Clearly, the local propagule pressure from colonizers of adjacent populations is considered independently from the habitat suitability of sites. Similar multivariate statistics can be used for fitting the data. Such hybrid models perform much better than basic SDMs alone as demonstrated for the invasion of Argentine ants in southern Europe (Roura-Pascual *et al.* 2009) and the annual weed *Ambrosia artemisiifolia* in eastern Europe (Smolik *et al.* 2010).

In these two setups of the spatial autoregressive model, the dependent variable y_i is related to the habitat suitability of its site i, but it is not related to the suitability of its neighbouring sites. In studies on animal and economic behaviour, this is equivalent to the utility theory where consumers (the species) choose/prefer to colonize sites that are highly suitable. Another framework for depicting the decision making in consumer and animal behaviour is prospect theory (Kahneman and Tversky 1979), where an animal tends to improve the quality of the habitat in which it lives by moving to a better one. In the utility theory, the probability of an animal moving to a site with suitability 0.9 to a site of 0.5 is equal to the probability of moving from a 0.3 suitability site to a 0.5 site, as only the suitability of the destiny site decides the probability of movement. In prospect theory, the latter option is more likely as the animal could gain a 0.2 increment in suitability, whereas the former option reduces its current site suitability by 0.4. To implement this habitat selection and preference in the spatial autoregressive model, we need to add the site gradient to the equation. Considering only sites that are previously unoccupied by the species ($y_{t,i} = 0$) in the first format of the spatial autoregressive model, the status of the site at time $t + 1$ is

$$y_{t+1,i} = \lambda \sum_{j=1}^{n} w_{ij} y_{t,j} + \sum_{k=1}^{s} \beta_k w_{ij} \left(x_{ik} - x_{jk} \right)$$
$$\left(1 - y_{t,j} \right) \left| y_{t+1,i} - y_{t+1,j} \right| + \varepsilon_i \qquad (5.8)$$

where the environmental characteristics in site i is compared with the characteristics of neighbouring sites j that have the opposite status (i.e. if site i is occupied during the period, j remains empty; if site i remains empty, j becomes occupied); the term $\left(1-y_{t,j} \right)$ specifies that we only consider empty sites at time t (because typical data records exclude the option for occupied sites to be become empty again); $\left| y_{t+1,i} - y_{t+1,j} \right|$ specifies that we only consider the scenario that sites i and j are in opposite status at time $t + 1$. For all three spatial autoregressive models, we suggest first transforming variables according to the model form before running the standard multivariate statistics.

The extra-limital colonization of many indigenous species provides an ideal natural experiment for clarifying the triggers and drivers of rapid spread of species with a previously rather static distribution. Extra-limital species (i.e. range-expanding native species) can colonize either by filling novel niches created by environmental changes (Guo and Ricklefs 2010) or by accumulating local propagule pressure (LPP; i.e. the accumulation of individuals dispersed from viable surrounding populations determines the colonization success of an empty habitat; Lockwood *et al.* 2005). As an example for this last regression approach, Horák and colleagues (2013) examined both the build-up of local propagule pressure and environmental gradients that could potentially drive the sudden extra-limital expansion of the rose chafer beetle, *Oxythyrea funesta*, an important pollinator at adulthood and scavenger at the larval stage in central Europe. Besides its widespread Palearctic distribution, *O. funesta* has previously been considered rare and close to extinction in the Czech Republic. Local propagule pressure (LPP) plays a key role in structuring natural communities and is also considered a key factor of invasion success (Richardson and Pyšek 2006; Catford *et al.* 2009; Lawrence and Cordell 2010). It depicts the effect of both the number and rate of incoming individuals dispersed from reachable populations on the colonization success of an empty habitat (Lockwood *et al.* 2005; Groom

et al. 2006). For a focal grid cell, Horák and colleagues (2013) first calculated the number of presences (p_1) and absences (a_1) in immediately adjacent cells $(n_1 = p_1 + a_1 \leq 8)$, and then the number of presences (p_2) and absences (a_2) in the secondary neighbouring cells $(n_2 = p_2 + a_2 \leq 16)$. The LPP was then calculated as $LPP = w_1(p_1 / n_1) + w_2(p_2 / n_2)$ $(0 \leq LPP \leq 1)$, where w_1 and w_2 are weights. Specifically, they chose $w_1 = 2/3$ and $w_2 = 1/3$ to reflect the fact that propagules released closer to the focal cell contribute more to colonization than propagules released further away. Since LPP follows a binomial distribution, they performed a logit transformation (Logit (LPP + 0.01)) in the analysis. This definition evidently considers not only the effect of local propagule pressure but also the spatial autocorrelation of existing propagules on the colonization of the focal cell.

To account for both the habitat preference of *O. funesta* and the spatial autocorrelation of environmental variables, Horák and colleagues (2013) transformed all environmental variables as follows. Let x be the value of a specific variable in the focal grid cell, x_0 and x_1 be the mean values of the variable in empty and occupied adjacent cells, respectively; let x_3 and x_4 be the mean values of the variable in empty and occupied secondary neighbouring cells. If the focal grid cell has been occupied during the period (i.e. the dependent variable is 1), we can transform the variable $X = w_1(x - x_0) + w_2(x - x_3)$;

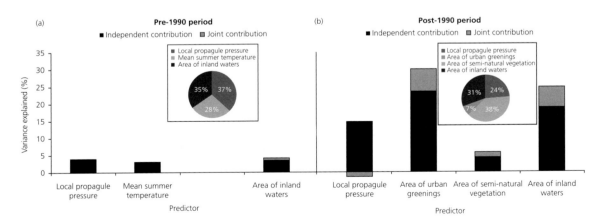

Fig. 5.11 Results of hierarchical partitioning showing the proportion of variance explained of selected predictors for the white-spotted rose beetle (*Oxythyrea funesta*) pre-1990 (a) and post-1990 (b). The pie diagrams are the results of the independent contributions of predictors as a proportion of total explained variance. From Horák *et al.* (2013). With permission from Springer.

if the focal cell remains empty (i.e. the dependent variable is 0), we have $X = w_1(x - x_1) + w_2(x - x_4)$. In this way, Horák and colleagues (2013) considered both the spatial autocorrelation of the variable and the habitat preference of the species ($X > 0$ indicates that the species prefers to colonize neighbouring cells with a higher level of the variable). They also standardized these transformed variables in the multivariate analyses.

They examined the changing role of LPP and other environmental variables in driving the colonization of *O. funesta* from 1981–90 to 1991–2000 (pre-1990), and from 1991–2000 to 2001–10 (post-1990). During the pre-1990 decade, the colonization of *O. funesta* was facilitated by LPP, mean summer temperature, and area of inland waters. During the post-1990 decade, the colonization was positively driven by LPP, the area of inland waters, and area of urban greening (Fig. 5.11). Evidently, this extralimital invader shifted to a fast-spreading phase after 1990, driven by the build-up of LPP and novel niches created by changing land use. The colonization *of O. funesta* was facilitated by high summer temperatures in the slow establishment phase pre-1990 but not in the fast-spreading phase post-1990. Instead, urban greening and inland waters became important for fast colonization by providing sufficient food supply for adults and larval development. The transition to the fast-spreading phase coincided with the collapse of the Iron Curtain in 1989, which drastically transformed the land use policy in Europe and influenced the regional spread dynamics of many invasive insect species (Roques *et al.* 2016). This serves to remind us that species distribution modelling, in the context of environmental changes, must recognize changing roles of ecological and environmental factors at different stages of biological invasion. Changes in the occupancy of a species not only offer the opportunity to elucidate the environmental drivers at play but also highlight the need to regularly revise the conservation or invasion status of species.

Taken together, the non-equilibrium dynamics of biological invasion provides an ideal model to deepen the discussion on the balance of nature, to reveal the density-dependent mechanisms especially when population densities are extremely low or high, and to understand how human disturbance

and a novel environment facing an introduced species could reshape and modify such regulating forces of its population dynamics. The density-dependent mechanisms and, perhaps more importantly, the mechanisms to release the population dynamics to be density-independent, could potentially be reflected by the spatial patterns and covariance between local populations. Management could first detect such varying mechanisms/processes at different phases of invasion using time series and autoregressive-type methods, and then target interceding key processes for efficient outcome. Here we emphasized the non-equilibrium dynamics of population densities and viability. Non-equilibrium dynamics can be multifaceted. In particular, diverse range dynamics of invasive species is another commonly documented non-equilibrium pattern (Chapter 2). Different mechanisms, from spatial selection to human-mediated dispersal, could drive boosted range expansion of many invasive species (Chapter 4). By escaping old and establishing new associations, invasive species experience non-equilibrium strengths of biotic interactions with resident species, posing dynamic impact on the recipient ecosystems (Chapter 6). Non-equilibrium invasion dynamics implies that invasion management should be flexible and employ different control strategies targeting spatially and temporally transient opportunities for optimal outcome (Chapter 9). This implies that a bigger picture of non-equilibrium invasion dynamics, in terms of temporal and spatial population variability, trait/niche adaptiveness, changing biotic interactions/impacts and recipient ecosystems, as well as flexible intervention strategies, should be taken into account when monitoring, assessing, and managing invasions.

References

Aanes, R., Sæther, B.E., and Øritsland, N.A. (2000) Fluctuations of an introduced population of Svalbard reindeer: the effects of density dependence and climatic variation. Ecography, 23, 437–43.

Allee, W.C. (1958) *The Social Life of Animals*. Revised edn. Boston, MA: Beacon Press.

Altieri, M.A. and Nicholis, C.I. (2003) Soil fertility management and insect pests: harmonizing soil and plant health in agroecosystems. Soil and Tillage Research, 72, 203–11.

Amarasekare, P. (1998) Allee effects in metapopulation dynamics. American Naturalist, 152, 298–302.

Anderson, J.T., Sparks, J.P., and Geber, M.A. (2010) Phenotypic plasticity despite source–sink population dynamics in a long-lived perennial plant. New Phytologist, 188, 856–67.

Andrewartha, H.G. (1957) The use of conceptual models in population ecology. Proceedings of the Cold Spring Harbor Symposia on Quantitative Biology, 22, 219–36.

Andrewartha, H.G. and Birch, L.C. (1954) *The Distribution and Abundance of Animals.* Chicago: University of Chicago Press.

Baranchikov, Y.N. (1989) Ecological basis of the evolution of host relationships in Eurasian gypsy moth populations. In: Wallner, W.E. and McManus, K.A. (eds) *Lymantriidae: A Comparison of Features of New and Old World Tussock Moths.* USDA General Technical Report NE-123. Pennsylvania, PA: United States Department of Agriculture Forest Service, pp. 319–38.

Beissinger, S.R. and Westphal, M.I. (1998) On the use of demographic models of population viability in endangered species management. Journal of Wildlife Management, 62, 821–41.

Berryman, A. and Turchin, P. (2001) Identifying the density-dependent structure underlying ecological time series. Oikos, 92, 265–70.

Betini, G.S., Griswold, C.K., and Norris, D.R. (2013) Carry-over effects, sequential density dependence and the dynamics of populations in a seasonal environment. Proceedings of the Royal Society B: Biological Sciences, 280, 20130110.

Bogdanowicz, S.M., Wallner, W.E., Bell, J., et al. (1993) Asian gypsy moths (Lepidoptera: Lymantriidae) in North America: evidence from molecular data. Annals of the Entomological Society of America, 86, 710–715.

Boivin, G., Fauvergue, X., and Wajnberg, E. (2004) Optimal patch residence time in egg parasitoids: innate versus learned estimate of patch quality. Oecologia, 138, 640–7.

Bossdorf, O., Auge, H., Lafuma, L., et al. (2005) Phenotypic and genetic differentiation between native and introduced plant populations. Oecologia, 144, 1–11.

Boukal, D.S. and Berec, L. (2002) Single-species models of the Allee effect: extinction boundaries, sex ratios and mate encounters. Journal of Theoretical Biology, 218, 375–94.

Boukal, D.S. and Berec, L. (2009) Modelling mate-finding Allee effects and populations dynamics, with applications in pest control. Population Ecology, 51, 445–58.

Box, G.E.P. and Jenkins, G.M. (1970) *Time Series Analysis: Forecasting and Control.* San Francisco, CA: Holen-Day.

Bradford, E. and Philip, J.R. (1970) Note on asocial populations dispersing in two dimensions. Journal of Theoretical Biology, 29, 27–33.

Breininger, D.R. and Carter, G.M. (2003) Territory quality transitions and source-sink dynamics in a Florida Scrub-Jay population. Ecological Applications, 13, 516–29.

Broennimann, O., Fitzpatrick, M.C., Pearman, P.B., et al. (2012) Measuring ecological niche overlap from occurrence and spatial environmental data. Global Ecology and Biogeography, 21, 481–97.

Broennimann, O., Mráz, P., Petitpierre, B., et al. (2014) Contrasting spatio-temporal climatic niche dynamics during the eastern and western invasions of spotted knapweed in North America. Journal of Biogeography, 41, 1126–36.

Broennimann, O., Treier, U.A., Müller-Schärer, H., et al. (2007) Evidence of climatic niche shift during biological invasion. Ecology Letters, 10, 701–9.

Brook, B.W. and Bradshaw, C.J. (2006) Strength of evidence for density dependence in abundance time series of 1198 species. Ecology, 87, 1445–51.

Chatfield, C. (1989) *The Analysis of Time Series.* New York, NY: Chapman and Hall.

Calabrese, J.M. and Fagan, W.F. (2004) Lost in time, and single: reproductive asynchrony and the Allee effect. American Naturalist, 164, 25–37.

Capiński, M. and Zastawniak, T. (2003) *Mathematics for Finance: An Introduction to Financial Engineering.* Berlin: Springer.

Cappuccino, N. (1995) Novel approaches to the study of population dynamics. In: Cappuccino, N. and Price, P.W. (eds) *Population Dynamics: New Approaches and Synthesis.* San Diego, CA: Academic Press, pp. 3–16.

Caswell, H., Nisbet, R.M., de Roos, A.M., et al. (1997) Structured-population models: many methods, a few basic concepts. In: Tuljaurkar, S. and Caswell, H. (eds) *Structured-Population Models in Marine, Terrestrial, and Freshwater Systems.* New York, NY: Chapman and Hall, pp. 3–17.

Catford, J.A., Jansson, R., and Nilsson, C. (2009) Reducing redundancy in invasion ecology by integrating hypotheses into a single theoretical framework. Diversity and Distributions, 15, 22–40.

Cattadori, I.M., Haydon, D.T., and Hudson, P.J. (2005) Parasites and climate synchronize red grouse populations. Nature, 433, 737–41.

Chen, L.L. and Hui, C. (2009) Habitat destruction and the extinction debt revisited: the Allee effect. Mathematical Biosciences, 221, 26–32.

Chesson, P.L. (1981) Models for spatially distributed populations: the effect of within-patch variability. Theoretical Population Biology, 19, 288–325.

Chesson, P.L. and Case, T.J. (1986) Overview: nonequilibrium community theories: chance, variability, and history. In: Diamond, J. and Case, T.J. (eds) *Community Ecology.* New York, NY: Harper and Row Publishers, pp. 229–39.

Connell, J.H. (1971) On the role of natural enemies in preventing competitive exclusion in some marine animals and in rain forest trees. In: Boer, P.J.D. and Gradwell, G.R. (eds) *Dynamics of Populations*. Wageningen : Center for Agriculture Publishing and Documentation, pp. 298–313.

Cooper, G. (2001) Must there be a balance of nature? Biology and Philosophy, 16, 481–506.

Cooper, W.S. (1923) The recent ecological history of Glacier Bay, Alaska: the present vegetation cycle. Ecology, 4, 223–46.

Crocker, R.L. and Major, J. (1955) Soil development in relation to vegetation and surface age at Glacier Bay, Alaska. Journal of Ecology, 43, 427–48.

Cumming, G.S., Buerkert, A., Hoffmann, E.M., *et al.* (2014) Implications of agricultural transitions and urbanization for ecosystem services. Nature, 515, 50–7.

DeAngelis, D.L. and Waterhouse, J.C. (1987) Equilibrium and nonequilibrium concepts in ecological models. Ecological Monographs, 57, 1–21.

Dellinger, A.S., Essl, F., Hojsgaard, D., *et al.* (2016) Niche dynamics of alien species do not differ among sexual and apomictic flowering plants. New Phytologist, 209, 1313–23.

Dennis, B. (1989) Allee effects: population growth, critical density and the chance of extinction. Natural Resource Modeling, 3, 481–538.

Dennis, B. (2002) Allee effects in stochastic populations. Oikos, 96, 389–401.

Deredec, A. and Courchamp, F. (2003) Extinction thresholds in host-parasite dynamics. Annales Zoologici Fennici, 40, 115–30.

Dias, P.C. (1996) Sources and sinks in population biology. Trends in Ecology and Evolution, 11, 326–30.

Drake, J.M. and Lodge, D.M. (2006) Allee effects, propagule pressure and the probability of establishment: risk analysis for biological invasions. Biological Invasions, 8, 365–75.

Drake, V.A. and Gatehouse, A.G. (1996) Population trajectories through space and time: a holistic approach to insect migration. In: Floyd, R.B., Sheppard, A.W., and De Barro, P.J. (eds) *Frontiers in Population Ecology*. Collingwood: CSIRO Publishing, pp. 399–408.

Early, R. and Sax, D.F. (2014) Climatic niche shifts between species' native and naturalized ranges raise concern for ecological forecasts during invasions and climate change. Global Ecology and Biogeography, 23, 1356–65.

Earn, D.J., Levin, S.A., and Rohani, P. (2000) Coherence and conservation. Science, 290, 1360–4.

Egerton, F.N. (1973) Changing concepts of the balance of nature. Quarterly Review of Biology, 48, 322–50.

Ehrlich, P.R. (1989) Attributes of invaders and the invading processes: vertebrates. In: Drake, J.A., Mooney, H.A., di

Castri, F., *et al.* (eds) *Biological Invasions: A Global Perspective.*. Chichester: Wiley, pp. 315–28.

Elith, J. and Leathwick, J.R. (2009) Species distribution models: ecological explanation and prediction across space and time. Annual Review of Ecology, Evolution, and Systematics, 40, 677–97.

Elkinton, J.S., Parry, D., and Boettner, G.H. (2006) Implicating an introduced generalist parasitoid in the invasive browntail moths enigmatic demise. Ecology, 87, 2664–72.

Elton, C.S. (1924) Periodic fluctuations in the numbers of animals: their causes and effects. Journal of Experimental Biology, 2, 119–63.

Elton, C.S. and Nicholson, M. (1942) The ten-year cycle in numbers of the lynx in Canada. Journal of Animal Ecology, 11, 215–44.

Fagan, W.F., Lewis, M.A., Neubert, M.G., *et al.* (2002) Invasion theory and biological control. Ecology Letters, 5, 1–10.

Farrow, R.A. and Daly, J.C. (1987) Long-range movements as an adaptive strategy in the genus *Heliothis* (Lepidoptera, Noctuidae): a review of its occurrence and detection in 4 pest species. Australian Journal of Zoology, 35, 1–24.

Fath, G. (1998) Propagation failure of traveling waves in a discrete bistable medium. Physica D, 116,176–90.

Fauvergue, X., Fleury, F., Lemaitre, C., and Allemand, R. (1999) Parasitoid mating structures when hosts are patchily distributed: field and laboratory experiments with Leptopilina boulardi and L. heterotoma . Oikos 86, 344–356.

Fauvergue, X., Malausa, J.C., Giuge, L., *et al.* (2007) Invading parasitoids suffer no Allee effect: a manipulative field experiment. Ecology, 88, 2392–403.

Feiner, Z.S., Aday, D.D., and Rice, J.A. (2012) Phenotypic shifts in white perch life history strategy across stages of invasion. Biological Invasions, 14, 2315–29.

Ferdy, J.B. and Molofsky, J. (2002) Allee effect, spatial structure and species coexistence. Journal of Theoretical Biology, 217, 413–24.

Figge, F. (2004) Bio-folio: applying portfolio theory to biodiversity. Biodiversity and Conservation, 13, 827–49.

Fitzpatrick, M.C., Weltzin, J.F., Sanders, N.J. *et al.* (2007) The biogeography of prediction error: why does the introduced range of the fire ant over-predict its native range? Global Ecology and Biogeography, 16, 24–33.

Fortuna, M.A., Gomez-Rodrıguez, C., and Bascompte, J. (2006) Spatial network structure and amphibian persistence in stochastic environments. Proceedings of the Royal Society B: Biological Sciences, 273, 1429–34.

Fowler, C.W. and Baker, J.D. (1991) A review of animal population dynamics at extremely reduced population levels. Report—International Whaling Commission, 41, 545–54.

Fowler, M.S. and Ruxton, G.D. (2002) Population dynamic consequences of Allee effects. Journal of Theoretical Biology, 215, 39–46.

Fox, G.A. and Kendall, B.E. (2002) Demographic stochasticity and the variance reduction effect. Ecology, 83, 1928–34.

Fox, J.W., Vasseur, D.A., Hausch, S., et al. (2011) Phase locking, the Moran effect and distance decay of synchrony: experimental tests in a model system. Ecology Letters, 14, 163–8.

Franklin, J. (2010) Moving beyond static species distribution models in support of conservation biogeography. Diversity and Distributions, 16, 321–30.

Frederiksen, M., Lebreton, J.-D., and Bregnballe, T. (2001) The interplay between culling and density-dependence in the great cormorant: a modelling approach. Journal of Applied Ecology, 38, 617–27.

Fukami, T. and Wardle, D.A. (2005) Long-term ecological dynamics: reciprocal insights from natural and anthropogenic gradients. Proceedings of the Royal Society B: Biological Sciences, 272, 2105–15.

Gallien L., Mazel F., Lavergne S., et al. (2015) Contrasting the effects of environment, dispersal and biotic interactions to explain the distribution of invasive plants in alpine communities. Biological Invasions, 17, 1407–23.

Gallien, L., Münkemüller, T., Albert, C.H., et al. (2010) Predicting potential distributions of invasive species: where to go from here? Diversity and Distributions, 16, 331–42.

Gallien, L., Saladin, B., Boucher, F.C., et al. (2016) Does the legacy of historical biogeography shape current invasiveness in pines? New Phytologist, 209, 1096–1105.

Gascoigne, J. and Lipcius, R.N. (2004) Allee effects in marine systems. Marine Ecology Progress Series, 269, 49–59.

Gatehouse, J.A. (2002) Plant resistance towards insect herbivores: a dynamic interaction. New Phytologist, 156, 145–69.

Glennon, K.L., Ritchie, M.E., and Segraves, K.A. (2014) Evidence for shared broad-scale climatic niches of diploid and polyploid plants. Ecology Letters, 17, 574–82.

Godfray, H.C.J. (1990) The causes and consequences of constrained sex allocation in haplodiploid animals. Journal of Evolutionary Biology, 3, 3–17.

Godfray, H.C.J. (1994) Parasitoids. Behavioral and Evolutionary Ecology. Princeton, NJ: Princeton University Press.

Greenman, J.V. and Benton, T.G. (2001) The impact of stochasticity on the behaviour of nonlinear population models: synchrony and the Moran effect. Oikos, 93, 343–51.

Grenfell, B.T., Wilson, K., Finkenstädt, B.F., et al. (1998) Noise and determinism in synchronized sheep dynamics. Nature, 394, 674–77.

Grevstad, F.S. (1999) Factors influencing the chance of population establishment: implications for release strategies in biocontrol. Ecological Applications, 9, 1439–47.

Groom, M.J. (1998) Allee effects limit population viability of an annual plant. American Naturalist, 151, 487–96.

Groom, M.J., Meffe, G.K., and Carroll, C.R. (2006) Principles of Conservation Biology. Sunderland, MA: Sinauer Associates.

Gruntfest, Y., Arditi, R., and Dombronsky, Y. (1997) A fragmented population in a varying environment. Journal of Theoretical Biology, 185, 539–47.

Guisan, A., Petitpierre, B., Broennimann, O., et al. (2014) Unifying niche shift studies: insights from biological invasions. Trends in Ecology and Evolution, 29, 260–9.

Guo, Q. and Ricklefs, R.E. (2010) Domestic exotics and the perception of invasibility. Diversity and Distributions, 16, 1034–9.

Gurney, W. and Nisbet, R.M. (1998) Ecological Dynamics. Oxford: Oxford University Press.

Gyllenberg, M., Hemminki, J., and Tammaru, T. (1999) Allee effects can both conserve and create spatial heterogeneity in population densities. Theoretical Population Biology, 56, 231–42.

Hadjiavgousti, D. and Ichtiaroglou, S. (2004) Existence of stable localized structures in population dynamics through the Allee effect. Chaos Solitons Fractals, 21, 119–31.

Hanski, I. (1998) Metapopulation dynamics. Nature, 396, 41–9.

Hanski, I. (1999) Habitat connectivity, habitat continuity, and metapopulations in dynamic landscapes. Oikos, 87, 209–19.

Hanski, I. and Gilpin, M. (1991) Metapopulation dynamics: brief history and conceptual domain. Biological Journal of the Linnean Society, 42, 3–16.

Hassell, M.P. (1975) Density-dependence in single-species populations. Journal of Animal Ecology, 44, 283–95.

Hassell, M.P., Latto, J., and May, R.M. (1989) Seeing the wood for the trees: detecting density dependence from existing life-table studies. Journal of Animal Ecology, 58, 883–92.

Hastings, A. (1996) Models of spatial spread: a synthesis. Biological Conservation, 78, 143–8.

Hedrick, P.W., Gadau, J., and Page, R.E. (2006) Genetic sex determination and extinction. Trends in Ecology and Evolution, 21, 55–7.

Hengeveld, R. and Walter, G.H. (1999) The two coexisting ecological paradigms. Acta Biotheoretica, 47, 141–70.

Henter, H.J. (2003) Inbreeding depression and haplodiploidy: experimental measures in a parasitoid and comparisons across diploid and haplodiploid insect taxa. Evolution, 57, 1793–1803.

Hermisson, J and Pennings, P.S. (2005) Soft sweeps molecular population genetics of adaptation from standing genetic variation. Genetics, 169, 2335–52.

Hixon, M.A. and Johnson, D.W. (2009) Density dependence and independence. eLS. DOI: 10.1002/9780470015902.a0021219.

Hopkins, G.A., Forrest, B.M., Jiang, W., et al. (2011) Successful eradication of a non-indigenous marine bivalve from a subtidal soft-sediment environment. Journal of Applied Ecology, 48, 424–31.

Hopper, K.R. and Roush, R.T. (1993) Mate finding, dispersal, number released, and the success of biological control introductions. Ecological Entomology, 18, 321–31.

Horak, J., Hui, C., Roura-Pascual, N., et al. (2013) Changing roles of propagule, climate, and land use during extralimital colonization of a rose chafer beetle. Naturwissenschaften, 100, 327–36.

Hugueny, B. (2006) Spatial synchrony in population fluctuations: extending the Moran theorem to cope with spatially heterogeneous dynamics. Oikos, 115, 3–14.

Hui, C. (2011) Forecasting population trend from the scaling pattern of occupancy. Ecological Modelling, 222, 442–6

Hui, C., Boonzaaier, C., and Boyero, L. (2012) Estimating changes in species abundance from occupancy and aggregation. Basic and Applied Ecology, 13, 169–77.

Hui, C. and Li, Z. (2003) Dynamical complexity and metapopulation persistence. Ecological Modelling, 164, 201–9.

Hui, C. and Li, Z. (2004) Distribution patterns of metapopulation determined by Allee effects. Population Ecology, 46, 55–63.

Hui, C., Richardson, D.M., Robertson, M.P., et al. (2011) Macroecology meets invasion ecology: linking the native distributions of Australian acacias to invasiveness. Diversity and Distributions, 17, 872–83.

Hui, C., Richardson, D.M., Visser, V., et al. (2014) Macroecology meets invasion ecology: performance of Australian acacias and eucalypts around the world revealed by features of their native ranges. Biological Invasions, 16, 565–76.

Hulme, P.E. (2006) Beyond control: wider implications for the management of biological invasions. Journal of Applied Ecology, 43, 835–47.

Hutchinson, G.E. (1948) Circular causal systems in ecology. Annals of the New York Academy of Sciences, 50, 221–46.

Janzen, D.H. (1970) Herbivores and the number of tree species in tropical forests. American Naturalist, 104, 501–28.

Johnson, D. M., Liebhold, A.M., Tobin, P.C., et al. (2006) Allee effects and pulsed invasion by the gypsy moth. Nature, 444, 361–3.

Johnson, D.J., Beaulieu, W.T., Bever, J.D., et al. (2012) Conspecific negative density dependence and forest diversity. Science, 336, 904–7.

Jonsen, I.D., Bourchier, R.S., and Roland, J. (2007) Influence of dispersal, stochasticity, and Allee effect on the persistence of weed biocontrol introductions. Ecological Modelling, 203, 521–6.

Jonsson, M., Wratten, S.D., Landis, D.A., et al. (2010) Habitat manipulation to mitigate the impacts of invasive arthropod pests. Biological Invasions, 12, 2933–45.

Kahneman, D. and Tversky, A. (1979) Prospect theory: an analysis of decision under risk. Econometrica, 47, 263–91.

Kakimoto, T., Fujisaki, K., and Miyatake, T. (2003) Egg laying preference, larval dispersion, and cannibalism in Helicoverpa armigera (Lepidoptera: Noctuidae). Annals of the Entomological Society of America, 96, 793–8.

Keeling, M.J., Jiggins, F.M., and Read, J.M. (2003) The invasion and coexistence of competing Wolbachia strains. Heredity, 91, 382–8.

Keena, M. A., Côté, M-J. Grinberg, P.S., et al. (2008) World distribution of female flight and genetic variation in Lymantria dispar (Lepidoptera: Lymantriidae). Environmental Entomology, 37, 636–49.

Keitt, T.H., Lewis, M.A., and Holt, R.D. (2001) Allee effects, invasion pinning, and species' borders. American Naturalist, 157, 203–16.

Keller, S.R. and Taylor, D.R. (2008) History, chance and adaptation during biological invasion: separating stochastic phenotypic evolution from response to selection. Ecology Letters, 11, 852–66.

Kendall, B.E., Bjørnstad, O.N., Bascompte, J., et al. (2000) Dispersal, environmental correlation, and spatial synchrony in population dynamics. American Naturalist, 155, 628–36.

Keszthelyi, S., Nowinszky, L., and Puskás, J. (2013) The growing abundance of Helicoverpa armigera in Hungary and its areal shift estimation. Central European Journal of Biology, 8, 756–64.

Kissling, W.D. and Carl, G. (2008) Spatial autocorrelation and the selection of simultaneous autoregressive models. Global Ecology and Biogeography, 17, 59–71.

Kogan, M. (1998) Integrated pest management: historical perspectives and contemporary developments. Annual Review of Entomology, 43, 243–70.

Koshio, C. (1996) Pre-ovipositional behaviour of female gypsy moth, Lymantria dispar L. (Lepidoptera: Lymantriidae). Applied Entomology and Zoology, 31, 1–10.

Kot, M., Lewis, M.A., and van den Driessche, P. (1996) Dispersal data and the spread of invading organisms. Ecology, 77, 2027–42.

Kuehn, C. (2013) Warning signs for wave speed transitions of noisy Fisher–KPP invasion fronts. Theoretical Ecology, 6, 295–308.

Lande, R. (1998) Demographic stochasticity and Allee effect on a scale with isotropic noise. Oikos, 83, 353–8.

Lande, R., Engen, S., and Sæther, B.E. (1999) Spatial scale of population synchrony: environmental correlation versus dispersal and density regulation. American Naturalist, 154, 271–81.

Lande, R. and Orzack, S.H. (1988) Extinction dynamics of age-structured populations in a fluctuating environment. Proceedings of the National Academy of Sciences of the USA, 85, 7418–21.

Landis, D.A., Wratten, S.D., and Gurr, G.M. (2000) Habitat management to conserve natural enemies of arthropod pests in agriculture. Annual Review of Entomology, 45, 175–201.

Lauzeral, C., Leprieur, F., Beauchard, O., et al. (2011) Identifying climatic niche shifts using coarse-grained occurrence data: a test with non-native freshwater fish. Global Ecology and Biogeography, 20, 407–14.

Lavergne, S., Evans, M.E., Burfield, I.J., et al. (2013) Are species' responses to global change predicted by past niche evolution? Philosophical Transactions of the Royal Society B: Biological Sciences, 368, 20120091.

Lawrence, D.J. and Cordell, J.R. (2010) Relative contributions of domestic and foreign sourced ballast water to propagule pressure in Puget Sound, Washington, USA. Biological Conservation, 143, 700–9

Levin, D.A., Kelley, C.D., and Sarkar, S. (2009) Enhancement of Allee effects in plants due to self-incompatibility alleles. Journal of Ecology, 97, 518–27.

Levins, R. (1969) Some demographic and genetic consequences of environmental heterogeneity for biological control. Bulletin of the Entomological Society of America, 15, 237–40.

Lewis, M.A. and Kareiva, P. (1993) Allee dynamics and the spread of invading organisms. Theoretical Population Biology, 43, 141–58.

Li, Z.Z., Gao, M., Hui, C., et al. (2005) Impact of predator pursuit and prey evasion on synchrony and spatial patterns in metapopulation. Ecological Modelling, 185, 245–54.

Liebhold, A. and Bascompte, J. (2003) The Allee effect, stochastic dynamics and the eradication of alien species. Ecology Letters, 6, 133–40.

Liebhold, A.M., Kamata, N., and Turcáni, M. (2008) Inference of adult female dispersal from the distribution of gypsy moth egg masses in a Japanese city. Agricultural and Forest Entomology, 10, 69–73.

Liebhold, A.M. and Tobin, P.C. (2008) Population ecology of insect invasions and their management. Annual Review of Entomology, 53, 387–408.

Liermann, M. and Hilborn, R. (1997) Depensation in fish stocks: a hierarchic Bayesian meta-analysis. Canadian Journal of Fisheries and Aquatic Sciences, 54, 1976–84.

Liu, Z.G., Zhang, F.P., and Hui, C. (2016) Density-dependent dispersal complicates spatial synchrony in tri-trophic food chains. Population Ecology, 58, 223–30.

Lockwood, J.L., Cassey, P., and Blackburn, T. (2005) The role of propagule pressure in explaining species invasions. Trends in Ecology and Evolution, 20, 223–8.

Lodge, D.M. (1993) Biological invasions: lessons for ecology. Trends in Ecology and Evolution, 8, 133–7.

Lorenzen, K. and Enberg, K. (2002) Density-dependent growth as a key mechanism in the regulation of fish populations: evidence from among-population comparisons. Proceedings of the Royal Society B: Biological Sciences, 26, 49–54.

Lucek, K., Sivasundar, A., and Seehausen, O. (2014) Disentangling the role of phenotypic plasticity and genetic divergence in contemporary ecotype formation during a biological invasion. Evolution, 68, 2619–32.

Markowitz, H. (1952) Portfolio selection. Journal of Finance, 7, 77–91.

May, R.M. (1974) Biological populations with nonoverlapping generations: stable points, stable cycles, and chaos. Science, 186, 645–7.

May, R.M. (1975) Biological populations obeying difference equations: stable points, stable cycles, and chaos. Journal of Theoretical Biology, 51, 511–24.

May, R.M. (1981) Patterns in multi-species communities. In: May, R.M. (ed.) Theoretical Ecology: Principles and Applications. Sunderland, MA: Sinauer Associates, pp. 197–227.

Medley, K.A. (2010) Niche shifts during the global invasion of the Asian tiger mosquito, Aedes albopictus Skuse (Culicidae), revealed by reciprocal distribution models. Global Ecology and Biogeography, 19, 122–33.

Memmott, J., Craze, P.G., Harman, H.M., et al. (2005) The effect of propagule size on the invasion of an alien insect. Journal of Animal Ecology, 74, 50–62.

Metz, J.A.J. and Diekmann, O. (1986) The Dynamics of Physiologically Structured Populations. New York, NY: Springer-Verlag.

Mikkola, K. (1971) The migratory habit of Lymantria dispar (Lepidoptera: Lymantriidae) adults of continental Eurasia in the light of a flight to Finland. Acta Entomologica Fennica, 28, 107–20.

Moore, J.W., McClure, M., Rogers, L.A., et al. (2010) Synchronization and portfolio performance of threatened salmon. Conservation Letters, 3, 340–8.

Moran, P.A.P. (1953) The statistical analysis of the Canadian lynx cycle. Australian Journal of Zoology, 1, 291–8.

Morris, W.F. and Doak, D.F. (2002) *Quantitative Conservation Biology; Theory and Practice in Conservation Biology.* Sunderland, MA: Sinauer Associates.

Murdoch, W.W. (1994) Population regulation in theory and practice. Ecology, 75, 271–87.

Myers, R.A., Barrowman, N.J., Hutchings, J.A., *et al.* (1995) Population-dynamics of exploited fish stocks at low population-levels. Science, 269, 1106–8.

Nelson, E., Mendoza, G., Regetz, J., *et al.* (2009) Modeling multiple ecosystem services, biodiversity conservation, commodity production, and tradeoffs at landscape scales. Frontiers in Ecology and the Environment, 7, 4–11.

Nicholson, A.J. (1957) The self-adjustment of populations to change. Proceeding of Cold Spring Harbor Symposia on Quantitative Biology, 22, 153–73.

Nicoll, M.A.C., Jones, C., and Norris, K. (2003) Declining survival rates in a reintroduced population of the Mauritius kestrel: evidence for non-linear density dependence and environmental stochasticity. Journal of Animal Ecology, 72, 917–26.

Novak, S.J. and Mack, R.N. (2005) Genetic bottlenecks in alien plant species. Influence of mating systems and introduction dynamics. In: Sax, D.F., Stachowicz, J.J., and Gaines, S.D. (eds) *Species Invasions: Insights into Ecology, Evolution and Biogeography.* Sunderland, MA: Sinauer Associates, pp 201–28.

Odum, H.T. and Allee, W.C. (1954) A note on the stable point of populations showing both intraspecific cooperation and disoperation. Ecology, 35, 95–7.

Ouyang, F., Hui, C., Ge, S., *et al.* (2014) Weakening density dependence from climate change and agricultural intensification triggers pest outbreaks: a 37-year observation of cotton bollworms. Ecology and Evolution, 4, 3362–74.

Peterson, A.T. (2011) Ecological niche conservatism: a time-structured review of evidence. Journal of Biogeography, 38, 817–27

Petitpierre, B., Kueffer, C., Broennimann, O., *et al.* (2012) Climatic niche shifts are rare among terrestrial plant invaders. Science, 335, 1344–8.

Philip, J.R. (1957) Sociality and sparse populations. Ecology, 38, 107–111.

Pianka, E.R. (1974) Niche overlap and diffuse competition. Proceedings of the National Academy of Sciences of the USA, 71, 2141–5.

Ponomarev, V.I. (1994) Population-genetic characteristics of outbreaks of mass propagation of gypsy moth (*Lymantria dispar* L.). Russian Journal of Ecology, 25, 380–5.

Prentis, P.J., Wilson, J.R.U., Dormontt, E.E., *et al.* (2008) Adaptive evolution in invasive species. Trends in Plant Science, 13, 288–94.

Price, P.W. (1980) *Evolutionary Biology of Parasites.* Princeton, NJ: Princeton University Press.

Pulliam, H.R. (1988) Sources, sinks, and population regulation. American Naturalist, 132, 652–61.

Quintero, I. and Wiens, J.J. (2013) Rates of projected climate change dramatically exceed past rates of climatic niche evolution among vertebrate species. Ecology Letters, 16, 1095–1103.

Ranta, E., Kaitala, V., Lindstrom, J., *et al.* H. (1995) Synchrony in population dynamics. Proceedings of the Royal Society B: Biological Sciences, 262, 113–18.

Ranta, E., Veijo, K., and Lindströom, J. (1999) Spatially autocorrelated disturbances and patterns in population synchrony. Proceedings of the Royal Society B: Biological Sciences, 266, 1851–6.

Richardson, D.M. and Pyšek, P. (2006) Plant invasions: merging the concepts of species invasiveness and community invasibility. Progress in Physical Geography, 30, 409–31.

Richardson, D.M., Hellmann, J.J., McLachlan, J.S., *et al.* (2009) Multidimensional evaluation of managed relocation. Proceedings of the National Academy of Sciences of the USA, 106, 9721–4.

Ripa, J. (2000) Analysing the Moran effect and dispersal: their significance and interaction in synchronous population dynamics. Oikos, 89, 175–87.

Ripa, J. and Ranta, E. (2007) Biological filtering of correlated environments: towards a generalised Moran theorem. Oikos, 116, 783–92.

Robinet, C., Lance, D.R., Thorpe, K.W., *et al.* (2008) Dispersion in time and space affect mating success and Allee effects in invading gypsy moth populations. Journal of Animal Ecology, 77, 966–73.

Robinet, C., Liebhold, A.M., and Gray, D. (2007) Variation in developmental time affects mating success and Allee effects. Oikos, 116, 1227–37.

Robinet, C. and Liebhold, A.M. (2009) Dispersal polymorphism in an invasive forest pest affects its ability to establish. Ecological Applications, 19, 1935–43.

Rödder, D. and Lötters, S. (2009) Niche shift versus niche conservatism? Climatic characteristics of the native and invasive ranges of the Mediterranean house gecko (*Hemidactylus turcicus*). Global Ecology and Biogeography, 18, 674–87.

Rodger, J.G., van Kleunen, M., and Johnson, S.D. (2013) Pollinators, mates and Allee effects: the importance of self-pollination for fecundity in an invasive lily. Functional Ecology, 27, 1023–33.

Rohde, K. (2006) *Nonequilibrium Ecology.* Cambridge: Cambridge University Press.

Roques, A., Auger-Rozenberg, M.-A., Blackburn, T.M., *et al.* (2016) Temporal and interspecific variation in rates of spread for insect species invading Europe during the last 200 years. Biological Invasions, 18, 907–20.

Rouget, M., Richardson, D.M., Nel, J.L., *et al.* (2004) Mapping the potential spread of major plant invaders in South Africa using climatic suitability. Diversity and Distributions, 10, 475–84.

Roura-Pascual, N., Bas, J.M., Thuiller, W., *et al.* (2009) From introduction to equilibrium: reconstructing the invasive pathways of the Argentine ant in a Mediterranean region. Global Change Biology, 15, 2101–15.

Royama, T., MacKinnon, W.E., Kettela, E.G., *et al.* (2005) Analysis of spruce budworm outbreak cycles in New Brunswick, Canada, since 1952. Ecology, 86, 1212–24.

Russell, T.L., Lwetoijera, D.W., Knols, B.G.J., *et al.* (2011) Linking individual phenotype to density-dependent population growth: the influence of body size on the population dynamics of malaria vectors. Proceedings of the Royal Society B: Biological Sciences, 278, 3142–51.

Salamin, N., Wüest, R.O., Lavergne, S., *et al.* (2010) Assessing rapid evolution in a changing environment. Trends in Ecology and Evolution, 25, 692–8.

Schaefer, P.W., Weseloh, R.M., Sun, X., *et al.* (1984) Gypsy moth, *Lymantria* (= *Ocneria*) *dispar* (L.) (Lepidoptera: Lymantriidae), in the People's Republic of China. Environmental Entomology, 13, 1535–41.

Scheuring, I. (1999) Allee effect increases the dynamical stability of populations. Journal of Theoretical Biology, 199, 407–14.

Schindler, D.E., Armstrong, J.B., and Reed, T.E. (2015) The portfolio concept in ecology and evolution. Frontiers in Ecology and the Environment, 13, 257–63.

Schindler, D.E., Hilborn, R., Chasco, B., *et al.* (2010) Population diversity and the portfolio effect in an exploited species. Nature, 465, 609–12.

Schöpf, R.J. and Sih, A. (2004) Dispersal behavior, boldness, and the link to invasiveness: a comparison of four Gambusia species. Biological Invasions, 6, 379–91.

Schreiber, S.J. (2004) On Allee effects in structured populations. Proceedings of the American Mathematical Society, 132, 3047–53.

Scott, K.D., Wilkinson, K.S., Lawrence, N., *et al.* (2005) Gene-flow between populations of cotton bollworm *Helicoverpa armigera* (Lepidoptera: Noctuidae) is highly variable between years. Bulletin of Entomological Research, 95, 381–92.

Sharov, A.A., Liebhold, A.M., and Ravlin, F.W. (1995) Prediction of gypsy moth (Lepidoptera: Lymantriidae) mating success from pheromone trap counts. Environmental Entomology, 24, 1239–44.

Shaw, A.K., Jalasvuori, M., and Kokko, H. (2014) Population-level consequences of risky dispersal. Oikos, 123, 1003–13.

Shea, K. and Possingham, H.P. (2000) Optimal release strategies for biological control agents: an application of stochastic dynamic programming to population management. Journal of Applied Ecology, 37, 77–86.

Sibly, R.M., Barker, D., Denham, M.C., *et al.* (2005) On the regulation of populations of mammals, birds, fish, and insects. Science, 309, 607–10.

Simberloff, D. (2014) The 'balance of nature'—evolution of a panchreston. PLoS Biology, 10, e1001963.

Smolik, M.G., Dullinger, S., Essl, F., *et al.* (2010) Integrating species distribution models and interacting particle systems to predict the spread of an invasive alien plant. Journal of Biogeography, 37, 411–22.

Soberón, A. and Peterson, A.T. (2005) Interpretation of models of fundamental ecological niches and species' distributional areas. Biodiversity Informatics, 2, 1–10.

Soboleva, T.K., Shorten, P.R., Pleasants, A.B., *et al.* (2003) Qualitative theory of the spread of a new gene into a resident population. Ecological Modelling, 163, 33–44.

Solberg, E.J., Saether, B.E., Strand, O., *et al.* (1999) Dynamics of a harvested moose population in a variable environment. Journal of Animal Ecology, 68, 186–204.

Solow, A.R. and Steele, J.H. (1990) On sample size, statistical power, and the detection of density dependence. Journal of Animal Ecology, 59, 1073–6.

South, A.B. and Kenward, R.E. (2001) Mate finding, dispersal distances and population growth in invading species: a spatially explicit model. Oikos, 95, 53–8.

Stearns, S.C. (1976) Life-history tactics: a review of the ideas. Quarterly Review of Biology, 51, 3–47.

Stephens, P.A., Sutherland, W.J., and Freckleton, R.P. (1999) What is the Allee effect? Oikos, 87, 185–90.

Strong, D.R. (1986) Density-vague population change. Trends in Ecology and Evolution, 1, 39–42.

Strubbe, D., Broennimann, O., Chiron, F., *et al.* (2013) Niche conservatism in non-native birds in Europe: niche unfilling rather than niche expansion. Global Ecology and Biogeography, 22, 962–70.

Suckling, D.M. and Brockerhoff, E.G. (2010) Invasion biology, ecology, and management of the light brown apple moth (Tortricidae). Annual Review of Entomology, 55, 285–306.

Sutcliffe, O.L., Thomas, C.D., and Moss, D. (1996) Spatial synchrony and asynchrony in butterfly population dynamics. Journal of Animal Ecology, 65, 85–95.

Taylor, C.M. and Hastings, A. (2005) Allee effects in biological invasions. Ecology Letters, 8, 895–908.

Tentelier, C., Desouhant, E., and Fauvergue, X. (2006) Habitat assessment by parasitoids: mechanisms for patch use behavior. Behavioral Ecology, 17, 515–21.

Theoharides, K.A. and Dukes, J.S. (2007) Plant invasion across space and time: factors affecting nonindigenous species success during four stages of invasion. New Phytologist, 176, 256–73.

Thiel, A. and Hoffmeister, T.S. (2004) Knowing your habitat: linking patch encounter rate and patch exploitation in parasitoids. Behavioral Ecology, 15, 419–25.

Thuiller, W., Münkemüller, T., Schiffers, K.H., *et al.* (2014) Does probability of occurrence relate to population dynamics? Ecography, 37, 1155–66.

Tingley, R., Vallinoto, M., Sequeira, F., *et al.* (2014) Realized niche shift during a global biological invasion. Proceedings of the National Academy of Sciences of the USA, 111, 10233–8.

Tobin, P.C., Berec, L., and Liebhold, A.M. (2011) Exploiting Allee effects for managing biological invasions. Ecology Letters, 14, 615–24.

Tobin, P.C., Robinet, C., Johnson, D.M., *et al.* (2009) The role of Allee effects in gypsy moth, *Lymantria dispar* (L.), invasions. Population Ecology, 51, 373–84.

Tobin, P.C., Whitmire, S.L., Johnson, D.M., *et al.* (2007) Invasion speed is affected by geographic variation in the strength of Allee effects. Ecology Letters, 10, 36–43.

Tokarska-Guzik, B., Brock, J.H., Brundu, G., *et al.* (2008) *Plant Invasions: Human Perception, Ecological Impacts and Management*. Leiden: Backhuys Publishers.

Tschinkel, W.R. (1981) Larval dispersal and cannibalism in a natural population of *Zophobas atratus* (Coleoptera: Tenebrionidae). Animal Behaviour, 29, 990–6.

Turchin, P. (1995) Population regulation: old arguments and a new synthesis. In: Cappuccino, N. and Price, P.W. (eds) *Population Dynamics: New Approaches and Synthesis*. San Diego, CA: Academic Press, pp. 19–40.

Turchin, P. (2003) *Complex Population Dynamics: A Theoretical/ Empirical Synthesis*. Princeton, NJ: Princeton University Press.

Vasseur, D.A. and Fox, J.W. (2009) Phase-locking and environmental fluctuations generate synchrony in a predator–prey community. Nature, 460, 1007–10.

Volterra, V. (1926) Fluctuations in the abundance of a species considered mathematically. Nature, 118, 558–60.

Wang, M.H. and Kot, M. (2001) Speeds of invasion in a model with strong or weak Allee effects. Mathematical Biosciences, 171, 83–97.

Wang, M.H., Kot, M., and Neubert, M.G. (2002) Integrodifference equations, Allee effects, and invasions. Journal of Mathematical Biology, 44, 150–68.

Weisberg, M. and Reisman, K. (2008) The robust Volterra principle. Philosophy of Science, 75, 106–31.

Whitmire, S.L. and Tobin, P.C. (2006) Persistence of invading gypsy moth populations in the United States. Oecologia, 147, 230–7.

Williamson, M., Dehnen-Schmutz, K., Kühn, I., *et al.* (2009) The distribution of range sizes of native and alien plants in four European countries and the effects of residence time. Diversity and Distributions, 15, 158–66.

Wilson, R.J., Thomas, C.D., Fox, R., *et al.* (2004) Spatial patterns in species distributions reveal biodiversity change. Nature, 432, 393–6.

Wright, D.H. (1991) Correlations between incidence and abundance are expected by chance. Journal of Biogeography, 18, 463–6.

Yamanaka, T. (2007) Mating disruption or mass trapping? Numerical simulation analysis of a control strategy for lepidopteran pests. Population Ecology, 49, 75–86.

Zayed, A. and Packer, L. (2005) Complementary sex determination substantially increases extinction proneness of haplodiploid populations. Proceedings of the National Academy of Sciences of the USA, 102, 10742–6.

PART II

Impact

CHAPTER 6

Biotic interactions

6.1 Pairwise interactions

6.1.1 A simple model

Alien species exert influence in recipient ecosystems partly by interacting directly with resident species, but biotic interactions are also indirect in many ways via other species or resources. Alien species, therefore, enter a complex web of interactions that involve many resident species—both native and alien taxa (Elton 1927; Paine 1966). Adding a species to such networks has repercussions that may ripple across ecosystems via pathways of multiple direct and mediated interactions. The sign (type) and strength of such biotic interactions depend on the system context and on the traits of the species involved. The biotic interactions we consider here are mainly pairwise interactions between two species, especially those between alien and native species. The type of pairwise interaction can be defined on the basis of the direction of the fitness change for the species participating in each interaction. For example, we can depict the population dynamics of a native species (N) and an alien invader (A) using a simplified Lotka–Volterra model:

$$\dot{N} = N(1 - N + \alpha_{12}A)$$
$$\dot{A} = A(1 - A + \alpha_{21}N) \qquad (6.1)$$

where the intrinsic population growth rate, carrying capacity, and negative density dependence are assumed, for simplicity, to be one; coefficient α_{12} depicts the interaction strength per capita of the alien species on the native species, and α_{21} indicates the interaction strength per capita of the native species on the alien species.

If both interaction strengths per capita are greater than zero, a mutualistic interaction (not necessarily a compulsory mutualism) exists. Also, to keep the dynamics within the feasible domain in the Lotka–Volterra model, the two interaction strengths per capita need to be kept to less than one (i.e. interspecific interaction needs to be weaker than negative density dependence). When both are less than zero, competition exists. When the two have opposite signs, we have an antagonistic interaction (e.g. resource consumer, predation, or parasitism, etc.). We could further define the cases of commensalism and amensalism when one of the strength coefficients is set to zero. The summarized dynamical behaviour of this simple model in Figure 6.1 assumes that the population size of the native species initially resides at its equilibrium whereas the alien species initially has a low population size, far below its potential equilibrium.

The simple conceptualization of interactions between native and alien species in Figure 6.1 provides a starting point for discussing the types of biotic interaction that are relevant when considering invasion dynamics. In their novel environments, all introduced species encounter other species—a motley collection of potential enemies, mutualists, and competitors with which the alien species shares little or no evolutionary history (Richardson et al. 2000; Levine et al. 2004; Parker and Gilbert 2004); likewise, resident species in recipient ecosystems encounter many new interactions. The performance of introduced species and the impacts they cause are fundamentally shaped by such interactions, many of which involve novel combinations of symbionts (Richardson et al. 2000). Mitchell and colleagues (2006) collated a list of 19

Invasion Dynamics. Cang Hui and David M. Richardson, Oxford University Press (2017).
© Cang Hui and David M. Richardson. DOI 10.1093/acprof:oso/9780198745334.001.0001

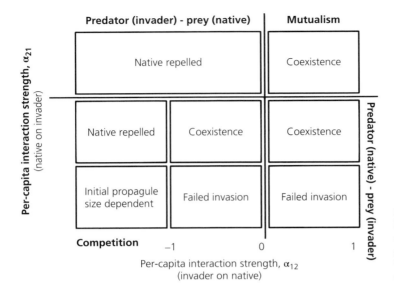

Fig. 6.1 The division of biotic interactions based on interaction strengths per capita. The qualitative results from the biotic interaction are given in the blocks; the zero lines separating the four biotic interactions indicate commensalism and amensalism.

hypotheses that capture the role of biotic interactions in determining the performance of alien plants (Table 6.1). Catford and colleagues (2009) compiled 29 hypotheses related to plant invasion, 21 of which involve biotic interactions (19 for direct interactions and two for mediated interactions). Many of these hypotheses also involve animals and all can be expanded to explore the success and invasion dynamics of alien animals. In what follows, we divide these hypotheses of pairwise direct biotic interactions into those related to antagonism, mutualism, and competition.

6.1.2 Antagonism

The loss of interactions with natural enemies in the novel range of a species following the human-mediated introduction of a species to a novel environment allows introduced populations of the species to achieve higher levels of fitness (Colautti et al. 2014). This often results in the alien species manifesting greater abundance or density than in their native ranges. The *enemy release* hypothesis (ERH) posits that alien species have a better chance of establishing and becoming dominant when released from the negative effects of natural enemies that cause high mortality rates and reduced productivity in their native range (Keane

and Crawley 2002). Many studies have shown lower pathogen and parasite loads and fewer herbivores on alien plants (Colautti et al. 2004; Hinz and Schwarzlaender 2004; Torchin and Mitchell 2004), although this pattern is not universal (e.g. Van der Putten et al. 2005). Although alien plants suffer less from antagonistic interactions than conspecific native species (DeWalt et al. 2004; Reinhart and Callaway 2004; Torchin and Mitchell 2004; Wolfe et al. 2005), within-community comparisons reveal exceptions (Colautti et al. 2004; Agrawal et al. 2005; Carpenter and Cappuccino 2005; Parker and Hay 2005).

Benefits due to release from the negative effects of natural enemies could be temporary and could weaken as the residence time of the alien species increases—as native species begin responding to the potential new hosts or prey provided by new arrivals. Many plant species can buffer damage from natural enemies through compensatory growth and reproduction, which complicates the assessment of the real effect of enemy release (Müller-Schärer et al. 2004). Also, trade-offs between resource acquisition and defence imply that fast-growing plant species adapted to high resource availability have a weaker defence against enemies and thus benefit more from enemy release in the novel range (Blumenthal et al. 2009). Moreover, to allow for

Table 6.1 Hypotheses that invoke enemies (E), mutualists (M), competitors (C), or favourable abiotic conditions (A) to explain the demographic performance of alien plant species. Each hypothesis either predicts a net increase (+) or decrease (−) of invader success. From Mitchell et al. (2006).

Hypothesis	E	M	C	A
Enemy release	+			
Evolution of increased competitive ability	+			
Biotic resistance from enemies	−			
New associations	−			
Mutualism facilitation		+		
Invasional meltdown		+		+
Biotic resistance from competitors			−	
Empty niche			+	
Novel weapons			+	
Mutualism−enemy release	+	+		
Competition−enemy release	+		+	
Mutualism−competition		+	+	
Enemy of my enemy	+		+	
Mutualism disruption		+	+	
Subsidized mutualism		+		
Resource−enemy release	+			+
Fluctuating resources			+	+
Opportunity windows	+	+	+	+
Naturalization	+	+	+	+

performance comparisons with congeners, species pairs selected for studies testing the ERH are subject to a lower effect size, which could be another reason for the fairly weak support for the ERH (Agrawal et al. 2005; Heger and Jeschke 2014).

The evolutionary consequences of enemy release could be an increase in resource allocation to growth and reproduction at the expense of allocation to defence. This has been termed the *evolution of increased competitive ability* (EICA) hypothesis (Blossey and Nötzgold 1995; Maron et al. 2004; Bossdorf et al. 2005). As generalist enemies are often widespread and abundant and are likely to appear in both native and novel ranges, the enemies that alien species escape in novel ranges are often specialists (Andow and Imura 1994; Hinz and Schwarzlaender 2004; Knevel et al. 2004; Torchin

and Mitchell 2004; Van der Putten et al. 2005). Consequently, alien species are primarily released from specialist enemies, and thus only reduce their production of defence compounds that specifically target specialist enemies but not generalist enemies (Müller-Schärer et al. 2004; Joshi and Vrieling 2005; Stastny et al. 2005).

The biotic resistance of the receiving ecosystem to invasion starts when resident enemies begin targeting introduced species, forming novel antagonistic interactions (Elton 1958; Mack 1996; Maron and Vilà 2001; Levine et al. 2004). For instance, recent evidence shows that native herbivores prefer non-native plants, thereby contributing to biotic resistance (Parker and Hay 2005). Such biotic resistance relies on the adaptability and flexibility of the foraging/host-selection behaviour of resident enemies and often emerges only after a delay during which time resident enemies learn the potential benefits of exploiting the introduced species (Mitchell and Power 2003; Zhang and Hui 2014). This means that introduced species as potential hosts and resources are engaging in apparent competition with native host/resource species by sharing resident natural enemies (Connell 1990; Mitchell and Power 2006; van Ruijven et al. 2003).

Due to the lack of shared co-evolutionary history with the alien species, those resident enemies in the novel range which start to target aliens could have a much stronger effect than those co-evolved enemies from the native range; this is known as the *new associations* hypothesis (NAH; Hokkanen and Pimentel 1989). This is because co-evolution often leads to better defence in hosts and reduced impact from natural enemies (e.g. stronger resistance and reduced virulence) (Jarosz and Davelos 1995; Parker and Gilbert 2004; Carroll et al. 2005; Parker and Hay 2005). Because generalist enemies are often promiscuous and thus likely to exploit potential new hosts or resources, alien species could evolve greater defences against the new associations with generalist enemies (Joshi and Vrieling 2005).

Together, these hypotheses imply that introduced species often experience temporary enemy release, followed by greater biotic resistance from new associations of resident enemies, which the introduced species could combat via the EICA. This suggests a temporal succession in the antagonistic interactions experienced by introduced species, although currently available evidence for such succession or enemy accumulation is not convincing (Andow and Imura 1994; Torchin and Mitchell 2004; Carpenter and Cappuccino 2005).

An important application of using antagonistic interactions is the suppression of many pests and alien species through biological control. The reduction of defence against specialist enemies (EICA) and the increase of defence against generalist enemies (NAH) suggest a component shift in the defence of alien species, giving opportunities to specialist biocontrol agents. Biological control typically follows four approaches (Frank *et al.* 2011; Pitcairn 2011). Firstly, augmentation (augmentative biocontrol) identifies enemies that already target the problematic species but with low efficiency and aims to enhance populations of the enemy to improve their regulation of the targeted species. Secondly, classical biocontrol involves the introduction of a co-evolved and preferably highly specific natural enemy from the native range of an alien species. For example, the parasitic wasp *Gonatocerus ashmeadi*, introduced as a biological control agent, drastically reduced the abundance of the invasive glassy-winged sharpshooter, *Homalodisca vitripennis*, in Tahiti with clear effects becoming evident in only seven months. Thirdly, conservation biocontrol, which emerged as a subdiscipline in the 1960s, aims to enhance resources that are important to key natural enemies, either native or introduced, thereby achieving greater reduction of population of the targeted species. For example, nectar source plants were provided to crabronid wasps, *Larra bicolor*, so that the wasps could suppress invasive *Scapteriscus* mole crickets in Florida. Fourthly, biopesticides are pathogens typically released in industrial quantities for suppressing pests, and normally involve the use of living viruses, bacteria, entomopathogenic nematodes, or fungi.

6.1.3 Mutualism

Many species require mutualists to complete parts of their life cycle or to increase their reproductive output. When such species are moved to new areas, progression along the introduction–naturalization–invasion continuum is contingent on the replacement of the lost services of co-evolved mutualists; this is known as the *mutualist facilitation* hypothesis (Richardson *et al.* 2000; Fig. 6.2). Introduced species often suffer from the loss of mutualistic partners, and the identity of potential new mutualists in the novel range can affect invasion success through the strength of species–specific interactions per capita with the aliens (Schemske and Horvitz 1984; Bever 2002; Klironomos 2003). Consequently, introduced species often experience high performance when the richness of mutualists is high (van der Heijden *et al.* 1998; Jonsson *et al.* 2001), when specialized mutualists are widespread, occurring in both native and invaded ranges (e.g. the convolvulus hawkmoth, *Agrius convolvuli*, occurs naturally in both tropical Asia, where the Asian Formosa lily, *Lilium formosanum*, is native, and in southern Africa, where the same *Lilium* species is invasive), or when specialized mutualists can be replaced by closely related similar taxa (e.g. South American hawkmoths, native pollinators of the moon vine, *Ipomoea alba*, were substituted by the local hawkmoths in the invasive range in South Africa; Johnson 2015). Because the benefit or payoff from a mutualistic partner is the product of interaction frequency and interaction strength per capita (Morris 2003; Vázquez *et al.* 2005), both abundance and the per-event pay off can affect the invasion success of introduced species.

Introduced plants often rely on both above-ground and below-ground mutualists that create synergies through their interactions with the plants. For example, arbuscular mycorrhizas in some plants can promote an increase in floral display and/or the quantity and quality of nectar, thereby directly increasing the flower visitation rates and effective pollination (Wolfe *et al.* 2005). Mutualism can further facilitate subsequent invasions through mediated interactions and the alteration of environments (Bruno *et al.* 2005) through, for example, increased soil nitrogen levels or fire frequency, a phenomenon termed invasional meltdown (Simberloff and Von Holle 1999; see Traveset and Richardson 2014 for other examples).

The origin of nitrogen-fixing symbionts utilized by invasive *Acacia* species in their new ranges has been intensively studied. Many successful invasive legumes, including acacias, form new mutualisms

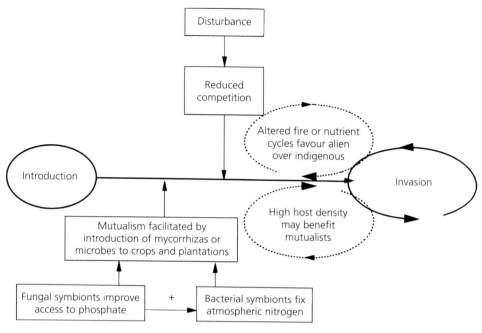

Fig. 6.2 Invasion of woody plants is often facilitated by co-introduced mutualists such as mycorrhizal fungi and microbes. Once alien species have invaded a community, they reinforce habitat change by further altering fire regimes, by accelerating nutrient cycling, or altering soil acidity and nutrient availability. Invasions of plants with nutrient-acquisition mutualisms are likely to benefit their mutualists, thus facilitating further invasions. From Richardson *et al.* (2000). Reproduced with permission from John Wiley & Sons.

with bacteria found in the introduced environment (the *host jumping* hypothesis), but symbionts can also be co-introduced with host plants, either intentionally as inoculants for species used in forestry or for other purposes, or accidentally by hitchhiking on introduced plant material and/or in soil (the *co-introduction* hypothesis).

Australian *Acacia* species have become a model group for elucidating the importance of mutualistic partners in plant invasions (Richardson *et al.* 2011). All studied Australian acacias are nodulated predominantly by strains of *Bradyrhizobium*, many of which were probably co-introduced with acacias to new geographical regions. Birnbaum and colleagues (2016) examined nitrogen-fixing bacterial communities associated with four *Acacia* spp. and a sister taxon, *Paraserianthes lophantha*, in their native ranges and outside the native ranges in Australia. They found similar bacterial communities in native and alien ranges (Fig. 6.3) and concluded that it is unlikely that the invasive success of species in areas

far removed from their native range are constrained by the absence of suitable symbionts.

A study of the taxonomic identity and diversity of root nodule and rhizospheric microbial symbionts associated with *A. pycnantha* in its native (Australian) and invasive (South African) range revealed that nodules in the invasive range contained both symbionts of South African origin and those that were likely to have been co-introduced with the plants from Australia (Ndlovu *et al.* 2013). The promiscuous nature of these and other woody legumes and the co-introductions of native symbionts are both important factors in explaining invasion success in this group (Rodríguez-Echeverría *et al.* 2011).

For three *Acacia* species adapted for seed dispersal by ants (myrmecochory), Wandrag and colleagues (2013) quantified seed removal probabilities associated with dispersal and predation in both native (Australian) and introduced (New Zealand) ranges. The probability of removed seed

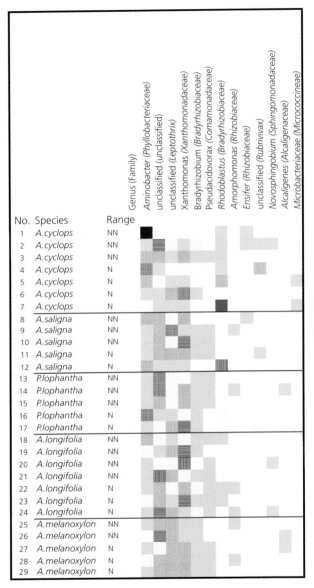

Fig. 6.3 Heat map showing the relative abundance of nitrogen-fixing bacterial communities detected in 29 nodule samples from five species of legume trees (*Acacia cyclops, A. longifolia, A. melanoxylon, A. saligna,* and *Paraserianthes lophantha*) in Australia. Operational taxonomic units that had ≤ 1 per cent abundance score were excluded. N—native range, NN—non-native range. From Birnbaum *et al.* (2016). Reproduced with permission from John Wiley & Sons.

being dispersed by invertebrates was similar in New Zealand and Australia, even though New Zealand has a relatively depauperate ant fauna. Therefore, differences in the invasion success of the three *Acacia* species in New Zealand could not be explained by differences in the strength of myrmecochorous interactions, suggesting that seed dispersers are unlikely to limit the spread of *Acacia* species in New Zealand. Overall, because many plant mutualists are host generalists, in most cases the absence of mutualists seldom limits the success of plant invasion (Richardson *et al.* 2000).

6.1.4 Competition

Competition between resident species and introduced species is frequently considered to be the chief mechanism for biotic resistance to invasion (although generalist herbivores or predators can also pose strong biotic resistance to invasion through antagonistic interactions). The accumulated negative effects of competition on invader performance are likely to increase with higher richness of resident species (Levine *et al.* 2004). The positive correlations between native and introduced species richness at large spatial scales (Stohlgren *et al.* 2003; Richardson *et al.* 2005) could be due to the confounding factor of spatial heterogeneity, which is positively associated with both native and invader richness (Davies *et al.* 2005). At local scales, the scale relevant to biotic interactions and population viability, high richness of resident species could exhaust resources and reduce the chance of invasion (Fargione and Tilman 2005). However, the validity of this *biotic resistance* hypothesis is currently disputed (Jeschke 2014).

The *novel weapons* hypothesis posits that some plant species become stronger competitors in novel environments than in their native ranges because resident competitors in the novel range lack the ability to tolerate allelopathic compounds from the invasive plants (Callaway and Ridenhour 2004). Moreover, as a form of invasional meltdown, the translocation of many co-evolved allelopathy-tolerant species (presumably from the same source community) to novel communities gives these introduced species a competitive advantage over the allelopathy-intolerant resident species (e.g. Callaway and Aschehoug 2000; Bais *et al.* 2003; Callaway and Ridenour 2004; Vivanco *et al.* 2004, but see Blair *et al.* 2005). The novel weapons hypothesis has received fairly strong support in empirical studies, although only relatively few species have been assessed; investigations of this hypothesis for a broader range of taxa are needed (Jeschke *et al.* 2012).

Alien species could encounter empty niches that are not exploited by resident species which will increase the chance of invasion success; this is known as the *empty niche* hypothesis (Elton 1958). The realized niche of a species in its native range is constrained by many competitors. When introduced to a novel environment, a species can benefit from the release from co-evolved competitors. Invaders can thus expand their realized range and access resources that cannot be exploited by native

species (Hierro *et al.* 2005; Traveset *et al.* 2015). The existence of empty niches in a community depends on the level of saturation in species packing, a character of evolutionary stability which is related to the trait composition and interaction strength of resident species (Hui *et al.* 2016).

The effect of competition on the establishment of introduced species was conceptualized by MacArthur and Levins (1967) in their *limiting similarity theory*. After two consumer species are established along a resource spectrum, each with a different optimal niche position, a third consumer species with an intermediate optimal niche position is then introduced to examine whether or not it can establish and coexist with the two resident species. The limiting similarity theory posits that the establishment of the third species depends critically on the level of niche similarity between the two established species (e.g. measured as the distance between their optimal niche positions). A minimum threshold for the level of niche similarity between the two established species exists, below which the third species cannot invade and will be competitively excluded (MacArthur and Levins 1967). The theoretical significance of limiting similarity theory and the principle of competitive exclusion has led to the advocacy for trait-based restoration and invasion management (e.g. Funk *et al.* 2008).

Abrams and Rueffler (2009) revisited the limiting similarity theory and found that when mortality is low, the combined resource consumption by two established species becomes high, leaving no chance for invasion by subsequently introduced species. High mortality will lead to an available niche for species that are introduced later to invade (Fig. 6.4A). Their results are consistent with the classic view that large niche separation creates an empty opportunity niche for the invasion of a species with an intermediate niche optimum, while small niche separation will preclude invasion (Fig. 6.4B).

Importantly, Abrams and Rueffler (2009) revealed that coexistence is most likely when the niche optimum of the latter introduced species is closer to one of the niche optima of the two established species (Fig. 6.4). In other words, although whether a species can invade depends on many factors (e.g. niche separation between established species and mortality from disturbance and poor habitat

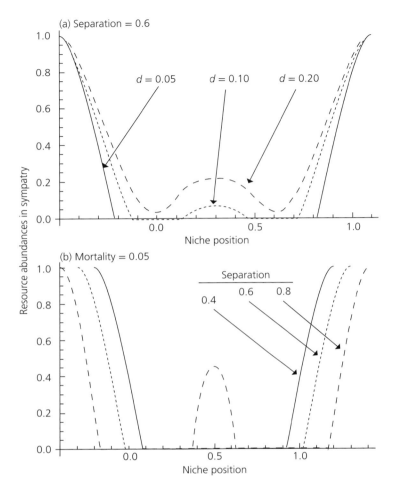

Fig. 6.4 The equilibrium of resource abundances when the two established consumer species are at equilibrium in sympatry. The two consumers are located symmetrically on either side of the niche position at the middle of the x axis (0.3 in Panel A and 0.5 in Panel B). In Panel A, only the mortality (d) of 0.05 produces a range for exclusion with a span > 1 on the resource axis. In Panel B, only a separation of 0.6 produces this result. From Abrams and Rueffler (2009). Reproduced with permission from John Wiley & Sons.

quality), an introduced species with a similar niche to a resident species might experience weak biotic resistance (Hui *et al.* 2016). A simple explanation for this phenomenon is that if a resident species can thrive in a community, an introduced species with a similar niche will be well equipped to flourish there as well. An introduced species, with its niche sitting between the niches of two resident species, faces an uncertain outcome: either there is an empty niche to allow the invasion, or no niches are available for the invasion; we call this the *niche resemblance* hypothesis (Hui *et al.* 2016). The application of the limiting similarity theory in assessing potential invasion performance needs to be revised to incorporate the situation of resembling niches between introduced and resident species (Price and Pärtel 2013).

6.2 Mediated interactions

No interaction is isolated. Not only will an alien species be affected by direct interactions, but perhaps more frequently by indirect interactions mediated by other species, native or alien, that are embedded in a network of interactions. Two species competing for a common abiotic resource is traditionally considered exploitative competition of direct interaction. However, host- or prey-mediated competition between two parasitic species or between two predators can strongly affect the performance of introduced parasites (Dunn *et al.* 2012). Parasites infecting the same individual can interact both through exploitative resource competition and through trait-mediated indirect effects via the host's immune response (Lello *et al.* 2004; Graham 2008), leading to changes in host physiology, parasite transmission, and virulence evolution, forming systemic-acquired resistance (Stout *et al.* 2006).

The most frequently studied indirect interaction is apparent competition mediated through a shared predator or mutualist. For instance, Recart and colleagues (2013) reported high fruit set for the native orchid *Bletia patula* with a low number of weevils, *Stethobaris polita*, when the alien orchid *Spathoglottis plicata* is absent, but an opposite pattern when the alien orchid is present. This is because the presence of alien orchids boosted the weevil population which then suppressed fruit set production of the native orchid. Consequently, the weevil population becomes a mediator of the strong apparent competition between the two orchid species. Brown and colleagues (2002) investigated the apparent competition between the native *Lythrum alatum* and the invasive *L. salicaria* for pollinators in five experimental treatments (Fig. 6.5): three treatments with pure natives (8/0, 16/0, and 24/0 plots) and two mixed treatments (8/8: 8 native plots plus 8 invasive plots; 8/16: 8 native plots plus 16 invasive plots). *Lythrum salicaria* is a self-incompatible tristylous plant with a showy floral display; it is native to Eurasia and is a major invader of wetlands and riverbanks in North America. Its congener *L. alatum* is a widespread distylous plant in North America. When the alien species was present, pollinator visitation and subsequent seed set of the native congener was reduced, and the effect size increased over time (Fig. 6.5). Reductions in both pollen quantity from competing for pollinators and pollen quality from heterospecific pollen transfer may have reduced *L. alatum* seed set.

Noonburg and Byers (2005) explored a more complicated case of apparent competition between alien and native species that was mediated by both shared resources and predators. They derived the risk of native extinction due to apparent competition from a successful invader as

$$\omega = 1 - \frac{ad\varepsilon_1}{ebr\varepsilon_2} \tag{6.2}$$

where *a* stands for the attack rate of alien species on the resource; *d* stands for the death rate of the predator in the absence of prey; ε_1 and ε_2 denote

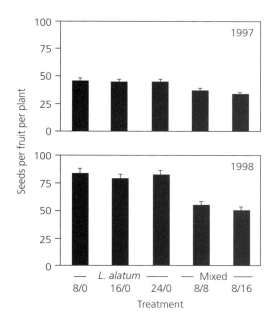

Fig. 6.5 Number of seeds per fruit per plant for the native winged loosestrife (*Lythrum alatum*) in 1997 and 1998. Treatment numbers on the *x* axis refer to the number of *L. alatum* plants and the number of alien purple loosestrife (*L. salicaria*) plants in each plot. From Brown *et al.* (2002). Reproduced with permission from John Wiley & Sons.

conversion efficiencies of the resource for the alien and native species, respectively; *e* denotes the conversion efficiency of the predator on the native species; *b* denotes the attack rate of the predator on the alien species; *r* denotes the intrinsic rate of increase of the resource. This formula divides these vital rates into two groups: increasing those in the numerator will reduce the negative effect of the alien species on the native species, while increasing those in the denominator will enhance the performance of the alien through apparent competition.

Mediated interactions often complicate the task of concluding whether biological invasions are a driver or a 'passenger' of ecosystem change, where the invasion is attributable to other driving forces such as habitat modification through land abandonment (Plieninger *et al.* 2014). A latent factor could drive the performance of both native and alien species, resulting in a strong correlation between invasion and the change of native assemblages

(Didham *et al.* 2005). If biological invasions are the driver of change (Fig. 6.6a), there are strong biotic interactions between alien species and resident species (indicated by double-headed arrows), and alien dominance is driven by competitive exclusion of native species. Removal of dominant alien species should result in native species recovery. If biological invasion only occurs as a result of other forces, say habitat disturbance (Fig. 6.6b), then biotic interactions between alien and native species are weak, and alien invaders can achieve dominance by 'filling the spaces' after disturbance. If biological invasion is both a driver and a passenger (Fig. 6.6c), a positive feedback loop could form between habitat disturbance and species invasion, as is the case in the 'invasional meltdown' of disturbed ecosystems

(Simberloff and von Holle 1999). After removing the alien species, some native species (e.g. species A and B) can recover, but not others (e.g. species C) (Didham *et al.* 2005). In all three cases, contemporary landscapes are dominated 90:10 by alien to native species, and it is impossible to distinguish which mechanism is operating simply through observation alone. Manipulative experiments involving the removal of alien species are needed to identify and separate causes and effects.

Some experimental data support the passenger model as the underlying cause of alien dominance. In their experiment of treatments to remove the dominant alien grass species *Dactylis glomerata* and *Poa pratensis* in deep-soil plots, MacDougall and Turkington (2005) found that the responses

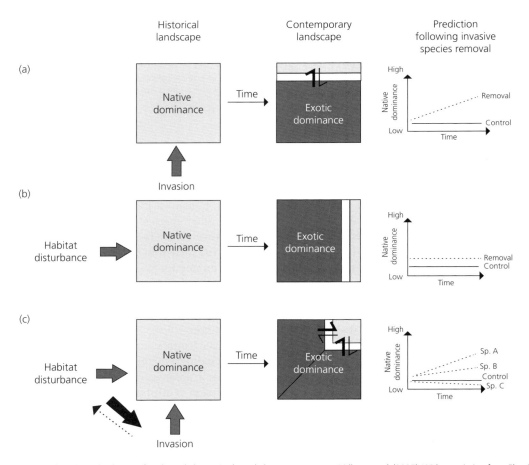

Fig. 6.6 Models describing the drivers of ecological change in degraded ecosystems. From Didham *et al.* (2005). With permission from Elsevier.

to the alien removal treatment were not the same for all native species. However, they did note that the combined suppressive and facilitative effects of habitat disturbance and invasion are substantial and that some native species did benefit from the removal of alien species. Such results are consistent with a growing body of work that shows that biological invasions cannot be explained solely by the competitive ability of the invader but must be understood within a broader context of environmental change, evolutionary adjustment, and life-history trade-offs (e.g. Callaway and Aschehoug 2000; Seabloom *et al.* 2003; Corbin and D'Antonio 2004).

6.3 Interaction strength

To determine the impact of invasion, we need to know more than the type of biotic interactions and whether the involved species are generalists or specialists. We also need to know the strength of biotic interactions so that the dynamics of multispecies systems can be investigated and the consequence of invasion predicted. However, measuring interaction strength is complicated because of the large number of species involved, the potential long-term feedbacks, multiple pathways of effects between species pairs, and possible non-linearities in interaction-strength functions (Wootton and Emmerson 2005).

Interaction strength is traditionally defined based on the consequences of removing species (Paine 1980; Morin *et al.* 1988; Menge *et al.* 1994), as in the case of defining network robustness as the amount of secondary extinctions from the removal of certain species (Dunne *et al.* 2002). We could borrow this concept and define the interaction strength between aliens and natives as the consequence of adding the alien species. However, such a definition has obvious shortcomings (Wootton and Emmerson 2005). The consequences of adding or removing species are highly context-dependent. As indirect effects and feedbacks could be part of the net effect of an alien species on native counterparts, estimates of the strength of interaction among species are likely to be strongly dependent on species composition, and therefore highly variable even for the same interaction (Gaertner *et al.* 2014).

At least 15 theoretical and empirical metrics of interaction strength have been used in food-web literature (Berlow *et al.* 2004). These include interaction matrix, community (Jacobian) matrix, inverse interaction matrix, non-linear functional response, relative prey preference, maximum consumption rate, biomass flux, change in population variability, link density, secondary extinctions, absolute prey response, Paine's index, log response ratio, statistical correlation, and frequency of consumption. Arguably the most basic and theoretically sound index is the per-capita interaction strength as measured by the community Jacobian matrix (Paine 1992; Laska and Wootton 1998; Parker *et al.* 1999). Assuming that the dynamics of a native species (N) and an alien species (A) are described by two differential equations:

$$\dot{N} = F(N, A)$$
$$\dot{A} = G(N, A) \tag{6.3}$$

we could define interaction strength as the partial derivatives of function F or G with respect to the other species' population density; that is, $\partial F/\partial A$ and $\partial G/\partial N$ at population equilibria of the two species define the impact of species A on N, and N on A, respectively. The two impacts can be of similar or different signs, and of different sizes. This index measures how one species in a system at equilibrium responds to the change of the other species as the result of a small perturbation, and is related to the concept of system resilience (linear stability) when assessed using the Jacobian matrix of the system (May 1973; Bender *et al.* 1984; Schoener 1993). The per-capita interaction strengths can then be simply defined as $\partial(F/N)/\partial A$ and $\partial(G/A)/\partial N$. Yodzis (1988) proposed using the elements of the inverted matrix of per-capita interaction strengths to describe the net effects of an individual of one species on another through both direct and indirect pathways.

These measures depict the system at equilibrium under small perturbations, which is different from the concept of robustness where an entire focal species is removed from the system (Dunne *et al.* 2002; Dambacher *et al.* 2003). Therefore, interaction strength can also be depicted by the removal method, by comparing changes in the abundance of a species at equilibrium before and after the removal of a species from the system. Interaction strength can also be measured as the functional response of energy or biomass flow (Hall *et al.* 2000), or the magnitudes of path coefficients in structural equation models (Wootton 1994; Gough and Grace 1999). An issue that remains is to define interaction strength for alien species. This is a challenge, as these species are normally yet to reach their equilibria in recipient environments; this means that the definition and measurement based on partial derivatives at the potential equilibrium of an invasive species might not be feasible.

To be able to model the essence of multispecies dynamics, the simplest non-linear form of functions *F* and *G* are second-order equations, such as the Lotka–Volterra model. This is equivalent to assuming that the interaction strength per capita is a constant. Whether it should be treated as a constant has been the subject of much debate (e.g. Abrams 2001; Sarnelle 2003). Species may modify the interaction per capita between individuals of other species, either by changing the traits of the interacting species ('trait-mediated indirect interactions') (Werner 1992; Abrams *et al.* 1996) or by changing the environmental context of an interaction (Crowder and Cooper 1982; Wootton 1992). Several approaches have been adopted in response to non-linear interactions, such as the maximum interaction strength (Ruesink 1998; Berlow *et al.* 1999), or parameters that describe the non-linear functions of interaction strengths (Abrams 2001). Another important aspect of non-linear, time-dependent interaction strength involves adaptive behaviour or learning in novel environments. Studies on experience-driven behaviour strategies suggest that the cognitive ability to remember past experience and comparing payoffs from different experiences in novel environments are essential to enable organisms to adjust their diets quickly to achieve payoffs as predicted from optimality theory (Zhang and Hui 2014). Such cognitive abilities and the potential species compositions in the novel environment will together determine how fast an alien species can adapt to the novel environment and the interaction strength with each potential partner.

To assess interaction strength accurately, field experiments are most appropriate (Bender *et al.* 1984; Paine 1992; Berlow *et al.* 1999), although full implementation poses substantial logistical challenges. Based on field experiments, Raffaelli and Hall (1996) found that interactions with high per-capita effects did not necessarily translate into large overall impacts on the system because of predator rarity. Short-term laboratory experiments may be used to provide estimates of interaction strength (Vandermeer 1969; Abrams 2001). Laboratory experiments have the advantage that species interactions can be examined in isolation from other species. Also, treatments and conditions can usually be more easily replicated, which allows for a more precise estimate of interaction strength between species pairs. Direct measures of biomass flow or consumption rate can be used to indicate the interaction strength between resource and consumer species. For instance, functional response experiments provide the means to compare impacts of different predatory species (Fig. 6.7) and to quantify their interaction strength on prey species (Dick *et al.* 2013, 2014; Paterson *et al.* 2015). However, designs of laboratory experiments often take species out of their natural context, which sometimes make it unclear whether estimates of interaction strength derived from these studies, as well as inferred competitive outcomes between natives and aliens from such estimates of interaction strength, can be realistically transferred to field conditions (Skelly 2002; Taylor *et al.* 2002). Moreover, effect size or effect size ratio from meta-analysis could also be used as indicators of relative interaction strengths (Gaertner *et al.* 2009, 2014; Vilà *et al.* 2013).

Observational information, such as rates of feeding, abundance, body size, and life-history traits, when appropriately transformed, can be used to estimate interaction strength (Yodzis and Innes 1992; Moore *et al.* 1993; Rafaelli and Hall 1996; Wootton 1997; Hall *et al.* 2000). The use of trait-based estimation for assessing interaction strength has been hotly debated recently. This approach could result in overgeneralization but is potentially useful for many purposes. A trait-based approach can, arguably, be used as a framework for restoring invaded systems (Funk *et al.* 2008). In particular, body size is one general property of organisms that might scale with interaction strength (Elton 1927; Yodzis and Innes 1992); investigating allometric relationships

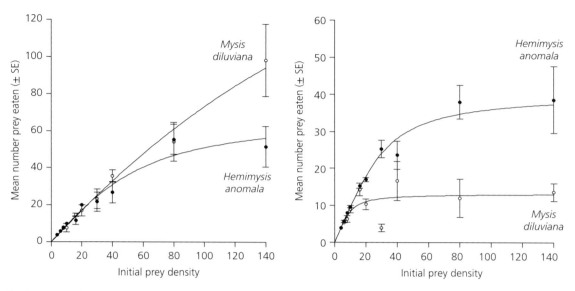

Fig. 6.7 Functional responses of the invasive Ponto–Caspian freshwater shrimp (*Hemimysis anomala*) compared to the North American native shrimp (*Mysis diluviana*) when tested separately using cladoceran prey *Ceriodaphnia quadrangula* (left) and *Daphnia pulex* (right), at 12 hours. From Dick *et al.* (2013). With permission from Elsevier.

between body size and interaction strength may therefore be profitable. Several allometric relationships relevant to various interaction-strength concepts or parameters in multispecies models in general are already well established. For example, metabolic rates and movement rates scale strongly with body size (Calder 1996; Harestad and Bunnell 1979; Peters 1983) and potentially affect organism death rates, encounter rates, and assimilation efficiencies. Foraging theory also suggests that consumption rate should be affected by the size of interacting species (e.g. Schluter 1982; Werner and Hall 1974). Several empirical datasets are available for a preliminary analysis of allometric relationships (e.g. Fig. 6.8; Emmerson and Raffaelli 2004; also see Wootton 1997; Sala and Graham 2002). When data from all these studies are combined, they tend to fall around the same relationship (Wootton and Emmerson 2005), (interaction strength per capita)$^{1/4}$ = 0.14 + 0.85 × (prey/predator mass)$^{1/4}$, which suggests a generally predictable pattern of interaction strength.

Analyses of linkage patterns in empirical food webs without reference to differences in the strength of the interactions that generate the links (e.g. Cohen and Briand 1984; Pimm *et al.* 1991) have

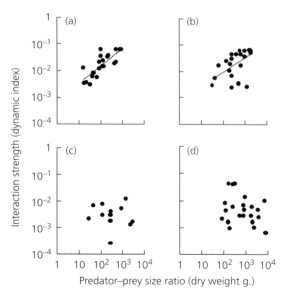

Fig. 6.8 Relationship between predator–prey body size ratio and interaction strength, with the mud shrimp *Corophium volutator* as the prey. Predators are (a) *Crangon crangon* (brown shrimp; $y \sim x^{0.66}$, $R^2 = 0.62$), (b) *Carcinus maenas* (green crab; $y \sim x^{0.54}$, $R^2 = 0.29$), (c) *Pomatoschistus microps* (common goby; not significant, $R^2 = 0.065$), and (d) *Platichthys flesus* (flounder; not significant, $R^2 = 0.03$). From Emmerson and Raffaelli (2004). Reproduced with permission from John Wiley & Sons.

been carried out extensively. Such studies have, however, generally not attempted to make specific predictions about experimental or environmental impacts. Dambacher and colleagues (2003) introduced methods for quantifying the strength and complexity of different feedback pathways in relatively complex systems. For example, linkage density (node-degree) could be related to interaction strength. Montoya and colleagues (2005) used the inverse community matrix derived from estimates of Emmerson and colleagues (2005) to assess the relative importance of species connectance within food webs. They found a negative relationship between linkage level of each species and the mean net indirect effect of that species on other species in the communities.

Empirical studies that have estimated interaction strength per capita among a number of species pairs have generally found distributions skewed towards many weak interactions and few strong interactions (Wootton and Emmerson 2005). This pattern is seen in the population level interactions (Jacobian elements) documented by de Ruiter and colleagues (1995), which clearly show a strong skew towards weaker interactions for both the negative effects of predators on prey and the positive effects of prey on predators. Strengths of pollinator effects on plants seem to exhibit a skewed distribution on a per-interaction basis (Schemske and Horvitz 1984; Pettersson 1991). Whether these correspond to interactions per capita is unclear because pollinator abundance was unknown and indices of pollination success (pollen deposition, fruit set) may not translate directly to population dynamics. Roxburgh and Wilson (2000) report the Jacobian matrix for a seven-species plant community, and the distribution of competitive interaction strengths also exhibits a skewed distribution. The skewed distribution of interaction strength in a community could reflect the result of assembly processes that stabilize community dynamics.

Spatial and temporal scales can affect estimates of interaction strength in several ways. Indirect effects and density-dependent feedback can change interaction-strength estimates over long periods of measurement. Strayer and colleagues (2006) argue that such processes are so widespread and important that ecologists need to adopt a long-term

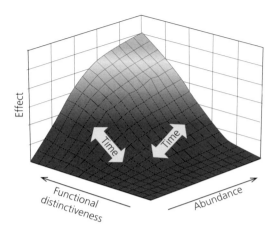

Fig. 6.9 The effects of an invading species increase with its abundance and functional distinctiveness (i.e. how much its functional characteristics differ from those of the species in the community that it is invading). Abundance and functional distinctiveness can increase or decrease through time, thereby altering the effect of the invader. From Strayer *et al.* (2006). With permission from Elsevier.

perspective on the effects of alien species. These processes (including evolution, shifts in species composition, accumulation of materials, and interactions with abiotic variables) can increase, decrease, or qualitatively change the impacts of an invader through time. Previous attempts to explain the ecological effects of an invader have focused on two attributes: its functional distinctiveness (i.e. the extent to which its characteristics such as nitrogen fixation, flammability, phenology, chemical defences, and diet differ from those of species already in the community) and its abundance (Vitousek 1990; Parker *et al.* 1999; Lovett *et al.* 2006). Evolutionary or ecological processes can change either of these attributes over time, thereby mediating the effects of the invader (Fig. 6.9). In the Galapagos Islands, Heleno and colleagues (2013) showed that the integration of alien plant species into seed dispersal networks was highest in areas with the longest history of invasion, suggesting a time lag between the arrival of invaders and their impacts on networks. Thus, a third factor (residence time) needs to be explicitly considered to understand the effects in many cases fully.

Spatial scale also affects our perception of interaction strength in several ways. One way is through changes in habitat heterogeneity if the performance of interacting species is habitat- dependent. Araújo and Rozenfeld (2014) found that co-occurrence arising from positive interactions, such as mutualism and commensalism, is manifested across scales. Negative interactions, such as competition and amensalism, are discernible at finer sampling grains but are indiscernible from random patterns at coarser spatial scales (Hui 2009). Scale dependence in consumer–resource interactions depends on the strength of positive dependencies between species. If the net positive effect is greater than the net negative effect, then interactions scale up similarly to positive interactions. In this case, fine-scale studies in one habitat would estimate strong interspecific interaction strengths among competitors consistent with competitive exclusion, whereas studies at larger scales that include multiple habitats would estimate lower interspecific interaction strengths, consistent with competitive coexistence.

Disturbance, especially human-driven disturbance, has a strong potential effect on the strength of biotic interactions. For instance, although it has its sceptics (e.g., Fox 2013), the *intermediate disturbance* hypothesis predicts a hump-shaped pattern between community diversity and disturbance intensity. Early stages of succession are most susceptible to invasion because resources and colonization opportunities are elevated after disturbance, a pattern that can be further accentuated by human-mediated dispersal that is biased towards translocating early successional and fast-growing species (Catford *et al*. 2012). This means that human disturbance, coupled with plant introductions, can extend the diversity–disturbance curve and shift peak diversity towards higher disturbance levels (Fig. 6.10a). However, invasive aliens can reduce native diversity at the community scale, especially in mid succession where competitive interactions structure communities (Fig. 6.10b). Certain invasive species may have greater impacts because they overcome some life-history trade-offs through their association with humans or novel evolutionary histories (e.g. enemy release). This may directly or indirectly (e.g. through plastic reallocation of resources from defence into growth) enable invasive plants to colonize earlier or persist into later stages of succession. By modifying disturbance regimes, invaders that transform the environment may also interfere with

succession and precipitate low-diversity communities. Low introduction rates of late-successional species may currently limit impacts of aliens under infrequent disturbance (Catford *et al*. 2012).

Other global change factors such as climate change can interact with alien species to affect the strength of biotic interactions. There are similarities and differences between invasive alien species and those native species that have expanded

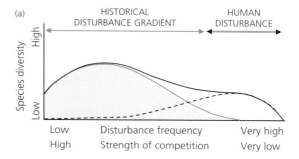

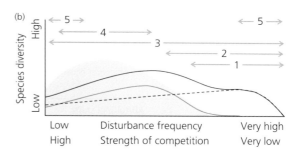

Fig. 6.10 Diversity-disturbance curves depicting the intermediate disturbance hypothesis extended for further considering human disturbance and biological invasion. (a) In early stages of invasion, the addition of alien species combined with human disturbance extends the *intermediate disturbance hypothesis* curve and augments species richness in high disturbance sites; it does not reduce native diversity. Grey shading indicates pre-invasion community species diversity under the historical disturbance regime; lines indicate diversity post-invasion and post-anthropogenic disturbance: solid black line, community diversity; dotted black line, alien diversity; solid grey line, native diversity. (b) Once established, invasive alien species are able to persist into late successional stages and may displace more than one native species causing a shift and reduction in peak diversity. Numbers indicate the mechanism that causes a reduction in diversity, 1: niche pre-emption, 2: apparent competition, 3: interference competition, 4: exploitative competition, 5: transformation of the environment; grey arrows indicate the position along the disturbance gradient where mechanisms have the greatest effect. From Catford *et al*. (2012). With permission from Elsevier.

their ranges due to climate change (Table 6.2; Van der Putten *et al.* 2010). Schweiger and colleagues (2010) proposed a framework for assessing how climate change and biological invasions together affect plant–pollinator interactions (Fig. 6.11). Effects of alien species may be disentangled into direct effects within trophic levels (plant–plant, pollinator–pollinator; arrows b and c in Fig. 6.11), direct effects across trophic levels (plant–pollinator and vice versa; arrows d and e in Fig. 6.11), and indirect effects via the corresponding trophic level (plant–pollinator–plant, pollinator–plant–pollinator; arrows e–a and d–a in Fig. 6.11). Within trophic levels, hybridization is an important force in the evolution of invasiveness in plant species (Abbott 1992; Vilà *et al.* 2000; Ellstrand and Schierenbeck 2000), and many widespread invasive species are allopolyploid hybrids (Lee 2002). Across trophic levels, alien species provide additional resources and can compensate for ecological/phenological mismatches due to climate change. Indirect effects of alien pollinators on native pollinators mediated via plants reflect resource competition rather than facilitation (Matsumura *et al.* 2004; Thomson 2006). These effects can be of diverse types (Bjerknes *et al.* 2007) and the interaction strength is likely to be density-dependent (Morales and Traveset 2009). Moreover, alien pollinators tend to preferentially visit alien plants (Stimec *et al.* 1997; Olesen *et al.* 2002; Goulson and Hanley 2004), potentially forming 'invader complexes'(Morales and Aizen 2006).

Interaction strength can also be mediated by abiotic environmental features, particularly resource availability and climate. Increased resource availability can increase or decrease resistance and tolerance to pathogen infection (Givnish 1999; Mitchell *et al.* 2003; Gilbert 2005). Increased resource availability can also interact with enemy release as introduced species with rapid growth rates could benefit from high resource availability and enemy release (the *resource–enemy release* hypothesis; Blumenthal 2005). Outbreaks of plant diseases are highly sensitive to climatic variation (Coakley *et al.* 1999; Harvell *et al.* 2002; Scherm 2004). The outcome of biotic interactions varies greatly across environmental gradients (Louda and Rodman 1996; Olff *et al.* 1997; DeWalt *et al.* 2004). Over half of all reported outcomes of competition between native and introduced plants summarized by Daehler

Table 6.2 Overview of similarities and differences between successful intracontinental range-expanding plants and intercontinental invasive plants (Van der Putten *et al.* 2010).

Process	Altitudinal shifts	Extra-limital range expansion	Invasion
Enemy release	Short-term, mainly from soil-borne enemies	Mid-long-term, probably stronger from soil-borne enemies than from above-ground enemies	Long-term, both above- and below-ground enemies
Dispersal barriers	No	Moderate	High
Range limitation	Probably temperature	Probably temperature and photoperiod	Probably temperature and photoperiod
Availability of symbiotic mutualists	Pollinators will be available; mycorrhizal fungi and nitrogen-fixing microbes close by or present	Pollinators may, or may not be available; mycorrhizal fungi and nitrogen-fixing microbes probably available	Pollinators will be available; mycorrhizal fungi and nitrogen-fixing microbes will be available
Continuity of gene flow from original populations	Yes	Yes	No
Hybridization with local species	Yes	Yes	Yes
Hybridization with other populations from same species	Yes	Yes	Depends on number of introductions and geographical spread of source populations

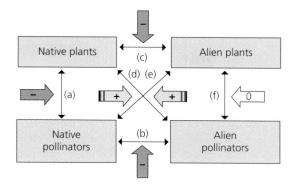

Fig. 6.11 Indirect effects of climate change (thick arrows) on interactions (thin arrows a–f) within a simplified pollination network of alien and native plants and pollinators. Dark grey arrows, negative effects; light grey arrows, positive effects; white arrow, no effects. From Schweiger *et al.* (2010). Reproduced with permission from John Wiley & Sons.

(2003) depended on the abiotic context. Nutrient-rich soil can greatly reduce the benefit of mycorrhizal fungi to host plants (Smith and Read 1997). Increased resource availability from natural or anthropogenic sources can decrease competition for those resources, facilitating the establishment and spread of introduced species (the *fluctuating resources theory of invasibility*; Davis *et al.* 2000). In general, invader success should be enhanced in situations where abiotic conditions reduce enemy impact or enhance the impact of mutualists on the invader relative to residents, or where they increase the competitive ability of invaders to residents. Introduced populations may be able to utilize spatial and temporal windows of beneficial biotic and abiotic opportunities to establish, and subsequently spread (the *invasion opportunity windows* hypothesis; Johnstone 1986; Agrawal *et al.* 2005).

6.4 Spatial interactions

Although recent studies suggest that the impact of biological invasions on native species richness is strongly scale-dependent (Gaertner *et al.* 2009), theoretical studies have confirmed that biotic resistance does have an effect on the rate of spread (Dunbar 1983, 1984; Okubo *et al.* 1989; Hui *et al.* 2011). For interspecific competition between invaders and native competitors, Okubo and colleagues (1989) considered a diffusion model of two competing species:

$$\frac{\partial N_1}{\partial t} = D_1 \frac{\partial^2 N_1}{\partial x^2} + r_1 N_1 \left(1 - \alpha_{11} N_1 - \alpha_{12} N_2\right)$$
$$\frac{\partial N_2}{\partial t} = D_2 \frac{\partial^2 N_2}{\partial x^2} + r_2 N_2 \left(1 - \alpha_{21} N_1 - \alpha_{22} N_2\right)$$

(6.4)

where α is the coefficient of intra- and interspecific competition. Okubo and colleagues (1989) found that the rate of spread for the invader (species 1) can be slowed down by its native competitor: $\hat{v} = 2\sqrt{r_1 D_1 (1 - \alpha_{12}/\alpha_{11})}$. A further expansion of this formula by considering the case of mutualism $(\alpha_{12}, \alpha_{21} < 0)$ suggests a higher rate of spread can be expected from mutualistic interactions. Dunbar (1983, 1984) examined the effect of predation on the rate of spread in a Lotka–Volterra model:

$$\frac{\partial N_1}{\partial t} = D_1 \frac{\partial^2 N_1}{\partial x^2} + r_1 N_1 \left(1 - \frac{N_1}{K}\right) - a N_1 N_2)$$
$$\frac{\partial N_2}{\partial t} = D_2 \frac{\partial^2 N_2}{\partial x^2} + \gamma a N_1 N_2 - \delta N_2$$

(6.5)

where a, γ, and δ are the rate of predation, the conversion rate of captured preys, and the death rate of the predators, respectively. If the predator is the invader, then it spreads according to the rate $\hat{v} = 2\sqrt{D_2 (\gamma a K - \delta)}$. Both studies suggest that biotic interactions such as competition and predation can reduce the spread rates of invasive species.

Besides the confirmation from such theoretical studies, real evidence of large-scale spatial competition is limited. Ferrer and colleagues (1991) reported that the range expansion of the common starling, *Sturnus vulgaris*, and the spotless starling, *S. unicolor*, in Spain slowed down due to competition between the two species. Similar evidence has also been reported for other bird species, for example, the decline of native endemic white-eyes (*Zosterops* spp.) due to interactions with the invasive red-whiskered bulbul, *Pycnonotus jocosus* (Clergeau and Mandon-Dalger 2001). However, such competitive resistance does not seem to be important for ants and wasps (Holway 1998; Roura-Pascual *et al.* 2010), and results are mixed for the studies of the spread of plants (Higgins *et al.* 2008). Other interspecific relationships such as pollination mutualisms and predation have also been suggested to influence the rate of spread (Lonsdale 1993; Richardson *et al.* 2000). Overall, biotic interactions can clearly affect rates of spread, and such impacts are taxon- and scale-dependent (Traveset and Richardson 2011).

Biotic interactions are thought to play a key role in shaping distributions at fine spatial scales, while other factors, such as climate play dominant roles at broader spatial extents (Pearson and Dawson 2003). Nevertheless, the question of how biotic interactions have shaped species distributions at broad spatial extents has been largely ignored. McGill (2010) illustrates such scale dependence of the effect of biotic interactions on species distributions from a tradition followed by a number of ecologists (e.g. Whittaker 1975) who concluded that the distribution of the terrestrial biomes of the world could be explained by the distribution of mean temperature and precipitation values alone (Fig. 6.12). Many aspects of this figure are debatable, but such far-reaching and multifaceted but testable hypotheses are needed in ecology.

By contrast, Wiens (2011) concluded that there is a paucity of good examples of large-scale patterns created by biotic interactions (e.g. García *et al.* 2011). Jablonski (2008) called for an integrative approach to incorporate biotic interactions in explanations of broad-scale ecological and evolutionary changes, despite the fact that many of these interactions are relatively ephemeral and play out at fine spatial scales in nature. Brooker and colleagues (2009), in replying to Ricklefs (2008), highlighted the need for studying biogeographic processes across a range

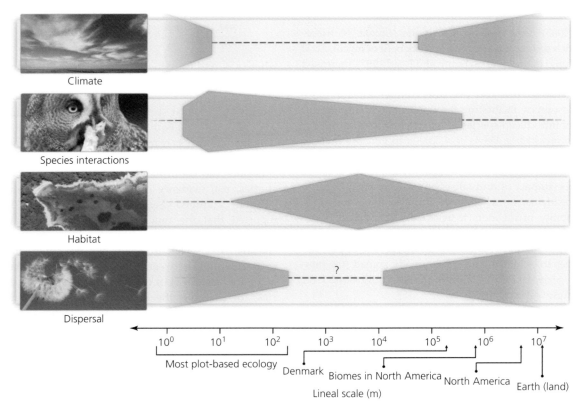

Fig. 6.12 Scale dependence of drivers controlling the distribution of species. Four main processes (vertical axis) are believed to control the distribution of organisms; their relative importance changes with scale (horizontal axis). The thickness of the bar for a given factor at a given scale indicates the importance of that factor at that scale. Ecologists began drawing such diagrams 30 years ago, but have only recently started to apply empirical studies to test the implied relationships. The question mark at intermediate scales of dispersal indicates that data on this process at these scales are scare. Climate is important at two scales, through two processes: microclimate (such as sun or shade) at fine scales and biogeography at broader spatial scales. From McGill (2010). Reprinted with permission from AAAS.

of temporal and spatial scales, and postulated that biotic interactions are important across all scales, although they are decreasingly influential at regional and continental scales.

Soberón (2007) argued for the distinction between the Grinnellian niche (defined by certain abiotic variables for which data are typically available at coarse resolution and broad spatial extents, e.g. average temperature, precipitation) and the Eltonian niche (defined by variables representing biotic interactions and resource consumer dynamics, typically measured at fine scales). He pointed out that the spatial structure of variables defining Grinnellian and Eltonian niches is a largely unexplored

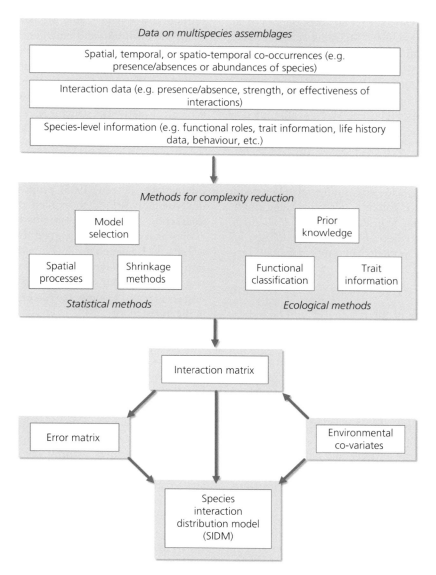

Fig. 6.13 A general framework for implementing species interaction distribution models (SIDMs). Data on multispecies assemblages are combined with methods for reducing complexity to estimate the interaction matrix. The interaction matrix together with environmental covariates and the error matrix are used to specify the final SIDM. From Kissling *et al.* (2012). Reproduced with permission from John Wiley & Sons.

research area, and that the degree of spatial structure will probably depend on the organisms under consideration.

Accounting for biotic interactions in species distribution models (SDMs) is still at its early stage, and Wisz and colleagues (2013) reviewed a few potential approaches in this regard. Firstly, some researchers have used surrogates for species interactions in SDMs, such as productivity, species richness, and traits (Michalet *et al.* 2006; Shipley *et al.* 2006; Mouillot *et al.* 2007; McGill *et al.* 2007; Meier *et al.* 2010; Guisan and Rahbek 2011; Roura-Pascual *et al.* 2011). However, biotic interactions are not considered in a reciprocal way as these interaction surrogates are simply considered explanatory variables in the SDMs. Hybrid SDMs have attempted to account for biotic interactions more explicitly (e.g. Midgley *et al.* 2010). Secondly, and more appropriately, biotic interactions can be directly integrated into species distribution models (SDMs) as they can influence both species–environmental relationship and co-occurrence patterns, normally using multivariate regressions (e.g. Roura-Pascual *et al.* 2010). In particular, the response variable for each species (Y_i) can be first modelled by environmental variables (X_i):

$$Y_i = \beta_i X_i + E_i \qquad (6.6)$$

and the structure of the residuals E_i can be indicative of biotic interactions. Specifically, the residuals of all species should follow a multivariate normal distribution MVN(0,**C**), with **C** the correlations between pairs of species. Ovaskainen and colleagues (2010) provided a clear example of such an approach in applying a multivariate logistic regression for modelling the community composition of wood-decaying fungal species for 22 species in Finland. Sebastian-Gonzalez and colleagues (2010) adopted the same approach for modelling community composition of waterbird species breeding in artificial irrigation ponds using seven species. Moreover, to account for spatial autocorrelation, multivariate autoregressive (MAR) models can be used (Hampton *et al.* 2013; see also Chapter 5).

Kissling and colleagues (2012) proposed a comprehensive way of modelling interactions between multiple species (Fig. 6.13) using these two methods and interaction matrices that quantify the effects of many species upon each other for each species pair, either as pairwise interaction coefficients or as functions that describe how pairwise interactions depend on other factors (e.g. environmental variables).

6.5 Interaction promiscuity and switching

Ehrlich and Raven (1964) argued that to understand the evolution and emergence of biotic interactions between different species we need to consider them in the light of co-evolution and adaptive radiation. This has led to a recent emphasis on the role of co-evolution in forming and shaping ecological network structures, especially mutualistic networks (Rezende *et al.* 2007; Urban *et al.* 2008; Guimarães *et al.* 2011; Nuismer *et al.* 2013; Minoarivelo *et al.* 2014). In his rebuttal to Ehrlich and Raven's (1964) co-evolutionary view of biotic interactions, Janzen (1985) proposed the concept of 'ecological fitting' to explain the multitude of interacting species in ecosystems that have no co-evolutionary history. He argued that no evolutionary change is needed to allow invading species to fit into local assemblages, and that ecological readjustment from resident native species to ecological fitting is sufficient to allow the establishment of novel biotic interactions.

Widespread species, including introduced species that extend their ranges as a result of the removal of dispersal barriers due to human actions, do not need to evolve in the novel habitat but only need to fit into the intricate web of biotic interactions. Many such species are generalists and have high levels of interaction fidelity and promiscuity, as well as sufficient trait plasticity to be able to establish novel interaction partners simply through ecological fitting. Janzen (1985) further argued that such widespread species might actually be quite maladapted to their habitats. These ideas on the origin of biotic interactions through co-evolution and ecological fitting are widely debated in the literature (Wilkinson 2004; Agosta 2006).

Many successful invasive alien species have promiscuous interactions and experience host jumps in their novel range. Biological invasions thus provide good evidence to support the notion of ecological fitting. Brooks and McLennan (2002) outlined three pathways of ecological fitting. Firstly, species traits can be co-opted (exapted) to perform novel functions so that novel species associations and interactions need not be precipitated by evolution or co-evolution. Secondly, parasites directly track host resources and not the host species themselves. This resource tracking is only indirectly related to phylogeny depending on how taxonomically widespread and phylogenetically conserved the resources being tracked are. Finally, the phylogenetic conservatism of traits related to resource use is a primary mechanism that will bias the formation of novel species associations. They argue that the ability of traits to be co-opted for novel functions and the evolutionary constraints on resource use make ecological fitting a potential force in the natural construction of species associations (Agosta 2006).

Adaptive interaction switches occur when the quantity and quality of available resources changes. Consumers prefer to select highly profitable resources rather than consuming all resources available to them as specified in optimal foraging theory (Stephens and Krebs 1986). Concurrently, they will exploit abundant resources over rare ones to avoid risk (Fossette *et al.* 2012; Zhang and Hui 2014). Psychologists explain the behaviour of decision making in animals by using concurrent schedules of reinforcement (e.g. using a Skinner box) and have proposed the momentary maximizing rules of hill climbing and melioration as possible explanations for diet choice (Staddon 2010). However, these rules are sensitive to the specific experimental environment, and their capacity in achieving optimization has not been verified (Herrnstein and Vaughan 1980; Hinson and Staddon 1983; Vaughan and Herrnstein 1987). Behavioural economists attempt to explain how animals make choices by analysing irrational decisions made by animals and humans (e.g. context-dependent choice, state-dependent behaviour, and heuristic decision making; Gigerenzer and Gaissmaier 2011; Rosati and Stevens 2009). Neuroscientists seek neural mechanisms of decision making in foraging (Basten *et al.* 2011; Kolling *et al.* 2012). Based on these observations, Zhang and Hui (2014) assume that a forager can (1) recognize the encountered resources (similar to the assumption in optimal diet model); (2) memorize

the profitability of food resources recently consumed (providing a reference point); and (3) perceive time elapsed since last feeding (an estimate of searching cost or hunger state); this leads to a recent experience-driven (RED) behaviour strategy for decision making during foraging. Importantly, the simple behavioural rule for decision making can cope with changing environment and predict an optimal energy intake rate.

Over short time scales, the nature and strength of cross-trophic interactions depend largely on preferences between prey and on predator behaviour patterns, a phenomenon known as prey switching. In prey switching, the number of attacks on a species is disproportionately large when the species is abundant relative to other prey, and disproportionately small when the species is relatively rare (Murdoch 1969). Some of the prey characteristics that influence predator preference are the nutritional quality of the prey and the ease of attack (Eubanks and Denno 2000). Predation on prey of highest nutritive value increases the predator's fitness. Capture success generally depends on prey mobility and access to a refuge (enemy-free space) (Fantinou *et al.* 2009). Jaworski and colleagues (2013) evaluated the prey switching of the generalist predator *Macrolophus pygmaeus* when it simultaneously encounters the silverleaf whitefly, *Bemisia tabaci*, and the leaf mining moth, *Tuta absoluta*. Prey switching was observed in both adult and juvenile predators (Jaworski *et al.* 2013): they over-attacked the most abundant prey when the mixed prey population was biased in favour of either prey species (Fig. 6.14).

Mounting evidence suggests that species often switch their interaction partners not only in antagonistic interactions (i.e. food webs; Murdoch 1969; Staniczenko *et al.* 2010) but also in mutualistic networks (Basilio *et al.* 2006; Fortuna and Bascompte 2006; Olesen *et al.* 2008; Petanidou *et al.* 2008). For instance, in a pollination network, pollinators continually switch the plant species with which they interact in response to environmental disturbances and the availability of resources (Whittall and Hodges 2007), whereas plants can also adjust phenology (e.g. flowering time) and morphology (e.g. flower heterostyly) to affect their pollinators (Aizen and Vazquez 2006; Barrett 2010; Kaiser-Bunbury *et al.* 2010). This implies a dynamic

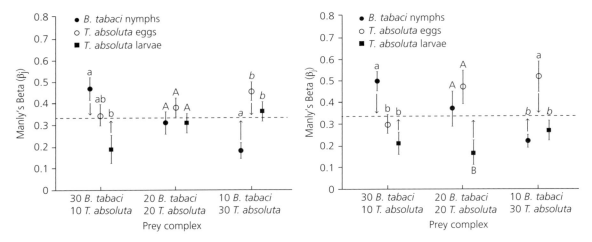

Fig. 6.14 Prey preference of *Macrolophus pygmaeus* adult (left) and juvenile (right) predators (based on Manly's beta values) depending on the initial relative abundance of prey (*Bemisia tabaci* and *Tuta absoluta*). The dotted line represents the expected β_j value against which calculated values for each prey are compared (significance difference with expected β_j values are indicated by arrows). Different indicate significantly different β_j values between the three prey types. From Jaworski *et al.* (2013).

nature of ecological networks; that is, both species abundance and species interaction could affect each other and change over time. Such interaction switches might not have exclusively ecological and environmental reasons (e.g. resource availability) but could also reflect adaptive behaviour of species for enhancing the efficiency of resource utilization (Zhang *et al.* 2011). Compared with fixed interactions between species, an interaction switch (or alternatively the rewiring of interactions between species) can lead to greater stability in food webs (Staniczenko *et al.* 2010) and pollination networks (Kaiser-Bunbury *et al.* 2010).

Mutualistic interactions contribute to the invasion of alien plants by enhancing plant growth, reproduction, and spread (Traveset and Richardson 2014). For instance, the invasion success of legumes can largely be attributed to their physiological adaptation to form symbioses with rhizobial bacteria for biological nitrogen fixation (Vilà *et al.* 2015; Inderjit and Cahill 2015; Vestergård *et al.* 2015), especially the ability to nodulate at low rhizobial abundance (Parker 2001; Perez-Fernanndez and Lamont 2003; Rodríguez-Echeverria *et al.* 2011). The *enhanced mutualism* hypothesis postulates that alien plants may benefit by forming novel mutualisms with native soil microbes that are capable of forming

associations with a wide range of hosts, thereby facilitating establishment and spread (Reinhart and Callaway 2006). Some alien woody legumes can readily nodulate (Lafay and Burdon 2006) and associate with novel bacterial communities in their alien ranges (Marsudi *et al.* 1999; Amrani *et al.* 2010; Callaway *et al.* 2011; Ndlovu *et al.* 2013).

Without co-evolved mutualists, symbiotic promiscuity is essential for invasion success, especially for compulsory mutualists (Aizen *et al.* 2012; Birnbaum *et al.* 2012; Heleno *et al.* 2013). Some invasive alien legumes can only form associations with bacteria from their native range (Chen *et al.* 2005). For instance, *Acacia longifolia* in Portugal associates with rhizobial communities similar to those in the native range of the plant in south-eastern Australia (Rodríguez-Echeverria 2010). However, the relationship between promiscuity and invasiveness is not universal, with evidence supporting both positive (Klock *et al.* 2015) and negative relationships (Marsudi *et al.* 1999; Mohamed *et al.* 2000; Lafay and Burdon 2006; Ndlovu *et al.* 2013). Birnbaum and colleagues (2016) found no difference in the nitrogen-fixing bacteria community of soils or nodules between native and introduced ranges, predominantly *Bradyrhizobium*, across the five widespread woody species (Fig. 6.3), suggesting

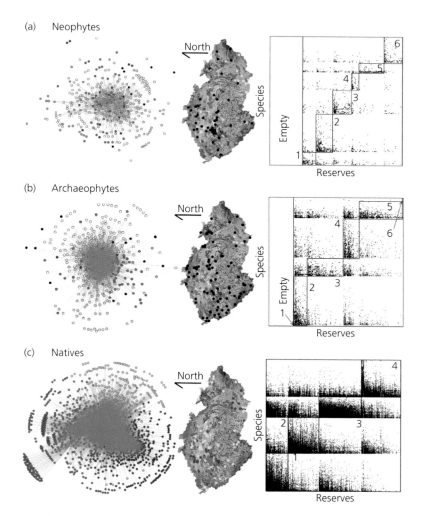

Plate 1 Network structures of vascular plants in the Czech Republic. Network expression, geographical location of reserves, and species-by-site matrices of modules identified for (a) neophytes, (b) archaeophytes, and (c) natives. In the network expression, open circles represent reserves. Blue, yellow, red, brown, black, and green points in the network expression and geographical maps indicate different modules identified in each of the three assemblages. Modules in the matrices are marked by the serial numbers and a rectangle, with points indicating the presence of a species (a row) occurring in a reserve (a column) and the rectangles of 'Empty' in neophytes and archaeophytes indicating reserves where these two species assemblages do not occur. From Hui *et al.* (2013). CC-BY. (see Figure 8.5 on page 201)

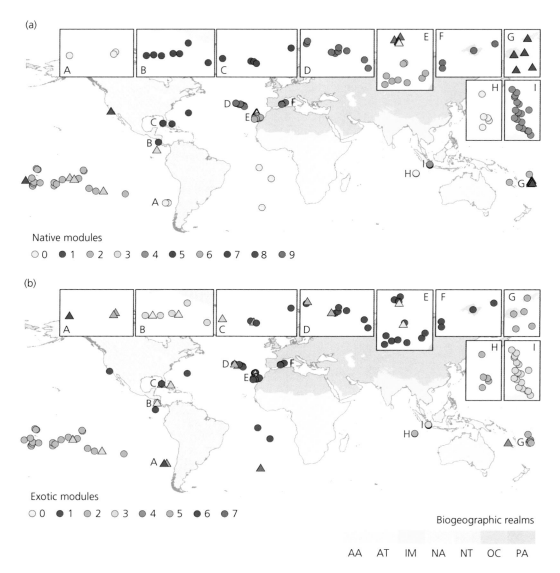

Plate 2 Distribution of (a) native and (b) alien modules for islands (i.e. distinct clusters of islands with a certain assemblage of species) derived from a modularity analysis. The colours indicate the module assigned by the modularity analysis, while the symbols indicate whether the conditional inference tree grouped that particular island in a node where the majority of islands have the same module (O) or a different one (Δ). The colours of the land masses show the biogeographic realms (corresponding to Australasia (AA), Afrotropic (AT), Indo-Malay (IM), Nearctic (NA), Neotropic (NT), Oceania (OC), Palearctic (PA)) (Olson *et al.* 2001). The inserts represent an expanded view of areas where symbols overlap. From Roura-Pascual *et al.* (2016). Reproduced with permission from John Wiley & Sons. (see Figure 8.13 on page 212)

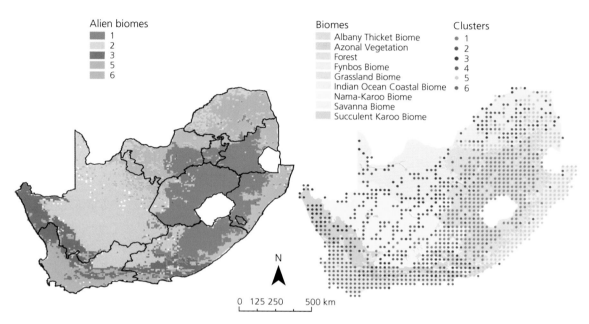

Plate 3 (Left) Invasive alien species clusters in relation to natural vegetation biomes (shading). Clusters were derived based on the presence/absence of invasive alien species in 15-minute grid cells (shown as circles). (Right) The potential distribution of alien biomes. Environmental determinants of the distribution of invasive alien species clusters were identified from a classification tree and mapped over the full range of environmental factors in South Africa. From Rouget *et al.* (2015). With permission from Elsevier. (see Figure 8.14 on page 213)

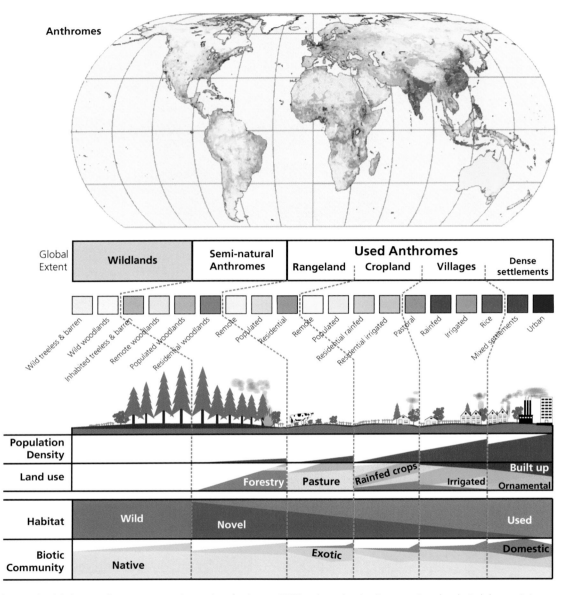

Plate 4 The global extent of contemporary anthromes (top; for the year 2000) and associated socio-economic and ecological characteristics, including human population density, land use, and integrity of habitat and biotic communities (bottom). From Martin *et al.* (2014). Reproduced with permission from John Wiley & Sons. (see Figure 11.2 on page 300)

that these legumes encounter similar bacteria communities in soils across Australia, supporting ecological fitting.

Plant mutualism ranges across the full spectrum of specialization and generalization (the number of mutualistic partner species) (Bascompte 2009). However, the relationship between specialization/generalization and the level of interaction promiscuity has not been thoroughly explored. To a large extent, generalist species are also considered promiscuous. We need to emphasize that these are two different concepts. As discussed in section 6.3, the specialization/generalization (node degree) is also related to interaction strength. We suspect that interaction strength, specialization/generalization, and interaction promiscuity are interlinked concepts worth further investigation. For instance, generalist hosts may be equally effective associating with a wide range of soil mutualists in both native and novel regions.

Interaction promiscuity and switching can have a profound influence on the complexity–stability relationship in ecological networks (Montoya et al. 2006). By incorporating the interaction switch (i.e. with foraging adaptation), Kondoh (2003) showed that the classic negative complexity–stability relationship in static food webs (May 1972) does not necessarily hold. Beckerman and colleagues (2006) and Petchey and colleagues (2008) were able to predict food-web complexity (connectivity) further using interaction switches. In mutualistic networks, allowing interaction switches can enhance the robustness of interaction strength against species loss over static networks (Kaiser-Bunbury et al. 2010). Thebault and Fontaine (2010) further demonstrated that a highly connected and nested architecture promotes stability in mutualistic networks. Moreover, Okuyama and Holland (2008) found that the resilience of mutualistic networks can be enhanced by increasing network size and the number of interactions and, thus, support a positive complexity–stability relationship. Specifically, they reported a weak effect of enhancing resilience by the nested architecture (based on the index matrix temperature; Atmar and Paterson 1993). Zhang and colleagues (2011) suggest that the interaction switch can increase the nestedness of a random network and thus could potentially enhance the stability of mutualistic networks (Okuyama and Holland 2008; Thebault and Fontaine 2010). Therefore, interaction switching could be an important process allowing a static community to become a dynamic (or adaptive) one and further foster ecological complexity and diversity (Nuwagaba et al. 2015). We develop this point further in Chapter 10.

6.6 Novel eco-evolutionary experiences

Biological invasions differ from other types of global change, such as global warming, nitrogen deposition, or pollutants, in that invasive species are not static selective agents. They can also evolve, and these evolutionary changes in invading species have the potential to either increase or decrease the strength of selection they impose on native community members. Introducing species to new regions further sets an ideal stage for novel eco-evolutionary experience and potential evolutionary diversification. Likewise, the altered selection regime, ecological opportunities, or possibilities for hybridization provided by alien species might promote diversification of some native species. For example, invasive garlic mustard has evolved reduced allelochemical concentrations over time (Lankau 2012) and several native species have evolved increased tolerance of garlic mustard allelochemicals (Lankau 2012; Lankau and Nodurft 2013). The evolution of reduced allelochemical production over time in garlic mustard is probably reducing both the negative consequences of invasion and the strength of selection imposed by garlic mustard on native species, while the evolution of increased tolerance to garlic mustard allelochemicals in natives is simultaneously reducing the fitness effects of garlic mustard invasion (Lankau 2012). In other cases, co-evolutionary responses may increase negative effects on the native. For example, native populations of big squirreltail grass, *Elymus multisetus*, have evolved increased competitive effects against invasive cheatgrass, *Bromus tectorum* (Rowe and Leger 2011), possibly exerting selection on cheatgrass for increased competitive ability.

Three categories of evolutionary diversification can be stimulated by invasions of alien species (Vellend et al. 2007). Firstly, diversification of alien species could occur in new regions (Fig. 6.15).

Bottlenecks and drift in the new allopatric population might cause it to diverge from populations in the native range (Berthouly-Salazar *et al.* 2013). Directional selection imposed by the novel environment might cause divergence from populations in the native range (Baker and Moeed 1987; Lomolino 2005). If environmental conditions are sufficiently heterogeneous, disruptive selection can lead to diversification within the introduced range (Johnston and Selander 1964, 1973). Across different invaded ranges, the introduced species could undergo similar phenotypic adaptation but with a different genetic basis (Huey *et al.* 2005).

Secondly, biological invasions could lead to the diversification of native species (Fig. 6.15; Vellend *et al.* 2007). Invaders could impose disruptive selection within populations of native species by providing an alternative host for herbivorous insects. Shifts of native phytophagous insects onto novel alien host plants provide the most striking and common empirical examples of native evolutionary diversification in response to invaders. The two best-documented cases are that of native apple maggot fly, *Rhagoletis pomonella*, which has evolved genetically differentiated ecotypes that feed on native hawthorns and introduced apples in North America (McPheron *et al.* 1988; Filchak *et al.* 2000),

and the North American soapberry bug, *Jadera haematoloma*, which has evolved different beak lengths to feed on the fruit from alien tree species (Carroll *et al.* 2001). Strauss and colleagues (2006) discuss the widespread phenomenon of native insect species undergoing genetic diversification in response to alien host plants.

Finally, diversification could happen through hybridization, either between native and alien species, between pairs of alien species brought into sympatry, or pairs of native species whose interaction is mediated by an alien species (Fig. 6.15; Vellend *et al.* 2007). Hybridization initiated by alien species invasions provides some of the most unambiguous cases of rapid evolutionary diversification (Ellstrand and Schierenbeck 2000). Aliens can hybridize with other aliens (Gaskin and Schaal 2002) or with native species (Rieseberg *et al.* 1990). Additionally, invasions of alien species can alter community structure and lead to hybridization between two formerly distinct native species (e.g. Schwarz *et al.* 2005). Results of any of these hybridization events can include the formation of new species, introgression of genetic material into one or both original species, or the formation of hybrid swarms, depending on the genetics and ecology of the two species (James and Abbott 2005).

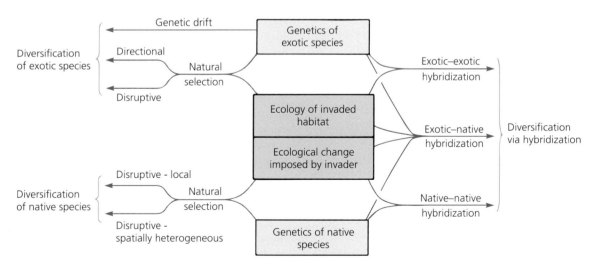

Fig. 6.15 Conceptual framework for understanding effects of invasions of alien species on evolutionary diversification. Characteristics of the genetics of alien and native species (top and bottom boxes) and the ecology of their shared habitats (middle boxes) interact via genetic drift, natural selection, and hybridization to result in evolutionary diversification of alien species, native species, or hybrid combinations of one or both of the alien and native species. From Vellend *et al.* (2007). With permission from Elsevier.

Sax and colleagues (2007) reviewed some of the major conceptual eco-evolutionary insights relevant to invasion ecology. Classic insights include that: (i) species are not optimally adapted to their environment (Gould 1997; Richardson and Higgins 1998) due to evolutionary constraints (Darwin 1859; Gould 1997); (ii) changes in geographical range can occur quickly (Darwin 1859); (iii) species are generally limited in their distribution by dispersal (Darwin 1859); (iv) speciation can occur sympatrically (Bolnick *et al.* 2007); (v) reproductive isolation can take millions of years as alien species that hybridize with native species can produce fertile offspring (Rice and Sax 2005; Burkhead and Williams 1991); (vi) individual plant species can transform ecosystems by altering nutrient availability and disturbance regimes (Vitousek *et al.* 1987; D'Antonio and Vitousek 1992). Emerging insights include: (i) ecological systems rarely show evidence of being saturated with species (Russell *et al.* 2006; Fridley *et al.* 2007); (ii) competition, unlike predation, seldom causes global extinction (Davis 2003; Gurevitch and Padilla 2004); (iii) community assembly often occurs through ecological sorting or species fitting (Janzen 1985; Agosta 2006); (iv) adaptive genetic change can occur rapidly (Carroll *et al.* 2005; Huey *et al.* 2005); (v) severe population bottlenecks do not preclude rapid adaptation (Novak and Mack 2005; Wares *et al.* 2005);

(vi) climate envelope approaches for predicting spread and ultimate distributions might be inadequate for many species (Broennimann *et al.* 2007; Fitzpatrick *et al.* 2007); (vii) variation in patterns of specialization in predators and mutualists influence community invasibility (Callaway *et al.* 2004; Sax *et al.* 2007). In particular, Callaway and colleagues (2004) suggested that invasion would be facilitated if native predators and pathogens tended to be specialists (and were thus unable to prey upon introduced species), and if native mutualists tended to be generalists (and were thus able to facilitate the invasion of alien species) (Fig. 6.16a). If most predators (including herbivores, parasites, and infectious diseases) were generalists and most mutualists were specialists then such communities should be difficult to invade (Fig. 6.16b; Parker *et al.* 2006; Gilbert and Parker 2006).

Although many studies have documented evolutionary responses of native species to biological invasions (reviewed in Strauss *et al.* 2006; Oduor 2013), we are only now beginning to understand the mechanisms involved. Many challenges to adaptation arise because biological invasions occur in complex species-rich communities in spatially and temporally variable environments. Consequently, many species may fail to adapt to biological invasions. Lau and terHorst (2015) reviewed these

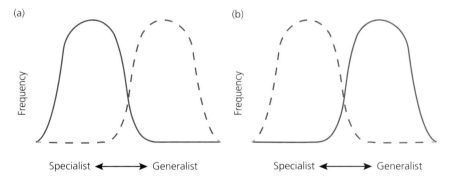

Fig. 6.16 Specialization in species interactions and invasibility. Individual ecosystems can vary such that the frequency distribution of species along a continuum from absolute specialists to absolute generalists differs between (solid lines) predators and pathogens, and (dashed lines) mutualists and facilitators. (a) In a high invasibility system, predators, and pathogens are more frequently specialists, whereas mutualists and facilitators are more frequently generalists. Such systems would be relatively easy to invade because few predators would be able to prey upon alien species (for which they would not be specialized), whereas many mutualists would be able to assist alien species. (b) A low invasibility system, with the opposite distribution and invasion outcome. The curves illustrated here are for heuristic purposes only; the actual shape of these curves is unknown empirically. Their impact on invasibility should operate as described here, however, as long as there is a difference in the mode of the two distributions, and as long as the frequency of interactions determines the average outcome of invasions. From Sax *et al.* (2007). With permission from Elsevier.

'ecological' constraints on adaptation, focusing on the complications that arise from the need to adapt simultaneously to multiple biotic agents and from temporal and spatial variation in both selection and demography. In particular, several ecological factors (top-row boxes) have the potential to influence the likelihood and extent of native species adaptation versus maladaptation (second-row boxes) to biological invasions (Fig. 6.17).

Introduced species can also respond to altered selection forces in novel environments through stochastic and adaptive phenotypic evolution (Keller and Taylor 2008). Experimental comparisons of species from their native and introduced range, or from the edge and the core populations in the invaded range, often reveal phenotypic changes (Bossdorf et al. 2005; also see Chapter 4) caused by prior evolutionary history, chance events, and response to selection (Fig. 6.18). For instance, 16 per cent of populations of the aquatic plant *Butomus umbellatus* in its native Europe are diploid, while 71 per cent are diploid in its invaded range of North

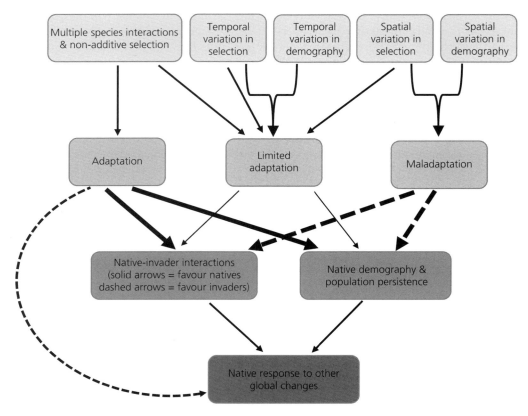

Fig. 6.17 Ecological constraints on native adaptation to biological invasions and resulting ecological effects of rapid adaptation (or failed adaptation). Multispecies interactions and non-additive selection can either facilitate or limit adaptation depending on the ecological scenario. Temporal variation in selection, especially when combined with temporal variation in demographic rates, can slow adaptation. Spatial variation coupled with modest gene flow can slow adaptation or, if coupled with spatial variation in demographic rates and asymmetric gene flow, cause maladaptation. The extent of native adaptation to invasion will influence competitive outcomes between native and invasive taxa and native population growth and persistence (third-row boxes). Altered native–invader competitive interactions that favour natives and increased native population growth rates in turn increase the likelihood that natives will respond less negatively (ecologically) and have a greater capacity to adapt to future global change (bottom box). The thickness of arrows indicates the hypothesized importance of that path or process. Solid arrows indicate positive effects on natives, dashed arrows indicate negative effects. From Lau and terHorst (2015). Reproduced with permission from John Wiley & Sons.

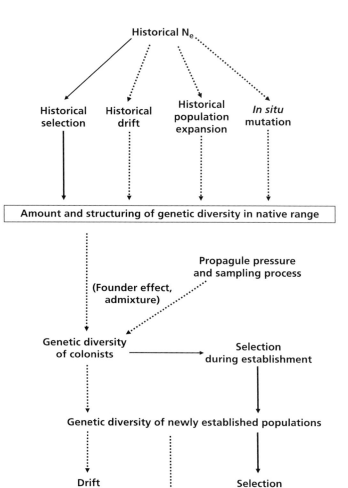

Fig. 6.18 Path diagram illustrating the contributions of evolutionary history, chance events, and natural selection to the genetics of introduced populations. The genetic diversity present among invasive populations has been shaped by a suite of historical (five historical factors above the top banner), stochastic (propagule pressure and sampling process, drift during expansion), and deterministic forces (selection during establishment and during expansion). Most path transitions leading up to and during invasion involve demographic events or intrinsic genetic effects (e.g. mutational input) that influence diversity at both neutral loci and quantitative traits (dotted arrows in diagram). These paths represent the contributions of historical and chance events that may influence quantitative trait evolution during invasion. Transitions involving selection (solid arrows) represent an additional influence on quantitative traits that may work in concert with or in opposition to chance and historical processes. Only a few stages of the invasion process are directly observable by empirical studies (boxes with outlines). Therefore, quantitative genetic studies must control statistically for the influence of unobserved stages in the invasion process when testing for adaptive evolution. From Keller and Taylor (2008). Reproduced with permission from John Wiley & Sons.

America as a result of natural selection for greater genetic variability and dispersal potential of sexually produced offspring (Brown and Eckert 2005). Consequently, adaptation in phenotypic evolution/change needs to be separated from the effects of history and chance. Keller and Taylor (2008) undertook this by proposing two experimental designs: ancestor–descendent comparisons using phylogenetics and multi-locus assignment methods; and comparisons of the genetic variance at neutral loci

(F_{ST}) relative to the variance in traits measured on pedigreed progeny (Q_{ST}).

Related, but different to phenotypic evolution during invasion, adaptive phenotypic plasticity can also greatly improve the establishment of introduced species in novel environments (Feiner *et al.* 2012; Lucek *et al.* 2012; Huang *et al.* 2015). Phenotypic plasticity could greatly improve the performance of alien plants (Richards *et al.* 2006), for example, through inducible prey defences (Engel *et al.* 2011). For instance, the widespread invasion of fountain grass *Pennisetum setaceum* in Hawai'i has been attributed to adaptive plasticity, not adaptive evolution (Williams *et al.* 1995). Phenotypic plasticity can be described as a reaction norm portraying how a phenotypic trait varies across environments. For instance, mammal species with a high variance in their adult body mass have a higher establishment success (González-Suárez *et al.* 2015). If the plasticity allows an introduced species to change its phenotypic trait in a novel environment still according to the reaction norm, and if the new phenotype allows individuals to have higher fitness, such plasticity becomes adaptive and can be important to adaptive evolution and invasion success in the new environment (Ghalambor *et al.* 2007). This means that phenotypic plasticity is often under strong selection during the invasion of new environment (e.g. the evolution of offspring size in Trinidadian guppies, *Poecilia reticulata*; Fitzpatrick *et al.* 2015; and in three-spined stickleback, *Gasterosteus aculeatus*; Bolnick *et al.* 2009; Lucek *et al.* 2014).

The eco-evolutionary experience of introduced species affects their establishment. Darwin (1859) proposed that introduced species that are more closely related to the resident species are less likely to become naturalized (*Darwin's naturalization* hypothesis; Daehler 2001). The chief mechanism proposed for this pattern is that more closely related species compete more strongly because resource use requirements are more similar among closer relatives. Available data support the hypothesis that greater phylogenetic relatedness increases the accumulation of pathogens and insect herbivores by introduced plants (Connor *et al.* 1980; Blaney and Kotanen 2001; Parker and Gilbert 2004). Several studies have reported this general pattern (e.g. Mack 1996; Wu *et al.* 2004), but others have found no support for this (Daehler 2001; Duncan and Williams 2002). For instance, the capacity for introduced species to form new associations with resident mutualists seems unlikely to depend strongly on phylogenetic relatedness (Richardson *et al.* 2000). These mechanisms are potentially countervailing. The success of more closely related invaders is expected to be reduced by greater accumulation of enemies and competitors, but enhanced by greater accumulation of mutualists and more

suitable abiotic conditions. The direction of predicted outcomes for invader success will depend on the relative phylogenetic dependence of the four factors, and the strength and form of the interactions between the factors (Mitchell *et al.* 2006).

To this end, Saul and colleagues (2013) advocate the adoption of an eco-evolutionary perspective on invasions. Such a view emphasizes the evolutionary antecedents of invasions, that is, the species' evolutionary legacy and its role in shaping novel biotic interactions that arise due to invasions. They present a conceptual framework comprising five hypothetical scenarios about the influence of eco-evolutionary experience in resident native and invading non-native species on invasion success. This framework integrates several of the key hypotheses in invasion ecology as discussed in this chapter and elsewhere in the book. Saul and Jeschke (2015) further consider the implications of different degrees of eco-evolutionary experience of interacting resident and alien species and define four qualitative risk categories for estimating the probability of successful establishment and impact of novel species (Fig. 6.19). Highly experienced non-resident species behave like native colonizing species, posing an intermediate level of risk, if they are functionally similar to resident species, so that other resident species also have a lot of experience in interacting with such type of species. In contrast, a high level of risk can be expected if the non-resident species are highly experienced and are functionally distinct to other resident species, so that the resident species have low degrees of experience in interacting with such type of species. If non-resident species have low degrees of experience with their resident interaction partners, risk expectations are again different, and interaction outcomes can be hard to predict (Fig. 6.19). Thus, the explicit consideration of the evolutionary legacy of species, as in terms of eco-evolutionary experience, highlights important differences between 'native colonizers' and 'introduced invaders' which are crucial for determining the outcome of novel species interactions (see also Strauss *et al.* 2006; Salo *et al.* 2007; Heger *et al.* 2013; Paolucci *et al.* 2013; Richardson and Ricciardi 2013).

Biotic interactions are an essential component, and one of the most intricate dimensions, of ecology. To elucidate their role in influencing invasion

Gilbert, G. and Parker, I. (2006) Invasion and the regulation of plant populations by pathogens. In: Cadotte, M.W., McMahon, S.M., and Fukami, T. (eds) *Conceptual Ecology and Invasion Biology: Reciprocal Approaches to Nature*. Berlin: Springer, pp. 289–305.

Gilbert, G.S. (2005) Dimensions of plant disease in tropical forests. In: Burslem, D., Pinard, M., and Hartley, S. (eds) *Biotic Interactions in the Tropics: Their Role in the Maintenance of Species Diversity*. Cambridge: Cambridge University Press, pp. 141–64.

Givnish, T.J. (1999) On the causes of gradients in tropical tree diversity. Journal of Ecology, 87, 193–210.

González-Suárez, M., Bacher, S., and Jeschke, J.M. (2015) Intraspecific trait variation is correlated with establishment success of alien mammals. American Naturalist, 185, 737–46.

Gough, L. and Grace, J.B. (1999) Effects of environmental change on plant species density: comparing predictions with experiments. Ecology, 80, 882–90.

Gould, S.J. (1997) Evolution: the pleasures of pluralism. New York Review of Books, 44, 47–52.

Goulson, D. and Hanley, M.E. (2004) Distribution and forage use of exotic bumblebees in South Island, New Zealand. New Zealand Journal of Ecology, 28, 225–32.

Graham, A.L. (2008) Ecological rules governing helminth–microparasite coinfection. Proceedings of the National Academy of Sciences of the USA, 105, 566–70.

Guimaraes Jr, P.R., Jordano, P., and Thompson, J.N. (2011) Evolution and coevolution in mutualistic networks. Ecology Letters, 14, 877–85.

Guisan, A. and Rahbek, C. (2011) SESAM–a new framework integrating macroecological and species distribution models for predicting spatio-temporal patterns of species assemblages. Journal of Biogeography, 38, 1433–44.

Gurevitch, J. and Padilla, D.K. (2004) Are invasive species a major cause of extinctions? Trends in Ecology and Evolution, 19, 470–4.

Hall, D.O., House, J., and Scrase, I. (2000) *An Overview of Biomass Energy*. London: Taylor and Francis.

Hampton, S.E., Holmes, E.E., Scheef, L.P., *et al.* (2013) Quantifying effects of abiotic and biotic drivers on community dynamics with multivariate autoregressive (MAR) models. Ecology, 94, 2663–9.

Harestad, A.S. and Bunnell, F.L. (1979) Home range and body weight—a re-evaluation. Ecology, 60, 389–402.

Harvell, C.D., Mitchell, C.E., Ward, J.R., *et al.* (2002) Climate warming and disease risks for terrestrial and marine biota. Science, 296, 2158–62.

Heger, T. and Jeschke, J.M. (2014) The enemy release hypothesis as a hierarchy of hypotheses. Oikos, 123, 741–50.

Heger, T., Saul, W.C., and Trepl, L. (2013) What biological invasions 'are' is a matter of perspective. Journal for Nature Conservation, 21, 93–6.

Heleno, R.H., Olesen, J.M., Nogales, M., *et al.* (2013) Seed dispersal networks in the Galápagos and the consequences of alien plant invasions. Proceedings of the Royal Society B: Biological Sciences, 280, 20122112.

Herrnstein, R.J. and Vaughan, W. (1980) Melioration and behavioral allocation. In: Staddon, J.E.R. (ed.) *Limits to Action: The Allocation of Individual Behavior*. Amsterdam: Elsevier, pp. 143–76.

Hierro, J.L., Maron, J.L., and Callaway, R.M. (2005) A biogeographical approach to plant invasions: the importance of studying exotics in their introduced and native range. Journal of Ecology, 93, 5–15.

Higgins, S.I., Flores, O., and Schurr, F.M. (2008) Costs of persistence and the spread of competing seeders and sprouters. Journal of Ecology, 96, 679–86.

Hinson, J.M. and Staddon, J.E.R. (1983) Matching, maximizing, and hill-climbing. Journal of the Experimental Analysis of Behavior, 40, 321–31.

Hinz, H.L. and Schwarzlaender, M. (2004) Comparing invasive plants from their native and exotic range: what can we learn for biological control? Weed Technology, 18, 1533–41.

Hokkanen, H.M. and Pimentel, D. (1989) New associations in biological control: theory and practice. Canadian Entomologist, 121, 829–40.

Holway, D.A. (1998) Effect of Argentine ant invasions on ground-dwelling arthropods in northern California riparian woodlands. Oecologia, 116, 252–8.

Huang, Q.Q., Pan, X.Y., Fan, Z.W., *et al.* (2015) Stress relief may promote the evolution of greater phenotypic plasticity in exotic invasive species: a hypothesis. Ecology and Evolution, 5, 1169–77.

Huey, R.B., Gilchrist, G.W., and Hendry, A.P. (2005) Using invasive species to study evolution. In: Sax, D.F., Stachowicz, J.J., and Gaines, S.D. (eds) *Species Invasions: Insights into Ecology, Evolution, and Biogeography*. Sunderland, MA: Sinauer Associates, pp. 139–64.

Hui, C. (2009) On the scaling patterns of species spatial distribution and association. Journal of Theoretical Biology, 261, 481–7.

Hui, C., Krug, R.M., and Richardson, D.M. (2011) *Modelling Spread in Invasion Ecology: A Synthesis. Fifty Years of Invasion Ecology: The Legacy of Charles Elton*. Chichester: Wiley-Blackwell, pp. 329–43.

Hui, C., Richardson, D.M., Landi, P., *et al.* (2016) Defining invasiveness and invasibility in ecological networks. Biological Invasions, 18, 971–83.

Inderjit and Cahill, J.F. (2015) Linkages of plant–soil feedbacks and underlying invasion mechanisms. AoB Plants, 7, plv022.

Jablonski, D. (2008) Biotic interactions and macroevolution: extensions and mismatches across scales and levels. Evolution, 62, 715–39.

James, J.K. and Abbott, R.J. (2005) Recent, allopatric, homoploid hybrid speciation: the origin of *Senecio squalidus* (Asteraceae) in the British Isles from a hybrid zone on Mount Etna, Sicily. Evolution, 59, 2533–47.

Janzen, D.H. (1985) Dan Janzen's thoughts from the tropics 1: on ecological fitting. Oikos, 45, 308–10.

Jarosz, A.M. and Davelos, A.L. (1995) Effects of disease in wild plant populations and the evolution of pathogen aggressiveness. New Phytologist, 129, 371–87.

Jaworski, C.C., Bompard, A., Genies, L., *et al.* (2013) Preference and prey switching in a generalist predator attacking local and invasive alien pests. PLoS One, 8, e82231.

Jeschke, J., Aparicio, L.G., Haider, S., *et al.* (2012) Support for major hypotheses in invasion biology is uneven and declining. NeoBiota, 14, 1–20.

Jeschke, J.M. (2014) General hypotheses in invasion ecology. Diversity and Distributions, 20, 1229–34.

Johnson, S. (2015) Invasive species in pollination networks. Quest, 11(2), 30–3.

Johnston, R.F. and Selander, R.K. (1964) House sparrows: rapid evolution of races in North America. Science, 144, 548–50.

Johnston, R.F. and Selander, R.K. (1973) Evolution in the house sparrow. III. Variation in size and sexual dimorphism in Europe and North and South America. American Naturalist, 107, 373–90.

Johnstone, I.M. (1986) Plant invasion windows: a time-based classification of invasion potential. Biological Review, 61, 369–94.

Jonsson, L.M., Nilsson, M.C., Wardle, D.A., *et al.* (2001) Context dependent effects of ectomycorrhizal species richness on tree seedling productivity. Oikos, 93, 353–64.

Joshi, J. and Vrieling, K. (2005) The enemy release and EICA hypothesis revisited: incorporating the fundamental difference between specialist and generalist herbivores. Ecology Letters, 8, 704–14.

Kaiser-Bunbury, C.N., Muff, S., Memmott, J., *et al.* (2010) The robustness of pollination networks to the loss of species and interactions: a quantitative approach incorporating pollinator behaviour. Ecology Letters, 13, 442–52.

Keane, R.M. and Crawley, M.J. (2002) Exotic plant invasions and the enemy release hypothesis. Trends in Ecology and Evolution, 17, 164–70.

Keller, S.R. and Taylor, D.R. (2008) History, chance and adaptation during biological invasion: separating stochastic phenotypic evolution from response to selection. Ecology Letters, 11, 852–66.

Kissling, W.D., Dormann, C.F., Groeneveld, J., *et al.* (2012) Towards novel approaches to modelling biotic interactions in multispecies assemblages at large spatial extents. Journal of Biogeography, 39, 2163–78.

Klironomos, J.N. (2003) Variation in plant response to native and exotic arbuscular mycorrhizal fungi. Ecology, 84, 2292–301.

Klock, M.M., Barrett, L.G., Thrall, P.H., *et al.* (2015) Host promiscuity in symbiont associations can influence exotic legume establishment and colonization of novel ranges. Diversity and Distributions, 21, 1193–203.

Knevel, I.C., Lans, T., Menting, F.B.J., *et al.* (2004) Release from native root herbivores and biotic resistance by soil pathogens in a new habitat both affect the alien *Ammophila arenaria in* South Africa. Oecologia, 141, 502–10.

Kolling, N., Behrens, T.E., Mars, R.B., *et al.* (2012) Neural mechanisms of foraging. Science, 336, 95–8.

Kondoh, M. (2003) Foraging adaptation and the relationship between food-web complexity and stability. Science, 299, 1388–91.

Lafay, B. and Burdon, J.J. (2006) Molecular diversity of rhizobia nodulating the invasive legume *Cytisus scoparius* in Australia. Journal of Applied Microbiology, 100, 1228–38.

Lankau, R.A. (2012) Coevolution between invasive and native plants driven by chemical competition and soil biota. Proceedings of the National Academy of Sciences of the USA, 109, 11240–5.

Lankau, R.A. and Nodurft, R.N. (2013) An exotic invader drives the evolution of plant traits that determine mycorrhizal fungal diversity in a native competitor. Molecular Ecology, 22, 5472–85.

Laska, M.S. and Wootton, J.T. (1998) Theoretical concepts and empirical approaches to measuring interaction strength. Ecology, 79, 461–76.

Lau, J.A. and Terhorst, C.P. (2015) Causes and consequences of failed adaptation to biological invasions: the role of ecological constraints. Molecular Ecology, 24, 1987–98.

Lee, C.E. (2002) Evolutionary genetics of invasive species. Trends in Ecology and Evolution, 17, 386–91.

Lello, J., Boag, B., Fenton, A., *et al.* (2004) Competition and mutualism among the gut helminths of a mammalian host. Nature, 428, 840–4.

Levine, J.M., Adler, P.B., and Yelenik, S.G. (2004) A meta-analysis of biotic resistance to exotic plant invasions. Ecology Letters, 7, 975–89.

Lomolino, M.V. (2005) Body size evolution in insular vertebrates: generality of the island rule. Journal of Biogeography, 32, 1683–99.

Lonsdale, W.M. (1993) Rates of spread of an invading species—*Mimosa pigra* in northern Australia. Journal of Ecology, 81, 513–21.

Louda, S.M. and Rodman, J.E. (1996) Insect herbivory as a major factor in the shade distribution of a native crucifer (*Cardamine cordifolia* A. Gray, bittercress). Journal of Ecology, 84, 229–37.

Lovett, G.M., Canham, C.D., Arthur, M.A., *et al.* (2006) Forest ecosystem responses to exotic pests and pathogens in eastern North America. BioScience, 56, 395–405.

Lucek, K., Sivasundar, A., and Seehausen, O. (2012) Evidence of adaptive evolutionary divergence during biological invasion. PLoS One, 7, e49377.

Lucek, K., Sivasundar, A., and Seehausen, O. (2014) Disentangling the role of phenotypic plasticity and genetic divergence in contemporary ecotype formation during a biological invasion. Evolution, 68, 2619–32.

MacArthur, R.H. and Levins, R. (1967) The limiting similarity convergence and divergence of coexisting species. American Naturalist, 101, 377–85.

MacDougall, A.S. and Turkington, R. (2005) Are invasive species the drivers or passengers of change in degraded ecosystems? Ecology, 86, 42–55.

Mack, R.N. (1996) Biotic barriers to plant naturalization. In: Moran, V.C. and Hoffman, J.H. (eds) *Proceedings of the IX International Symposium on Biological Control of Weeds*. Cape Town: University of Cape Town, pp. 39–46.

Maron, J.L. and Vila, M. (2001) When do herbivores affect plant invasion? Evidence for the natural enemies and biotic resistance hypotheses. Oikos, 95, 361–73.

Maron, J.L., Vila, M., and Arnason, J. (2004) Loss of enemy resistance among introduced populations of St. John's Wort (*Hypericum perforatum*). Ecology, 85, 3243–53.

Marsudi, N.D.S., Glenn, A.R., and Dilworth, M.J. (1999) Identification and characterization of fast- and slow-growing root nodule bacteria from South-Western Australian soils able to nodulate *Acacia saligna*. Soil Biology and Biochemistry, 31, 1229–38.

Matsumura, C., Yokoyama, J., and Washitani, I. (2004). Invasion status and potential ecological impacts of an invasive alien bumblebee, Bombus terrestris L. (Hymenoptera: Apidae) naturalized in Southern Hokkaido, Japan. Global Environmental Research, 8, 51–66.

May, R.M. (1972) Will a large complex system be stable? Nature, 238, 413–14.

May, R.M. (1973) Time-delay versus stability in population models with two and three trophic levels. Ecology, 54, 315–25.

McGill, B.J. (2010) Matters of scale. Science, 328, 575–6.

McGill, B.J., Etienne, R.S., Gray, J.S., et al. (2007) Species abundance distributions: moving beyond single prediction theories to integration within an ecological framework. Ecology Letters, 10, 995–1015.

McPheron, B.A., Smith, D.C., and Berlocher, S.H. (1988) Genetic differences between host races of *Rhagoletis pomonella*. Nature, 336, 64–6.

Meier, E.S., Kienast, F.,Pearman, P.B., et al. (2010) Biotic and abiotic variables show little redundancy in explaining tree species distributions. Ecography, 33, 1038–48.

Menge, B.A., Berlow, E.L., Blanchette, C.A., et al. (1994) The keystone species concept: variation in interaction strength in a rocky intertidal habitat. Ecological Monographs, 64, 249–86.

Michalet, R., Brooker, R.W., Cavieres, L.A., et al. (2006) Do biotic interactions shape both sides of the humped-back model of species richness in plant communities? Ecology Letters, 9, 767–73.

Midgley, G.F., Davies, I.D., Albert, C.H., et al. (2010) BioMove–an integrated platform simulating the dynamic response of species to environmental change. Ecography, 33, 612–16.

Minoarivelo, H.O., Hui, C., Terblanche, J.S., et al. (2014) Detecting phylogenetic signal in mutualistic interaction networks using a Markov process model. Oikos, 123, 1250–60.

Mitchell, C.E., Agrawal, A.A., Bever, J.D., et al. (2006) Biotic interactions and plant invasions. Ecology Letters, 9, 726–40.

Mitchell, C.E. and Power, A.G. (2003) Release of invasive plants from fungal and viral pathogens. Nature, 421, 625–7.

Mitchell, C.E. and Power, A.G. (2006) Plant communities and disease ecology. In: Collinge, S.K. and Ray, C. (eds) *Disease Ecology: Community Structure and Pathogen Dynamics*. Oxford: Oxford University Press, pp. 58–72.

Mitchell, C.E., Reich, P.B., Tilman, D., et al. (2003) Effects of elevated CO_2, nitrogen deposition, and decreased species diversity on foliar fungal plant disease. Global Change Biology, 9, 438–51.

Mohamed, S.H., Smouni, A., Neyra, M., et al. (2000) Phenotypic characteristics of root-nodulating bacteria isolated from Acacia spp. grown in Libya. Plant and Soil, 224, 171–83.

Montoya, J.M., Emmerson, M.C., Solé, R.V., et al. (2005) Perturbations and indirect effects in complex food webs. In: De Ruiter, P.C., Wolters, V., Moore, J.C. (eds) *Dynamic Food Webs: Multispecies Assemblages, Ecosystem Development, and Environmental Change*. Amsterdam: Elsevier, pp. 369–80.

Montoya, J.M., Pimm, S.L., and Solé, R.V. (2006) Ecological networks and their fragility. Nature, 442, 259–64.

Moore, J.C., De Ruiter, P.C., and Hunt, H.W. (1993) Influence of productivity on the stability of real and model ecosystems. Science, 261, 906–8.

Morales, C.L. and Aizen, M.A. (2006) Invasive mutualisms and the structure of plant–pollinator interactions in the temperate forests of north-west Patagonia, Argentina. Journal of Ecology, 94, 171–80.

Morales, C.L. and Traveset, A. (2009) A meta-analysis of impacts of alien vs. native plants on pollinator visitation and reproductive success of co-flowering native plants. Ecology Letters, 12, 716–28.

Morin, P.J., Lawler, S.P., and Johnson, E.A. (1988) Competition between aquatic insects and vertebrates: interaction strength and higher order interactions. Ecology, 69, 1401–9.

Morris, W.F. (2003) Which mutualists are most essential? Buffering of plant reproduction against the extinction of pollinators. In: Kareiva, P. and Levin, S.A. (eds) *The Importance of Species: Perspectives on Expendability and Triage*. Princeton, NJ: Princeton University Press, pp. 260–80.

Mouillot, D., Dumay, O., and Tomasini, J.A. (2007) Limiting similarity, niche filtering and functional diversity in coastal lagoon fish communities. Estuarine, Coastal and Shelf Science, 71, 443–56.

Müller-Schärer, H., Schaffner, U., and Steinger, T. (2004) Evolution in invasive plants: implications for biological control. Trends in Ecology and Evolution, 19, 417–22.

Murdoch, W.W. (1969) Switching in general predators: experiments on predator specificity and stability of prey populations. Ecological Monographs, 39, 335–54.

Ndlovu, J., Richardson, D.M., Wilson, J.R.U., et al. (2013) Co-invasion of South African ecosystems by an Australian legume and its rhizobial symbionts. Journal of Biogeography, 40, 1240–51.

Noonburg, E.G. and Byers, J.E. (2005) More harm than good: when invader vulnerability to predators enhances impact on native species. Ecology, 86, 2555–60.

Novak, S.J. and Mack, R.N. (2005) Genetic bottlenecks in alien plant species: influence of mating systems and introduction dynamics. In: Sax, D.F., Stachowitz, J.J., and Gaines, S.D. (eds) Species Invasions: Insights into Ecology, Evolution, and Biogeography. Sunderland, MA: Sinauer Associates, pp. 201–28.

Nuismer, S.L., Jordano, P., and Bascompte, J. (2013) Coevolution and the architecture of mutualistic networks. Evolution, 67, 338–54.

Nuwagaba, S., Zhang, F., and Hui, C. (2015) A hybrid behavioural rule of adaptation and drift explains the emergent architecture of antagonistic networks. Proceedings of the Royal Society B: Biological Sciences, 282, 20150320.

Oduor, A.M.O. (2013) Evolutionary responses of native plant species to invasive plants: a review. New Phytologist, 200, 986–92.

Okubo, A., Maini, P.K., Williamson, M.H., et al. (1989) On the spatial spread of the grey squirrel in Britain. Proceedings of the Royal Society B: Biological Sciences, 238, 113–25.

Okuyama, T. and Holland, J.N. (2008) Network structural properties mediate the stability of mutualistic communities. Ecology Letters, 11, 208–16.

Olesen, J.M., Bascompte, J., Elberling, H., et al. (2008) Temporal dynamics in a pollination network. Ecology, 89, 1573–82.

Olesen, J.M., Eskildsen, L.I., and Venkatasamy, S. (2002) Invasion of pollination networks on oceanic islands: importance of invader complexes and endemic super generalists. Diversity and Distributions, 8, 181–92.

Olff, H., De Leeuw, J., Bakker, J.P., et al. (1997) Vegetation succession and herbivory in a salt marsh: changes induced by sea level rise and silt deposition along an elevational gradient. Journal of Ecology, 85, 799–814.

Ovaskainen, O., Hottola, J., and Siitonen, J. (2010) Modeling species co-occurrence by multivariate logistic regression generates new hypotheses on fungal interactions. Ecology, 91, 2514–21.

Paine, R.T. (1966) Food web complexity and species diversity. American Naturalist, 100, 65–75.

Paine, R.T. (1980) Food webs: linkage, interaction strength and community infrastructure. Journal of Animal Ecology, 49, 667–85.

Paine, R.T. (1992) Food-web analysis through field measurement of per capita interaction strength. Nature, 355, 73–5.

Paolucci, E.M., MacIsaac, H.J., and Ricciardi, A. (2013) Origin matters: alien consumers inflict greater damage on prey populations than do native consumers. Diversity and Distributions, 19, 988–95.

Parker, I.M. and Gilbert, G.S. (2004) The evolutionary ecology of novel plant–pathogen interactions. Annual Reviews in Ecology, Evolution and Systematics, 35, 675–700.

Parker, J.D. and Hay, M.E. (2005) Biotic resistance to plant invasions? Native herbivores prefer non-native plants. Ecology Letters, 8, 959–67.

Parker, M.A. (2001) Mutualism as a constraint on invasion success for legumes and rhizobia. Diversity and Distributions, 7, 125–36.

Parker, M.A., Malek, W., and Parker, I.M. (2006) Growth of an invasive legume is symbiont limited in newly occupied habitats. Diversity and Distributions, 12, 563–71.

Parker, I.M., Simberloff, D., Lonsdale, W.M., et al. (1999) Impact: toward a framework for understanding the ecological effects of invaders. Biological Invasions, 1, 3–19.

Paterson, R.A., Dick, J.T., Pritchard, D.W., et al. (2015) Predicting invasive species impacts: a community module functional response approach reveals context dependencies. Journal of Animal Ecology, 84, 453–63.

Pearson, R.G. and Dawson, T.P. (2003) Predicting the impacts of climate change on the distribution of species: are bioclimate envelope models useful? Global Ecology and Biogeography, 12, 361–71.

Pérez-Fernández, M.A. and Lamont, B.B. (2003) Nodulation and performance of exotic and native legumes in Australian soils. Australian Journal of Botany, 51, 543–53.

Petanidou, T., Kallimanis, A.S., Tzanopoulos, J., et al. (2008) Long-term observation of a pollination network: fluctuation in species and interactions, relative invariance of network structure and implications for estimates of specialization. Ecology Letters, 11, 564–75.

Petchey, O.L., Beckerman, A.P., Riede, J.O., et al. (2008) Size, foraging, and food web structure. Proceedings of the National Academy of Sciences of the USA, 105, 4191–6.

Peters, R.H. (1983) The Ecological Implications of Body Size. Cambridge: Cambridge University Press.

Pettersson, M.W. (1991) Pollination by a guild of fluctuating moth populations: option for unspecialization in *Silene vulgaris*. Journal of Ecology, 79, 591–604.

Pimm, S.L., Lawton, J.H., and Cohen, J.E. (1991) Food web patterns and their consequences. Nature, 350, 669–74.

Pitcairn, M.J. (2011) Biological control, of plants. In: Simberloff, D. and Rejmánek, M. (eds) *Encyclopedia of Biological Invasions*, Berkeley, CA: University of California Press, pp. 63–70.

Plieninger, T., Hui, C., Gaertner, M., *et al.* (2014) The impact of land abandonment on species richness and abundance in the Mediterranean Basin: a meta-analysis. PLoS One, 9, e98355.

Price, J.N. and Pärtel, M. (2013) Can limiting similarity increase invasion resistance? A meta-analysis of experimental studies. Oikos, 122, 649–56.

Raffaelli, D.G. and Hall, S.J. (1996) Assessing the relative importance of trophic links in food webs. In: Polis, G.A. and Winemiller, K.O. (eds) *Food Webs: Integration of Patterns and Dynamics*. New York, NY: Chapman and Hall, pp. 185–91.

Recart, W., Ackerman, J.D., and Cuevas, A.A. (2013) There goes the neighborhood: apparent competition between invasive and native orchids mediated by a specialist florivorous weevil. Biological Invasions, 15, 283–93.

Reinhart, K.O. and Callaway, R.M. (2004) Soil biota facilitate exotic *Acer* invasions in Europe and North America. Ecological Applications, 14, 1737–45.

Reinhart, K.O. and Callaway, R.M. (2006) Soil biota and invasive plants. New Phytologist, 170, 445–57.

Rezende, E.L., Lavabre, J.E., Guimarães, P.R., *et al.* (2007) Non-random coextinctions in phylogenetically structured mutualistic networks. Nature, 448, 925–8.

Rice, W.R. and Sax, D.F. (2005) Testing fundamental evolutionary questions at large spatial and demographic scales: species invasions as an underappreciated tool. In: Sax, D.F., Stachowitz, J.J., and Gaines, S.D. (eds) *Species Invasions: Insights into Ecology, Evolution, and Biogeography*. Sunderland, MA: Sinauer Associates, pp. 291–308.

Richards, C.L., Bossdorf, O., Muth, N.Z., *et al.* (2006) Jack of all trades, master of some? On the role of phenotypic plasticity in plant invasions. Ecology Letters, 9, 981–93.

Richardson, D.M., Allsopp, N., D'Antonio, C.M., *et al.* (2000) Plant invasions—the role of mutualisms. Biological Reviews, 75, 65–93.

Richardson, D.M., Carruthers, J., Hui, C., *et al.* (2011) Human-mediated introductions of Australian acacias—a global experiment in biogeography. Diversity and Distributions, 17, 771–87.

Richardson, D.M. and Higgins, S.I. (1998) Pines as invaders in the southern hemisphere. In: Richardson, D.M. (ed.) *Ecology and Biogeography of Pinus*. Cambridge: Cambridge University Press, pp. 450–73.

Richardson, D.M. and Ricciardi, A. (2013) Misleading criticisms of invasion science: a field guide. Diversity and Distributions, 19, 1461–7.

Richardson, D.M., Rouget, M., Ralston, S.J., *et al.* (2005) Species richness of alien plants in South Africa: environmental correlates and the relationship with indigenous plant species richness. EcoScience, 12, 391–402.

Ricklefs, R.E. (2008) Disintegration of the ecological community. American Naturalist, 172, 741–50.

Rieseberg, L.H., Beckstrom-Sternberg, S., and Doan, K. (1990) *Helianthus annuus* ssp. texanus has chloroplast DNA and nuclear ribosomal RNA genes of *Helianthus debilis* ssp. cucumerifolius. Proceedings of the National Academy of Sciences of the USA, 87, 593–7.

Rodríguez-Echeverria, S. (2010) Rhizobial hitchhikers from Down Under: invasional meltdown in a plant-bacteria mutualism? Journal of Biogeography, 37, 1611–22.

Rodríguez-Echeverría, S., Le Roux, J.J., Crisóstomo, J.A., *et al.* (2011) Jack-of-all-trades and master of many? How does associated rhizobial diversity influence the colonization success of Australian *Acacia* species? Diversity and Distributions, 17, 946–57.

Rosati, A.G. and Stevens, J.R. (2009) Rational decisions: the adaptive nature of context-dependent choice. In: Watanabe, A.P., Blaisdell, L., Huber, L., *et al.* (eds) *Rational Animals, Irrational Humans*. Tokyo: Keio University Press, pp. 101–17.

Roura-Pascual, N., Krug, R.M., Richardson, D.M., *et al.* (2010) Spatially-explicit sensitivity analysis for conservation management: exploring the influence of decisions in invasive alien plant management. Diversity and Distributions, 16, 426–38.

Roura-Pascual, N., Richardson, D.M., Chapman, R.A., *et al.* (2011) Managing biological invasions: charting courses to desirable futures in the Cape Floristic Region. Regional Environmental Change, 11, 311–20.

Rowe, C.L. and Leger, E.A. (2011) Competitive seedlings and inherited traits: a test of rapid evolution of *Elymus multisetus* (big squirreltail) in response to cheatgrass invasion. Evolutionary Applications, 4, 485–98.

Roxburgh, S.H. and Wilson, J.B. (2000) Stability and coexistence in a lawn community: mathematical prediction of stability using a community matrix with parameters derived from competition experiments. Oikos, 88, 395–408.

Ruesink, J.L. (1998) Variation in per capita interaction strength: thresholds due to nonlinear dynamics and nonequilibrium conditions. Proceedings of the National Academy of Sciences of the USA, 95, 6843–7.

Russell, R., Wood, S.A., Allison, G., *et al.* (2006) Scale, environment, and trophic status: the context dependency of community saturation in rocky intertidal communities. American Naturalist, 167, E158–70.

Sala, E. and Graham, M.H. (2002) Community-wide distribution of predator–prey interaction strength in kelp forests. Proceedings of the National Academy of Sciences of the USA, 99, 3678–83.

Salo, P., Korpimaki, E., Banks, P.B., *et al.* (2007) Alien predators are more dangerous than native predators to prey populations. Proceedings of the Royal Society B: Biological Sciences, 274, 1237–43.

Sarnelle, O. (2003) Nonlinear effects of an aquatic consumer: causes and consequences. American Naturalist, 161, 478–96.

Saul, W.C. and Jeschke, J.M. (2015) Eco-evolutionary experience in novel species interactions. Ecology Letters, 18, 236–45.

Saul, W.-C., Jeschke, J.M., and Heger, T. (2013) The role of eco-evolutionary experience in invasion success. NeoBiota, 17, 57–74.

Sax, D.F., Stachowicz, J.J., Brown, J.H., *et al.* (2007) Ecological and evolutionary insights from species invasions. Trends in Ecology and Evolution, 22, 465–71.

Schemske, D.W. and Horvitz, C.C. (1984) Variation among floral visitors in pollination ability: a precondition for mutualism specialization. Science, 225, 519–21.

Scherm, H. (2004) Climate change: can we predict the impacts on plant pathology and pest management? Canadian Journal of Plant Pathology, 26, 267–73.

Schluter D. (1982) Seed and patch selection by Galapagos Ecuador ground finches Geospiza in relation to foraging efficiency and food supply. Ecology, 63, 1106–20.

Schoener, T.W. (1993) On the relative importance of direct versus indirect effects in ecological communities. In: Kawanabe, H., Cohen, J.E., and Iwasaki, K. (eds) *Mutualism and Community Organization: Behavioral, Theoretical and Food Web Approaches*. Oxford: Oxford University Press, pp. 365–411.

Schwarz, D., Matta, B.M., Shakir-Botteri, N.L., *et al.* (2005) Host shift to an invasive plant triggers rapid animal hybrid speciation. Nature, 436, 546–9.

Schweiger, O., Biesmeijer, J.C., Bommarco, R., *et al.* (2010) Multiple stressors on biotic interactions: how climate change and alien species interact to affect pollination. Biological Reviews, 85, 777–95.

Seabloom, E.W., Harpole, W.S., Reichman, O.J., *et al.* (2003) Invasion, competitive dominance, and resource use by exotic and native California grassland species. Proceedings of the National Academy of Sciences of the USA, 100, 13384–9.

Sebastián-González, E., Sánchez-Zapata, J.A., and Botella, F. (2010) Agricultural ponds as alternative habitat for waterbirds: spatial and temporal patterns of abundance and management strategies. European Journal of Wildlife Research, 56, 11–20.

Shipley, B., Vile, D., and Garnier, É. (2006) From plant traits to plant communities: a statistical mechanistic approach to biodiversity. Science, 314, 812–14.

Simberloff, D. and Von Holle, B. (1999) Positive interactions of nonindigenous species: invasional meltdown? Biological Invasions, 1, 21–32.

Skelly, D.K. (2002) Experimental venue and estimation of interaction strength. Ecology, 83, 2097–101.

Smith, S.E. and Read, D.J. (1997) *Mycorrhizal Symbiosis*. San Diego, CA: Academic Press.

Soberón, J. (2007) Grinnellian and Eltonian niches and geographic distributions of species. Ecology Letters, 10, 1115–23.

Staddon, J.E.R. (2010) *Adaptive Behaviour and Learning*. Cambridge: Cambridge University Press.

Staniczenko, P., Lewis, O.T., Jones, N.S., *et al.* (2010) Structural dynamics and robustness of food webs. Ecology Letters, 13, 891–9.

Stastny, M., Schaffner, U., and Elle, E. (2005) Do vigour of introduced populations and escape from specialist herbivores contribute to invasiveness? Journal of Ecology, 93, 27–37.

Stephens, D.W. and Krebs, J.R. (1986) *Foraging Theory*. Princeton, NJ: Princeton University Press.

Stimec, J., ScottDupree, C.D., and McAndrews, J.H. (1997) Honey bee, *Apis mellifera*, pollen foraging in southern Ontario. Canadian Field-Naturalist, 111, 454–6.

Stohlgren, T.J., Barnett, D.T., and Kartesz, J.T. (2003) The rich get richer: patterns of plant invasions in the United States. Frontiers in Ecology and the Environment, 1, 11–14.

Stout, M.J., Thaler, J.S., and Thomma, B.P. (2006) Plant-mediated interactions between pathogenic microorganisms and herbivorous arthropods. Annual Reviews in Entomology, 51, 663–89.

Strauss, S.Y., Lau, J.A., and Carroll, S.P. (2006) Evolutionary responses of natives to introduced species: what do introductions tell us about natural communities? Ecology Letters, 9, 357–74.

Strayer, D.L., Eviner, V.T., Jeschke, J.M., *et al.* (2006) Understanding the long-term effects of species invasions. Trends in Ecology and Evolution, 21, 645–51.

Taylor, B.W., McIntosh, A.R., and Peckarsky, B.L. (2002) Reach-scale manipulations show invertebrate grazers depress algal resources in streams. Limnology and Oceanography, 47, 893–9.

Thébault, E. and Fontaine, C. (2010) Stability of ecological communities and the architecture of mutualistic and trophic networks. Science, 329, 853–6.

Thomson, D. M. (2006) Detecting the effects of introduced species: a case study of competition between Apis and Bombus. Oikos, 114, 407–18.

Torchin, M.E. and Mitchell, C.E. (2004) Parasites, pathogens, and invasions by plants and animals. Frontiers in Ecology and the Environment, 2, 183–90.

Traveset, A., Olesen, J.M., Nogales, M., *et al.* (2015) Bird–flower visitation networks in the Galápagos unveil a widespread interaction release. Nature Communications, 6, 6376.

Traveset, A. and Richardson, D.M. (2011) Mutualisms—key drivers of invasions . . . key casualties of invasions. In: Richardson, D.M. (ed.) *Fifty Years of Invasion Ecology. The Legacy of Charles Elton.* Oxford: Wiley-Blackwell, pp. 143–60.

Traveset, A. and Richardson, D.M. (2014) Mutualistic interactions and biological invasions. Annual Review of Ecology, Evolution, and Systematics, 45, 89–113.

Urban, M.C., Leibold, M.A., Amarasekare, P., *et al.* (2008) The evolutionary ecology of metacommunities. Trends in Ecology and Evolution, 23, 311–17.

Van der Heijden, M.G., Klironomos, J.N., *et al.* (1998) Mycorrhizal fungal diversity determines plant biodiversity, ecosystem variability and productivity. Nature, 396, 69–72.

Vandermeer, J.H. (1969) The competitive structure of communities: an experimental approach with protozoa. Ecology, 50, 362–71.

Van der Putten, W.H., Macel, M., and Visser, M.E. (2010) Predicting species distribution and abundance responses to climate change: why it is essential to include biotic interactions across trophic levels. Philosophical Transactions of the Royal Society B: Biological Sciences, 365, 2025–34.

Van Der Putten, W.H., Yeates, G.W., Duyts, H., *et al.* (2005) Invasive plants and their escape from root herbivory: a worldwide comparison of the root-feeding nematode communities of the dune grass *Ammophila arenaria* in natural and introduced ranges. Biological Invasions, 7, 733–46.

Van Ruijven, J., De Deyn, G.B., and Berendse, F. (2003) Diversity reduces invasibility in experimental plant communities: the role of plant species. Ecology Letters, 6, 910–18.

Vaughan, W. and Herrnstein, R.J. (1987) Choosing among natural stimuli. Journal of the Experimental Analysis of Behavior, 47, 5–16.

Vázquez, D.P., Morris, W.F., and Jordano, P. (2005) Interaction frequency as a surrogate of population-level effects of animal mutualists on plants. Ecology Letters, 8, 1088–94.

Vellend, M., Harmon, L.J., Lockwood, J.L., *et al.* (2007) Effects of exotic species on evolutionary diversification. Trends in Ecology and Evolution, 22, 481–8.

Vestergård, M., Rønn, R., and Ekelund, F. (2015) Above–belowground interactions govern the course and impact of biological invasions. AoB Plants, 7, plv025.

Vilà, M., Carrillo-Gavilán, A., Vayreda, J., *et al.* (2013) Disentangling biodiversity and climatic determinants of wood production. PLoS One, 8, e53530.

Vilà, M., Rohr, R.P., Espinar, J.L., *et al.* (2015) Explaining the variation in impacts of non-native plants on local-scale species richness: the role of phylogenetic relatedness. Global Ecology and Biogeography, 24, 139–46.

Vilà, M., Weber, E., and D'Antonio, C. (2000) Conservation implications of invasion by plant hybridization. Biological Invasions, 2, 207–17.

Vitousek, P.M. (1990) Biological invasions and ecosystem processes: towards an integration of population biology and ecosystem studies. Oikos, 57, 7–13.

Vitousek, P.M., Walker, L.R., Whiteaker, L.D., *et al.* A. (1987) Biological invasion by *Myrica faya* alters ecosystem development in Hawaii. Science, 238, 802–4.

Vivanco, J.M., Bais, H.P., Stermitz, F.R., *et al.* (2004) Biogeographical variation in community response to root allelochemistry: novel weapons and exotic invasion. Ecology Letters, 7, 285–92.

Wandrag, E.M., Sheppard, A., Duncan, R.P., *et al.* (2013) Reduced availability of rhizobia limits the performance but not invasiveness of introduced Acacia. Journal of Ecology, 101, 1103–13.

Wares, J.P., Hughes, A.R., and Grosberg, R.K. (2005) Mechanisms that drive evolutionary change: insights from species introductions and invasions. In: Sax, D.F., Stachowitz, J.J., and Gaines, S.D. (eds) *Species Invasions: Insights into Ecology, Evolution, and Biogeography.* Sunderland, MA: Sinauer Associates, pp. 229–57.

Werner, E.E. (1992) Individual behavior and higher-order species interactions. American Naturalist, 140 Supplement, S5–32.

Werner, E.E. and Hall, D.J. (1974) Optimal foraging and the size selection of prey by the bluegill sunfish *Lepomis macrochirus*. Ecology, 55, 1042–52.

Whittaker, R.H. (1975) *Communities and Ecosystems.* New York, NY: Macmillan.

Whittall, J.B. and Hodges, S.A. (2007) Pollinator shifts drive increasingly long nectar spurs in columbine flowers. Nature, 447, 706–9.

Wiens, J.J. (2011) The niche, biogeography and species interactions. Philosophical Transactions of the Royal Society B: Biological Sciences, 366, 2336–50.

Wilkinson, D.M. (2004) The parable of Green Mountain: Ascension Island, ecosystem construction and ecological fitting. Journal of Biogeography, 31, 1–4.

Williams, D.G., Mack, R.N., and Black, R.A. (1995) Ecophysiology of introduced *Pennisetum setaceum* on Hawaii: the role of phenotypic plasticity. Ecology, 76, 1569–80.

Wisz, M.S., Pottier, J., Kissling, W.D., *et al.* (2013) The role of biotic interactions in shaping distributions and realised assemblages of species: implications for species distribution modelling. Biological Reviews, 88, 15–30.

Wolfe, B.E., Husband, B.C., and Klironomos, J.N. (2005) Effects of a belowground mutualism on an aboveground mutualism. Ecology Letters, 8, 218–23.

Wootton, J.T. (1992) Indirect effects, prey susceptibility, and habitat selection: impacts of birds on limpets and algae. Ecology, 73, 981–91.

Wootton, J.T. (1994) The nature and consequences of indirect effects in ecological communities. Annual Review of Ecology and Systematics, 25, 443–66.

Wootton, J.T. (1997) Estimates and tests of per capita interaction strength: diet, abundance, and impact of intertidally foraging birds. Ecological Monographs, 67, 45–64.

Wootton, J.T. and Emmerson, M. (2005) Measurement of interaction strength in nature. Annual Review of Ecology, Evolution, and Systematics, 419–44.

Wu, S.H., Hsieh, C.F., Chaw, S.M., *et al.* (2004) Plant invasions in Taiwan: insights from the flora of casual and naturalized alien species. Diversity and Distributions, 10, 349–62.

Yodzis, P. (1988) The indeterminacy of ecological interactions as perceived through perturbation experiments. Ecology, 69, 508–15.

Yodzis, P. and Innes, S. (1992) Body size and consumer-resource dynamics. American Naturalist, 139, 1151–75.

Zhang, F. and Hui, C. (2014) Recent experience-driven behaviour optimizes foraging. Animal Behaviour, 88, 13–19.

Zhang, F., Hui, C., and Terblanche, J.S. (2011) An interaction switch predicts the nested architecture of mutualistic networks. Ecology Letters, 14, 797–803.

CHAPTER 7

Regime shifts

7.1 Regime and bifurcation

The biosphere is in a critical transition driven by ongoing human-induced changes to the environment (Barnosky *et al.* 2012). Biological invasions are a major contributor to such changes and have driven many ecosystems towards and beyond critical transitions to cause major degradation, suggesting the existence of alternative stable states or regimes in ecosystems. A theory of multiple stable states in ecosystems was proposed by Holling (1973), Lewontin (1969), and May (1977) to explain bistable patterns and catastrophic regime shifts in coral reefs, semi-arid ecosystems, lakes, and fisheries (Scheffer *et al.* 2001; Schröder *et al.* 2005). Theoretical work suggests that even simple systems can possess alternative regimes and unpredictable behaviours as long as appropriate non-linearities and positive feedbacks exist. Although it is straightforward to document the critical transition from one regime to another, determining whether a system is approaching, or has already crossed, tipping points is non-trivial. To achieve such predictive power we need models that capture full knowledge of the system. There has been lively debate regarding theories pertaining to early warning signals for regime shifts.

The magnitude of biological invasions is increasing worldwide: more species are invading, and the area affected by invasions and the types and overall extent and complexity of impacts are increasing (Pyšek and Richardson 2010). Limited resources mean that not all invasions can be, or need to be, managed. There is increasing pressure on managers to apply objective and defendable protocols for deciding where to use available resources (Hulme *et al.* 2013). Impacts are too often assumed rather than proven and quantified, and efforts may be wasted on managing invasive species that have little or no impact (Hejda and Pyšek 2006; Meffin *et al.* 2010). Some invasions generate only community-level changes whereas others can fundamentally alter the structure and functioning of ecosystems (Levine *et al.* 2003; Vilà *et al.* 2011; Hui *et al.* 2013), with major consequences for native biodiversity and ecosystem functioning, and for the delivery of ecosystem services (Pejchar and Mooney 2009). To improve management efficacy we need to separate those species that have the potential to inflict ecosystem-level impacts from those that have less impact (Hulme *et al.* 2013). Species that have the potential to cause regime shifts—that is, to restructure and transform ecosystems fundamentally—should be considered the most high-impact species and given priority for intervention.

A system's regime is a hysteretic behaviour of the system, with its resilience (and local stability) maintained by mutually balancing processes; a regime shift is associated with the loss of the current regime resilience largely driven by reinforcing processes or feedbacks (Scheffer *et al.* 2001; Bennett *et al.* 2005). This means that regime shifts are often large and abrupt changes in ecosystem structure and functioning (Rietkerk *et al.* 2004; Scheffer *et al.* 2012) that are triggered either by the reorganization of internal feedback mechanisms or by the introduction of new feedbacks (Fig. 7.1). Regime-shift phenomena have been reported in many types of ecosystems, notably in freshwater lakes, coral reefs, semi-arid systems and savannas (Scheffer *et al.* 2001). They may be triggered by many factors, including extreme weather, human disturbance, or biological invasions. Restoring a system to its previous regime after a drastic shift is often difficult or impossible due to the 'sticky' behaviour of the system (MA 2005).

Invasion Dynamics. Cang Hui and David M. Richardson, Oxford University Press (2017).
© Cang Hui and David M. Richardson. DOI 10.1093/acprof:oso/9780198745334.001.0001

(a) **Low-impact invader**

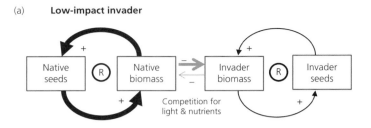

(b) **Post-regime shift for a high-impact invader**

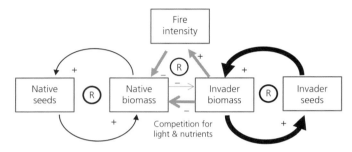

Fig. 7.1 A simplified conceptualization of the change in dominant system feedbacks that may accompany a regime shift driven by a high-impact invader (the example is based on tree invasions in South African fynbos vegetation). The thickness of curved arrows indicates the relative amount of energy and resources entrained in the competing feedback loops. The grey arrows connecting the different loops indicate the size of the negative effect of the invader on the native species and vice versa. R indicates a reinforcing feedback. a) Feedback configuration of an ecosystem for a low-impact invader. The negative effects of the dominant native species (through competition) control population numbers of the invader. b) Feedback configuration of an ecosystem after a regime shift driven by a high-impact invader has occurred. The invader introduces a new reinforcing feedback (fire intensity) that destabilizes the ecosystem and shifts it to a regime where the invasive species has the dominant influence. From Gaertner *et al.* (2014). Reproduced with permission from John Wiley & Sons.

Elucidation of regime shifts, including those generated by biological invasions, demands an understanding of the determinants of stability of a dynamical system. Dynamical systems are models describing the change rate of state variables in a system using differential or difference equations (Dercole and Rinaldi 2012). In most cases, state variables of a system are ecologically meaningful and measurable components of an ecosystem. The most common state variable would be the density or biomass of a particular population, although other variables can also be included if a mechanistic understanding is sufficient for depicting its dynamics, for instance the amount of harvest driven by both ecological and economic contexts. In the most common scenario, each system is described by s number of state equations using ordinary differential equations in the general form:

$$\frac{dn_i}{dt} = f_i\left(n_1[t], \ldots, n_s[t]\right) \tag{7.1}$$

with $i = 1, \ldots, s$. The left-hand side dn_i/dt depicts the time derivative of state variable $n_i[t]$, which is the slope or gradient of n_i along time t (in some studies it is also denoted as $\dot{n}_i$). In the case of population dynamics, this time derivative often means population change rate. For instance, in the classic Lotka–Volterra model, dn_i/dt is the rate of population change of species i in a community. The function f_i depicts the exact form of the population change rate, often as a density-dependent function of fecundity and mortality.

Given an initial state, the state equations will uniquely define a trajectory of the system. A time-invariant trajectory defines equilibrium. A reinforcing feedback loop between two state variable n_1 and n_2 ($n_1 \rightarrow n_2 \rightarrow n_1$; e.g. representing the loop between invader biomass and invader seeds in Fig. 7.1) can be depicted by their positive interaction strengths on each other; that is, partial derivatives $\partial f_1 / \partial n_2 > 0$ (representing interaction from n_2 on n_1)

and $\partial f_2 / \partial n_1 > 0$ (representing interaction from n_1 on n_2) calculated at a given equilibrium. Note, having both negative partial derivatives also defines a reinforcing feedback loop, although often uninterested in ecology. A reinforcing feedback loop can be identified from any number of state variables (e.g. $n_1 \to n_2 \to n_3 \to n_1$) if their interaction strengths along the loop are not trivial and if the multiplication of these interaction strengths is positive (e.g. among fire intensity, native and invader biomass in Fig. 7.1). We explore the characteristics of an introduced species that are more likely to cause such positive reinforcing feedbacks in invaded ecosystems later in the chapter.

The stability of the equilibrium point can be examined by calculating the eigenvalues of the linearization matrix of the system, known as the Jacobian. Typical research questions when examining a dynamical system are often related to the behaviour (shape) of these trajectories. The equilibria with all adjacent trajectories moving closer are called attractors of the system (Dercole and Rinaldi 2012). Each attractor of a dynamical system portrays a unique way of balancing forces through negative feedbacks (i.e. the multiplication of the interaction strengths along the feedback loop is negative). We can define the basin of attraction for each attractor as the range of initial values with all trajectories starting within the range and eventually converging towards the attractor. The type of the attractor (e.g. node or limit cycle) defines the regime, and the size and shape of the basin defines the resilience of a regime (i.e. its capacity to revert to its initial stable state following a perturbation).

When a system has multiple attractors, the phase space will be subdivided into many basins, each corresponding to one attractor. Regime shifts require internal or external forces to break down the current balance, often by establishing positive feedbacks, shifting the system away from the current basin to a new one reinforced by altered stabilizing negative feedbacks (Scheffer *et al.* 2001; Bennett *et al.* 2005). Invasive species can establish positive reinforcing feedbacks in invaded ecosystems in three different ways (D'Antonio and Vitousek 1992), by altering (i) resource supply (e.g. *Myrica faya, Mesembryanthemum crystallinum, Carpobrotus edulis*), (ii) trophic structure (e.g. through introducing the brown tree snake as a new top predator to Guam), and/or

(iii) disturbance regimes (e.g. feral pigs *Sus scrofa* altering decomposition, cheatgrass *Bromus tectorum* in North American prairies and *Acacia* trees in South African fynbos altering fire regimes, thereby promoting their invasions).

A system under perturbations could shift the state variables (e.g. by reducing the biomass of native species but increasing that of invasive species) but will return to its original attractor as long as the shift remains within the basin. Systems suffering large perturbations that shock the state variables to move outside their current basin will experience a jump across basins; that is, a regime shift due to bistability (the existence of alternative regimes). A key question when addressing regime shifts is to identify the boundaries of current and potential regimes which are often defined as thresholds for the state variables (e.g. invader biomass less than a certain amount); these thresholds are known as tipping points. When numerically modelling a dynamical system, attractors can be identified by plotting trajectories to infinite time, whereas the boundaries of attractors by reversing time to negative infinity (i.e. tipping points are the attractors of the system when time is reversed).

A system's regime can change its type due to perturbations or changes in the system parameters known as bifurcations (e.g. changes in nutrient cycling leading to alterations of reinforcing feedback loops); this depicts the structural stability of the system in response to changes in system parameters (Kuznetsov 2010). Bifurcations can be catastrophic and can induce hysteresis. Supposing a slight change of a parameter from p to $p + \varepsilon$ were to trigger a catastrophic regime change, a system can be called hysteretic if changing the parameter back to p does not push the system back to its original attractor; for example, if removal of the invader does not necessarily trigger a shift to the state that existed prior to invasion. A dynamical system often experiences one of seven bifurcations (Dercole and Rinaldi 2012). In particular, a saddle-node (fold) bifurcation indicates a catastrophic regime change and is often associated with regime shifts and hysteresis in ecological systems (Fig. 7.2; Scheffer *et al.* 2001; Kuehn 2011)—this happens when an attractor (stable fixed point) collides with an unstable fixed point.

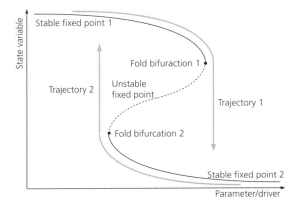

Fig. 7.2 A dynamic system with hysteresis generated by two saddle-node bifurcations. From Lade *et al.* (2013). With permission from Springer.

It is necessary to point out that system regimes and resilience, as discussed earlier, and early warning signals for regime shifts, as will be discussed later, are largely developed based on the concept of Lyapunov stability (resilience to small perturbations). There are other concepts of system stability, such as D-stability, sign stability, Volterra dissipativeness, and many others (Logofet 2005); their relationships can be visualized using a 3D Venn diagram (Fig. 7.3), with the three biggest grey petals indicating the subsets of community (Jacobian) matrices conforming to Lyapunov stability. Evidently, Lyapunov stability is only a weak stability notion. To fully understand the community-level response to perturbations and system-level adaptive evolution, we need to acknowledge and clarify the relationships between these stability concepts (Borrelli *et al.* 2015). This means that we need to differentiate these stability concepts and their related stability measures, such as resilience (response to perturbation), robustness (response to species removal; Dunne *et al.* 2002), feasibility (likelihood of persistence), and invasibility (response to species introduction) (see later in this chapter and Chapter 10). For instance, invasibility could be the flip side of robustness and could reflect the structural and evolutionary instability of a system (Minoarivelo and Hui 2016a). Major theoretical progress is needed in this area (see Hui *et al.* 2016 for some ideas in this regard).

7.2 System reshuffling

For systems comprising multiple species, identifying the exact regimes and their boundaries can be difficult. However, the possibility of losing system stability under the prevailing regime can be used as indicative of the presence of alternative regimes (Scheffer *et al.* 2012), or at least considering system collapse as an alternative regime. As mentioned earlier, positive feedback loops typically drive bistability and regime boundaries in ecological systems and thus represent a major source of system instability. For instance, positive density dependence—that is, the Allee effect—separates the feasible region of a single species system, $dN / dt = rN(1 - N / K)(N - a)$, into two regimes ($N = 0$ or K), divided by the Allee effect threshold (a). In a simple system with two competing species, $dN_i / dt = N_i(1 - \mu N_i - \alpha N_j)$ (where $i = 1$ or 2), stronger interspecific competition (destabilizing force) greater than intraspecific density dependence (stabilizing force), $\alpha > \mu$, creates a positive reinforcing feedback loop and the possibility of alternative outcomes from competitive exclusion that depends on the initial population sizes (see Chapter 6).

For a multispecies community, May (1972) derived a condition for the current regime to lose its stability: $\sigma\sqrt{SC} > \mu$, where σ represents the standard deviation of non-zero interaction strength in the community, C the connectance, μ the intraspecific negative density dependence, and S the number of species in the community (updated by Allesina and Tang 2012). The left of this condition signals the strength of destabilizing force, while the right denotes the stabilizing force (see Chapter 5). When the destabilizing force is greater than the stabilizing force, a positive reinforcing feedback loop is created, amplifying and repeatedly reinforcing the effect of any perturbations on the system resilience until the collapse of current regimes. The community has to reorganize and reshuffle after the collapse which creates opportunities for including alien species in the reshuffled assemblages. The value of $\sigma\sqrt{SC} / \mu$ is a generic indicator for the potential loss of the standing ecosystem regime or, strictly speaking, a sign of system reshuffling. Whether a particular alien species will benefit from such reshuffling critically depends on its traits relative to those of the

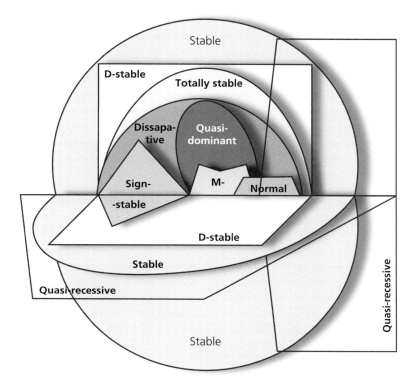

Fig. 7.3 Different concepts of system stability illustrated in a 3D Venn diagram. Let a community be depicted by a Jacobian matrix A = {a$_{ij}$}. The Venn diagram contains three half-planes: the upper vertical one represents Jacobian matrices with no zero entries on the main diagonal; the horizontal one with all the diagonal entries non-positive and at least one zero; the lower vertical half-plane with at least one positive diagonal element. Lyapunov stability requires the real parts of all eigenvalues of the matrix being negative (three 'stable' petals). From Borrelli *et al.* (2015), originally adapted from Logofet (2005). With permission from Elsevier.

resident species and opportunity niches that have emerged during the reshuffling (e.g. Minoarivelo and Hui 2016a).

This index of system reshuffling identifies four factors that could trigger the regime shift and initiate assemblage reshuffling: stronger interaction strength (σ), higher species richness (S), more connected pairs of interacting species or generalists (C), and weaker density dependence (μ). These four factors are largely congruent with the factors that facilitate the establishment and impact of alien species in recipient ecosystems. In particular, propagule pressure (Ricciardi *et al.* 2011), trait distinctiveness (Daehler 2001; Moles *et al.* 2012), resource availability (Guo *et al.* 2015), interaction strength (Williamson and Fitter 1996; Lonsdale 1999), system connectivity (Ives and Carpenter 2007), and disturbance (Davis *et al.* 2000) have

been shown to contribute substantially to the susceptibility of an ecosystem to invasion. The potential for regime shifts will increase with higher propagule pressure (increasing σ and C, reducing μ), unique invader traits (increasing σ), more invasive species (increasing S and C, invasional meltdown), more resources and stronger disturbance (reducing μ). Besides biological invasions, other drivers of change could further complicate and even aggravate an already fragile ecosystem, thereby triggering regime shifts.

7.3 Evolutionary regimes

Evolutionary bistability—the existence of two alternative evolutionary attractors (i.e. evolutionarily stable strategies) –is a phenomenon during trait evolution and appears more frequently where trait evolution is under two opposing selective forces for adaptation. For example, the evolutionary dynamics of pathogenic virulence can exhibit bistability when multiple modes of transmission are available in the system, leading to opposing selection pressures (Roche *et al.* 2011). Besides such bistability, many evolutionary systems show a bifurcation of punctuated equilibria where slight changes in system parameters lead to a drastic jump of the evolutionary trajectory in trait space from one to another evolutionary attractor (Dercole *et al.* 2003; Ferrière and Legendre 2013); that is, an evolutionary regime shift.

The assembly (succession) dynamics of an ecological community or network are governed by both its evolutionary pathway and the coupling of ecological and evolutionary processes (Minoarivelo and Hui 2016b). The eco-evolutionary coupling can create a strong priority effect in, for example, a mutualistic community (May 1977; Young *et al.* 2001; Fukami 2005), thereby opening alternative evolutionary/successional pathways which are sensitive to the initial trait composition of pioneer species (Dieckmann *et al.* 1995; Dercole *et al.* 2006; Landi *et al.* 2015). Biological invasions that add new distinct traits to a system can strongly affect system stability (Ricciardi and Atkinson 2004), often leading to chaotic evolutionary scenarios, especially in the case of evolutionary bistability and bifurcation. For instance, perturbations could push the adaptive trait onto the trajectory of evolutionary suicide (negative fitness) or towards an evolutionary trap (local fitness valley but with no escape) (Rankin *et al.* 2007; Zhang *et al.* 2013). Therefore, we cannot predict the evolutionary or successional trajectory of a co-evolving community perturbed by biological invasions based only on the knowledge of ancestral conditions and eco-evolutionary processes.

Successful invasions often create a strong directional selection and shift the system into an alternative evolutionary trajectory. Factors that could affect the strength of directional selection, such as reduced pollination success from phenological mismatches due to rising temperatures, may result in a change in the evolutionary trajectory (Ferrière and Legendre 2013). This means that trait evolution in a community cannot depend solely on the evolutionary history of ancestral or pioneer species; resilience of the predetermined evolutionary trajectory in the face of biotic perturbations also matters (Smallegange and Coulson 2013). Disruption of interactions due to invasion can have not only ecological impacts but can also alter selective pressures that drive evolutionary responses in native species as well as the path of community succession (Traveset and Richardson 2014; also see Chapters 8 and 10). This is a joint result of biotic interactions and environmental perturbations from biological invasions and other drivers of change.

7.4 Early warning signals

7.4.1 Critical slowing down

Regime shifts often result in changes in ecosystem functioning that detrimentally affect the capacity of the ecosystem to deliver certain services, with negative impacts on human well-being. Consequently, management strategies for adaptation and mitigation, or even the avoidance of these unfavourable shifts, are often sought (Boettiger *et al.* 2013). The cost of implementing such strategies needs to be weighed against the costs of the regime shift. Consequently, the ability to foresee and avoid a regime shift is appealing to managers. Avoidance depends on both the ability to predict regime shifts in advance and the time scale of the response of the system. Adaptation and mitigation might require the ability to predict a shift in advance if the time scale of implementation is long relative to the rate at which damages occur. For this purpose, certain early warning signals (EWS) potentially associated with pending regime shifts have been identified (Scheffer *et al.* 2009, 2012). The phenomenon of critical slowing down (CSD) has recently enjoyed considerable discussion in this regard.

CSD is generic system behaviour that manifests especially when the system nears a tipping point, when the variance or autocorrelation and return time of the state variables of the system are greatly increased. CSD has been of great interest to theorists because of its role in the stability of complex dynamical systems (Carpenter and Brock 2006; Dakos *et al.* 2011; Boettiger and Hastings 2012a; Lade and Gross 2012). Insights in this regard are also seen as a priority for precautionary biodiversity management (Barnosky *et al.* 2012; Essl *et al.* 2015). CSD occurs when the dominant eigenvalue of the system

Jacobian approaches zero; this could be the result of either abiotic or biotic disturbance such as adding a species to an ecosystem. A negative dominant eigenvalue suggests that small perturbations to the system will be dampened (i.e. the system is resilient to small perturbations), whereas a positive one amplifies the perturbations the system receives (i.e. the current system configuration is not stable and will collapse or change towards a new regime). Consequently, a zero-dominant eigenvalue indicates the tipping point of the current system regime. When the eigenvalue approaches zero, the ability of a stable system to subdue the perturbation is greatly weakened and slowed, and a CSD can be flagged.

Rapid catastrophic regime shifts, where the behaviour of the system changes abruptly, is a scenario intertwined with the aforementioned concepts of bifurcation and CSD (Fig. 7.4) (Boettiger *et al.* 2013). Recent research on EWS addresses all these concepts, and signals of increased variance and coefficient of variation (Carpenter and Brock 2006), autocorrelation (Dakos *et al.* 2008), and skewness (Guttal and Jayaprakash 2008) are directly deductible from the changing eigenvalue in a saddle-node (fold) bifurcation. For instance, Dai and colleagues (2012) observed slowed recovery time

and increased variance and autocorrelation in a microcosm with yeast when the density was reduced to near the threshold for a fold bifurcation from Allee effects. Similar results were found in a system of cyanobacteria (Veraart *et al.* 2012). A classic and large-scale example was presented by Carpenter and colleagues (2011) on the rapid regime shift in lake ecosystems triggered by the introduction of largemouth bass, *Micropterus salmoides*, with trophic interactions from adult fish eating the juveniles of their competitors acting as a trigger to push the system to undergo a saddle-node (fold) bifurcation (Carpenter *et al.* 2012).

Establishing the mechanism behind a saddle-node bifurcation is essential for assessing the feasibility of using CSD as a warning signal. Although manipulative experiments are useful for simple systems (Dai *et al.* 2012; Veraart *et al.* 2012), they are impractical for most natural systems. We could certainly assume that the saddle-node bifurcation applies to a limited set of systems that resemble well-studied examples such as lakes undergoing eutrophication (Scheffer *et al.* 2001), lakes with a trophic cascade (Carpenter and Kitchell 1996; Walters and Kitchell 2001; Carpenter *et al.* 2008), and forest/savanna transitions (Hirota *et al.* 2011; Staver

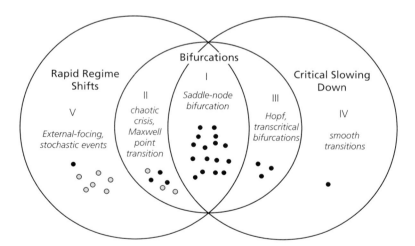

Fig. 7.4 The intersecting domains of rapid regime shifts, bifurcations, and critical slowing down (CSD). Labels below the Roman numerals are examples of phenomena that occur in each domain. Each dot represents a study in the domain. Dots in grey represent literature that do not explicitly test early warning signals (EWS) but which demonstrate phenomena related to EWS. The centre domain (I), where all three phenomena intersect, is the most extensively researched domain in the early warning signal field. Literature outside this charted research is less extensive, but hints at how existing signals based on CSD may be insufficient or misleading. From Boettiger *et al.* (2013). With permission from Springer.

et al. 2011). However, making such assumptions for less well-understood systems could lead to false positives and exaggeration of the frequency of such fold bifurcation in ecosystems (Boettiger and Hastings 2012b).

There are seven main types of bifurcations in dynamical systems, not all of which lead to catastrophic regime shifts. These other bifurcations could cause long-term changes in system behaviours without approaching the zero eigenvalue and therefore lacking the phenomenon of CSD (Schreiber 2003; Schreiber and Rudolf 2008; Hastings and Wysham 2010). As those typical EWS of enhanced variance and autocorrelation are mainly developed with fold bifurcations in mind, they could be misleading when the system experiences other types of bifurcations. For instance, a class of bifurcations that shows no CSD before a regime shift, known as *crises*, are characterized by the sudden appearance of chaotic attractors (Grebogi *et al.* 1983; Boettiger *et al.* 2013). Regime shifts are not always catastrophic and smooth transitions do exist (Fig. 7.5),

especially when the system is under the influence of demographic stochasticity combined with realistic features such as limited mobility and spatial heterogeneity (Martín *et al.* 2015).

Some rapid regime shifts are not attributable to bifurcations at all. An internal stochastic event may switch a system between dynamical regimes. Large environmental shocks may change the behaviour of a system without any warning; this is commonly recognized as bistability (Scheffer *et al.* 2001, 2009, 2012; Barnosky *et al.* 2012). For instance, Schooler and colleagues (2011) found that lakes invaded by the floating fern, *Salvinia molesta*, and herbivorous weevils alternated between low- and high-*Salvinia* states driven by disturbances from regular external flooding events. System stochasticity could also shift systems from one state to another even when environmental conditions remain constant (Hastings and Wysham 2010). In such exogenous regime shifts with no underlying bifurcations, identifying EWS becomes irrelevant, and the CSD undetectable.

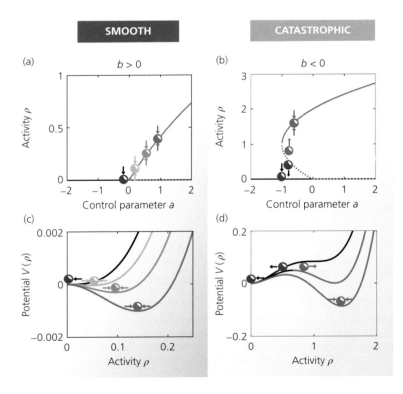

Fig. 7.5 Bifurcation diagram and deterministic potential V(p) for continuous (a and c) and abrupt (b and d) regime shifts. Lines in (a) and (b) represent the steady-state solutions of $\dot{p} = ap - bp^2 - cp^3$, with spheres indicating four specific steady states. Lines in (c) and (d) represent the effective potential, defined as $\partial V / \partial p = -\dot{p}$. From Martín *et al.* (2015).

7.4.2 Other early warning signals

As mentioned earlier, recent studies have advocated the use of early warning signals (EWS) such as increased variance and autocorrelation that normally coincide prior to regime shifts between alternate attractors (van Nes and Scheffer 2007; Carpenter *et al.* 2008; Scheffer *et al.* 2009; Brock and Carpenter 2012). However, these indicators do not increase simultaneously if regime shifts are driven by exogenous shocks (Ditlevsen and Johnsen 2010; Wang *et al.* 2012). These indicators will also not respond to step changes in control variables. For instance, the introduction of largemouth bass to a lake may suddenly push the system to a new attractor without gradual changes in internal feedback mechanisms (Carpenter *et al.* 2011). Moreover, using these indicators often requires making arbitrary calculations; for example, the power of lag-1 autocorrelation could be modified by changing methods of data aggregation, detrending, and filtering signal bandwidth (Lenton *et al.* 2012). Although these choices may be optimized with enough calibration data, they are not applicable for many ecological systems. Sophisticated metrics such as multiple-method (Lindegren *et al.* 2012) and composite indices (Drake and Griffen 2010) have been proposed, but their power has yet to be tested.

Delayed processes with time lags are a destabilizing force and can induce a variety of interesting behaviours (see the minnow-bath case in section 7.6). A linear dependence on a delayed variable may lead to instabilities, oscillations, and Hopf bifurcations in complex systems, while the interaction of delayed dynamics with stochasticity may induce noise-sustained fluctuations (D'Odorico *et al.* 2013). In a univariate delayed system driven by additive noise, the variance of the state variable will still increase when the system is close to a critical transition—it can thus still be used as an EWS (D'Odorico *et al.* 2013). However, in the case of systems with multiplicative noise, rising variance is not necessarily a warning of an imminent regime shift (Dakos *et al.* 2012). Consequently, care is needed when considering regime shifts in systems involving time lags (Guttal *et al.* 2013).

As most ecosystems are embedded in highly stochastic environments, they may start to 'flicker' between basins of attraction long before system bifurcation. Under such noisy conditions, one would fail to detect CSD when using generic leading indicators, (Contamin and Ellison 2009; Scheffer *et al.* 2009; Perretti and Munch 2012). CSD can be identified either directly after perturbation (Veraart *et al.* 2012; Fig. 7.6a), or by comparing generic indicators to a control treatment (Drake and Griffen 2010; Fig. 7.6b), although such noise-free time series are rare in practice. By contrast, Carpenter and colleagues (2011) found that the abundance of largemouth bass flickered between alternative states one year before the regime shift to piscivore dominance in the lake (Fig. 7.6c). Other paleo-isotope data of abrupt shifts between warming and cooling episodes over the past 6000 years, known as Dansgaard–Oeschgaard events (Dansgaard *et al.* 1993), have also shown stochastically induced jumps ('flickering') between alternative attractors (Ganopolski and Rahmstorf 2002; Ditlevsen and Johnsen 2010; Fig. 7.6d). Variance will rise as flickering occurs (Carpenter and Brock 2006; Carpenter *et al.* 2008), but temporal autocorrelation may either increase or decrease depending on the interval of time series (Dakos *et al.* 2012; Lenton *et al.* 2012; Wang *et al.* 2012). Under extreme stochasticity, neither variance nor autocorrelation will work as an EWS (Contamin and Ellison 2009; Perretti and Munch 2012).

It is difficult to use leading indicators of EWS (variance and autocorrelation) to differentiate between CSD and flickering (Scheffer *et al.* 2012), especially for variance (Fig. 7.7). When flickering, indicators of such time series will exhibit much larger variation than the smooth curves of CSD, and only variance can still serve as a good indicator of upcoming transitions (Fig. 7.7; Dakos *et al.* 2013). Again, strong stochasticity weakens the detectability of EWS for imminent critical transitions (Perretti and Munch 2012). A different set of indicators might be especially suitable for detecting CSW, flickering, or other extreme stochastic regimes in the system. In any case, the most interesting feature of flickering is that it gives us a preview of the size and shape of alternative attractor (i.e. a glimpse of future) (Livina *et al.* 2010; Dakos *et al.* 2013; Scheffer *et al.* 2012).

There have been some exciting recent developments regarding EWS, in particular the generalized modelling-based early warning signal (Lade and Gross 2012) which shares a mathematical basis with both the qualitative theory of differential equations (Kelley and Peterson 2010) and catastrophe theory

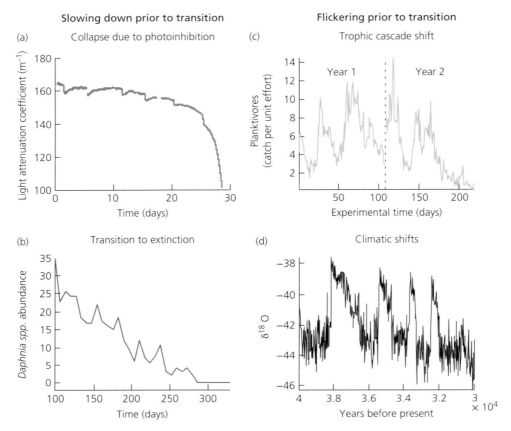

Fig. 7.6 Typical time series supporting the use of critical slowing down (CSD) (a and b) or flickering (c and d) as EWS for regime shifts. (a) CSD in a plankton chemostat (Veraart *et al.* 2012); (b) CSD in a zooplankton experiment (Drake and Griffen 2010); (c) flickering from trophic cascade in a lake experiment (Carpenter *et al.* 2011); (d) flickering between warming and cooling episodes of Dansgaard–Oeschgaard events (Svensson *et al.* 2008). From Dakos *et al.* (2013). With permission from Springer.

(Zeeman 1977). Unlike the variance and autocorrelation approaches, this approach requires further knowledge of the system structure. Once a generalized model is constructed, the Jacobian matrix can be estimated, and the time series observations will then be used to estimate the derivatives in the Jacobian matrix (thus the resilience of the system; Gross and Feudel 2006; Kuehn *et al.* 2013). Generalized modelling thus can be used to parameterize the Jacobian matrix directly and connect the model structure to properties of real systems. This makes generalized modelling perform equally well to other systems dynamics approaches (Sterman 2000; Lade *et al.* 2013) such as structural equation modelling (Kline 2011) and flexible functional forms (Chambers 1988).

7.5 Minnow-bass bistability

Bistable dynamics, the existence of alternative regimes, emerge typically in non-linear systems due to positive feedbacks with processes that control resource availability or the disturbance regime (e.g. Scheffer *et al.* 2009). To test whether a system possesses alternative regimes, predictions from models with and without multiple attractors can be compared using long-term time series (Scheffer and Carpenter 2003; Mumby *et al.* 2007; Ives and Carpenter 2007; Schooler *et al.* 2011). A classic example is the examination of potential of multiple attractors and regime shifts in a small Michigan lake before and after the introduction of the predatory

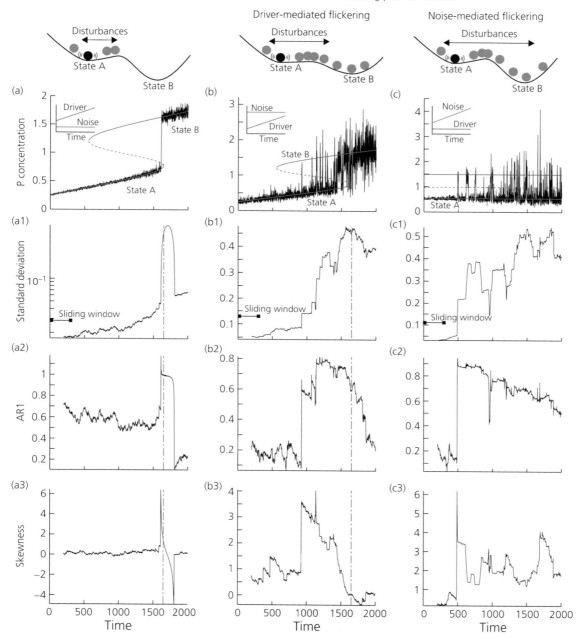

Fig. 7.7 Critical slowing down (CSD) versus flickering. In the CSD scenario, the system state stays within the basin of attraction of the current state under weak disturbances (upper left panel), while in the flickering scenarios stronger disturbances can push the system state across the basin of attraction when the basin shrinks (driver-mediated flickering middle panel) or when noise intensity increases (noise-mediated flickering left panel). An example of a typical time series derived from transient simulations under CSD (a), driver-mediated flickering (b), and noise-mediated flickering (c) scenarios. Generic leading indicators (variance measured as standard deviation (a1, b1, c1), autocorrelation at lag 1 (AR1, a2, b2, c2), and skewness (a3, b3, c3)) were estimated for each scenario within sliding windows of 200 points across the time series (in a, b, and c, solid lines denote stable equilibria, dotted lines unstable equilibria; dash dot lines in a1 to a3 and b1 to b3 denote the threshold at which the critical transition to the eutrophic state would occur in the deterministic case). From Dakos et al. (2013). With permission from Springer.

largemouth bass, *Micropterus salmoides* (Carpenter *et al.* 2001; Seekell *et al.* 2013). Before the introduction of a small number of adult largemouth bass, the ecosystem was dominated by minnows (planktivorous fish). The minnow's dominance was maintained because the growth and recruitment of largemouth bass were limited by predation and competition affecting juvenile largemouth bass (Walters and Kitchell 2001; Carpenter *et al.* 2008, 2011). As the number of adult bass increased beyond a critical point, competition pressure from other species on juvenile largemouth bass declined due to the predation from adult largemouth bass, and adult recruitment increased, causing a regime shift to bass dominance (Carpenter *et al.* 2008).

Before the first addition of bass to the lake, small prey fishes were abundant and the dynamics showed high variability (Fig. 7.8, top panel). The catches declined immediately after the first addition of bass, and the variability shifted to lower frequencies of longer-term oscillations. After the

last addition of bass, daily catches were extremely low and showed low variability (Fig. 7.8, top panel). Seekell and colleagues (2013) used the standardized residuals from the generalized autoregressive conditional heteroscedasticity (GRACH) model to detect regime shifts during the experiment (Fig. 7.8, bottom panel); evidently, the residuals show larger oscillations during the transition between two point attractors, which could be attributed to non-linear dynamics.

The regime shift between minnow dominance and bass dominance can be clearly illustrated in the phase portrait, where the minnow catch declines during the study (Fig. 7.9a). Prior to the first addition of bass, the trajectory circles around an attractor. The trajectory during the transition period indicates a limit cycle. After the last bass addition, the trajectory follows a new attractor close to the extirpation of minnows. The standardized GARCH residuals (Fig. 7.9b) make the circular pattern of a limit cycle more evident during the transition period (the

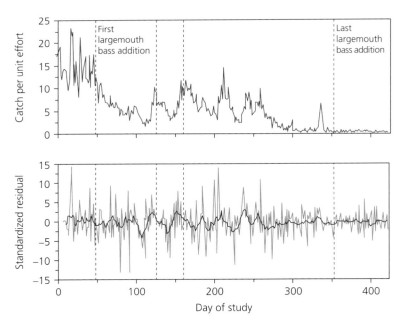

Fig. 7.8 Top: Mean catch per day for minnow traps in the manipulation lake (Peter Lake, 2.6 ha in extent, in Michigan, USA) for four summers. Additions of largemouth bass (*Micropterus salmoides*) are indicated with dashed vertical lines. The dynamics before the first addition and after the last addition are considered to be stable. The time between these additions is considered the transition. Bottom: The lighter grey line is the time series of standardized residuals from the generalized autoregressive conditional heteroscedasticity time series model fitted to the mean catch per-day time series. The darker grey line is the standardized residuals smoothed using a seven-point moving average. From Seekell *et al.* (2013). With permission from Springer.

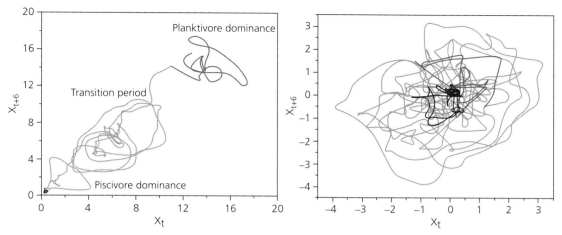

Fig. 7.9 Phase portrait of minnow trap time series (left) and standardized residuals from the generalized autoregressive conditional heteroscedasticity model. The top-right darker grey trajectory (planktivore dominance) is the period prior to the first addition of largemouth bass in the manipulated Peter Lake in Michigan, USA. The lighter grey trajectory is the transition period. The bottom-left darker grey trajectory (piscicore dominance) is the period after the last largemouth bass addition. The system is initially at a point attractor, but enters into a limit cycle after the first largemouth bass addition. The system has returned to a new point attractor by the time of the last largemouth bass addition. From Seekell *et al.* (2013). With permission from Springer.

grey trajectory). This limit cycle during the transition phase is driven by the interaction of fast and slow processes in the system. The population size of bass is a much slower variable than the population of minnows, due to reproductive cycles and foraging behaviours (Carpenter *et al.* 2011; Walker *et al.* 2012). The oscillations could have resulted from delays due to annual reproductive cycles. The magnitude of variability during the transition is greater than the variability around the attractors (Fig. 7.9b), suggesting that variance is a robust EWS for regime shifts in lake food webs (Carpenter *et al.* 2008).

There are three different ways for invasive species to establish positive reinforcing feedbacks and thus potentially trigger regime shifts: by altering resource supply, trophic structure, or disturbance regimes (D'Antonio and Vitousek 1992). The introduction of largemouth bass clearly altered the trophic structure in the lake food web discussed earlier. There are several good examples of regime shifts caused by aquatic invasions triggered by the other two categories. The dense zebra mussel (*Dreissena polymorpha*) population in Lake St Clair (USA/Canada) has shifted a turbid lake into a clear, weedy lake with concomitant changes in the lake food web through the filtration activity that alters resource

supply (Vanderploeg *et al.* 2002). By interfering with both resource supply and disturbance regimes, introduced goldfish, *Carassius auratus*, and common carp, *Cyprinus carpio*, can shift shallow ponds and lakes into turbid systems with only a few rooted macrophytes (Richardson *et al.* 1995; Lougheed *et al.* 1998). These ecosystem engineering processes, often triggered by invaders at low trophic levels, establish positive reinforcing feedbacks that favour the progress of the invasion. A well-known case is the dense cover created by water hyacinth, *Eichhornia crassipes*, in Lake Victoria that reduces atmospheric oxygen exchange with surface waters, thereby promoting an accumulation of decaying vegetation on bottom sediments and causing anoxic conditions that release phosphate which then stimulates further proliferation of water hyacinth (Ntiba *et al.* 2001).

7.6 Reinforcing feedbacks during invasion

The potential of reinforcing feedbacks to drive regime shifts in plant invasion ecology has been explored, especially for grass-fire feedbacks (D'Antonio and Vitousek 1992; Brooks *et al.* 2004) and plant-soil feedbacks (Suding *et al.* 2013; van der

Putten *et al.* 2013). However, quantifying reinforcing feedbacks is done largely through manipulative experiments by comparing effect size with treatment size, making the results highly contextual and generalization difficult. Gaertner and colleagues (2014) provide a comprehensive overview and synthesis of potential reinforcing feedback mechanisms that could drive regime shifts in landscapes dominated by invasive plants. More importantly, they use the ratio of effect size between treatment and control experiments measured from a meta-analysis of the existing literature, which allows them to detect key reinforcing feedbacks associated with different types of plant invasion. We provide a short review of their methodology and results regarding these reinforcing feedbacks during plant invasion and ways of quantifying them to indicate potential regime shifts.

Many frameworks for conceptualizing impacts of alien invasions have been proposed (Parker *et al.* 1999; Blackburn *et al.* 2014). Vitousek (1990) argued that invaders can have large effects on ecosystem processes if they can alter disturbance regimes. Many studies have demonstrated that the impacts of invasive species are strongly context-dependent and can have substantially different outcomes depending on the type of invader and the invaded habitat (Pyšek *et al.* 2012; Ricciardi *et al.* 2013). This means that potential changes in feedback mechanisms or regime shifts to alternate states must be carefully considered when assessing the impact of invasion (Suding *et al.* 2004; Chapin *et al.* 2011; Richardson and Gaertner 2013). Due to the complications of EWS (see section 7.4), a complementary approach has been proposed. This uses systems-analysis tools to analyse the feedback structure of a system based on knowledge of ecosystem drivers, processes, and impacts, which helps us to understand whether a particular ecosystem may be susceptible to specific regime shifts (Scheffer *et al.* 2009; Biggs *et al.* 2012).

Gaertner and colleagues (2014) applied such a systems-analysis approach combined with a meta-analysis of the literature to identify potential high-impact plant invaders. Based on feedbacks identified in 443 studies on effects of alien plant invasion involving 173 species, they identified five generic types of reinforcing feedback processes that are associated with alien plant invasions (Fig. 7.10): changes in seed bank composition; fire regime; soil nutrient cycling; litter quantity and/or quality; and soil biota structure and function. These feedbacks favour the invader in competition with native species for resources (light, nutrients, water), or change the soil environment so as to suppress the germination of native seedlings. As such, these feedbacks can promote the accumulation of invader biomass and thus amplify the competitiveness of the invader in several ways:

- By producing large numbers of seeds, many non-native plants can establish a reinforcing feedback loop that promotes their own abundance (R1 in Fig. 7.10; e.g. Australian *Acacia* species in South African fynbos; Holmes *et al.* 1987).
- By altering fire frequencies and intensities through changing fuel load, invasive plants can transform ecosystems (R2 in Fig. 7.10; D'Antonio and Vitousek 1992; Brooks *et al.* 2004; Rossiter *et al.* 2003; Rossiter-Rachor *et al.* 2008; Stevens and Beckage 2009). For example, *Chromolaena odorata*, invasions in South African savannas introduce 'ladder' fuels that carry fires into the crowns of trees, thereby transforming a regime of low-intensity surface fires to one characterized by high-intensity fires that extend into the canopy and kill trees (te Beest *et al.* 2012).
- Through nitrogen fixation (R3 in Fig. 7.10), some invasive plants can affect decomposition (Ehrenfeld 2003), enhance conditions for themselves and other weeds (Richardson *et al.* 2000; Vinton and Goergen 2006), and suppress recruitment of native plants (e.g. Marchante *et al.* 2008). Prominent examples include the invasion of Australian acacias in South African fynbos (Yelenik *et al.* 2004; Le Maitre *et al.* 2011), and the invasion of young volcanic soils in Hawai'i by the Fire tree (*Morella faya*) (Vitousek *et al.* 1987).
- High litter volumes produced by some invasive species can inhibit native species growth while promoting the growth of the invader (R4 in Fig. 7.10; Farrer and Goldberg 2009). For example, reed canary grass (*Phalaris arundinacea*) induces litter feedbacks in invaded habitats, causing profound ecosystem-level changes (potentially a regime shift) to a high-litter, invader-dominated state (Eppinga *et al.* 2011; Eppinga and Molofsky 2013).

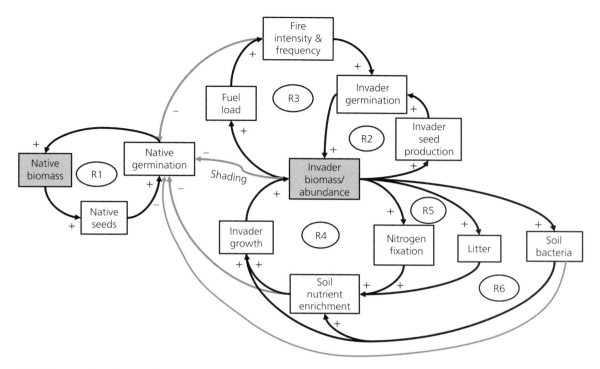

Fig. 7.10: Causal loop diagram of five main feedback mechanisms implicated in plant invasions (R1: seed-biomass (native); R2: seed-biomass (invasive); R3: fire feedback; R4, 5 and 6: soil-nutrient, litter, and soil biota feedback). Note, feedbacks can also operate in the opposite direction (e.g. fire may be enhanced or suppressed). From Gaertner *et al.* (2014). Reproduced with permission from John Wiley & Sons.

- By affecting soil biota by releasing specific secondary compounds, invasive plants can facilitate their own growth (R5 in Fig. 7.10; e.g. Wolfe and Klironomos 2005; te Beest *et al.* 2009; Felker-Quinn *et al.* 2011; Wolfe *et al.* 2008). Many species can also affect soil nutrient levels, thereby indirectly changing the composition and abundance of mycorrhizal fungi necessary for other plants (Sanon *et al.* 2009).

Based on a meta-analysis of the literature on invasive plant species that have caused significant ecosystem impacts, potentially through affecting different reinforcing feedback loops, Gaertner and colleagues (2014) assessed the presence of amplified ecosystem effects by comparing the response effect size of the characteristic change in recipient ecosystems (e.g. the standardized change in soil nutrients) with the cause–effect size of status change in the invasive species (e.g. the standardized change in the invader's biomass) (Table 7.1). An *effect size ratio* (ESR) of greater than 1 is taken as indicating a high probability that the species could trigger or change one or more reinforcing feedback process. The meta-analysis revealed that the soil-nutrient feedback loop was the most commonly recorded feedback, and was most often observed in forest ecosystems and grasslands. Other frequently described feedback loops were those involving fire, mainly in shrublands, and soil biota, mainly in greenhouse studies. Feedbacks involving litter and seed production were the least frequently recorded type, but occurred in a variety of different ecosystems. A few invasive plant species can simultaneously affect multiple feedback loops (Gaertner *et al.* 2014).

Despite the dearth of studies on regime-shift phenomena related to invasive plants, Gaertner and colleagues (2014) found that about 25 per cent of studies which reported ecosystem changes in fact

Table 7.1 A meta-analysis of amplified effects in ecosystems from plant invasions (Gaertner *et al.* 2014). The effect size ratio (ESR) was calculated for each growth form–ecosystem–impact combination. D indicates whether the response was negative or positive; μ and SD are the mean and standard deviation of the logarithmic of the absolute effect size ratio (ln|ESR|). P1 is the *p* value for the one tail *t* test against on whether the absolute ESR is less than 1. P1 < 0.05 means significantly less than 1, i.e. dampened feedbacks. P1 > 0.05 means the species–ecosystem combination could indicate an amplified feedback. NA stands for *n* = 1. P is the *p* value for the two tail *t* test on whether the mean of the logarithmic of absolute ESR is different from expected from a permutation test. PC: plant composition; FC: fauna composition.

Species–Ecosystem–Impact Combination	D	μ	SD	P1	P
Grass–forests–fire	−	−1.65	0.95	0.06	0.34
Grass–forests–nutrients	−	−1.26	0.84	0.09	0.80
Grass–forests–PC	−	−2.05	1.07	0.04	0.05
Grass–grasslands–biota	+	−2.23	2.12	0.15	<0.01
Grass–grasslands–litter	+	−1.34	0.83	0.06	0.47
Grass–grasslands–PC	+	−0.25	1.03	0.41	<0.01
Grass–grasslands–seed	−	−0.24	0.52	0.33	0.09
Grass–riparian–PC	−	−2.79	2.87	0.19	0.04
Grass–shrublands–FC	−	0.50	0.00	NA	0.31
Grass–wetlands–nutrients	+	0.34	1.38	0.41	0.01
Herb–agriculture–PC	−	−1.12	0.15	0.01	0.98
Herb–grasslands–biota	+	−0.48	1.11	0.35	0.50
Herb–wetlands–biota	+	−1.11	0.00	NA	0.99
Herb–wetlands–FC	−	−1.38	0.93	0.12	0.74
Herb–wetlands–litter	−	−1.44	0.81	0.06	0.51
Herb–wetlands–nutrients	+	−0.28	0.69	0.35	0.08
Herb–wetlands–PC	−	0.51	0.11	<0.01	0.02
Herb–wetlands–PC	−	−0.72	0.45	0.09	0.64
Shrub–agriculture–PC	−	−0.97	0.00	NA	0.95
Shrub–forests–nutrients	−	−0.55	0.57	0.19	0.42
Shrub–forests–PC	−	−0.55	1.17	0.33	0.48
Shrub–shrublands–FC	+	0.17	0.00	NA	0.42
Tree–forests–litter	+	−1.69	3.15	0.32	0.57
Tree–forests–nutrients	−	−1.35	2.20	0.27	0.26
Tree–forests–PC	−	−1.20	1.33	0.20	0.86
Tree–grass–fire	−	0.19	0.00	NA	0.41
Tree–riparian–PC	+	−0.85	0.06	<0.01	0.82
Tree–wetlands–PC	−	−0.09	0.01	0.01	0.35

cause regime shifts. It is often difficult to determine whether an invasion has caused, or will cause, a regime shift until its impact is detectable; at this stage the impact is often unmanageable or irreversible. It is worth noting that strong impacts (i.e. those that are dramatic or obvious) are not always those that trigger reinforcing feedback loops and regime shifts.

Using a meta-analysis to identify species–ecosystem–impact combinations that could cause amplifying feedbacks (e.g. Gaertner *et al.* 2014), ecologists and managers could connect invasiveness of alien species and invasibility of recipient ecosystems and identify most vulnerable feedback loops in ecosystems potentially affected by biological invasions. Such an integrated perspective may show, for example, that regime shifts could be triggered either by increased propagule pressure or by affecting feedback loops related primarily to resources and disturbance. Unlike many invasive species with strong competitive ability, some invaders penetrate the biotic resistance in recipient communities through ecosystem engineering that accumulatively amplifies their impact, initially causing little impact but eventually driving irreversible regime shifts (Crooks 2002; Gonzalez *et al.* 2008).

The potential for using early warning signals (EWS) to flag imminent regime shifts associated with biological invasions is uncharted territory that deserves attention. It is likely to be difficult, given the diverse patterns of invasion and the many socio-economic factors involved. A clear depiction of the system, including all key features of the dominant and potential characteristics of disturbance, is a fundamental prerequisite. Several of the EWS metrics discussed in section 7.4 and multifaceted indices (e.g. $\sigma\sqrt{SC} / \mu$) could be used to detect the potential for the loss of system resilience and regime shifts. The quasi-stable phase of CSD at the tipping point of a regime shift, if detected, provides a window for effective management intervention, after which the system will rapidly converge towards an alternative regime. Management could then aim to mitigate these reinforcing positive feedbacks by breaking chains of interactions along the reinforcing loops, or by strengthening the stabilising force of dampening feedbacks in the system (i.e. enhancing the resistance to invasion).

Biological invasions can potentially reorganize the internal feedback mechanisms in recipient ecosystems, abruptly changing the structure and functioning of ecosystems (Biggs *et al.* 2012; Gaertner *et al.* 2014). Recipient ecosystems often exhibit hysteresis towards invasion, showing a certain ability to resist or even repel biological intrusions under certain conditions (see Chapters 6 and 8); such ability is part of the concept of invasibility which defines the vulnerability of an ecosystem to invasion (Richardson and Pyšek 2006). Factors such as high propagule pressure, the distinctiveness of invaders relative to resident species, the fullness of niche space, and the availability of limiting resources to the invasive species are key determinants

of ecosystem invasibility, although a predictive understanding of how these factors interact to influence invasibility remains elusive (Minoarivelo and Hui 2016a; Hui *et al.* 2016). Elucidation of system regimes and the potential for regime shifts need to be considered when assessing ecosystem invasibility.

Along the introduction–naturalization–invasion continuum and between different phases of spread, both the invasive species and the recipient ecosystems face potentially altered regulating forces (see Chapter 5) and assembly processes (see Chapter 8) which create windows of opportunity for potential regime shifts. Factors such as the regulating mechanism of the invasion dynamics that can determine the position of the tipping points (i.e. the boundary of the basin of attraction for a particular regime) and the slope of the potential surface near the tipping points are crucial targets for management to boost the resilience of the system against biological invasions. Even in the absence of clear EWSs, managers need to take actions to reduce the possibility of regime shifts (Fischer *et al.* 2009; Polasky *et al.* 2011). For instance, improved coordination between pre- and post-border management efforts is needed to prevent sudden bursts of high propagule pressure which greatly increase the likelihood of regime shifts (Takimoto 2009). A critical task for research on EWS is to connect robust indicators of regime shifts with invasion monitoring schemes so that data required for EWSs are available when needed (see Chapter 9).

References

Allesina, S. and Tang, S. (2012) Stability criteria for complex ecosystems. Nature, 483, 205–8.

Barnosky, A.D., Hadly E.A., Bascompte, J., *et al.* (2012) Approaching a state shift in earth's biosphere. Nature, 486, 52–8.

Bennett, E.M., Cumming, G.S., and Peterson, G.D. (2005) A systems model approach to determining resilience surrogates for case studies. Ecosystems, 8, 945–57.

Biggs, R., Blenckner, T., Folke, C., *et al.* (2012) Regime shifts. In: Hastings, A. and Gross, L. (eds) *Encyclopaedia of Theoretical Ecology*. Berkeley, CA: University of California Press, pp. 609–17.

Blackburn, T.M., Essl, F., Evans, T., *et al.* (2014) A unified classification of alien species based on the magnitude of their environmental impacts. PLoS Biology, 12, e1001850.

Boettiger, C. and Hastings, A. (2012a) Quantifying limits to detection of early warning for critical transitions. Journal of the Royal Society Interface, 9, 2527–39.

Boettiger, C. and Hastings, A. (2012b) Early warning signals and the prosecutor's fallacy. Proceedings of the Royal Society B: Biological Sciences, 279, 4734–9.

Boettiger, C., Ross, N., and Hastings, A. (2013) Early warning signals: the charted and uncharted territories. Theoretical Ecology, 6, 255–264.

Borrelli, J.J., Allesina, S., Amarasekare, P., *et al.* (2015) Selection on stability across ecological scales. Trends in Ecology and Evolution, 30, 417–25.

Brock, W.A. and Carpenter, S.R. (2012) Early warnings of regime shift when the ecosystem structure is unknown. PLoS One 7, e45586.

Brooks, M.L., D'Antonio, C.M., Richardson, D.M., *et al.* (2004) Effects of invasive alien plants on fire regimes. BioScience, 54, 677–88.

Carpenter, S.R. and Brock, W.A. (2006) Rising variance: a leading indicator of ecological transition. Ecology Letters, 9, 311–18.

Carpenter, S.R., Brock, W.A., Cole, J.J., *et al.* (2008) Leading indicators of trophic cascades. Ecology Letters, 11, 128–38.

Carpenter, S.R., Cole, J.J., Hodgson, J.R., *et al.* (2001) Trophic cascades, nutrients, and lake productivity: whole-lake experiments. Ecological Monographs, 71, 163–86.

Carpenter, S.R., Cole, J.J., Pace, M.L., *et al.* (2011) Early warnings of regime shifts: a whole-ecosystem experiment. Science, 332, 1079–82.

Carpenter, S.R., Folke, C., Norström, A., *et al.* (2012) Program on ecosystem change and society: an international research strategy for integrated social–ecological systems. Current Opinion in Environmental Sustainability, 4, 134–38.

Carpenter, S.R. and Kitchell, J.F. (1996) *The Trophic Cascade in Lakes*. Cambridge: Cambridge University Press.

Chambers, R.G. (1988) *Applied Production Analysis: A Dual Approach*. Cambridge: Cambridge University Press.

Chapin III, F.S., Matson, P.A., and Vitousek, P. (2011) *Principles of Terrestrial Ecology*. New York, NY: Springer.

Contamin, R. and Ellison, A.M. (2009) Indicators of regime shifts in ecological systems: what do we need to know and when do we need to know it. Ecological Applications, 19, 799–816

Crooks, J.A. (2002) Characterizing ecosystem-level consequences of biological invasions: the role of ecosystem engineers. Oikos, 97, 153–66.

Daehler, C.C. (2001) Darwin's naturalization hypothesis revisited. American Naturalist, 158, 324–30.

Dai, L., Vorselen, D., Korolev, K.S., *et al.* (2012) Generic indicators for loss of resilience before a tipping point leading to population collapse. Science, 336, 1175–77.

Dakos, V., Scheffer, M., van Nes, E.H., *et al.* (2008) Slowing down as an early warning signal for abrupt climate change. Proceedings of the National Academy of Science of the USA, 105, 14308–12.

Dakos, V., Carpenter, S.R., Brock, W.A., *et al.* (2012) Methods for detecting early warnings of critical transitions in time series illustrated using simulated ecological data. PLoS One, 7, e41010.

Dakos, V., Kefi, S., Rietkerk, M., *et al.* (2011) Slowing down in spatially patterned ecosystems at the brink of collapse. American Naturalist, 177, E153–66.

Dakos, V., van Nes, E.H., and Scheffer, M., (2013) Flickering as an early warning signal. Theoretical Ecology, 6, 309–17.

Dansgaard, W., Johnsen, S.J., Clausen, H.B., *et al.* (1993) Evidence for general instability of past climate from a 250-kyr ice-core record. Nature, 364, 218–20.

D'Antonio, C.M. and Vitousek, P.M. (1992) Biological invasions by exotic grasses, the grass/fire cycle, and global change. Annual Review of Ecology and Systematics, 23, 63–87.

Davis, M.A., Grime, J.P., and Thompson, K. (2000) Fluctuating resources in plant communities: a general theory of invasibility. Journal of Ecology, 88, 528–34.

Dercole, F., Irisson, J.-O., and Rinaldi, S. (2003) Bifurcation analysis of a prey–predator coevolution model. SIAM Journal of Applied Mathematics, 63, 1378–91.

Dercole, F., Ferriere, R., Gragnani, A., *et al.* (2006) Coevolution of slow–fast populations: evolutionary sliding, evolutionary pseudo-equilibria and complex Red Queen dynamics. Proceedings of the Royal Society B: Biological Sciences, 273, 983–90.

Dercole, F. and Rinaldi, S. (2012) Bifurcations. In: Hastings, A. and Gross, L. (eds) *Encyclopaedia of Theoretical Ecology*. Berkeley, CA: University of California Press, pp. 88–95.

Dieckmann, U., Marrow, P., and Law, R. (1995) Evolutionary cycling in predator–prey interactions: population dynamics and the Red Queen. Journal of Theoretical Biology, 176, 91–102.

Ditlevsen, P.D. and Johnsen, S.J. (2010) Tipping points: early warning and wishful thinking. Geophysical Research Letters, 37, 2–5.

D'Odorico, P., Ridolfi, L., and Laio, F. (2013) Precursors of state transitions in stochastic systems with delay. Theoretical Ecology, 6, 265–70.

Drake, J.M. and Griffen, B.D. (2010) Early warning signals of extinction in deteriorating environments. Nature 467, 456–7.

Dunne, J.A., Williams, R.J., and Martinez, N.D. (2002) Network structure and biodiversity loss in food webs: robustness increases with connectance. Ecology Letters, 5, 558–67.

Ehrenfeld, J.G. (2003) Effects of exotic plant invasions on soil nutrient cycling processes. Ecosystems, 6, 503–523.

Eppinga, M.B., Kaproth, M.A., Collins, A.R., *et al.* (2011) Litter feedbacks, evolutionary change and exotic plant invasion. Journal of Ecology, 99, 503–14.

Eppinga, M.B. and Molofsky, J. (2013) Eco-evolutionary litter feedback as a driver of exotic plant invasion. Perspectives in Plant Ecology, Evolution and Systematics, 15, 20–31.

Essl, F., Bacher, S., Blackburn, T.M., *et al.* (2015) Crossing frontiers in tackling pathways of biological invasions. BioScience, 65, 769–82.

Farrer, E.C. and Goldberg, D.E. (2009) Litter drives ecosystem and plant community changes in cattail invasion. Ecological Applications, 19, 398–412.

Felker-Quinn, E., Bailey, J.K., and Schweitzer, J.A. (2011) Soil biota drive expression of genetic variation and development of population-specific feedbacks in an invasive plant. Ecology, 92, 1208–14.

Ferriere, R. and Legendre, S. (2013) Eco-evolutionary feedbacks, adaptive dynamics and evolutionary rescue theory. Philosophical Transactions of the Royal Society B: Biological Sciences, 368, 20120081.

Fischer, J., Peterson, G.D., Gardner, T.A., *et al.* (2009) Integrating resilience thinking and optimisation for conservation. Trends in Ecology and Evolution, 24, 549–54.

Fukami, T. (2005) Integrating internal and external dispersal in metacommunity assembly: preliminary theoretical analyses. Ecological Research, 20, 623–31.

Gaertner, M., Biggs, R., Te Beest, M., *et al.* (2014) Invasive plants as drivers of regime shifts: identifying high-priority invaders that alter feedback relationships. Diversity and Distributions, 20, 733–44.

Ganopolski, A. and Rahmstorf, S. (2002) Abrupt glacial climate changes due to stochastic resonance. Physical Review Letters, 88, 038501.

Gonzalez, A., Lambert, A., and Ricciardi, A. (2008) When does ecosystem engineering cause invasion and species replacement? Oikos, 117, 1247–57.

Grebogi, C., Ott, E., and Yorke, J.A. (1983) Crises, sudden changes in chaotic attractors and transient chaos. Physica D, 7, 181–200.

Gross, T. and Feudel, U. (2006) Generalized models as a universal approach to the analysis of nonlinear dynamical systems. Physical Review E, 73, 016205.

Guo, Q., Fei, S., Dukes, J.S., *et al.* (2015) A unified approach for quantifying invasibility and degree of invasion. Ecology, 96, 2613–21.

Guttal, V. and Jayaprakash, C. (2008) Changing skewness: an early warning signal of regime shifts in ecosystems. Ecology Letters, 11, 450–60.

Guttal, V., Jayaprakash, C., and Tabbaa, O.P. (2013) Robustness of early warning signals of regime shifts in time-delayed ecological models. Theoretical Ecology, 6, 271–83.

Hastings, A. and Wysham, D.B. (2010) Regime shifts in ecological systems can occur with no warning. Ecology Letters, 13, 464–72.

Hejda, M. and Pyšek, P. (2006) What is the impact of Impatiens glanduliferaon species diversity of invaded riparian vegetation? Biological Conservation, 132, 143–52.

Hirota, M., Holmgren, M., Van Nes, E.H., *et al.* (2011) Global resilience of tropical forest and savanna to critical transitions. Science, 334, 232–5.

Holling, C.S. (1973) Resilience and stability of ecological systems. Annual Review of Ecology and Systematics, 4, 1–23.

Holmes, P.M., Macdonald, A.W., and Juritz, J. (1987) Effects of clearing treatment to seed banks of the alien invasive shrubs *Acacia saligna* and *Acacia cyclops* in the southern and south-western cape, South Africa. Journal of Applied Ecology, 24, 1045–51.

Hui, C., Richardson, D.M., Landi, P., *et al.* (2016) Defining invasiveness and invasibility in ecological networks. Biological Invasions, 18, 971–83.

Hui, C., Richardson, D.M., Pyšek, P., *et al.* (2013) Increasing functional modularity with residence time in the co-distribution of native and introduced vascular plants. Nature Communications, 4, 2454.

Hulme, P.E., Pyšek, P., Jarošík, V.C., *et al.* (2013) Bias and error in understanding plant invasion impacts. Trends in Ecology Evolution, 28, 212–18.

Ives, A.R. and Carpenter, S.R. (2007) Stability and diversity of ecosystems. Science, 317, 58–62.

Kelley, W.G. and Peterson, A.C. (2010) *The Theory of Differential Equations: Classical and Qualitative*. Upper Saddle River: Pearson Prentice Hall.

Kline, R.B. (2011) *Principles and Practice of Structural Equation Modeling*. New York, NY: The Guilford Press.

Kuehn, C. (2011) A mathematical framework for critical transitions: bifurcations, fast–slow systems and stochastic dynamics. Physica D, 240, 1020–35.

Kuehn, C., Siegmund, S., and Gross, T. (2013) Dynamical analysis of evolution equations in generalized models. IMA Journal of Applied Mathematics, 78, 1051–77.

Kuznetsov, Y. (2010) *Elements of Applied Bifurcation Theory*. New York, NY: Springer.

Lade, S.J. and Gross, T. (2012) Early warning signals for critical transitions: a generalized modeling approach. PLoS Computer Biology, 8, e1002360.

Lade, S.J., Tavoni, A., Levin, S.A., *et al.* (2013) Regime shifts in a social-ecological system. Theoretical Ecology, 6, 359–72.

Landi, P., Hui, C., and Dieckmann, U., 2015. Fisheries-induced disruptive selection. Journal of Theoretical Biology, 365, 204–16.

Le Maître, D.C., Gaertner, M., Marchante, E., *et al.* (2011) Impacts of invasive Australian acacias: implications for management and restoration. Diversity and Distributions, 17, 1015–29.

Lenton, T.M., Livina, V.N., Dakos, V., *et al.* (2012) Early warning of climate tipping points from critical slowing down: comparing methods to improve robustness. Philosophical Transactions of the Royal Society A: Mathematical Physical and Engineering Sciences, 370, 1185–204.

Levine, J.M., Vilà, M., D'Antonio, C.M., *et al.* (2003) Mechanisms underlying the impacts of exotic plant invasions. Proceedings of the Royal Society B: Biological Sciences, 270, 775–81.

Lewontin, R.C. (1969) The meaning of stability. In: Woodwell G.W. and Smith H.H. (eds) *Brookhaven Symposia in Biology*. Upton, NY: Brookhaven National Laboratory, pp. 13–25.

Lindegren, M., Dakos, V., Groger, J.P., *et al.* (2012) Early detection of ecosystem regime shifts: a multiple method evaluation for management application. PLoS One, 7, e38410.

Livina, V.N., Kwasniok, F., and Lenton, T.M. (2010) Potential analysis reveals changing number of climate states during the last 60 kyr. Climate of the Past, 6, 77–82.

Logofet, D.O. (2005) Stronger-than-Lyapunov notions of matrix stability, or how 'flowers' help solve problems in mathematical ecology. Linear Algebra and its Applications, 398, 75–100.

Lonsdale, W.M. (1999) Global patterns of plant invasions and the concept of invasibility. Ecology, 80, 1522–36.

Lougheed, V.L., Crosbie, B., and Chow-Fraser, P. (1998) Predictions on the effect of common carp (*Cyprinus carpio*) exclusion on water quality, zooplankton, and submergent macrophytes in a Great Lakes wetland. Canadian Journal of Fisheries and Aquatic Sciences, 55, 1189–97.

MA (2005) *Ecosystems and Human Well-Being: Synthesis. Report of the Millennium Ecosystem Assessment*. Washington, DC: Island Press.

Marchante, E., Kjoller, A., Struwe, S., *et al.* (2008) Short- and long-term impacts of Acacia longifolia invasion on the below ground processes of a Mediterranean coastal dune ecosystem. Applied Soil Ecology, 40, 210–17.

Martín, P.V., Bonachela, J.A., Levin, S.A., *et al.* (2015) Eluding catastrophic shifts. Proceedings of the National Academy of Sciences of the USA, 112, E1828–36.

May, R.M. (1972) Will a large complex system be stable? Nature, 238, 413–14.

May, R.M. (1977) Thresholds and breakpoints in ecosystems with a multiplicity of stable states. Nature, 269, 471–7.

Meffin, R., Miller, A.L., Hulme, P.E., *et al.* (2010) Experimental introduction of the alien weed Hieracium lepidulum reveals no significant impact on montane plant communities in New Zealand. Diversity and Distributions, 16, 804–15.

Minoarivelo, H.O. and Hui, C. (2016a) Invading a mutualistic network: to be or not to be similar. Ecology and Evolution, 6, 4981–96.

Minoarivelo, H.O. and Hui, C. (2016b) Trait-mediated interaction leads to structural emergence in mutualistic networks. Evolutionary Ecology, 30, 105–21.

Moles, A.T., Flores-Moreno, H., Bonser, S.P., *et al.* (2012) Invasions: the trail behind, the path ahead, and a test of a disturbing idea. Journal of Ecology, 100, 116–27.

Mumby, P.J., Hastings, A., and Edwards, H.J. (2007) Thresholds and the resilience of Caribbean coral reefs. Nature, 450, 98–101.

Ntiba, M.J., Kudoja, W.M., and Mukasa, C.T. (2001) Management issues in the Lake Victoria watershed. Lakes and Reservoirs: Research and Management, 6, 211–16.

Parker, I.M., Simberloff, D., Lonsdale, W.M., *et al.* (1999) Impact: toward a framework for understanding the ecological effects of invaders. Biological Invasions, 1, 3–19.

Pejchar, L. and Mooney, H.A. (2009) Invasive species, ecosystem services and human well-being. Trends in Ecology and Evolution, 24, 497–504.

Perretti, C.T. and Munch, S.B. (2012) Regime shift indicators fail under noise levels commonly observed in ecological systems. Ecological Applications, 22, 1772–9.

Polasky, S., Carpenter, S.R., Folke, C., *et al.* (2011) Decision-making under great uncertainty: environmental management in an era of global change. Trends in Ecology and Evolution, 26, 398–404.

Pyšek, P., Jarošík, V., Hulme, P.E., *et al.* (2012) A global assessment of invasive plant impacts on resident species, communities and ecosystems: the interaction of impact measures, invading species' traits and environment. Global Change Biology, 18, 1725–37.

Pyšek, P. and Richardson, D.M. (2010) Invasive species, environmental change and management, and ecosystem health. Annual Review of Environment and Resources, 35, 25–55.

Rankin, D.J., Bargum, K., and Kokko, H. (2007) The tragedy of the commons in evolutionary biology. Trends in Ecology and Evolution, 22, 643–51.

Ricciardi, A. and Atkinson, S.K. (2004) Distinctiveness magnifies the impact of biological invaders in aquatic ecosystems. Ecology Letters, 7, 781–84.

Ricciardi, A., Hoopes, M.F., Marchetti, M.P., *et al.* (2013) Progress toward understanding the ecological impacts of nonnative species. Ecological Monographs, 83, 263–82.

Ricciardi, A., Jones, L.A., Kestrup, Å.M., *et al.* (2011) Expanding the propagule pressure concept to understand the impact of biological invasions. In: Richardson, D.M.

(ed.) *Fifty Years of Invasion Ecology: The Legacy of Charles Elton*. Oxford: Wiley-Blackwell, pp. 225–35.

Richardson, D.M., Allsopp, N., D'Antonio, C.M., *et al.* (2000) Plant invasions—the role of mutualisms. Biological Reviews, 75, 65–93.

Richardson, D.M. and Gaertner, M. (2013) Plant invasions as builders and shapers of novel ecosystems. In: Hobbs, R.J., Higgs, E.S., and Hall, C.M. (eds) *Novel Ecosystems: Intervening in the New Ecological World Order*. Chichester: John Wiley & Sons, pp. 102–14.

Richardson, D.M. and Pyšek, P. (2006) Plant invasions—merging the concepts of species invasiveness and community invasibility. Progress in Physical Geography, 30, 409–31.

Richardson, M.J., Whoriskey, F.G., and Roy, L.H. (1995) Turbidity generation and biological impacts of an exotic fish *Carassius auratus*, introduced into shallow seasonally anoxic ponds. Journal of Fish Biology, 47, 576–85.

Rietkerk, M., Dekker, S.C., de Ruiter, P.C., *et al.* (2004) Self-organized patchiness and catastrophic shifts in ecosystems. Science, 305, 1926–9.

Roche, B., Drake, J.M., and Rohani, P. (2011) The curse of the Pharaoh revisited: evolutionary bi-stability in environmentally transmitted pathogens. Ecology Letters, 14, 569–75.

Rossiter, N.A., Setterfield, S.A., Douglas, M.M., *et al.* (2003) Testing the grass-fire cycle: alien grass invasion in the tropical savannas of northern Australia. Diversity and Distributions, 9, 169–76.

Rossiter-Rachor, N.A., Setterfield, S.A., Douglas, M.M., *et al.* (2008) Andropogon gayanus (Gamba grass) invasion increases fire-mediated nitrogen losses in the tropical savannas of northern Australia. Ecosystems, 11, 77–88.

Sanon, A., Beguiristain, T., Cebron, A., *et al.* (2009) Changes in soil diversity and global activities following invasions of the exotic invasive plant, Amaranthus viridis L., decrease the growth of native sahelian Acacia species. FEMS Microbiology Ecology, 70, 118–31.

Scheffer, M., Bascompte, J., Brock, W.A., *et al.* (2009) Early warning signals for critical transitions. Nature, 461, 53–9.

Scheffer, M. and Carpenter, S.R. (2003) Catastrophic regime shifts in ecosystems: linking theory to observation. Trends in Ecology and Evolution, 18, 648–656.

Scheffer, M., Carpenter, S.R., Foley, J.A., *et al.* (2001) Catastrophic shifts in ecosystems. Nature, 413, 591–6.

Scheffer, M., Carpenter, S.R., Lenton, T.M., *et al.* (2012) Anticipating critical transitions. Science, 338, 344–8.

Schooler, S.S., Salau, B., Julien, M.H., *et al.* (2011) Alternative stable states explain unpredictable biological control of *Salvinia molesta* in Kakadu. Nature, 470, 86–9.

Schreiber, S.J. (2003) Allee effects, extinctions, and chaotic transients in simple population models. Theoretical Population Biology, 64, 201–9.

Schreiber, S. and Rudolf, V.H.W. (2008) Crossing habitat boundaries: coupling dynamics of ecosystems through complex life cycles. Ecology Letters, 11, 576–87.

Schröder, A., Persson, L., and de Roos, A.M. (2005) Direct experimental evidence for alternative stable states: a review. Oikos, 110, 3–19.

Seekell, D.A., Cline, T.J., Carpenter, S.R., *et al.* (2013) Evidence of alternate attractors from a whole-ecosystem regime shift experiment. Theoretical Ecology, 6, 385–94.

Smallegange, I.M. and Coulson, T. (2013) Towards a general, population-level understanding of eco-evolutionary change. Trends in Ecology and Evolution, 28, 143–8.

Staver, A.C., Archibald, S., and Levin, S. (2011) Tree cover in sub-Saharan Africa: rainfall and fire constrain forest and savanna as alternative stable states. Ecology, 92, 1063–72.

Sterman, M.B. (2000) EEG markers for attention deficit disorder: pharmacological and neurofeedback applications. Child Study Journal, 30, 1–23.

Stevens, J.T. and Beckage, B. (2009) Fire feedbacks facilitate invasion of pine savannas by Brazilian pepper (*Schinus terbinthifolius*). New Phytologist, 184, 365–75.

Suding, K.N., Gross, K.L., and Houseman, G.R. (2004) Alternative states and positive feedbacks in restoration ecology. Trends in Ecology and Evolution, 19, 46–53.

Suding, K.N., Stanley Harpole, W., Fukami, T., *et al.* (2013) Consequences of plant-soil feedbacks in invasion. Journal of Ecology, 101, 298–308.

Svensson, A., Andersen, K.K., Bigler, M., *et al.* (2008) A 60 000 year Greenland stratigraphic ice core chronology. Climate of the Past, 4, 47–57.

Takimoto, G. (2009) Early warning signals of demographic regime shifts in invading populations. Population Ecology, 51, 419–26.

te Beest, M., Cromsigt, J.G.M., Ngobese, J., *et al.* (2012) Managing invasions at the cost of native habitat? An experimental test of the impact of fire on the invasion of *Chromolaena odorata* in a South African savanna. Biological Invasions, 14, 607–18.

te Beest, M., Stevens, N., Olff, H., *et al.* (2009) Plant-soil feedback induces shifts in biomass allocation in the invasive plant Chromolaena odorata. Journal of Ecology, 97, 1281–90.

Traveset, A. and Richardson, D.M. (2014) Mutualistic interactions and biological invasions. Annual Review of Ecology, Evolution, and Systematics, 45, 89–113.

van der Putten, W.H., Bardgett, R.D., Bever, J.D., *et al.* (2013) Plant-soil feedbacks: the past, the present and future challenges. Journal of Ecology, 101, 265–76.

Vanderploeg, H.A., Nalepa, T.F., Jude, D.J., *et al.* (2002) Dispersal and emerging ecological impacts of Pont-Caspian species in the Laurentian Great Lakes. Canadian Journal of Fisheries and Aquatic Sciences, 59, 1209–28.

Van Nes, E.H. and Scheffer, M. (2007) Slow recovery from perturbations as a generic indicator of a nearby catastrophic shift. American Naturalist, 169, 738–47.

Veraart, A.J., Faassen, E.J., Dakos, V., *et al.* (2012) Recovery rates reflect distance to a tipping point in a living system. Nature, 481, 357–9.

Vilà, M., Espinar, J.L., Hejda, M., *et al.* (2011) Ecological impacts of invasive alien plants: a meta-analysis of their effects on species, communities and ecosystems. Ecology Letters, 14, 702–8.

Vinton, M.A. and Goergen, E.M. (2006) Plant–soil feedbacks contribute to the persistence of Bromus inermis in tallgrass prairie. Ecosystems, 9, 967–76.

Vitousek, P.M. (1990) Biological invasions and ecosystem processes: towards an integration of population biology and ecosystem studies. Oikos, 57, 7–13.

Vitousek, P.M., Walker, L., Whiteaker, L., *et al.* (1987) Biological invasion of *Myrica faya* alters ecosystem development in Hawaii. Science, 238, 802–4.

Walker, B.H., Carpenter, S.R., Rockstrom, J., *et al.* (2012) Drivers, 'slow' variables, 'fast' variables, shocks, and resilience. Ecology and Society, 17, 30.

Walters, C. and Kitchell, J.F. (2001) Cultivation/depensation effects on juvenile survival and recruitment: implications for the theory of fishing. Canadian Journal of Fisheries and Aquatic Sciences, 58, 39–50.

Wang, R., Dearing, J.A., Langdon, P.G., *et al.* (2012) Flickering gives early warning signals of a critical transition to a eutrophic lake state. Nature, 492, 419–22.

Williamson, M. and Fitter, A., 1996. The varying success of invaders. Ecology, 77, 1661–6.

Wolfe, B.E. and Klironomos, J.N. (2005) Breaking new ground: soil communities and exotic plant invasion. BioScience, 55, 477–87.

Wolfe, B.E., Rodgers, V.L., Stinson, K.A., *et al.* (2008) The invasive plant *Alliaria petiolata* (garlic mustard) inhibits ectomycorrhizal fungi in its introduced range. Journal of Ecology, 96, 777–83.

Yelenik, S.G., Stock, W.D., and Richardson, D.M. (2004) Ecosystem level impacts of invasive *Acacia saligna* the South African fynbos. Restoration Ecology, 12, 44–51.

Young, T.P., Chase, J.M., and Huddleston, R.T. (2001) Community succession and assembly comparing, contrasting and combining paradigms in the context of ecological restoration. Ecological Restoration, 19, 5–18.

Zeeman, E.C. (1977) *Catastrophe Theory: Selected Papers, 1972–1977*. Boston: Addison-Wesley.

Zhang, F., Hui, C., and Pauw, A. (2013) Adaptive divergence in Darwin's race: how coevolution can generate trait diversity in a pollination system. Evolution, 67, 548–60.

CHAPTER 8

Community assembly and succession

8.1 Ecological community invaded

Human-mediated introductions of species to areas far removed from their native ranges have forged a global-scale natural experiment that offers many opportunities to improve our understanding of the factors that determine the composition of ecological communities. Invasions interact with other drivers of global environmental change such as climate change and altered land use to influence, and sometimes dismantle, the stabilizing forces that permit species coexistence in communities. Research in community ecology, especially work that has shifted the focus away from an emphasis on community assembly rules to address assembly processes and patterns more generally, has contributed much to our understanding of how multiple species interact and coexist. Increasing attention is being given to understanding the effect of dispersal between local communities on species coexistence in regional meta-communities. Such studies are casting new light on the processes that determine biodiversity patterns at different spatial and temporal scales.

Research on community assembly has expanded in scope recently, following decades of debate over the relevance of Diamond's (1975) assembly rules (see section 8.3.1). The objective of studying assembly rules as currently accepted in the field is to explain and predict which subset of species from regional pools could occur in a given habitat (Keddy 1992). Progress has been made by clarifying the key assembly processes and constraints at work; for example, facilitation, tolerance, and inhibition (Connell and Slatyer 1977), and the similarities between concepts that are often considered distinct (e.g. succession and community assembly) (Young

et al. 2001); hereafter we use these terms interchangeably. Processes similar to the fundamental ones that drive evolution (selection, drift, speciation, and dispersal) act upon community dynamics (Belyea and Lancaster 1999; Vellend 2010). These processes can be further elucidated as the main drivers of community assembly: the structure of the environment, species morphology, the economics of its behaviour, population dynamics, dispersal, and evolutionary processes (MacArthur 1972; Vellend 2010).

The phenomenon of biological invasions and the increasing occurrence of alien species in many ecosystems have reopened debates on several key issues pertaining to community assembly rules. What set of processes and rules governs the composition and abundance of co-occurring species at a locality? Is there such a thing as community integrity? Is there a maximum number of species that a community can hold? Such questions have long intrigued community ecologists, and biological invasions clearly offer opportunities to gain crucial new insights on the determinants of community assembly. Following the discussion of regime shifts triggered by invasive species in Chapter 7, we now examine how biological invasions shed light on these and other questions in community ecology and how insights from community ecology could help solve some conundrums in invasion ecology.

Community ecology focuses largely on three interrelated assembly-level issues: coexistence, succession, and diversity patterns (see Chase and Leibold 2003; Verhoef and Morin 2010; Vellend 2016, for recent overviews). We summarize key points of these issues but focus more on links between these aspects and recent studies in invasion ecology, our main aim being to draw novel insights at the interface between community ecology and invasion

Invasion Dynamics. Cang Hui and David M. Richardson, Oxford University Press (2017).
© Cang Hui and David M. Richardson. DOI 10.1093/acprof:oso/9780198745334.001.0001

ecology. As community assembly is driven largely by forces that foster or inhibit species coexistence, we first discuss a list of mechanisms that mediate species coexistence through stabilizing and equalizing forces (section 8.2). We then explore succession by elucidating links between secondary succession and community assembly/disassembly at local and regional scales (section 8.3). The diversity issue is addressed in section 8.4 (at local and metacommunity level) and in 8.5 (regional community and insular assembly patterns).

8.2 Species coexistence

Resident species in a community often need to reach some level of ecological compromise to be able to coexist with each other (Chesson 2000; Leibold *et al.* 2004). Based on the standard Lotka–Volterra competition model (see Chapter 6), coexisting species in a local community have to meet one fundamental requirement: the impact of competition from a species on itself must exceed the competitive impact from other species; that is, intraspecific competition must exceed interspecific competition. Alien species, on the other hand, can often take advantage of being released from the original ecological compromise in their native communities (see Chapter 6) and thus often enjoy a competitive edge over resident species in newly invaded communities. Consequently, to understand the performance of an alien species in a community fully we need to elucidate the mechanisms that foster coexistence between resident species and then to assess whether such coexistence mechanisms are susceptible or resistant to invasions (Melbourne *et al.* 2007).

The criterion for coexistence between two competing species can be derived based on the condition for *reciprocal invasibility* (Armstrong and McGehee 1980), which is known as *mutual invasibility* in biomathematics (e.g. Dieckmann and Law 1996). Coexistence of two species requires each species to be able to increase when rare. Based on this criterion, Chesson (2000) derived a common approximation of the long-term low-density growth rate of a species i, $\bar{r}_i$, in such a multispecies community with n species:

$$\frac{\bar{r}_i}{b_i} \approx \left(k_i - \bar{k}\right) + \frac{(1-\rho)D}{n-1}$$

where b_i is the mortality, k fitness, ρ niche overlap, and D a positive constant. The term $(1-\rho)$ measures the level of niche difference between alien species i and the resident species in the community. Niche difference reduces competition intensity and facilitates the establishment and stable coexistence of the invader and resident species in the community. Because it promotes stable coexistence, mechanisms that ensure niche difference are hereafter labelled stabilizing forces. By contrast, the term $(k_i - \bar{k})$ denotes the difference between the fitness of invasive species i and the average fitness of all resident species. Invasive species with a higher fitness $(k_i > \bar{k})$ have a competitive advantage over resident species and could exclude the latter. Fitness difference leads to competitive exclusion and is thus detrimental to species coexistence. Because equal fitness between species could result in neutral coexistence of species in a community, mechanisms that minimize fitness difference are hereafter labelled equalizing forces. Stabilizing forces are necessary for coexistence; without this, equalizing force can only slow down competitive exclusion (Chesson 2000).

Eight mechanisms of invasion and coexistence have been identified: classic niche partitioning, storage effect, fitness-density covariance, relative non-linearity, patch dynamics, species sorting (spatial niche partitioning), mass effects (source-sink dynamics and rescue effect), and neutral models (Hubbell 2001); arguably a model for co-occurrence rather than coexistence. Whereas the neutral model explores coexistence via equalizing force (i.e. reducing $k_i - \bar{k}$), the other seven are stabilizing mechanisms (i.e. they modify the second term by changing the way in which different concepts of niches (spatial, temporal, environmental, and biotic) are partitioned).

Classic niche partitioning requires coexisting species to differ in the types of resources or habitats (Tilman 1982; Holt *et al.* 1994); these resources/habitats do not need to be temporally or spatially heterogeneous. Most other stabilizing mechanisms require resources that vary in space or time. In particular, *storage effect* is a strategy for invaders to take the opportunities and gains in time or space that are created by chance or disturbance to compensate for loss during harsh times or in unfavourable environments (Chesson 1994). With spatial heterogeneity, competitive exclusion at local scales can switch to species coexistence at regional scales in meta-communities due to trade-offs and spatial heterogeneity; that is, a spatial storage effect (Chase and Leibold 2003; Kneitel and Chase 2004).

Relative non-linearity depicts competitive coexistence through shared biotic resources (e.g. a common prey). Different levels of non-linearity in functional response create conditions for stable coexistence (Armstrong and McGehee 1980). With less non-linearity in its functional response than the resident species, an invader will experience an advantage that may allow it to establish. When comparing the functional response of the invasive Ponto–Caspian freshwater shrimp, *Hemimysis anomala*, with that of the native shrimps *Mysis diluviana* and *M. salemaai*, Dick and colleagues' (2013) results support relative non-linearity, with the invader showing a more linear form of functional response. *Patch dynamics* depicts the possibility of coexistence in meta-communities through competition-colonization trade-off (Hastings 1980; Tilman 1994; Yu and Wilson 2001). Successful invaders are often good colonizers that take advantage of empty patches created by disturbance (see section 8.3). *Species sorting* requires each species to have different favourable habitats so that species distributions are spatially separated due to habitat heterogeneity; that is, spatial niche partitioning (Tilman 1982; Chase and Leibold 2003).

These mechanisms of coexistence rely on trade-offs of life-history strategies (e.g. the three-way competitive, stress tolerant, and ruderal (CSR) trade-offs) and across scales (Kneitel and Chase 2004). In particular, Grime's (1974, 1997) scheme of three-way trade-offs in plants among competitive (C), stress tolerant (S), and ruderal (R) strategies (Fig. 8.1) has received increasing support (Diaz *et al.* 2016; Kunstler *et al.* 2016). Through such trade-offs, niches can be partitioned among species which reduces interspecific competition to allow the coexistence of species, with consistent evidence in plant communities across the globe (Kunstler *et al.* 2016). Consequently, the classic equilibrium view of species coexistence is based on the niche

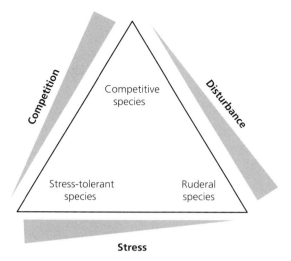

Fig. 8.1 An illustration of Grime's (1974, 1997) three-way trade-offs in plants among competitive, tolerant, and ruderal strategies.

theory that relies on life-history trade-offs (Tilman 2004). Competitors excel in habitats with high productivity and low disturbance; stress-tolerant species survive in habitats with low productivity; ruderal species occupy habitats that experience frequent disturbance. Plant species can thus allocate their resources to traits constrained by the CSR trade-off for survival, reproduction, and growth. Most invasive species are ruderals whose establishment and proliferation is facilitated by high levels of disturbance, while a few highly competitive invaders or stress-tolerant species introduced to harsh environments are also highly invasive. Communities facing altered assembly processes create opportunities for invasion when there is a mismatch between resident traits and habitat characteristics.

We have discussed the features/traits of successful invaders (Chapter 4), niche evolution and shift in invasive species (Chapter 5), and a list of hypotheses related to competition and other types of biotic interactions, such as limiting similarity, Darwin's naturalization hypothesis, apparent competition, and the evolution of increased competitive ability (Chapter 6). These chapters all discuss niche-based theories for the coexistence of natives and aliens and thus the emergence of a community assembled from native and alien species. In addition, many factors (e.g. disturbance, stress, and shared resources/natural enemies) can affect the nature of interactions between competing species, thereby altering the likelihood of competitive exclusion. For instance, according to Connell's (1978) *intermediate disturbance* hypothesis (IDH), species diversity should be highest at intermediate levels of disturbance. Together with trade-offs (Chesson and Huntly 1997; Fox 2013), low disturbance can result in the dominance of competitive species; high disturbance leads to the extinction and extirpation of many species. The *stress-gradient* hypothesis predicts that the frequency of facilitation will increase while the frequency of competition will decline along the stress gradient (Bertness and Callaway 1994; Pugnaire and Luque 2001). Again, low-stress environments can lead to competitive dominance, while extreme high stress will eventually reduce diversity even with the increasing facilitation due to biotic interactions (Michalet *et al.* 2006; Maestre *et al.* 2009). Apparent competition and prey switching from shared natural enemies could also facilitate coexistence by reducing impacts on rare species due to density-dependent foraging (the relative non-linearity mechanism). We discuss their relevance of biotic interactions in biological invasions in more detail in Chapter 6.

Fitness-density covariance describes the scenario where population density can build up in high-quality habitats so as to reduce each individual's fitness due to intense intraspecific competition (Chesson 2000), resulting in a positive covariance between environmental quality and intraspecific competition intensity. Along the gradient of habitat quality, individual fitness will thus follow a hump-shaped curve, different from the linear relationship assumed in species sorting. Alternatively, this covariance can be considered a joint result of species sorting and density dependence. With this covariance, a species experiences the strongest intraspecific competition when in its most favourable environment, and suffers the strongest interspecific competition when in environments that are preferred by its competitors the most. The existence of ecologically similar resident species in a community signals the availability of quality habitat for an invader. Due to this positive covariance, the individual fitness of these ecologically similar resident species could have been hampered from intensified

intraspecific competition, giving rare invaders a competitive edge over abundant resident species and thus opportunities to establish in their favourable habitat. In unfavourable habitat, rare invaders suffer from the combined effect of poor habitat quality and strong interspecific competition from resident species, making establishment difficult. Because invasive species often start with small populations, they can escape such covariance that reduces individual fitness when population density is higher. The advantage of being rare, together with the spatial storage effect, allows populations of invaders to grow quickly in favourable habitats. This mechanism has been further developed to incorporate stochastic drift into stabilizing mechanisms through trade-offs and niche partitioning (Tilman 2004).

Both the *mass effect* and *neutral models* emphasize the role of dispersal in enhancing persistence, thus the potential of coexistence. The former emphasizes the role of directional dispersal from source to sink populations in enhancing local population persistence and thus coexistence (Mouquet and Loreau 2003; Amarasekare and Nisbet 2001), whereas the latter depicts transient unstable coexistence of identical species and emphasizes the dispersal from regional species pool to counterbalance local extirpation (Hubbell 2001). The application of neutral theory to describe and explain the human-mediated invasion of communities by alien species is still in its infancy (Daleo *et al.* 2009; Herben 2009). Daleo and colleagues (2009) reasoned that both neutral and niche processes contribute to invasions, but they did not demonstrate this empirically or theoretically. Using neutral simulation models, Herben (2009) showed that several general empirical invasion patterns are explained well by such models. Recent formulations recognize that species assemblages in unsaturated local communities are at least partly shaped by neutral forcing via the continuous influx of 'invaders'—species from the regional pool and, recently, alien species (Stohlgren *et al.* 2003). Despite contrasting opinions on the applicability of neutral theory to real-world communities (Chase 2005; Clark 2012; Rosindell *et al.* 2012), it is now widely accepted that deterministic and stochastic processes, or roughly neutral and niche processes, or drift and adaptive processes, interact to structure species assemblages (Chase and Myers 2011; Bar-Massada *et al.* 2014; Nuwagaba *et al.* 2015).

The current debate on the importance of niche-based versus neutral processes in driving community assembly clearly hinges on the role of stabilizing mechanisms versus equalizing mechanisms in supporting species coexistence (Chesson 2000; Adler *et al.* 2007). In an attempt to clarify the concept of the neutral–niche continuum has been proposed—this view posits that natural communities lie along a continuum defined on the basis of the relative importance of these two types of processes, with patterns of pure neutral and pure niche communities at the two extremes of the continuum (e.g. Gravel *et al.* 2006; Leibold and McPeek 2006; Mutshinda and O'Hara 2011; Bar-Massada *et al.* 2014). The potential correlation between the neutrality (i.e. the position of a community along this continuum (Gravel *et al.* 2006; Haegeman and Loreau 2011; Bar-Massada *et al.* 2014) and invasibility of a community could be related but this question is yet to be fully explored. The fact that highly disturbed communities (mostly anthropogenically driven) are often also highly invaded (Hobbs and Huenneke 1992) provides some support for the potential neutrality–invasibility correlation. Neutral processes (e.g. distance-limited dispersal) could reinforce niche processes (species filtering) in spatially autocorrelated environments, creating community patterns near and even beyond the

two ends of the neutral–niche continuum (Latombe *et al.* 2015). A complete list of combinations of assembly processes that could drive the shift of a community towards the neutral end of the continuum, thereby making the community more susceptible to invasion, is an enticing aim. It will require further dialogue between these two sub-disciplines of ecology.

The question of whether equalizing mechanisms (neutral and stochastic) are more or less important than stabilizing mechanisms (niche-based and deterministic) in driving community assembly and species coexistence has no simple answer. The relative roles of neutral and niche processes have been shown to differ across spatio-temporal scales (Bell 2005; Cadotte 2007; Stokes and Archer 2010). Both mechanisms are scale- and context-dependent—this is known as MacArthur's paradox (Schoener 1983; Loreau and Mouquet 1999). Recent efforts to incorporate coexistence mechanisms in invasion ecology (Shea and Chesson 2002; Melbourne *et al.* 2007; MacDougall *et al.* 2009; Mayfield and Levine 2010) allow us to conceptualize these mechanisms into a simple diagram based on Chesson's (2000) formula on invasion performance through both equalizing force (fitness difference) and stabilizing force (niche difference) (Fig. 8.2). We first describe the two axes and then four corners of the diagram:

- **Horizontal axis**: The absolute value of fitness difference measures equalizing force which determines the speed/pace of invasion dynamics. It is

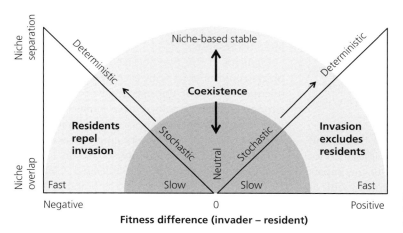

Fig. 8.2 Multiple coexistence mechanisms for explaining the invasion performance in a community.

related to the invasiveness of an invader and describes how fast its population grows, or whether it can competitively exclude the resident species, or be expelled by the resident species due to biotic resistance. **Vertical axis**: By contrast, niche overlap/separation measures stabilizing force and indicates whether the outcome is driven by neutral or niche processes.

- **Lower centre**: Closeness to the origin (equal fitness plus similar niches) means that the fate of invasion is governed by stochastic processes, with even successful invaders only having low invasiveness. Some coexistence mechanisms (e.g. fitness-density covariance), density-dependence (e.g. the Allee effect), and low propagule pressure could critically shift the origin and thus affect invasion performance (see Chapters 5 and 6). Since initial propagule pressure is low compared to the population size of resident species, fitness metrics need to reflect the intrinsic population growth rate when the invader is rare; this is known as the *invasion fitness* (Chapter 10). Successful invaders of this kind are strongly influenced by introduction history and propagule pressure.

- **Lower left or right**: With overlapped niches and big fitness differences, coexistence becomes unlikely. Successful invaders of this kind normally have superior competitive abilities and can rapidly exclude resident species, causing big impacts in the community. Invaders with lower fitness than the resident species, however, cannot establish and will be expelled by the resident species.

- **Upper centre**: With similar fitness but separated niches, establishment, and coexistence are assured, even if invaders have lower fitness than resident species. However, the invasiveness of the invaders is low. Such invaders usually possess distinct traits but have low impact.

- **Upper left or right**: With different fitness and separated niches, the establishment is assured for invaders having higher fitness, with the impact immediately felt. Invaders either coexist with or expel the resident species. Invaders with lower fitness, however, could still establish but have little impact. Successful invaders of this kind

normally possess high invasiveness and distinct traits and potentially have big impact, such as ecosystem engineers.

8.3 Assembly succession via invasion

8.3.1 Community succession

Studies on community assembly are central to ecology, their key aims being to predict assembly composition and to clarify constraints and patterns of species coexistence (Götzenberger *et al.* 2012). This approach concentrates on the possibility of multiple sets of species in a community and the potential conditions and trajectories leading to a specific set. However, to understand fully how resident communities absorb invasions we need to address important aspects of community assembly: the process of community assembly through the sequential establishment of colonists/invaders and the fluctuation of their populations (Fukami 2010). This dynamic view of community assembly pre-dated the discussion of assembly rules in the literature. It grew from discussions of ecological succession as early as the mid-eighteenth century when the French naturalist Georges-Louis Leclerc (Comte de Buffon; 1707–1788) deliberated on aspects of forest succession, and the concept of vegetation type in biogeography as set out in the work of Alexander von Humboldt (1769–1859) and Jean-Baptiste Lamarck (1744–1824). However, it was only in the early twentieth century that the concept of succession was given prominence by Frederic Clements (1916) with his proposition on secondary succession after disturbance.

Clements (1916) hypothesized that earlier colonizers could interact with each other and also the micro-environment through feedbacks that allow stronger competitors to replace earlier colonizers until the community reaches a 'climax'. This super-organism view of community emphasizes a deterministic destiny of community succession. Gleason (1927) and Tansley (1935) challenged Clements' view by arguing that species in a community behave individually with no regard for such a 'climax' or community integrity. Tansley (1935) coined the term 'ecosystem' to replace community and argued for the potential of multiple climaxes (or multiple community assemblages; i.e. alternative stable states) and trajectories to these climaxes (Egler 1954; see also Chapter 7). As such, the history and timing of species arrival to a community could have a priority effect on community succession, potentially altering final trajectories (Chase 2003; Fukami 2015).

Different histories in a community can diverge into different assemblages whereas a single climax implies the convergence in community succession; that is, the final assemblage is insensitive to the initial condition and history (May 1977). The existence of multiple alternative assemblages often requires the existence of positive reinforcing feedbacks that foster system bistability and tipping points in separating basins of attraction (Scheffer *et al.* 2001; also see Chapter 7), potentially triggered by biological invasions of alien species (some act as ecosystem engineers) that can alter nutrient cycling and disturbance regimes (Ricciardi *et al.* 2013; Gaertner *et al.* 2014). A community could be both convergent and divergent at different levels of community organization, where both convergence (communities proceed towards a pre-disturbance state regardless of historical conditions) and divergence (historical factors continue to affect the long-term trajectory of community development) are present in nature (Young *et al.* 2001). In studies of experimentally assembled plant communities it has been reported that while the identities of individual species remain unique across different community replicates, species traits generally became more similar (Fukami 2005; Gao *et al.* 2015).

Jared Diamond (1975) introduced the term *assembly rules* based on his work on the birds of New Guinea and satellite islands to define a set of empirical patterns generalized for depicting community compositional structure. In particular, he discussed permissible and forbidden species pairs in a checkerboard pattern of species co-occurrence using a species-by-site matrix. However, these rules are descriptions of patterns, not the processes that could drive and derive these patterns, leaving the link between pattern and process open to debate. For instance, checkerboard patterns that show that two species tend to avoid each other are widely considered to have been overgeneralized by Diamond (1975) to be the outcome of interspecific

competition. Connor and Simberloff (1979) criticized such pattern-to-process inference and argued that chance alone as generated by null models can produce similar checkerboard patterns, thus questioning Diamond's overemphasis of competition in structuralizing community assemblages (Gotelli and Graves 1996; Gotelli and McCabe 2002). More sophisticated tests have been developed to detect the underdispersion in co-occurrence patterns, taking into account factors such as functional traits (McGill *et al.* 2006), phylogeny (Webb *et al.* 2002), environmental heterogeneity (Heino and Grönroos 2013), and the distance decay of similarity (Soininen *et al.* 2007), and to explain the regional co-occurrence patterns of native and alien assemblages wth different residence times using these factors (Hui *et al.* 2013; Roura-Pascual *et al.* 2016).

Several authors have bemoaned the fact that invasion ecology has sought explanations for processes associated with introductions and invasions without due cognisance of insights from ecology in general. For example, Davis and colleagues (2001) argued that 'invasion ecology has largely dissociated itself from other sub-disciplines of ecology, particularly succession ecology'. Although human-mediated introductions do change key aspects of ecology and therefore require certain special rules, it is clear that community ecology and invasion ecology could both benefit by drawing insights from each other. Much work has been done recently to exploit opportunities offered by the experiments that invasions have created at scales that could never be achieved in formal experiments for exploring the potential trajectories of regional changes in species co-distribution. As with resident native species during community assembly, introduced species also need to cross a series of filters to become established and to spread. For alien species, this trajectory has been conceptualized as the introduction–naturalization–invasion continuum (Richardson *et al.* 2000; Blackburn *et al.* 2011; Richardson and Pyšek 2012). The stochastic component of 'random' introduction is gradually diminished through multiple dispersal and environmental filters, with the remnant species emerging as 'winners'. These filters along the introduction–naturalization–invasion continuum thus define the direction in both human-mediated and natural selections.

8.3.2 Local-scale succession

Given the increasing levels of human-mediated disturbance affecting local and regional communities, a framework based on successional stages after major disturbance could offer conservation managers a more powerful tool for monitoring and prioritizing invasions for management. Understanding how native species establish differentially during stages of succession could help in assessing the potential susceptibility of communities to invasions from non-native species. For instance, closed-canopy and late-successional forests appear to be fairly resistant to invasion by alien plants (DeFerrari and Naiman 1994; Rejmánek et al. 2005; Martin et al. 2009). This is probably because most invasive plants are early-successional species (Martin et al. 2008) and because alien species assemblages follow similar succession dynamics to native assemblages (McLane et al. 2012). Invasibility might be better understood with the successional stages of community assembly in mind (Meiners et al. 2002; Campagnoni and Halpern 2009; McLane et al. 2012).

Secondary community succession driven by external perturbations can facilitate the establishment of alien species by creating temporarily available opportunity niches, while increased biotic resistance could be experienced by invaders at later succession stages. For instance, Corenblit and colleagues (2014) documented the alien and native plant species in two habitats along the Tech River in south-eastern France: herbaceous communities on the alluvial bars that are exposed to strong hydrogeomorphic disturbances, and woody vegetation on the river margins that escape annual hydrogeomorphic disturbances (Fig. 8.3). Positive correlations between alien and native plant species richness were observed in both habitats. The higher number of alien species in the herbaceous community than in the woody vegetation was attributed to changes in local exposure to hydrogeomorphic disturbances driven by woody ecosystem-engineer plant species and to vegetation succession. Succession has caused an increase of γ plant diversity at the scale of the riparian corridor due to the stabilization and construction of new habitats by pioneer engineering plants, a local decrease of a plant diversity during the succession related to the exclusion of several native and alien ruderals from shading by the native tree species *Alnus glutinosa, Populus nigra,* and *Salix alba,* and an increase in resistance to invasion during secondary succession.

Early invaders can have strong impacts on the invasion/colonization success of species that arrive later (priority effect and invasion meltdown; Simberloff and Von Holle 1999), especially when the introduction rate and/or the succession rate is low enough to allow species to become established before replacement (Drake 1991). For example, *Tamarix* species have successfully invaded many rivers in the western United States and have replaced populations of the native pioneer trees *Populus fremontii* and *Salix gooddingii* in reaches with intermittent stream flows (Friedman et al. 2005; Stromberg et al. 2007). The implication of *Tamarix* replacing *Populus* and *Salix* species in terms of ecosystem functioning in a human-impacted and hydrogeomorphically altered environment has been considered in the planning of restoration (Stromberg et al. 2009). Gao and colleagues (2015) found that the herbaceous plant community was highly structured and shaped by the presence of *T. chinensis*. At the local scale, two functional traits (plant height and leaf area) were found to be significantly convergent. Dispersal, environmental stress, and interspecific competition had trivial effects on the local community assembly. The facilitating effect of *T. chinensis* on the pioneering herbaceous plants by acting as a wind shelter was proposed as the dominant community assembly process.

Alien species can exploit transient opportunities during secondary succession and can also drive secondary succession by disassembling existing resident assemblages as the flip side of an assembly rule, thereby leading to irreversible successional trajectories. For instance, many native ant communities exhibit significant spatial segregation as a likely means of avoiding interspecific resource competition. Through their ability to exploit resources and interfere with competitors, invasive Argentine ants, *Linepithema humile,* can drive a shift from species segregation in intact communities to species aggregation in invaded communities (Fig. 8.4), thereby rapidly dismantling the mechanism of coexistence and

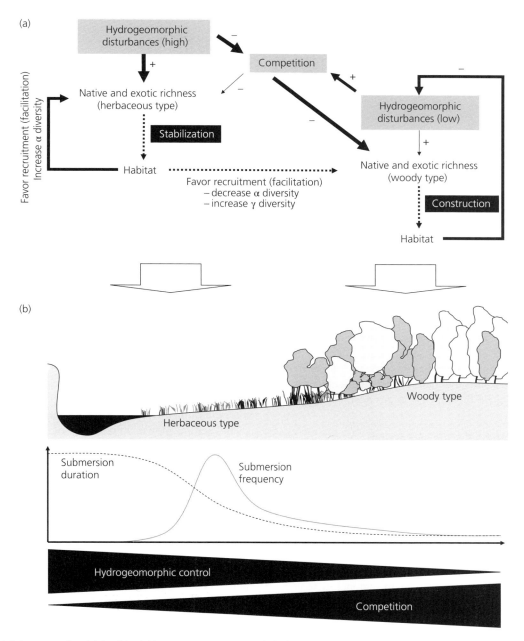

Fig. 8.3 A conceptual model describing (a) how ecosystem engineering by herbaceous and woody pioneer vegetation controls native and alien richness through the modification of the balance between extrinsic and intrinsic controls, hydrogeomorphic disturbances and competition; and (b) the localization of the processes along the transverse gradient of hydrogeomorphic connectivity. In (a), extrinsic and intrinsic controls are figured in grey boxes and ecosystem engineer effects in black boxes. 'Construction' corresponds to ecosystem engineering (*sensu* Jones *et al.* 1994). Increasing or decreasing effects are denoted by plus and minus signs. Dotted arrows express processes related to the biogeomorphic succession; solid arrows express primary and secondary controls according to their thickness. From Corenblit *et al.* (2014). Reproduced with permission from John Wiley & Sons.

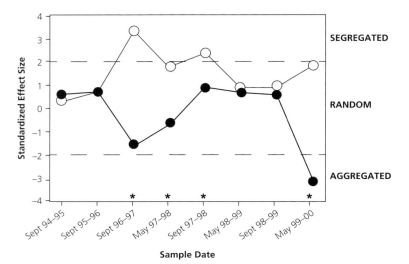

Fig. 8.4 A comparison of native ant community organization in sample plots immediately before and after invasion by Argentine ants (*Linepithema humile*). Paired symbols indicate the same plots sampled before (open circles) and one year after invasion (black dots) (Sanders *et al.* 2003). Copyright (2003) National Academy of Sciences, U.S.A.

co-occurrence (Sanders *et al.* 2003). Such community disassembly as indicated by the shift of co-occurrence patterns occurs not only because the spatial heterogeneity of the invader's density (thus impact) but more importantly also because native ant species are differentially affected by Argentine ants. Habitat generalists can persist temporarily after the invasion, while habitat specialists could be competitively expelled; some other native species appear to coexist with *L. humile* through a storage effect.

8.3.3 Regional-scale succession

Regional meta-communities are often unsaturated and absorbent to colonizers either from regional species pool or as human facilitated non-indigenous species (Cornell and Lawton 1992; Hubbell 2001; Loreau 2000; Whittaker *et al.* 2001; Wang and Loreau 2014). Species gain membership of regional assemblages by passing through multiple ecological and environmental filters. To capture the potential trajectory of structural changes in regional meta-communities driven by biological invasions, one can categorize species pools into assemblages of different residence times. Assemblages that have resided for longer periods in a region could have experienced more environmental filters than newly arrived species; consequently, these older assemblages should become more functionally ordered and structured. Although species with different residence times do interact, the role of interspecific interactions within

the same trophic level is relatively trivial at the regional scale compared with top-down regional processes which are driven largely by habitat suitability and dispersal barriers. Two specific hypotheses have been proposed to help in unravelling the potential trajectories of compositional and structural changes in regional assemblages driven by biological invasions (Hui *et al.* 2013): the *settling-down* hypothesis of diminishing stochasticity with residence time, and the *niche-mosaic* hypothesis of inlaid neutral modules in regional meta-communities.

The settling-down hypothesis of diminishing stochasticity assumes that because species in older assemblages are survivors of longer selection, stronger signals of matching between their habitat requirements and the characteristics of inhabited sites should be expected (i.e. a lock-and-key relationship), with groups of species likely to inhabit non-random subsets of sites that reflect this match (i.e. through species sorting). In contrast, more recent introductions should have a poorer match, as many species are initially randomly introduced to suboptimal sites. As a regional meta-community is an open-ended constantly evolving system, where propagules exchange either naturally or through mediation by humans between local communities and beyond the regional boundaries, it is rather difficult to predict the invasion dynamics of a particular introduced species given such ecological complexity. Instead, the succession dynamics of the regional meta-community could be better reflected by changes in system structures and orders such as those measured by co-occurrence and network metrics.

The niche-mosaic hypothesis of inlaid neutral modules recognizes that biodiversity maintenance and species coexistence can be achieved by balancing ecologically identical (equalizing) and distinctive (stabilizing) forces (*sensu* Chesson 2000; see section 8.2). The stabilizing force of niche separation could form niche-differentiated modules (or communities), whereas the equalizing force could ensure that species within a module possess rather similar niches, analogous to the concept of emergent neutrality (e.g. Vergnon *et al.* 2012). The modules will become more functionally distinctive

as residence time increases; that is, a shift from an initially neutral or stochastic assemblage to a niche-based functional-driven multi-module assemblage along the introduction–naturalization–invasion continuum. These two hypotheses regarding regional-scale succession were supported by results from a modularity analysis of three different-aged assemblages (neophytes, introduced after 1500 AD; archaeophytes, introduced before 1500 AD, and natives), comprising 2054 vascular plant species in 302 reserves in central Europe (Fig. 8.5).

The roles of species' traits and niche functions in structuring species assemblages have been hotly debated, largely due to the strong dichotomy between neutral-stochastic and niche-based models

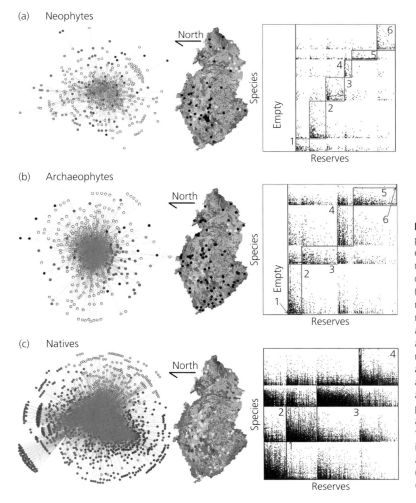

(a) Neophytes

(b) Archaeophytes

(c) Natives

Fig. 8.5 Network structures of vascular plants in the Czech Republic. Network expression, geographical location of reserves, and species-by-site matrices of modules identified for (a) neophytes, (b) archaeophytes, and (c) natives. In the network expression, open circles represent reserves. Blue, yellow, red, brown, black, and green points in the network expression and geographical maps indicate different modules identified in each of the three assemblages. Modules in the matrices are marked by the serial numbers and a rectangle, with points indicating the presence of a species (a row) occurring in a reserve (a column) and the rectangles of 'Empty' in neophytes and archaeophytes indicating reserves where these two species assemblages do not occur. From Hui *et al.* (2013). Distributed under a Creative Commons CC-BY license. (see Plate 1) (e.g. Cadotte *et al.* 2011; Flynn *et al.* 2011). Placing the genesis of communities into one of these categories is often done by comparing the similarity of assemblage patterns generated from these models with observations. Different processes can lead to similar patterns, and a single process can lead to multiple patterns. For this reason, comparisons between such patterns cannot provide conclusive support for the mechanisms embedded in the model (Grimm *et al.* 2005; Chase and Myers 2011). As modularity analysis relies only on species co-distribution, additional data on the characteristics of sites and species are needed. Further tests show little phylogenetic differentiation between species of different neophyte modules, in contrast to the high phylogenetic distinctiveness of modules comprising natives. This provides further support that the transition from neutral/stochastic processes to niche-/functional-based processes governs the regional meta-community succession and invasion (Hui *et al.* 2013). Arguably, the difference between the largely randomly assembled neophytes and the regional assemblage of natives with unique functional clusters points to the direction of changing roles of assembly processes during regional succession—decreasing the role of stochastic forces and increasing the importance of deterministic processes. The long-term structural changes in meta-communities could offer a temporal perspective for reconciling the debate between neutral and niche-based schools of thought. This suggests that alien species follow rather similar assembly processes to native species.

Refined conservation plans could be designed for each module (see also section 8.5). This module-based risk assessment and planning is consistent with trait- and function-based conservation. As these matrices depict only the co-distribution pattern of species association, species co-distribution could be more informative than species distribution for quantifying species invasiveness and performance in novel environments (Paini *et al.* 2010; see also section 8.4). Non-random species associations emerge along the introduction–naturalization–invasion continuum, and these co-distribution patterns of species association reflect the match between species' functional roles and their habitat requirements. The increasing modularity from young to mature assemblages not only identifies a

specific facet of the directional change in regional assemblages, but also suggests a transition from an assemblage driven by a stochastic process to functional-driven multi-module assemblages along the invasion pathway (Hui *et al.* 2013).

8.4 Diversity signature

8.4.1 Invasion paradox

Species diversity is often positively related to community function—for example, primary productivity, soil fertility, resistance to disturbance, or speed of recovery (resilience) (Cardinale *et al.* 2012; Tilman *et al.* 2014). Two hypotheses have been proposed to explain these relations based on the degree of strength and overlap of the ecological function of species (Cain *et al.* 2014). Firstly, as species richness increases, each species could add an equal effect, resulting in a linear increase in community function (*complementarity* hypothesis). Secondly, when there is overlap in their function, the community function will gradually flatten off with the addition of species (*redundancy* hypothesis). In addition, if there is a large variation in ecological function among species, community function might not rise smoothly with the addition of species. For these reasons, different diversity-function curves could help to explain how community assembly responds to biological invasions.

Invasibility is a community-level feature of ecosystems and is intuitively related to the richness of resident species. Elton's (1958) niche-based hypothesis proposed that species- and functionally rich communities are less susceptible to invasion because high-diversity assemblages limit establishment opportunities for invaders by reducing access to resources. Resource availability, the strength of resource competition, and disturbance that could interfere with resource accessibility and competition, have subsequently been identified as key determinants of community invasibility (Davis *et al.* 2000; Davis and Pelsor 2001; Tilman 2004; Renne *et al.* 2006). Is there a negative diversity–invasibility curve? The proposition that high levels of species richness increase the resistance of the community to invasion has received strong support from fine-scale experiments (Fridley *et al.* 2007), where species richness is normally negatively correlated with resource availability (Tilman 1997; Levine 2000, 2001; Naeem *et al.* 2000; Kennedy *et al.* 2002).

There are counterarguments and evidence to support a positive diversity–invasibility relationship; this is known as the *rich get richer* hypothesis (Palmer and Maurer 1997; Stohlgren *et al.* 2003). There is thus an *invasion paradox* in that there is support for both a negative and a positive relationship between native biodiversity and invasibility (Fridley *et al.* 2007). Structurally complex communities may increase microhabitat diversity and thus favour certain invaders, especially when communities include ecosystem engineers that facilitate the establishment of other species (Bruno 2000; Bruno *et al.* 2003; Bulleri *et al.* 2008). More importantly, processes affecting biotic interactions and species distributions are scale-dependent (McGill 2010; also see Chapter 6). Plot-scale experiments are not supported by observations at the scale of landscapes or larger areas; this could be because the strong effects of competition that are evident at the plot scale are overridden by landscape processes at broader scales (Levine 2000; Brown and Peet 2003; Stohlgren *et al.* 2003).

Fridley and colleagues (2007) identified mechanisms that could affect the correlation of native and alien species richness. Three of these, statistical artefact, Eltonian biotic resistance, and invasional meltdown, produce negative correlations. Mechanisms that produce positive correlations are: neutral process plus spatial variance in immigration rates or in disturbance rates; spatial environmental heterogeneity; biotic acceptance plus non-equilibrium conditions; and facilitation. No single mechanism can produce the pattern of a negative correlation at small scales and a positive correlation at large scales. However, when two or more mechanisms underlie the community assembly rules, the invasion paradox can be resolved. Studies show that the sign of diversity–invasibility relation, depicted as the correlation of native and alien species richness, is indeed scale-dependent, potentially due to processes at work at different scales (Fig. 8.6; Fridley *et al.* 2007).

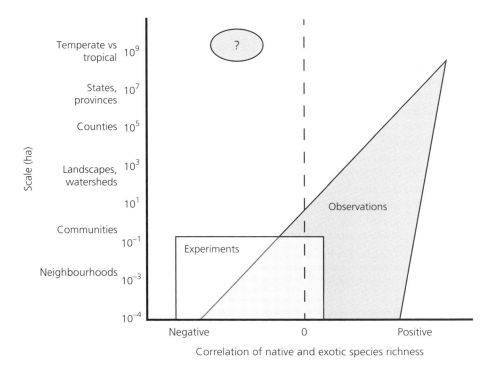

Fig. 8.6 The invasion paradox, illustrated as the contrasting signs of correlation between native and alien species richness from experimental and observational studies, could be potentially explained by the scale of study (Fridley *et al.* 2007). Reproduced with permission from John Wiley & Sons.

Based on explanations from Shea and Chesson (2002) and Davies and colleagues (2005) on the invasion paradox, Melbourne and colleagues (2007) showed how biotic interactions and environmental heterogeneity together produce the slipping sign of correlation of native and alien species richness when shifting from plot to larger scales (Fig. 8.7). At small scales, density dependence and resource competition within sites places constraints on both native and alien species richness (Levine 2000), inevitably leading to a negative correlation. At large scales, resource heterogeneity between sites (Shea and Chesson 2002) and within sites (Davies *et al.* 2005) contributes to the capability of accommodating more species, regardless of whether they are native or alien. Environmental heterogeneity could increase invasion success but also reduce the impact of invasion by promoting coexistence that is otherwise impossible in homogenous environments (Fig. 8.7; Melbourne *et al.* 2007). This means that community assemblages at large spatial scales reflect regional habitat heterogeneity, with species that have been locally extirpated from competition still remaining through a spatial storage effect. These co-varying environmental factors and regional processes (e.g. immigration) at large scales will thus promote both native and non-native diversity (Levine and D'Antonio 1999; Brown and Peet 2003; Shea and Chesson 2002; Stohlgren *et al.* 2003; Melbourne *et al.* 2007). The most useful spatial scale, however, for discussing community assembly and issues relating to species coexistence is the landscape scale (roughly from 10^2 to 10^4 ha) and/or the scale of meta-communities (Whittaker *et al.* 2001). These scales straddle finer scales (where deterministic processes often dominate) and larger scales where stochastic forces operate (Kneitel and Chase 2004; Rosindell *et al.* 2012; Heino and Grönroos 2013) and at which management of biodiversity and invasions is most often done (Hobbs *et al.* 2006; Pauchard and Shea 2006).

Besides these explanations, that are primarily based on features of recipient environments and ecosystems, characteristics of alien species also contribute to the invasion paradox. For instance, a positive correlation between the richness of native and alien species could emerge with weak invaders (species that persist but do not dominate habitats), while a negative correlation could result for invaders that dominate/transform the landscape (Ortega and Pearson 2005). Biotic resistance does not necessarily require a high number of resident species; sometimes, one or two native species could pose strong resistance across the landscape. For instance, two common native perennial grass species in North America, *Calamagrostis nutkaensis* and *Festuca rubra*, pose strong resistance through competition for light with the European perennial grass invader *Holcus lanatus*, while another common native grass, *Bromus carinatus* var. *maritimus*, has little effect on the performance of *H. lanatus* (Thomsen and D'Antonio 2007). Different responses of resident functional groups to invasion could be more important to community invasibility than simply the richness of resident species.

Importantly, biological invasions are driven by human activities (e.g. introduction efforts and disturbance), and factors associated with human actions could be more direct drivers of any relationships between native and alien species richness than any inherent features of recipient ecosystems. That is, human activity could dictate the relationship between the richness of aliens and natives, especially in deeply transformed ecosystems. For instance, global freshwater fish distributions and communities have been fundamentally transformed by habitat alteration (e.g. the construction of hydroelectric dams and inter-basin water transfer), water overuse (from irrigation and industrial need), pollution (e.g. from waste water), disturbance (e.g. waterway traffic), freshwater fisheries and fishing (Vörösmarty *et al.* 2010). Using global freshwater fish distributions, Leprieur and colleagues (2008) tested the 'human activity' hypothesis (human activities facilitate the establishment of non-native species by disturbing natural landscapes and by increasing propagule pressure); the 'biotic resistance' hypothesis (species-rich communities will readily impede the establishment of non-native species); and the *biotic acceptance* hypothesis (environmentally suitable habitats for native species are also suitable for non-native species). Only the 'human activity' was supported, suggesting that a high level of human disturbance could override the presence or absence of invasion paradox (Fig. 8.8).

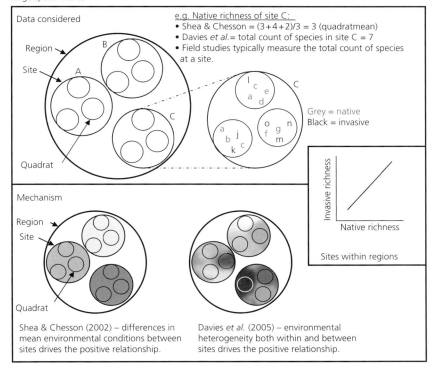

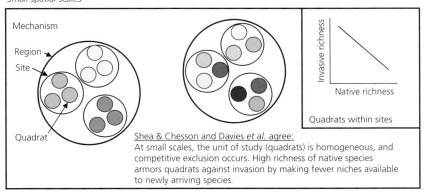

Fig. 8.7 Differences between the models of Shea and Chesson (2002) and Davies *et al.* (2005) regarding the diversity–invasibility paradox. Small spatial scales are those at which individuals interact (e.g. experience inter- and intraspecific competition). Large spatial scales are those greater than the scale of individual interactions. Shading represents variation in an environmental (exogenous) factor. Scales of species richness: alpha diversity is the mean diversity of quadrats within a site; beta diversity is the difference in species composition between quadrats within a site; gamma diversity is the diversity of a site (i.e. total count of species). From Melbourne *et al.* (2007). Reproduced with permission from John Wiley & Sons.

(a)

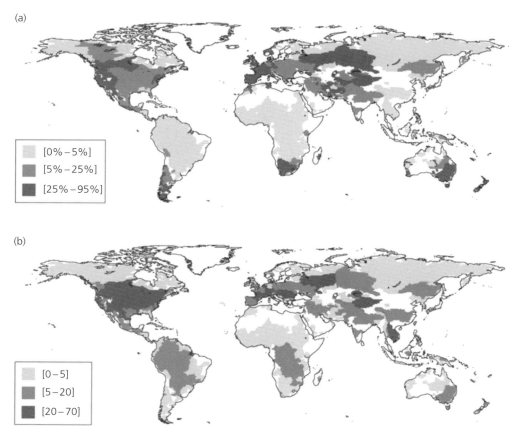

[0% − 5%]
[5% − 25%]
[25% − 95%]

(b)

[0 − 5]
[5 − 20]
[20 − 70]

Fig. 8.8 The global distribution of non-native freshwater fish species. (a) The percentage of non-native species per basin (i.e. the ratio of non-native species richness/total species richness) and (b) the non-native species richness per basin. The maps were drawn using species occurrence data for 9968 species in 1055 river basins covering more than 80 per cent of continental areas worldwide. Invasion hotspots are defined as areas where more than a quarter of the species are non-native (dark areas on map (a)), leading to six invasion hotspots: the Pacific coast of North and Central America, southern South America, western and southern Europe, central Eurasia, South Africa and Madagascar, southern Australia, and New Zealand. From Leprieur *et al.* (2008).

There are potentially more issues regarding using the correlation of native and alien species richness to test the assembly-level mechanism on the *diversity-invasibility* hypothesis, pointing towards several future research priorities regarding the invasion paradox. Firstly, although many authors have questioned the potential statistical artefacts in producing certain signs of the correlation (e.g. Rejmánek 2003), a more interesting question is whether the correlation is linear or non-linear; that is, could the native–alien richness curve eventually level off? This brings the discussion back to the two basic diversity-function hypotheses (complementarity versus redundancy). Secondly, we are essentially equating the diversity–invasibility relationship to the correlation of native and alien species richness; that is, the invasibility of a community is measured by the number of alien species that can establish. The introduction rate of alien species and propagule pressure are largely subject to human activity. Because humans prefer to inhabit high-productivity land (which often coincides with high biodiversity), introduction biased towards species-rich areas preferred by humans could introduce an additional confounding effect to the correlation. Moreover, the number of established alien species is a transient property as invasion dynamics have hardly reached equilibrium. How will the final richness of alien species, given sufficient residence time, correlate with the native richness? This question is yet to be answered. Certainly, invasibility of the recipient community needs to be explained more fundamentally than simply by relating it to the richness of native species (also see Chapter 10).

8.4.2 Species turnover

If species richness oversimplifies the concept of community invasibility, other diversity signatures, in particular species turnover in space (e.g. Martin and Wilsey 2015), could be more indicative of secondary successional dynamics after disturbance and invasibility of the community. Indeed, when inferring assembly process from biodiversity patterns, we could drastically reduce the chance of committing type I errors by considering multiple diversity signatures or patterns simultaneously (e.g. Grimm *et al.* 2005). If human-mediated disturbance that triggers secondary succession does not affect native assemblage but only facilitates aliens, we could see a low number of aliens in undisturbed communities. In this scenario, species richness might not be an informative indicator of community assemblage responding to invasion. Instead, assembly composition and turnover, especially of particular kinds of species (e.g. common and rare, generalists and specialists), could better indicate successional stages and cross-site differences of invasibility. For instance, a lower alien species cover within the woody vegetation type than within the herbaceous type suggests an increase of resistance to invasion during biogeomorphic succession (Corenblit *et al.* 2014): the engineering effects of woody vegetation through landform construction resulted in a decrease of alpha (α) diversity at the patch scale but caused an increase in gamma (γ) diversity at the scale of river segment. Species compositional turnover could reveal the invasibility and community assembly processes more explicitly.

Based on a recent proposal of using both alpha (α) and beta (β) diversity (i.e. diversity signature) to differentiate neutral (stochastic) community assembly from those driven by deterministic (niche-based) processes (Jost 2007; Chase and Myers 2011), we

could detect both community assembly processes and the kind of native species (common or rare) that are susceptible to the invasion based on changes in diversity signature due to invasion (Fig. 8.9), in driving variation in species composition across spatial and environmental gradients, using species turnover measured as beta (β) diversity (Jost 2007; Chase and Myers 2011). This approach can be applied to discern alternative alien community assembly mechanisms. Firstly, the pattern of isolation-by-distance (also known as the distance decay of similarity; Nekola and White 1999) implies a positive correlation between β diversity (measured as dissimilarity) and the distance between sites if the community assembly is driven by a neutral force (e.g. dispersal limitation), but a null correlation in the case of a niche-based community (Fig. 8.9a, assuming no strong spatial autocorrelation in habitat heterogeneity). The pattern of isolation by resistance, instead, implies a positive correlation between β diversity and environmental difference between sites if the community assembly is driven by niche-based processes, but a null correlation in the case of a neutral community (Fig. 8.9b). If alien species are passengers of human activity, we would expect the relationship to shift towards neutral lines, showing opposite directions in these two plots (arrows Fig. 8.9a and b). If alien species are drivers of change in the resident community, the observed relationship should shift towards niche lines. When β diversity is plotted against α diversity (Fig. 8.9d), such a diversity signature could also indicate processes driving alien community assembly.

Venn diagrams can be used to illustrate three scenarios regarding the impact of invasions: (i) if invasion affects common and rare species equally ((i) in Fig. 8.9c), then α diversity will decline but β diversity will stay roughly the same ((i) in Fig. 8.10d), suggesting a gradual increase in deterministic forces in the community assembly mechanism; (ii) if invasion has a disproportionally larger impact on rare species ((ii) in Fig. 8.9c), both α and β diversity will decline ((ii) in Fig. 8.9d), suggesting a drastic increase in deterministic forces in the community assembly mechanism; (iii) if invasion has a disproportionally large impact on common/widespread species ((iii) in Fig. 8.9c), α diversity will decline but β diversity will increase ((iii) in Fig. 8.9d),

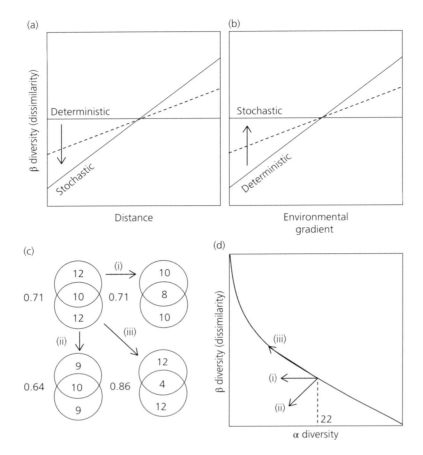

Fig. 8.9 (a) Isolation by distance and (b) isolation by resistance in a hypothetical community. Neutral and niche lines set the extremes, while observed dashed lines shift to opposite directions when biological invasion functions as a stochastic force in disassembling the resident community. (c) Venn diagrams of α and β diversity before (top left) and after the invasion (the other three diagrams). Numbers in the diagram indicate species richness and outside β diversity (Jaccard's dissimilarity); six species were lost due to invasion, which could be affecting both common and rare species (top right), only affecting rare species (bottom left), or only affecting common species (bottom right). (d) diversity signature of a purely stochastic driven assembly (solid curve), with arrows indicate the three scenarios of invasion impact listed in (c). Inspired by Chase and Myers (2011) and modified to apply to biological invasion.

suggesting that community assembly is still driven by stochastic forces.

Species turnover between pairwise β diversity metrics, together with α diversity, could potentially signal whether a community has deviated from certain assembly processes after invasion. However, species turnover measured by β diversity is largely driven by the difference in the composition of rare species between sites. More importantly, β diversity does not allow the direction estimation of γ diversity, or of the spatial scaling between α and γ, because they fail to account for patterns of co-occurrence among

more than two sites (Baselga 2013; Socolar *et al.* 2016). A new method called zeta (ζ) diversity (ζ_i is simply the mean number of species shared by i sites) was proposed by Hui and McGeoch (2014) to generalize β diversity to examine overlap in trios, quartets, and larger collections of samples. Zeta diversity ζ_i declines monotonically with the number of sites concerned i, with close to 60 per cent of real cases following a power law and 30 per cent an exponential decline (Fig. 8.10). The exponential and power law forms of zeta diversity decline are underpinned by distinct hypotheses about ecological processes; that is, they

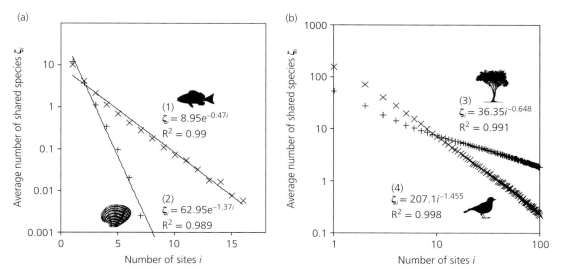

Fig 8.10 Two dominant forms of decline in zeta diversity: the exponential (a) and the power law (b) in four communities (freshwater fish, marine fouling organisms, tropical trees, and birds). From Hui and McGeoch (2014). With permission from the University of Chicago Press.

represent species turnover as either largely stochastic (exponential zeta) or driven principally by niche differentiation processes (power law zeta). A null model with all the species having the same probability of occurring in a site, regardless of the heterogeneity across sites, will produce this exponential form of zeta diversity decline; that is, a common and a rare species will have an equal chance to occur in an additional site. A null model with species differing in their probability of occupying a site (e.g. species have different site or habitat preferences) commonly produces this power law form of the zeta diversity decline; that is, a common species is more likely to occur in an extra site than a rare species.

Other diversity and macroecological patterns, such as species–area relationship and species accumulation curves provide additional examples of diversity signatures of species turnover and community assembly patterns. For instance, Powell and colleagues (2013) showed that invasive species reduce the intercept and increase the slope of the species–area relationship (Fig. 8.11) through a combination of neutral (e.g. by reducing individual density) and non-neutral processes (e.g. local species extinctions), and environmental heterogeneity. They explored four mechanisms that could have caused such a universal pattern: neutral sampling; non-neutral shifts in relative abundance; local

extirpation; and aggregation. Neutral sampling and local extirpation were identified as the main drivers. Importantly, common species were found to be more influenced by invasion than rare species. This highlights the fact that common and rare species

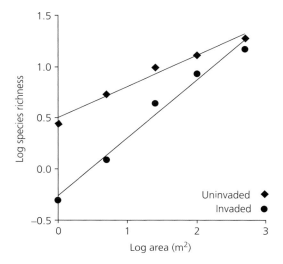

Fig 8.11 Species–area relationship for invaded and uninvaded sites from Highlands Hammock State Park, Florida, highlighting how increases in the slopes in invaded communities lead to smaller species richness declines with increasing spatial scale. From Powell *et al.* (2013). Reprinted with permission from AAAS.

respond to invasion differently, and that methods that are able to assess the changes in common species due to invasion (e.g. higher order ζ diversity) are needed to capture the full impact of invasion. Pairwise β diversity overemphasizes the role of rare species in turnover and thus could potentially underestimate the impact of invasion. Community assembly processes and invasibility, as well as the kind of resident species that could be susceptible to invasion, can be evaluated by diversity signatures of invaded communities, especial using the turnover metrics of β and ζ diversity.

8.5 Large-scale assembly patterns

8.5.1 Alien island biogeography

Island biogeography holds a special place in community ecology (MacArthur and Wilson 1963, 1967). Islands are considered ideal models for testing many biogeographical hypotheses that seek to explain the determinants of biodiversity through experiments (Simberloff and Wilson 1969) or observations (Losos and Ricklefs 2010). Trade-marked by the well-known species–area relationship, island biogeography has revealed a list of factors behind insular diversity and composition (Kreft et al. 2008; Whittaker et al. 2008; Triantis et al. 2012) including primarily island size and its temporal and spatial isolation from other islands and mainlands, plus habitat quality and heterogeneity and human activity. These factors mainly affect the rate of immigration and extinction, and thus determine the species richness (and composition) when immigration counterbalances local extinction; the rate of speciation can also be incorporated. Alien species have been continuously invading islands across the globe, and some have disrupted networks of insular interactions of resident species (Borges et al. 2006; Blackburn et al. 2008; Traveset et al. 2014; Whittaker et al. 2014). We focus on two studies that have explored the biodiversity patterns of native and alien species on islands for plants and birds (Blackburn et al. 2016) and for ants (Roura-Pascual et al. 2016).

Using data on ant species diversity from 102 islands around the world, Roura-Pascual and colleagues (2016) found that island area is the most

important predictor of both native and alien species richness, followed by distance to the nearest continent in the case of native species and human population size for alien species. The global patterns of native and alien species composition also appeared to be constrained to certain geographic regions by the same factors related to natural isolation. Using data on plants from 62 islands and birds from 68 islands, Blackburn and colleagues (2016) found that species richness for both alien birds and plants was strongly correlated with island area. Alien species richness was strongly positively related to native species richness (reflecting habitat heterogeneity), with additional effects of human population size (determining the number of alien species that arrive on an island). Distance to continent was only considered as an indirect but still strong effect on alien species richness. However, the structure equation model used could be slightly overcomplicated and could produce a spurious negative direct effect of island area on alien plant richness, albeit giving an overall positive effect (also incorporating indirect effects). In both cases, human population accounts for considerable variation in alien species richness because of the role of humans as dispersal agents that facilitate the establishment of alien species across the world (Rizali et al. 2010; Morrison 2014).

There are some fundamental differences between these taxa regarding their introduction and establishment on islands. Alien birds and plants are primarily introduced to islands by humans for ornamental, cultural, and agricultural reasons, whereas alien ants are almost invariably introduced unintentionally, normally as hitch-hikers. By contrast, native birds could have arrived on islands by flying, while native plants and ants often arrive via various vectors as hitch-hikers. Ants and plants could establish on extremely small islands, whereas birds normally require bigger islands to establish. These differences could have led to the differences in the observed species–area relationships. The species–area relationships for birds and plants seem stronger than those for ants (Fig. 8.12), although this could be because larger islands were included in Blackburn and colleagues' (2016) analysis than in that of Roura-Pascual and colleagues' (2016). Species richness, especially for native birds and plants on islands ranging in size between 10^4 km^2

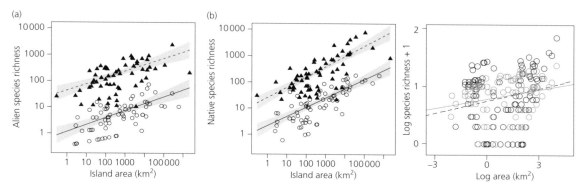

Fig 8.12 Species–area relationships for native and alien species on global islands (Blackburn *et al.* 2016; Roura-Pascual *et al.* 2016). (a): alien plants (triangles; z = 0.24) and alien birds (circles; z = 0.27); (b): native plants (triangles; z = 0.39) and native birds (circles; z = 0.36); (c): native ants (black circles; z = 0.18) and alien ants (grey circles; z = 0.18). Reproduced with permission from John Wiley & Sons.

and 10^6 km², needs to consider the contribution of potential speciation events. Island area is likely to increase habitat heterogeneity and thus species diversity, but it can also act as a determinant of species speciation/extinction rate, and it is a crucial variable for colonization (Rosenzweig 1995). The detectability and sampling efforts could differentially affect species lists, and more reliable checklists and surveys of island fauna and flora are needed.

As mentioned in section 8.4, species richness tells us nothing about the compositional difference between the assemblages of different islands. With the species-by-island lists, we should explore the reasons for species turnover. Roura-Pascual and colleagues (2016) used two different approaches to examine the drivers of species turnover in assemblages of insular ants. Firstly, they calculated the zeta diversity (Hui and McGeoch 2014) and found that zeta diversity of both native and alien ants declined exponentially with the increase of number of islands, signalling stochastically driven assemblage patterns. Secondly, using modularity analysis (e.g. Newman and Girvan 2004; Hui *et al.* 2013), they identified the compositional modules of islands based on species co-distributions (Fig. 8.13), and then used conditional inference forests (a recursive partitioning method) to infer these identified compositional modules.

Ocean current was identified as the most informative connectivity variable for explaining the differences between modules for both native and alien species. Poor dispersal ability in ants confines them

to adjacent islands connected by prevailing winds and ocean currents (Fisher 2010; Morrison 2014). Alien ants represent only a quarter of the total pool of species, but a few of these species occur on a large number of islands. The most prominent species is *Paratrechina longicornis*, followed by *Monomorium floricola*, *Tetramorium bicarinatum*, *Tapinoma melanocephalum*, and *Pheidole megacephala*. Biotic interactions and other variables (including environmental heterogeneity through island area and climate, distance to the nearest land masses such as geographic coordinates and nearest continents/islands, and human population size) do not seem to influence the modular structure of ant assemblages (Fig. 8.13). The importance of ocean currents and regions in explaining the modularity of ant assemblages could indicate the complexity and directionality of movement of people and commodities that are responsible for these geographic patterns, acting as surrogates of both past natural and current anthropogenic isolation (Fisher 2010), while alien modules would be more dependent on economic factors that influence the present-day transport of humans and commodities among islands and therefore the arrival of alien ants (Helmus *et al.* 2014; Morrison 2014).

8.5.2 Alien biomes

Humans have intentionally or accidentally moved organisms around the world for centuries (Elton 1958). This has resulted in many species establishing

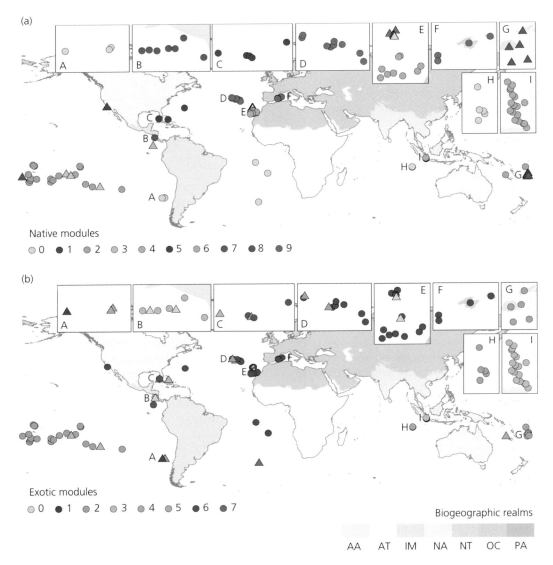

Fig 8.13 Distribution of (a) native and (b) alien modules for islands (i.e. distinct clusters of islands with a certain assemblage of species) derived from a modularity analysis. The colours indicate the module assigned by the modularity analysis, while the symbols indicate whether the conditional inference tree grouped that particular island in a node where the majority of islands have the same module (○) or a different one (△). The colours of the land masses show the biogeographic realms (corresponding to Australasia (AA), Afrotropic (AT), Indo-Malay (IM), Nearctic (NA), Neotropic (NT), Oceania (OC), Palearctic (PA)) (Olson *et al.* 2001). The inserts represent an expanded view of areas where symbols overlap. From Roura-Pascual *et al.* (2016). Reproduced with permission from John Wiley & Sons. (see Plate 2)

self-sustaining populations outside their native ranges, that is, in regions separated from the natal ranges by substantial biogeographic barriers (Thuiller *et al.* 2006; van Kleunen *et al.* 2010; Richardson 2011), with many becoming invasive and having detrimental effects in recipient ecosystems (e.g. Gaertner *et al.* 2009; Vilà *et al.* 2011). Many aspects of the biogeography and ecology of invasions have been well studied (Pyšek and Richardson 2006; van Kleunen *et al.* 2010; Richardson 2011). The spatial patterns and dynamics of alien species assemblages (i.e. how alien species assemble and co-occur across landscapes) are receiving increasing attention (Pyšek *et al.* 2005; Hui *et al.* 2013; Rouget *et al.* 2015).

Plant assemblages define vegetation patterns at different scales, from plant communities at local scales to biomes at regional and global scales. Biomes are typically defined on the basis of broad vegetation types and the biophysical features that exercise fundamental control on the distribution of plants (Cox and Moore 2000). The widespread alteration of ecosystems by humans caused by agriculture, urbanization, and other land uses has led to the formation of modified ecological patterns across the globe. Anthropogenic biomes ('anthromes') have been defined to reflect this new ecological order (Ellis and Ramankutty 2008). Biological invasions add a new layer of complexity to natural biomes and anthromes. In particular, invasive alien plant species have transformed regional landscapes and now dominate vegetation in many parts of the world, although their invasion dynamics and distributions can usually be explained by assessing life-history traits and introduction history (e.g. introduction and dissemination pathways, propagule pressure, and residence time; Richard-

son *et al.* 2011, 2014; Wilson *et al.* 2007). The short time-span and human-driven nature of invasions suggest that the distribution of invasive species might follow the distribution of anthromes.

One of the fundamental reasons for the success of some invasive species is the lack of a shared evolutionary history with the components of recipient ecosystems (Cox and Moore 2000). Biological invasions can be used as a natural experiment for exploring the determinants of vegetation boundaries. However, there is little understanding of how assemblages of alien plants are collectively affected by biotic and abiotic features. This is because few, if any, invasive species have sampled all potentially invasible habitats in their introduced ranges. Consequently, the roles of traits and introduction histories typically override those of the fundamental biological/physical processes in shaping the current distributions of invasive alien plants (Thuiller *et al.* 2006). Different sets of assembly rules could therefore apply in alien assemblages. Based on potential drivers that could drive alien assemblages (Catford *et al.* 2009), Rouget and colleagues (2015) derived six biogeographical hypotheses based on

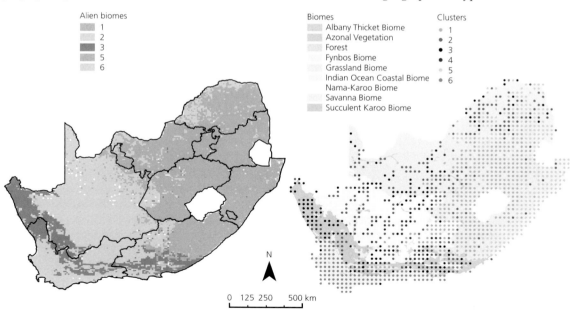

Fig 8.14 (Left) Invasive alien species clusters in relation to natural vegetation biomes (shading). Clusters were derived based on the presence/absence of invasive alien species in 15-minute grid cells (shown as circles). (Right) The potential distribution of alien biomes. Environmental determinants of the distribution of invasive alien species clusters were identified from a classification tree and mapped over the full range of environmental factors in South Africa. From Rouget *et al.* (2015). With permission from Elsevier. (see Plate 3)

Table 8.1 Hypotheses to explain the current distributions of invasive alien plant (IAP) species. The hypotheses vary in the degree to which IAP distributions are expected to correspond to biomes defined on the basis of native plant assemblages. The hypotheses also vary in the importance of propagule pressure, abiotic characteristics, biotic characteristics, and human influence of humans. From Rouget *et al.* (2015).

Hypothesis	Mechanism	Explanation	Predicted pattern	Management implication
The weed-shaped hole	Disturbance (Abiotic)	Invasions are facilitated by certain levels of disturbance; these may be anthropogenic or natural (and so an inherent function of the invasibility of an ecosystem).	IAP species clustering tends to follow anthromes and/or biomes where fire is a key driver of ecosystem dynamics (in fynbos, savanna, and grassland). Current distributions of IAPs are largely a function of the spatial patterns of disturbance regimes.	Disturbance is a key driver of establishment and spread of IAP populations. Need to focus on managing disturbance regimes to control invasions, or plan management actions around disturbance events.
The biome decides	Competition (Biotic)	Biomes differ inherently in invasibility, such that there is selectivity regarding which invasive species can and cannot invade.	IAP species clustering will correspond with biome boundaries, with some invasive species negatively associated with certain biomes despite opportunities for such species to invade and seemingly suitable physiological conditions.	Provides insights on how species interact and coexist within biomes. Given the importance of context and species interactions, it might be expected that invasiveness elsewhere in the world would be a less important driver. There might be a large difference between taxa observed to be invasive in transformed areas from those that invade natural ecosystems. Need to separate strategic plans for IAP risk assessment and management for each biome.
Goldilocks	Climate (Abiotic)	Biomes and invasive species share broadly similar abiotic requirements, such that they occupy the same niches.	IAP species clusters will correspond to particular climatic niches. Expect natural biome boundaries to also follow climatic boundaries, and potentially biomes in different parts of the world show similar climatic boundaries.	Insights on environmental factors limiting the spread of IAPs. Bioclimatic models of species distributions based on native ranges will provide accurate estimates of potential ranges in new regions. Management should be based on climatic zones and expect shifts in invasion with changing climate.
A new world order	Global change (Humans)	A new set of boundaries is formed by suites of interacting native and alien species. These new associations lead to novel ecosystems that need not be similar in nature to those in previous biomes.	Invasions cluster together though not in relation to factors mentioned above. New boundaries are formed but IAP clusters do not correspond with biomes. Likely to see invasional meltdown.	The concept of a 'biome' will have little meaning for IAPs, but how such novel ecosystems form will provide insights on a potentially new set of drivers of distributions. It will be hard to predict the trajectories of invasions, and might need to manage novel ecosystems or species on a case-by-case basis.
Something in the way you move	Introduction dynamics (Propagule pressure)	Distribution patterns in invasions are dominated by drivers associated with introduction histories of IAPs.	Species over-represented close to points of introduction; and human activities. Species clustering tend to follow anthromes. These patterns may be expected to decrease as residence time increases (i.e. the 'settling-down' hypothesis in section 8.3), but equally such initial stochasticity might last, leading to novel ecosystems (see earlier).	Current distribution patterns are strongly influenced by historical socio-economic forces and so require a multi-disciplinary approach. Current invasive distributions are likely to be poor predictors of future (or potential) distributions. Control should focus around points of introductions and historical and future pathways.

Hypothesis	Mechanism	Explanation	Predicted pattern	Management implication
Random tessellation	Geometry (none)	The distributions of IAP species are inherently idiosyncratic. When many species are considered, it appears that the relative clustering of distributions is simply constrained by geometry.	There will be no significant correlation to any individual biotic or abiotic variables. Clustering patterns will correspond to a null model where distributions are allocated randomly to the map with the constraint that overall distributions in extent and arrangement are preserved for each species. Different levels of species clustering no different from null models, though such clustering may, at first glance, look as though patterns are present.	Each species must be studied and managed individually and no particular mechanistic clustering of distributions.

main drivers of patterns, including disturbance, competition, climate, global change, introduction dynamics, and physical/topological geometry (Table 8.1), with human activities and propagule pressure included as part of factors of global change and introduction dynamics, unique to alien biomes.

Many invasive alien plant species in South Africa were introduced more than two centuries ago and have been widely disseminated within the region. The assemblage of these alien plant species provides an opportunity to test whether the same fundamental factors that define native plant assemblages and biomes also define the composition of alien assemblages. Using 69 well established and invasive alien plant species with a median residence time of around 150 years, Rouget and colleagues (2015) derived several 'alien biomes' based on the current alien plant assemblages and associated environmental drivers (Fig. 8.14). They tested the six hypotheses shown in Table 8.1. The alien biomes were largely defined by the same factors that separate natural vegetation biomes in southern Africa (Fig. 8.14), and hypotheses relating to anthropogenic factors and/or random assemblage of species were rejected.

Although anthropogenic drivers clearly influence the establishment and spread of alien species (Pyšek and Richardson 2006; Pyšek et al. 2010), the broad-scale distribution of assemblages of alien species is largely explained by the same environmental factors that separate biomes based on native vegetation. These broad environmental factors

seem to operate as barriers, dividing the environmental space into distinct clusters where native and alien species cannot easily cross cluster boundaries (Cowling and Pressey 2001; Rouget et al. 2003). Similar factors might constrain the distribution of alien plant species in South Africa (Fig. 8.14). However, more causal tests are needed to discern whether the biome boundaries drive the alien biomes (the 'biome decides' hypothesis) or whether the both alien species and native species respond to the same external drivers (the 'Goldilocks' hypothesis).

The spread of invasive alien species can result in biotic homogenization where the native biodiversity is suppressed and replaced by widespread invasive species (McKinney and Lockwood 1999). More studies in different regions of the world, ideally with different levels of urbanization and agricultural development, are needed to test these hypotheses. Because predicting the current and future impacts of invasive species is a daunting task (Blackburn et al. 2014), the existence of such alien biomes could provide managers with the means for coordinating strategies to target the alien assemblage processes within each alien biome (Rouget et al. 2015). Most current frameworks for assessing the impact of biological invasions are species-specific (i.e. each alien species is scored separately), making it difficult to quantify the cumulative impacts of invaded areas which are usually affected by multiple species. Knowledge of alien species assemblages and potential alien biomes could provide a tool for assessing the cumulative impacts of groups of invasive species and prioritizing management effort to target specific areas (see Chapter 9).

References

Adler, P.B., HilleRisLambers, J., and Levin, J.M. (2007) A niche for neutrality. Ecology Letters, 10, 95–104.

Amarasekare, P. and Nisbet, R.M. (2001) Spatial heterogeneity, source-sink dynamics, and the local coexistence of competing species. American Naturalist, 158, 572–84.

Armstrong, R.A. and McGehee, R. (1980) Competitive exclusion. American Naturalist, 115, 151–70.

Bar-Massada, A., Kent, R., and Carmel, Y. (2014) Environmental heterogeneity affects the location of modelled communities along the niche–neutrality continuum. Proceedings of the Royal Society B: Biological Sciences, 281, 20133249.

Baselga, A. (2013) Multiple site dissimilarity quantifies compositional heterogeneity among several sites, while average pairwise dissimilarity may be misleading. Ecography, 36, 124–8.

Bell, G. (2005) The co-distribution of species in relation to the neutral theory of community ecology. Ecology, 86, 1757–70.

Belyea, L.R. and Lancaster, J. (1999) Assembly rules within a contingent ecology. Oikos, 86, 402–16.

Bertness, M.D. and Callaway, R. (1994) Positive interactions in communities. Trends in Ecology and Evolution, 9, 191–3.

Blackburn, T.M., Cassey, P., Duncan, R.P., et al. (2008) Threats to avifauna on oceanic islands revisited. Conservation Biology, 22, 492–4.

Blackburn, T.M., Delean, S., Pyšek, P., et al. (2016) On the island biogeography of aliens: a global analysis of the richness of plant and bird species on oceanic islands. Global Ecology and Biogeography, 25, 859–68.

Blackburn, T.M., Essl, F., Evans, T., et al. (2014) A unified classification of alien species based on the magnitude of their environmental impacts. PLoS Biology, 12, e1001850.

Blackburn, T.M., Pyšek, P., Bacher, S., et al. (2011) A proposed unified framework for biological invasions. Trends in Ecology and Evolution, 26, 333–9.

Borges, P.A.V., Lobo, J.M., de Azevedo, E.B., et al. (2006) Invasibility and species richness of island endemic arthropods: a general model of endemic vs. exotic species. Journal of Biogeography, 33, 169–87.

Brown, R.L. and Peet, R.K. (2003) Diversity and invasibility of southern Appalachian plant communities. Ecology, 84, 32–39.

Bruno, J.F. (2000) Facilitation of cobble beach plant communities through habitat modification by Spartina alterniflora. Ecology, 81, 1179–92.

Bruno, J.F., Stachowicz, J.J., and Bertness, M.D. (2003) Inclusion of facilitation into ecological theory. Trends in Ecology and Evolution, 18, 119–25.

Bulleri, F., Bruno, J.F., and Benedetti-Cecchi, L. (2008) Beyond competition: incorporating positive interactions between species to predict ecosystem invasibility. PLoS Biology, 6, e162.

Cadotte, M.W. (2007) Concurrent niche and neutral processes in the competition-colonization model of species coexistence. Proceedings of the Royal Society B: Biological Sciences, 274, 2739–44.

Cadotte, M.W., Carscadden, K., and Mirotchnick, N. (2011) Beyond species: functional diversity and the maintenance of ecological processes and services. Journal of Applied Ecology, 48, 1079–87.

Cain, M.L., Bowman, W.D., and Hacker, S.D. (2014) Ecology, 3rd edn. Sunderland, MA: Sinauer Associates.

Cardinale, B.J., Duffy, E., Gonzalez, A., et al. (2012) Biodiversity loss and its impact on humanity. Nature, 486, 59–67.

Catford, J.A., Jansson, R., and Nilsson, C. (2009) Reducing redundancy in invasion ecology by integrating hypotheses into a single theoretical framework. Diversity and Distributions, 15, 22–40.

Chase, J.M. (2003) Community assembly: when should history matter? Oecologia, 136, 489–98.

Chase, J.M. (2005) Towards a really unified theory for metacommunities. Functional Ecology, 19, 182–6.

Chase, J.M. and Leibold, M.A. (2003) Ecological Niches: Linking Classical and Contemporary Approaches. Chicago, IL: University of Chicago Press.

Chase, J.M. and Myers, J.A. (2011) Disentangling the importance of ecological niches from stochastic processes across scales. Philosophical Transactions of the Royal Society B: Biological Sciences, 366, 2351–63.

Chesson, P. (1994) Multispecies competition in variable environments. Theoretical Population Biology, 45, 227–76.

Chesson, P. (2000) Mechanisms of maintenance of species diversity. Annual Review of Ecology and Systematics, 31, 343–66.

Chesson, P. and Huntly, N. (1997) The roles of harsh and fluctuating conditions in the dynamics of ecological communities. American Naturalist, 150, 519–53.

Clark, J.S. (2012) The coherence problem with the unified neutral theory of biodiversity. Trends in Ecology and Evolution, 27, 198–202.

Clements, F.E. (1916) Plant Succession: An Analysis of the Development of Vegetation. Washington, DC: Carnegie Institution of Washington.

Compagnoni, A. and Halpern, C.B. (2009) Properties of native plant communities do not determine exotic success during early forest succession. Ecography, 32, 449–58.

Connell, J.H. (1978) Diversity in tropical rain forests and coral reefs. Science, 199, 1302–10.

Connell, J.H. and Slatyer, R.O. (1977) Mechanisms of succession in natural communities and their role in community stability and organization. American Naturalist, 111, 1119–44.

Connor, E.F. and Simberloff, D. (1979) The assembly of species communities: chance or competition? Ecology, 60, 1132–40.

Corenblit, D., Steiger, J., González, E., et al. (2014) The biogeomorphological life cycle of poplars during the fluvial biogeomorphological succession: a special focus on Populus nigra L. Earth Surface Processes and Landforms, 39, 546–63.

Cornell, H.V. and Lawton, J.H. (1992) Species interactions, local and regional processes, and limits to the richness of ecological communities: a theoretical perspective. Journal of Animal Ecology, 61, 1–12.

Cowling, R.M. and Pressey, R.L. (2001) Rapid plant diversification: planning for an evolutionary future. Proceedings of the National Academy of Sciences of the USA, 98, 5452–7.

Cox, C.B. and Moore, P.D. (2000) Biogeography: An Ecological and Evolutionary Approach, 7th edn. Oxford: Wiley-Blackwell.

Daleo, P., Alberti, J., and Iribarne, O. (2009) Biological invasions and the neutral theory. Diversity and Distributions, 15, 547–53.

Davies, K.F., Chesson, P., Harrison, S., et al. (2005) Spatial heterogeneity explains the scale dependence of the native-exotic diversity relationship. Ecology, 86, 1602–10.

Davis, M.A., Grime, J.P., and Thompson, K. (2000) Fluctuating resources in plant communities: a general theory of invasibility. Journal of Ecology, 88, 528–34.

Davis, M.A. and Pelsor, M. (2001) Experimental support for a resource-based mechanistic model of invasibility. Ecology Letters, 4, 421–8.

Davis, M.A., Thompson, K., and Grime, J.P. (2001) Charles S. Elton and the dissociation of invasion ecology from the rest of ecology. Diversity and Distributions, 7, 97–102.

DeFerrari, C.M. and Naiman, R.J. (1994) A multi-scale assessment of the occurrence of exotic plants on the Olympic Peninsula, Washington. Journal of Vegetation Science, 5, 247–58.

Diamond, J.M. (1975) Assembly of species communities. In: Cody, M.L. and Diamond, J. (eds) Ecology and Evolution of Communities. Cambridge, MA: Harvard University Press, pp. 342–444.

Díaz, S., Kattge, J. Cornelissen, J.H.C., et al. (2016) The global spectrum of plant form and function. Nature, 529, 167–71.

Dick, J.T.A., Gallapher, K., Avlijas, S., et al. (2013) Ecological impacts of an invasive predator explained and predicted by comparative functional responses. Biological Invasions, 15, 837–46.

Dieckmann, U. and Law, R. (1996) The dynamical theory of coevolution: a derivation from stochastic-ecological processes. Journal of Mathematical Biology, 34, 579–612.

Drake, J.A. (1991) Community-assembly mechanics and the structure of an experimental species ensemble. American Naturalist, 137, 1–26.

Egler, F.E. (1954) Vegetation science concepts. I. Initial floristic composition, a factor in old-field vegetation development. Vegetatio, 4, 412–7.

Ellis, E.C. and Ramankutty, N. (2008) Putting people in the map: anthropogenic biomes of the world. Frontiers in Ecology and the Environment, 6, 439–47.

Elton, C.S. (1958) The Ecology of Invasions by Animals and Plants. London: Methuen.

Fisher, B.L. (2010) Biogeography. In: Lach, L., Parr, C.L., and Abbott, K.L. (eds) Ant Ecology. Oxford: Oxford University Press, pp. 18–37.

Flynn, D.F.B., Mirotchnick, N., Jain, M., et al. (2011) Functional and phylogenetic diversity as predictors of biodiversity–ecosystem-function relationships. Ecology, 92, 1573–81.

Fox, J.W. (2013) The intermediate disturbance hypothesis should be abandoned. Trends in Ecology and Evolution, 28, 86–92.

Fridley, J.D., Stachowicz, J.J., Naeem, S., et al. (2007) The invasion paradox: reconciling pattern and process in species invasions. Ecology, 88, 3–17.

Friedman, J.M., Auble, G.T., Shafroth, P.B., et al. (2005) Dominance of non-native riparian trees in western USA. Biological Invasions, 7, 747–51.

Fukami, T. (2005) Integrating internal and external dispersal in metacommunity assembly: preliminary theoretical analyses. Ecological Research, 20, 623–31.

Fukami, T. (2010) Community assembly dynamics in space. In: Verhoef, H.A. and Morin, P.J. (eds) Community Ecology: Processes, Models, and Applications. Oxford: Oxford University Press, pp. 45–54.

Fukami, T. (2015) Historical contingency in community assembly: integrating niches, species pools, and priority effects. Annual Review of Ecology, Evolution, and Systematics, 46, 1–23.

Gaertner, M., Biggs, R., teBeest, M., et al. (2014) Invasive plants as drivers of regime shifts: identifying high priority invaders that alter feedback relationships. Diversity and Distributions, 20, 733–44.

Gaertner, M., Den Breeÿen, A., Hui, C., et al. (2009) Impacts of alien plant invasions on species richness in Mediterranean-type ecosystems: a meta-analysis. Progress in Physical Geography, 33, 319–38.

Gao, M., Wang, X.X., Hui, C., et al. (2015) Assembly of plant communities in coastal wetlands—the role of saltcedar

Tamarix chinensis during early succession. Journal of Plant Ecology, 8, 539–48.

Gleason, H.A. (1927) Further views on the succession-concept. Ecology, 8, 299–326.

Gotelli, N.J. and Graves, G.R. (1996) *Null Models in Ecology*. Washington, DC: Smithsonian Institute Press.

Gotelli, N.J. and McCabe, D.J. (2002) Species co-occurrence: a meta-analysis of JM Diamond's assembly rules model. Ecology, 83, 2091–6.

Götzenberger, L., de Bello, F., Bråthen, K.A., *et al.* (2012) Ecological assembly rules in plant communities—approaches, patterns and prospects. Biological Reviews, 87, 111–27.

Gravel, D., Canham, C.D., Beaudet, M., *et al.* (2006) Reconciling niche and neutrality: the continuum hypothesis. Ecology Letters, 9, 399–409.

Grime, J.P. (1974) Vegetation classification by reference to strategies. Nature, 250, 26–31.

Grime, J.P. (1997) Biodiversity and ecosystem function: the debate deepens. Science, 277, 1260–1.

Grimm, V., Revilla, E., Berger, U., *et al.* (2005) Pattern-oriented modeling of agent-based complex systems: lessons from biology. Science, 310, 987–91.

Haegeman, B. and Loreau, M. (2011) A mathematical synthesis of niche and neutral theories in community ecology. Journal of Theoretical Biology, 269, 150–65.

Hastings, A. (1980) Disturbance, coexistence, history, and competition for space. Theoretical Population Biology, 18, 363–73.

Heino, J. and Grönroos, M. (2013) Does environmental heterogeneity affect species co-occurrence in ecological guilds across stream macroinvertebrate metacommunities? Ecography, 36, 926–36.

Helmus, M.R., Mahler, D.L., and Losos, J.B. (2014) Island biogeography of the Anthropocene. Nature, 513, 543–46.

Herben, T. (2009) Invasibility of neutral communities. Basic and Applied Ecology, 10, 197–207.

Hobbs, R.J., Arico, S., Aronson, J., *et al.* (2006) Novel ecosystems: theoretical and management aspects of the new ecological world order. Global Ecology and Biogeography, 15, 1–7.

Hobbs, R.J. and Huenneke, L.F. (1992) Disturbance, diversity, and invasion: implications for conservation. Conservation Biology, 6, 324–37.

Holt, R.D., Grover, J., and Tilman, D. (1994) Simple rules for interspecific dominance in systems with exploitative and apparent competition. American Naturalist, 144, 741–71.

Hubbell, S.P. (2001) *The Unified Neutral Theory of Biodiversity and Biogeography*. Princeton, NJ: Princeton University Press.

Hui, C. and McGeoch, M.A. (2014) Zeta diversity as a concept and metric that unifies incidence-based biodiversity patterns. American Naturalist, 184, 684–94.

Hui, C., Richardson, D.M., Pyšek, P., *et al.* (2013) Increasing functional modularity with residence time in the co-distribution of native and introduced vascular plants. Nature Communications, 4, 2454.

Jost, L. (2007) Partitioning diversity into independent alpha and beta components. Ecology, 88, 2427–39.

Keddy, P.A. (1992) Assembly and response rules: two goals for predictive community ecology. Journal of Vegetation Science, 3, 157–64.

Kennedy, T.A., Naeem, S., Howe, K.M., *et al.* (2002) Biodiversity as a barrier to ecological invasion. Nature, 417, 636–8.

Kneitel, J.M. and Chase, J.M. (2004) Trade-offs in community ecology: linking spatial scales and species coexistence. Ecology Letters, 7, 69–90.

Kreft, H., Jetz, W., Mutke, J., *et al.* (2008) Global diversity of island floras from a macroecological perspective. Ecology Letters, 11, 116–27.

Kunstler, G., Falster, D., Coomes, D.A., *et al.* (2016) Plant functional traits have globally consistent effects on competition. Nature, 529, 204–7.

Latombe, G., Hui, C., and McGeoch, M.A. (2015) Beyond the continuum: a multi-dimensional phase space for neutral-niche community assembly. Proceedings of the Royal Society B: Biological Sciences, 282, 20152417.

Leibold, M.A., Holyoak, M., Mouquet, N., *et al.* (2004) Themetacommunity concept: a framework for multi-scale community ecology. Ecology Letters, 7, 601–13.

Leibold, M.A. and McPeek, M.A. (2006) Coexistence of the niche and neutral perspectives in community ecology. Ecology, 87, 1399–410.

Leprieur, F., Beauchard, O., Blanchet, S., *et al.* (2008) Fish invasions in the world's river systems: when natural processes are blurred by human activities. PLoS Biology, 6, e28.

Levine, J.M. (2000) Species diversity and biological invasions: relating local process to community pattern. Science, 288, 852–4.

Levine, J.M. (2001) Local interactions, dispersal, and native and exotic plant diversity along a California stream. Oikos, 95, 397–408.

Levine, J.M. and D'Antonio, C.M. (1999) Elton revisited: a review of evidence linking diversity and invasibility. Oikos, 87, 15–26.

Loreau, M. (2000) Biodiversity and ecosystem functioning: recent theoretical advances. Oikos, 91, 3–17.

Loreau, M. and Mouquet, N. (1999) Immigration and the maintenance of local species diversity. American Naturalist, 154, 427–40.

Losos, J.B. and Ricklefs, R.E. (eds) (2010) *The Theory of Island Biogeography Revisited*. Princeton, NJ: Princeton University Press.

MacArthur, R.H. (1972) *Geographical Ecology. Patterns in the Distribution of Species*. New York, NY: Harper and Row.

MacArthur, R.H. and Wilson, E.O. (1963) An equilibrium theory of insular zoogeography. Evolution, 17, 373–87.

MacArthur, R.H. and Wilson, E.O. (1967) *The Theory of Island Biogeography*. Princeton, NJ: Princeton University Press.

MacDougall, A.S., Gilbert, B., and Levine, J.M. (2009) Plant invasions and the niche. Journal of Ecology, 97, 609–15.

Maestre, F.T., Callaway, R.W., Valladares, F., *et al.* (2009) Refining the stress-gradient hypothesis for competition and facilitation in plant communities. Journal of Ecology, 97, 199–205.

Martin, L.M. and Wilsey, B.J. (2015) Differences in beta diversity between exotic and native grasslands vary with scale along a latitudinal gradient. Ecology, 96, 1042.

Martin, M.R., Tipping, P.W., and Sickman, J.O. (2009) Invasion by an exotic tree alters above and belowground ecosystem components. Biological invasions, 11, 1883–94.

Martin, P.H., Canham, C.D., and Marks, P.L. (2008) Why forests appear resistant to exotic plant invasions: intentional introductions, stand dynamics, and the role of shade tolerance. Frontiers in Ecology and the Environment, 7, 142–9.

May, R.M. (1977) Thresholds and breakpoints in ecosystems with a multiplicity of stable states. Nature, 269, 471–7.

Mayfield, M.M. and Levine, J.M. (2010) Opposing effects of competitive exclusion on the phylogenetic structure of communities. Ecology Letters, 13, 1085–93.

McGill, B.J. (2010) Matters of scale. Science, 328, 575–6.

McGill, B.J., Enquist, B., Weiher, E., *et al.* (2006) Rebuilding community ecology from functional traits. Trends in Ecology and Evolution, 21, 178–85.

McKinney, M.L. and Lockwood, J.L. (1999) Biotic homogenization: a few winners replacing many losers in the next mass extinction. Trends in Ecology and Evolution, 14, 450–3.

McLane, C.R., Battaglia, L.L., Gibson, D.J., *et al.* (2012) Succession of exotic and native species assemblages within restored floodplain forests: a test of the parallel dynamics hypothesis. Restoration Ecology, 20, 202–10.

Meiners, S.J., Pickett, S.T.A., and Cadenasso, M.L. (2002) Exotic plant invasions over 40 years of old field successions: community patterns and associations. Ecography, 25, 215–23.

Melbourne, B.A., Cornell, H.V., Davies, K.F., *et al.* (2007) Invasion in a heterogeneous world: resistance, coexistence or hostile takeover? Ecology Letters, 10, 77–94.

Michalet, R., Brooker, R.W., Cavieres, L.A., *et al.* (2006) Do biotic interactions shape both sides of the humped-back model of species richness in plant communities? Ecology Letters, 9, 767–77.

Morrison, L.W. (2014) The small-island effect: empty islands, temporal variability and the importance of species composition. Journal of Biogeography, 41, 1007–17.

Mouquet, N. and Loreau, M. (2003) Community patterns in source-sink metacommunities. American Naturalist, 162, 544–57.

Mutshinda, C.M. and O'Hara, R.B. (2011) Integrating the niche and neutral perspectives on community structure and dynamics. Oecologia, 166, 241–51.

Naeem, S., Knops, J.M., Tilman, D., *et al.* (2000) Plant diversity increases resistance to invasion in the absence of covarying extrinsic factors. Oikos, 91, 97–108.

Nekola, J.C. and Whilte, P.S. (1999) The distance decay of similarity in biogeography and ecology. Journal of Biogeography, 26, 867–78.

Newman, M.E.J and Girvan, M. (2004) Finding and evaluating community structure in networks. Physical Review E, 69, 026113.

Nuwagaba, S., Zhang, F., and Hui, C. (2015) A hybrid behavioural rule of adaptation and drift explains the emergent architecture of antagonistic networks. Proceedings of the Royal Society B: Biologica Sciences, 282, 20150320.

Olson, D.M., Dinerstein, E., Wikramanayake, E.D., *et al.* (2001). Terrestrial ecoregions of the world: a new map of life on earth. BioScience, 51, 933–8.

Ortega, Y.K. and Pearson, D.E. (2005) Weak vs. strong invaders of natural plant communities: assessing invasibility and impact. Ecological Applications, 15, 651–61.

Paini, D.R., Worner, S.P., Cook, D.C., *et al.* (2010) Using a self organising map to predict invasive species: sensitivity to data errors and a comparison with expert opinion. Journal of Applied Ecology, 47, 290–8.

Palmer, M.W. and Maurer, T.A. (1997) Does diversity beget diversity? A case study of crops and weeds. Journal of Vegetation Science, 8, 235–40.

Pauchard, A. and Shea, K. (2006) Integrating the study of non-native plant invasions across spatial scales. Biological Invasions, 8, 399–413.

Powell, K.I., Chase, J.M., and Knight, T.M. (2013) Invasive plants have scale-dependent effects on diversity by altering species-area relationships. Science, 339, 316–18.

Pugnaire, F.I. and Luque, M.T. (2001) Changes in plant interactions along a gradient of environmental stress. Oikos, 93, 42–9.

Pyšek, P., Jarošík, V., Chytrý, M., *et al.* (2005) Alien plants in temperate weed communities: prehistoric and recent invaders occupy different habitats. Ecology, 86, 772–85.

Pyšek, P., Jarošík, V., Hulme, P.E., *et al.* (2010) Disentangling the role of environmental and human pressures on biological invasions across Europe. Proceedings of the National Academy of Sciences of the USA, 107, 12157–62.

Pyšek, P. and Richardson, D.M. (2006) The biogeography of naturalization in alien plants. Journal of Biogeography, 33, 2040–50.

Rejmánek, M. (2003) The rich get richer—responses. Frontiers in Ecology and the Environment, 1, 123.

Rejmánek, M., Richardson, D.M., Higgins, S.I., *et al.* (2005) Ecology of invasive plants: state of the art. In: Mooney, H.A., Mack, R.N., McNeely, J.A., *et al.* (eds) *Invasive Alien Species: A New Synthesis*. Washington, DC: Island Press, pp. 104–61.

Renne, I.J., Tracy, B.F., and Colonna, I.A. (2006) Shifts in grassland invasibility: effects of soil resources, disturbance, composition, and invader size. Ecology, 87, 2264–77.

Ricciardi, A., Hoopes, M.F., Marchetti, M.P., *et al.* (2013) Progress toward understanding the ecological impacts of nonnative species. Ecological Monographs, 83, 263–82.

Richardson, D.M. (ed.) (2011) *Fifty Years of Invasion Ecology: The Legacy of Charles Elton*. Oxford: Wiley-Blackwell.

Richardson, D.M., Carruthers, J., Hui, C., *et al.* (2011) Human-mediated introductions of Australian acacias—a global experiment in biogeography. Diversity and Distributions, 17, 771–87.

Richardson, D.M., Hui, C., Nuñez, M.A., *et al.* (2014) Tree invasions: patterns, processes, challenges and opportunities. Biological invasions, 16, 473–81.

Richardson, D.M. and Pyšek, P. (2012) Naturalization of introduced plants: ecological drivers of biogeographic patterns. New Phytologist 196, 383–96

Richardson, D.M., Pyšek, P., Rejmánek, M., *et al.* (2000) Naturalization and invasion of alien plants—concepts and definitions Diversity and Distributions, 6, 93–107.

Rizali, A., Lohman, D.J., Buchori, D., *et al.* (2010) Ant communities on small tropical islands: effects of island size and isolation are obscured by habitat disturbance and 'tramp'ant species. Journal of Biogeography, 37, 229–36.

Rosenzweig, M.L. (1995) *Species Diversity in Space and Time*. Cambridge: Cambridge University Press.

Rosindell, J., Hubbell, S.P., He, F., *et al.* (2012) The case for ecological neutral theory. Trends in Ecology and Evolution, 27, 203–8.

Rouget, M., Cowling, R.M., Pressey, R.L., *et al.* (2003) Identifying spatial components of ecological and evolutionary processes for regional conservation planning in the Cape Floristic Region, South Africa. Diversity and Distributions, 9, 191–210.

Rouget, M., Hui, C., Renteria, J., *et al.* (2015) Plant invasions as a biogeographical assay: vegetation biomes constrain the distribution of invasive alien species assemblages. South African Journal of Botany, 101, 24–31.

Roura-Pascual, N., Sanders, N.J., and Hui, C. (2016) The distribution and diversity of insular ants: do exotic species play by different rules? Global Ecology and Biogeography, 25, 642–54.

Sanders, N.J., Gotelli, N.J., Heller, N.E., *et al.* (2003) Community disassembly by an invasive species. Proceedings of the National Academy of Sciences of the USA, 100, 2474–77.

Scheffer, M., Carpenter, S., Foley, J.A., *et al.* (2001) Catastrophic shifts in ecosystems. Nature, 413, 591–6.

Schoener, T.W. (1983) Rate of species turnover decreases from lower to higher organisms—a review of the data. Oikos, 41, 372–7.

Shea, K. and Chesson, P. (2002) Community ecology theory as a framework for biological invasions. Trends in Ecology and Evolution, 17, 170–6.

Simberloff, D. and von Holle, D. (1999) Positive interactions of nonindigenous species: invasion meltdown? Biological Invasions, 1, 21–32.

Simberloff, D.S. and Wilson, E.O. (1969) Experimental zoogeography of islands: the colonization of empty islands. Ecology, 50, 278–96.

Socolar, J.B., Gilroy, J.J., Kunin, W.E., *et al.* (2016) How should beta-diversity inform biodiversity conservation? Trends in Ecology and Evolution, 31, 67–80.

Soininen, J., McDonald, R., and Hillebrand, H. (2007) The distance decay of similarity in ecological communities. Ecography, 30, 3–12.

Stohlgren, T.J., Barnett, D.T., and Kartesz, J.T. (2003) The rich get richer: patterns of plant invasions in the United States. Frontiers in Ecology and the Environment, 1, 11–14.

Stokes, C.J. and Archer, S.R. (2010) Niche differentiation and neutral theory: an integrated perspective on shrub assemblages in a parkland savanna. Ecology, 91, 1152–62.

Strömberg, J.C., Chew, M.K., Nagler, P.L., *et al.* (2009) Changing perceptions of change: the role of scientists in tamarix and river management. Restoration Ecology, 17, 177–86.

Tansley, A.G. (1935) The use and abuse of vegetational concepts and terms. Ecology, 16, 284–307.

Thomsen, M.A. and D'Antonio, C.M. (2007) Mechanisms of resistance to invasion in a California grassland: the roles of competitor identity, resource availability, and environmental gradients. Oikos, 116, 17–30.

Thuiller, W., Richardson, D.M., Rouget, M., *et al.* (2006) Interactions between environment, species traits, and human uses describe patterns of plant invasions. Ecology, 87, 1755–69.

Tilman, D. (1982) *Resource Competition and Community Structure*. Princeton, NJ: Princeton University Press.

Tilman, D. (1997) Community invasibility, recruitment limitation, and grassland biodiversity. Ecology, 78, 81–92.

Tilman, D. (1994) Competition and biodiversity in spatially structured habitats. Ecology, 75, 2–16.

Tilman, D. (2004) Niche tradeoffs, neutrality, and community structure: a stochastic theory of resource competition, invasion, and community assembly. Proceedings of the National Academy of Sciences of the USA, 101, 10854–61.

Tilman, D., Isbell, F., and Cowles, J.M. (2014) Biodiversity and ecosystem functioning. Annual Review of Ecology, Evolution, and Systematics, 45, 471–493.

Traveset, A., Kueffer, C., and Daehler, C.C. (2014) Global and regional nested patterns of non-native invasive floras on tropical islands. Journal of Biogeography, 41, 823–32.

Triantis, K.A., Hortal, J., Amorim, I., *et al.* (2012) Resolving the Azorean knot: a response to Carine and Schaefer (2010). Journal of Biogeography, 39, 1179–87.

Van Kleunen, M., Dawson, W., Schlaepfer, D., *et al.* (2010) Are invaders different? A conceptual framework of comparative approaches for assessing determinants of invasiveness. Ecology Letters, 13, 947–58.

Vellend, M. (2010) Conceptual synthesis in community ecology. Quarterly Review of Biology, 85, 183–206.

Vellend, M. (2016) *The Theory of Ecological Communities.* Princeton, NJ: Princeton University Press.

Vergnon, R., van Nes, E.H., and Scheffer, M. (2012) Emergent neutrality leads to multimodal species abundance distributions. Nature Communications, 3, 663.

Verhoef, H.A. and Morin, P.J. (2010) *Community Ecology: Processes, Models, and Applications.* Oxford: Oxford University Press.

Vilà, M., Espinar, J.L., Hejda, M., *et al.* (2011) Ecological impacts of invasive alien plants: a meta-analysis of their effects on species, communities and ecosystems. Ecology Letters, 14, 702–8.

Vörösmarty, C.J., McIntyre, P.B., Gessner, M.O., *et al.* (2010) Global threats to human water security and river biodiversity. Nature, 467, 555–61.

Wang, S. and Loreau, M. (2014) Ecosystem stability in space: α, β and γ variability. Ecology Letters, 17, 891–901.

Webb, C.O., Ackerly, D.D., McPeek, M.A., *et al.* (2002) Phylogenies and community ecology. Annual Review of Ecology and Systematics, 33, 475–505.

Whittaker, R.J., Rigal, F., Borges, P.A.V., *et al.* (2014) Functional biogeography of oceanic islands and the scaling of functional diversity in the Azores. Proceedings of the National Academy of Sciences of the USA, 111, 13709–14.

Whittaker, R.J., Triantis, K.A., and Ladle, R.J. (2008) A general dynamic theory of oceanic island biogeography. Journal of Biogeography, 35, 977–94.

Whittaker, R.J., Willis, K.J., and Field, R. (2001) Scale and species richness: towards a general, hierarchical theory of species diversity. Journal of Biogeography, 28, 453–70.

Wilson, J.R., Richardson, D.M., Rouget, M., *et al.* (2007) Residence time and potential range: crucial considerations in modelling plant invasions. Diversity and Distributions, 13, 11–22.

Young, T.P., Chase, J.M., and Huddleston, R.T. (2001) Community succession and assembly comparing, contrasting and combining paradigms in the context of ecological restoration. Ecological Restoration, 19, 5–18.

Yu, D.W. and Wilson, H.B. (2001) The competition-colonization trade-off is dead; long live the competition-colonization trade-off. American Naturalist, 158, 49–63.

CHAPTER 9

Monitoring and management

9.1 Introduction

Management plans and actions for problematic invasive species are diverse, with attention primarily given to pre-border prevention by establishing rules and norms, at-border screening, post-border early detection, and rapid response. However, for established species, a new set of management options is needed, largely to address constraints relating to budgets and other resources and the scope of the problem. This chapter focuses on ten issues related to post-border invasion monitoring and management. Coordinated management efforts across regions and the globe demand a clear, objective focus for the accurate and efficient monitoring different aspects of invasions. This chapter discusses a minimum set of essential variables (section 9.2) and the monitoring/sampling strategies that are needed to collate required information (section 9.3). These monitored variables must allow us to estimate the risk and impact of targeted species (section 9.4) and to provide the foundation for assigning priorities among species, pathways, and sites (section 9.5). As it is an ongoing non-equilibrium problem, the extent and impact of invasions will continue to grow in the absence of efficient control measures, creating a debt for future management (section 9.6). Optimal control models of bioeconomics have been used to identify the optimum control strategies within the constraints due to aspects of political commitment (section 9.7) and budget and management efficacy (section 9.8). Such models often require a reliable estimate of the economic costs (e.g. opportunity versus utility costs) of the management effort and socio-ecological

impact (section 9.9). This often relies on sophisticated methods for framing and solving problems (section 9.10), especially aspects concerning the optimal allocation of management efforts across landscapes and regions (9.11).

9.2 Essential variables for monitoring

Many countries have launched strategies for dealing with biological invasions in response to the rapid increase in the overall magnitude of negative ecological and economic impacts attributable to invasive species. International strategies are also in place, notably the Convention on Biological Diversity (CBD). Aichi Target 9 of the CBD calls for 'invasive alien species and pathways [to be] identified and prioritized, [and for] priority species [to be] controlled or eradicated, and [for] measures [to be] in place to manage pathways to prevent their introduction and establishment' by 2020. There has been a substantial increase in efforts to prioritize and manage those invasive species that have the greatest negative impact on biodiversity and ecosystem functioning (Pyšek and Richardson 2010; Tittensor *et al.* 2014; McGeoch *et al.* 2015; van Kleunen *et al.* 2015). International cooperation is essential to reduce new incursions and to ensure the availability of information on alien species' distributions and their characteristics so that the risks associated with expanding trade pathways can be assessed (Essl *et al.* 2015). Administrative borders often slow or block the efficient sharing of data on invasions. However, data from multiple stakeholders, agencies, and governments are crucial for determining

Invasion Dynamics. Cang Hui and David M. Richardson, Oxford University Press (2017).
© Cang Hui and David M. Richardson. DOI 10.1093/acprof:oso/9780198745334.001.0001

the status and trends of invasions. There is a huge disparity in the infrastructure, availability of resources, and accessibility of data among countries (Pyšek *et al.* 2008; Nunez and Pauchard 2010; McGeoch *et al.* 2010). Even within countries, different methodologies for quantifying invasions make it difficult to determine the extent of invasions and changes over time (Guo 2011). A global monitoring and repository system based on standardized regional reporting is urgently needed to guide the formulation and implementation of effective strategies for managing biological invasions (Latombe *et al.* 2016).

Monitoring of invasions requires information on the status of introduced species that is updated at intervals that are relevant for management. Conservation assessments normally use a decadal scale for estimating population trends, and projections are made at the scale of centuries (Mace *et al.* 2008). What scales are appropriate for assessing trends in biological invasions? Assessments made over months run the risk of flagging species that may show initial rapid population expansions but which may subsequently decline in abundance without the need for intervention (boom-and-bust invaders; Simberloff and Gibbons 2004; see section 2.7). Such short-interval assessments may trigger interventions that are unnecessary and which may waste resources and cause more harm than benefit to ecosystems. On the other hand, assessments at decadal intervals run the risk of resulting in missed opportunities to deal with major invasive species early in the invasion process when eradication is still feasible. A standardized global baseline to guide such assessments is clearly desirable. Information for such a baseline needs to be captured relatively quickly and inexpensively, but procedures should be flexible enough to ensure that data are appropriate for a wide range of situations and applications. As invasion dynamics are context-dependent, such baseline data could be used together with contextual information such as data on features of species and ecosystems and introduction history to elucidate aspects of invasiveness and invasibility required for policy development. Standardized baseline data, collected using common protocols, would pave the way for more accurate meta-analyses to discern global and regional trends (van Kleunen *et al.* 2010).

Many indicators and metrics for quantifying the level of invasion have been proposed. For instance, Catford and colleagues (2012) discussed the pros and cons of 12 potential indicators of ecosystem invasion levels and the feasibility of each being incorporated in management. These indicators include: the presence/absence of alien species; thresholds of alien species' richness; thresholds of alien species' abundance; alien species' richness; relative alien species' richness (relative to resident species' richness); relative alien species' richness for different functional groups; relative richness of transformer species; alien species' total abundance; relative species abundance; relative alien species' abundance for different functional groups; relative abundance of transformer species; alien species' diversity; alien species' diversity relative to native species' diversity; alien species' evenness. Clearly, these indicators are not all independent of each other; some are basic whereas others can be expressed and estimated from the basic indicators. These basic indicators are true metrics obtained from ground surveys by management teams, whereas others require the use of specific statistical methods or mathematical models to estimate and could be model-/method-dependent.

The Group on Earth Observations Biodiversity Observation Network (GEO BON) recently proposed the elements required for a global observation and monitoring system for biological invasions (McGeoch and Squires 2015; Latombe *et al.* 2016). It proposed a minimum set of essential variables as the basis for measuring and monitoring invasions. Key variables, including essential biodiversity variables, are the minimum information set needed for the study, reporting, and management of an environmental problem (Pereira *et al.* 2013; Geijzendorffer *et al.* 2015; Nativi *et al.* 2015). Together with context- and taxa-dependent variables, the essential variables for invasion monitoring provide the minimum information required for assessing and tracking the status of biological invasions and their biodiversity impacts at multiple scales. The three essential variables are: (i) the presence or absence (occurrence) of alien species over defined spatial units; (ii) information on the alien status of species within their current geographic ranges; and (iii) a measure of alien species' impact (McGeoch and Squires 2015; Latombe *et al.* 2016) (Fig. 9.1).

Firstly, spatially explicit presence/absence records are the fundamental unit for quantifying the geography and movement of species and for monitoring changes in their distribution (McGeoch and Latombe 2016). Quantifying the size, extent, and nature of biological invasions globally can be greatly improved by improving the availability of standardized occurrence data across countries for alien species from multiple taxonomic groups (McGeoch

et al. 2010). The occurrence or 'occupancy' (Azaele *et al.* 2012) of alien species at appropriate levels of resolution provides the foundation for quantifying several derived variables and indicators of invasion. For instance, presence/absence records can be used to estimate species occupancy which then can be used to estimate Catford and colleagues' (2012) abundance-related indicators using different mathematical models (Hui *et al.* 2009). Co-occurrence matrices developed from these presence/absence records can be used to develop different macroecological patterns (Hui and McGeoch 2014) which can then be used for testing null models (Gotelli and Graves 1996). Moreover, tracking the level of

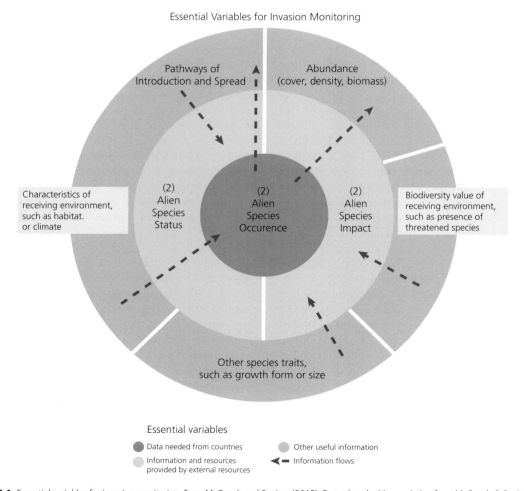

Fig. 9.1 Essential variables for invasion monitoring. From McGeoch and Squires (2015). Reproduced with permission from McGeoch & Squires.

invasion and evaluating the success of policies and interventions can be achieved by repeated measurement of relevant records of the presence and abundance of alien species.

The second essential variable for invasion monitoring is the confirmation of the presence of a species during routine surveillance or by incidental observation, categorized according to the criteria proposed by Pyšek and colleagues (2004). Such confirmation will contribute data to national inventories of alien species and possibly initiate risk assessments to limit detrimental impacts. To this end, advances in molecular ecology provide exciting opportunities to confirm the presence of target species using new molecular and genetic approaches. For instance, next generation sequencing and DNA barcoding have been used to identify the presence of tick-borne pathogen *Ixodes ricinus* in Europe (Vayssier-Taussat *et al.* 2013), the presence of the American bullfrog, *Lithobates catesbelanus*, in south-western France (Dejean *et al.* 2012), and to differentiate the presence of the zebra mussel (*Dreissena polymorpha*) and the quagga mussel (*D. rostriformis bugensis*) in the Meuse river in Europe (Marescaux and Van Doninck 2013), albeit with potentially high type-I errors due to contamination from upstream source populations in rivers, or from dead individuals.

The third essential variable for invasion monitoring is the per-capita impact of an alien species on the environment, which is critical for prioritizing efforts to prevent future introductions and to contain the spread of the most harmful species (Blackburn *et al.* 2014; see Section 9.4). The concept of per-capita impact, initially outlined by Parker and colleagues (1999) in their compound impact index (impact = range size × abundance × per-capita impact), underpins the different lists currently used to report on biodiversity targets and inform policies for the management of biological invasions (McGeoch *et al.* 2010). Prioritizing where to invest in action is a pivotal part of effective policy and management (McGeoch *et al.* 2016), as emphasized by the CBD's Strategic Plan for Biodiversity 2020 and Aichi Target 9 (UNEP 2011). The three essential variables provide essential information for a global observation system for biological invasions (McGeoch and Squires 2015; Latombe *et al.* 2016). This is not to say that other variables are not important (e.g. Catford *et al.* 2012); information on abundance, pathways, and characteristics of the receiving environment is also useful for monitoring and managing invasive species. Many of these other variables can be derived from the three essential variables together with context- and taxa-dependent information. As an example, Wilson and colleagues (2014) proposed a list of fundamental metrics for monitoring tree invasions that allow for progress towards management goals to be assessed, thereby increasing compatibility across administrative borders and between invasions (Box 9.1).

Information on these essential variables (McGeoch and Squires 2015; Latombe *et al.* 2016) and

---

Box 9.1 Essential variables for monitoring tree invasions

For efficient monitoring the status of tree invasions, six metrics were proposed (Wilson *et al.* 2014): (a) status per region (including an interpretation for trees following Blackburn and colleagues' (2011) unified invasion framework); (b) potential distribution (using a species distribution model to project potential range based on climatic suitability; see Chapter 3); (c) the number of foci requiring management; (d) compressed canopy area (i.e. area of occupancy or net infestation) as a measure of abundance; (e) range size (i.e. extent of occurrence or gross infestation) as an estimate of the total affected area that needs to be considered for management; and (f) observations of current and potential impact. These metrics can be grouped into four clusters: (i) current and potential status; (ii) abundance and population growth rate; (iii) extent and spread; and (iv) impacts and threats. Issues relating to these four clusters are discussed later, drawing on experience with tree invasions.

Firstly, the presence of an alien taxon is the primary information used for guiding biosecurity policy and invasion management (Randall 2007; Wilson *et al.* 2014). Although there are many problems with such presence records and species lists, such complications are generally less severe for

continued

Box 9.1 *Continued*

trees than for many other groups since most species were intentionally introduced which means that good introduction records often exist, and trees are relatively easy to detect (Richardson *et al.* 2014). Sub-specific identity is important for documenting tree invasions due to polyploid hybridization (Petit 2004; Thompson *et al.* 2011). Molecular methods are useful for this purpose (Thompson *et al.* 2015) and for identifying the provenance of invasive populations, hybridization, and for resolving taxonomic misclassifications (Le Roux and Wieczorek 2009). Standard criteria (e.g. spread over at least 100 m within 50 years; Richardson *et al.* 2000) have paved the way for the development of a standardized global list of invasive trees and shrubs (Richardson and Rejmánek 2011; Rejmánek and Richardson 2013). More informative metrics, such as measures of climatic suitability (Hulme 2012), can be calculated based on confirmed presence records. Species distribution models (SDMs) provide a good first estimate of the potential for a species to establish (Thuiller *et al.* 2005; see Chapter 3), though the temptation to overstate the meaning of the quantitative results needs to be tempered by the various methodological and theoretical limitations to the approach (Elith and Leathwick 2009). Refinements to SDMs, including the integration of biotic interactions (see Chapter 6), genetic considerations (Scoble and Lowe 2010) and the quantification of levels of climate matching (van Wilgen *et al.* 2009), will increase the usefulness of such methods in risk assessment. A standard metric could integrate such information to assess the probability and status of invasion (Leung *et al.* 2012).

Secondly, abundance and population trends are crucial metrics to consider in conservation assessments. For taxa with complex life cycles, age or stage structure could be important additional information for assessing abundance and population trends, for example, alien trees with large seed banks, where the size and longevity of seed banks profoundly influence management options and outcomes (Richardson and Kluge 2008; Panetta *et al.* 2011; Wilson *et al.* 2011), or insects with different instars having different functions in ecosystems (e.g. juvenile and adult harlequin ladybirds, *Harmonia axyridis*, have different diets and pathogens; Roy *et al.* 2016). Measurements of abundance, mortality, and fecundity over time are typically needed to calculate growth rates, although proxies of population trends can also be derived from occupancy patterns across scales (Wilson *et al.* 2004; Hui *et al.* 2014). Given the size and age structure of invasive tree populations, matrix population models (Caswell 2001) are useful for estimating population growth rates, and have been applied for invasive species *Ardisia elliptica*

(Koop and Horvitz 2005), *Gleditsia triacanthos* (Marco and Paez 2000), *Pinus nigra* (Buckley *et al.* 2005), *Prosopis* spp. (Pichancourt *et al.* 2012), and *Prunus serotina* (Sebert-Cuvillier *et al.* 2007).

Thirdly, extent and spread are two measures adopted by the IUCN to describe the status of species' distributions (IUCN 2012) as they provide distinct, but equally valuable, information. A raster-type approach can be used to describe the area of occupancy for a particular unit (Gaston 2003), while a vector-type approach—for example, using convex and alpha hulls—can be used to estimate the extent of occurrence and area of occupancy (Gaston and Fuller 2009; Hui *et al.* 2011). Spread rate can be estimated by monitoring the extent of change over time (see Chapter 2) or based on mechanistic models of spread (See Chapter 3). Various dating techniques (e.g. growth rings, morphometric measures, radiocarbon dating) can be used to age individuals in populations of woody plants which allows for the reconstruction of invasions at broad spatial scales (e.g. Münzbergová *et al.* 2013), but also for elucidating links between disturbance events and stages in population growth (e.g. Richardson 1988).

Finally, the impact of an invasive species has been defined as the product of geographical range, abundance, and the per-capita impact per unit of invaded area (Parker *et al.* 1999). As discussed earlier, range and abundance are objective, but per-capita impact is more difficult to quantify and varies dramatically between species, for example, for cost of control (see Hulme 2012; also see section 9.9). Any measurement of impact or threat is also complicated by the subjectivity of definitions and the choice of metrics used (Jeschke *et al.* 2014). Effect per individual (or unit area) allows for a quantitative measure but, despite many useful conceptual models, a detailed quantification of impact is often precluded by data requirements and the non-linearity of impacts (Kumschick *et al.* 2015; see also section 9.4). The negative effects of many invasions are likely underappreciated (e.g. poorly studied, difficult to detect or demonstrate) or inaccurately quantified (e.g. due to the lag between the beginning of an invasion and the manifestation of impact) (Simberloff 2011). Also, some positive and negative impacts are difficult to detect or measure objectively and metrics for their quantification are controversial in some cases (Richardson and Ricciardi 2013). Wilson and colleagues (2014) proposed that for an initial assessment of impact for invasive trees, two observations can be used: height in relation to the native vegetation (see Chapter 8), and whether a species has the traits for triggering regime shifts (Rejmánek *et al.* 2013; see Chapter 7).

supplementary site- and taxa-dependent variables (Catford *et al.* 2012; Wilson *et al.* 2014) have already formed a key part of monitoring or management programs in some regional monitoring and reporting schemes (e.g. Stokes *et al.* 2006; Allen *et al.* 2009). To maximize the value of investment in monitoring biological invasions, the information generated from monitoring schemes must be shared to ensure that assessments of the status of biological invasion are comparable across local, regional and global scales (e.g. Katsanevakis and Roy 2015; van Kleunen *et al.* 2015). Moreover, occurrence data from the first essential variable are readily scalable (Vilà *et al.* 2010) and are suitable for incremental increases in survey efforts over time. Even at coarse scales, alien species occurrence data can be combined with information gathered for the other two essential variables to inform actions at the country level (Blackburn *et al.* 2014).

Digital infrastructure for big data on essential variables for biodiversity conservation and invasion monitoring is being developed (Jetz *et al.* 2012; Canhos *et al.* 2015; Nativi *et al.* 2015). In particular, the Global Biodiversity Information Facility (GBIF) is an international network for organizing evidence of species presence, and the Map of Life provides species' range information and inventories by location (Jetz *et al.* 2012). Such data infrastructure provides the flexible platforms necessary for compiling data from both citizen science and professional ecological surveys. The Global Invasive Alien Species Information Partnership (GIASIP) that was initiated in 2012 already delivers a freely accessible mechanism for sharing and integrating data, including an Information Gateway that can accommodate all EBVs for monitoring biological invasions (UNEP 2014). The Global Invasive Species Database (GISD) which is managed by the IUCN Species Survival Commission's Invasive Species Specialist Group (ISSG) contains information on the ecology of introduced and invasive species, although it requires further work to remove taxon bias and expert subjectivity (Pyšek *et al.* 2008; Bellard *et al.* 2016). Knowledge products such as a Global Register of Introduced and Invasive Species (GRIIS) will provide access to verified national inventories (UNEP 2014). This register links the species name and geographic reference to the country or site of occurrence with primary data

sources. It aims to soon include over 120 verified national inventories, made available through the Partnership (UNEP 2014). Overall, of the three EBVs for invasion monitoring (McGeoch and Squires 2015; Latombe *et al.* 2016), the first two can be incorporated to the global effort and infrastructures for biodiversity monitoring (see section 9.3), and the schemes for assessing the last one has emerged (see section 9.4).

9.3 Monitoring strategies

Monitoring the abundance and spatial distribution of invasive alien populations is important for designing and measuring the efficacy of long-term management strategies. Methods for monitoring over large areas with minimum sampling effort, but with sufficient accuracy, are crucial for effective management. Together with proposed essential variables for invasion monitoring, the GEO BON team further proposed a modular approach for monitoring alien and invasive species (Latombe *et al.* 2016) for countries at different stages of developing their invasion monitoring systems (not necessarily equivalent to the developing/developed country dichotomy). At the global level, this modular approach allows for the combination of information from countries with different monitoring infrastructures by only delivering a national species inventory of aliens (Fig. 9.2). In particular, the Parties to the CBD have identified the need for such inventories, and several countries and some regions now have inventories for various taxa, and for management and policy purposes (Canhos *et al.* 2015; UNEP 2014). More detailed information on the status and distribution of alien species at national levels could enhance the commitment of cross-border management actions through controlled trading and human movement, as well as custom practice (van Kleunen *et al.* 2015).

Within a country, this modular approach can allow for the further development of fine-scale priority sites and species (Fig. 9.2), such as for protected areas with high conservation value or hubs of new incursions (McGeoch *et al.* 2015). For countries at the early phase of developing their monitoring infrastructures, range extent, and occupancy can be estimated for selected taxa. For countries that already have observation and monitoring systems, the inclusion of a network of long-term monitoring sites where alien species occurrence is recorded at regular intervals allows for accurate evaluation of population trends and devising effective management plans (Latombe *et al.* 2016). In many countries, invasion monitoring can be improved by integrating such efforts with existing biodiversity monitoring schemes (e.g. see Box 9.2 for examples from South Africa's Kruger National Park). Countries may also capitalize on emerging citizen science initiatives, and online and remote technologies for data capture (Skidmore *et al.* 2015) to improve records of invasions. For instance, Visser and colleagues (2014) demonstrated the potential of Google Earth for monitoring the spread of invasive trees into certain vegetation types, and proposed the use of this freely available technology as the basis for the establishment of networks of sentinel sites. For all countries, the goal should be to provide, at regular intervals (at least every five years), data on alien species occurrence corresponding to their maximum level of resolution, for their national inventory, either for priority sites or selected taxa (McGeoch and Squires 2015; Latombe *et al.* 2016).

As discussed earlier, the two most important variables that need to be accurately estimated in alien species monitoring programmes are spatial distribution and abundance. Large-scale monitoring programmes have obvious advantages over local schemes. Firstly, sources of invasions can be included and identified in large-scale programmes, which is often impossible for local studies. Secondly, spatial stochasticity is often minimized to facilitate the recognition of trends and patterns. Finally, environmental complexity and heterogeneity are often included to facilitate robust inference and projection that are compatible with policy making (Johnson 1993; Urquhart *et al.* 1998). Nevertheless, the implementation of large-scale programmes is often constrained by cost (e.g. Bottrill *et al.* 2009). Consequently, designing cost-efficient methods of inference and sampling schemes (protocols) that are accurate enough to provide appropriate data for a range of management and planning activities is crucial for improving the effectiveness of such programmes.

The effectiveness of large-scale monitoring programmes can be improved through the careful design of sampling strategies (Carlson and Schmie-

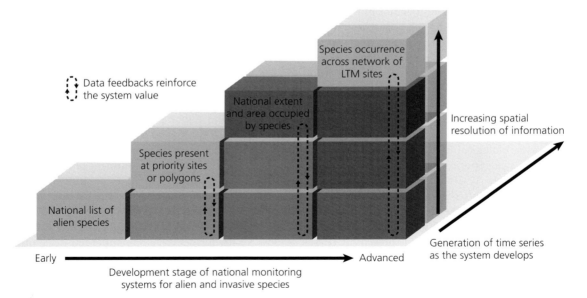

Fig. 9.2 Modular approach to the development of national observation and monitoring system for biological invasions, built from the variable 'alien species occurrence', which is one of the three essential variables for invasion monitoring and also an essential biodiversity variable. LTM: Long-term monitoring. From McGeoch and Squires (2015). Reproduced with permission from McGeoch & Squires.

Box 9.2 Monitoring plant invasions in Kruger National Park

The Kruger National Park (KNP) in South Africa, almost 20 000 km² in extent, is one of the largest protected areas in the world that is actively managed primarily for biodiversity conservation. More than 370 alien plant species have been recorded in KNP (Foxcroft *et al.* 2008). The goals of invasive species management in KNP are to detect new invasions at an early stage and to respond rapidly to these to maintain current invasions at low abundances where eradication is not feasible, and to undertake relevant actions to prevent further invasions (Foxcroft and Richardson 2003; Foxcroft and Freitag-Ronaldson 2007). The CyberTracker system (Fig. 9.3) is used by rangers during patrols to record features they observe. The system can be programmed to take geographical coordinates at predefined time intervals. When recording a specific observation, other features, if present,

will also be captured. Thus, timed points and records of other features can be used as 'absence' point data when analysing alien plant invasions (Foxcroft *et al.* 2009). Between 2004 and 2007 the system collected almost 2.4 million records (including 27 777 presences of invasive alien plants and 2.3 million inferred absence records), with most records representing *Opuntia stricta* (72.1 per cent) and *Lantana camara* (8.4 per cent) (Hui *et al.* 2011).

Presence and absence records can be used for projecting the potential distributions of alien plants in the park using correlative species distribution models, although the low percentage (roughly 1 per cent) of presence records and the non-equilibrium of ongoing invasions create major challenges to the validity of such models (see Chapter 3). As a monitoring programme, the landscape can be divided into

Records/Cell (0.5 km):
0
1–2
3–7
8–24
25–20 460

N

40   0   40   80 Kilometers

Fig. 9.3 The CyberTracker system has been used since 2003 in the Kruger National Park to achieve management goals. The system, a user-friendly interface for Palm OS computers linked to a GPS, provides high-precision data, including information on animals, plants, water holes, poaching activity, fence condition, fire scars, and numerous other features. Figure copyright: DST-NRF Centre of Excellence for Invasion Biology. Reprinted with permission from John Wiley & Sons.

continued

Box 9.2 *Continued*

grids at a resolution for each cell to be a potential management unit, and the presence/absence records can be used to estimate simultaneously occupancy, abundance, and detection rate of alien plants within each cell (e.g. MacKenzie *et al.* 2003; Royle *et al.* 2007; Hui *et al.* 2009, 2011; Barwell *et al.* 2014). Large-scale monitoring programmes need cost-efficient methods for estimating the abundance and distribution of target species. Traditional mensuration methods of estimating abundance, such as mark-recapture techniques, are only useful at local scales (e.g. $0.1–10$ km^2 for complete counts) due to the method of data collection and the associated costs. This has led to an increased interest in the use of binary (presence/absence) data for large-scale monitoring (Brotons *et al.* 2004; Joseph *et al.* 2006). In this regard, two categories of abundance estimation models have been developed. The first is designed for true presence/absence binary data and includes the intraspecific occupancy-abundance relationship that is grounded in the ubiquitous positive correlation between species abundance and range size (Hui *et al.* 2009), and the scaling pattern of occupancy describes how adjacent occupied cells merge with increasing grain (Hartley and Kunin 2003; Hui *et al.* 2006; Hui 2009). A multi-criteria test suggests the supremacy of the models of the scaling pattern of occupancy over those of the occupancy abundance relationship in estimating abundance and yielding macroecological patterns (Hui *et al.* 2009; Barwell *et al.* 2014).

To improve the efficiency of the programme, five monitoring strategies for abundance and distribution estimation of invasive alien plants in KNP were examined (Hui *et al.* 2011). These including weighted (the monitoring/sampling effort is allocated according to observations from the current CyberTracker data), uniform (i.e. equal weighted cells), addictive (the ranger tends to visit the cells having more presence records), elusive (the ranger will try to avoid visiting the cells with presence records), and random-walk options (the ranger will randomly choose a cell adjacent to the cell visited at the last time). Although there are other sophisticated sampling options for increasing the sampling efficiency (e.g. adaptive cluster sampling; Dryver and Thompson 2005), they are difficult to implement in practice, especially given the species-specific spatial variation and the multi-faceted targets in initiatives such as the KNP's monitoring programme. All five monitoring schemes are robust because the estimates approached the known abundance and distribution for simulated species with the increasing number of records, and the uniform sampling scheme performed best (Hui *et al.* 2011). A strong interspecific power law relationship between the optimal sampling effort for abundance (OSEA) and the population density emerged (Fig. 9.4a), and also between the optimal sampling effort for distribution (OSED) and the population densities (Fig. 9.4b). Overall, a spatially random, rare species required a much larger monitoring effort to achieve a satisfactory level of accuracy. About one record per hectare is needed to have reliable abundance and distribution estimates for widespread invasive species (Hui *et al.* 2011). Overall, presence records should be used efficiently to address key challenges facing invasion management (Richardson and Whittaker 2010), and reliable estimates of abundance and range of invasive plants then pave the way for improved prioritization of various interventions to address problems associated with biological invasions.

---

gelow 2002; Thompson 2002; Fortin and Dale 2005). Issues that must be considered include the determination of an appropriate sampling effort, extent, unit size (grain), and sampling strategy (scheme, or spatial layout of the sampling unit); all these factors affect the potential of the initiative to detect different spatial patterns (Dungan *et al.* 2002; Hui *et al.* 2010, 2011). Importantly, monitoring programmes could be coordinated with control strategy (Hauser and McCarthy 2009), with management decisions for surveillance and eradication in a robust fashion (Rout *et al.* 2009). For instance, adaptive cluster sampling enhances the efficiency compared to simple random sampling for estimating population densities of aggregated species (Thompson 1990). As a

rule of thumb, sampling designs should be guided by the minimum requirement for adequate spatial analysis and inference (Fortin and Dale 2005), an issue which is likely to be difficult in a large-scale monitoring programme (Goodman 2003). As a commonly used and cost-efficient format for data recording (Brotons *et al.* 2004; Joseph *et al.* 2006), presence/absence data and ad hoc records have been adopted as the standard format in many monitoring programmes. In such cases, the choice of extent and grain is not an issue. Instead, we face the problem of false-negatives and pseudo-absences in the model inference (e.g. MacKenzie *et al.* 2002; Royle and Nichols 2005). An efficient sampling design thus requires an adequate sampling effort for

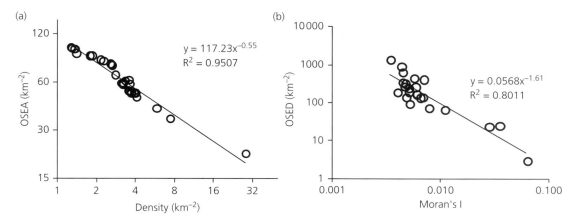

Fig. 9.4 The relationship between the density of invasive plants and the optimal sampling effort for abundance estimation (OSEA) (a), between the aggregation, as measured by Moran's I, and the optimal sampling effort for distribution estimation (OSED) (b), for the 29 most widespread and abundant alien plant species in Kruger National Park, South Africa. From Hui *et al.* (2011). Reprinted with permission from John Wiley & Sons.

an efficient and practical sampling scheme that can provide enough information for accurate inference of abundance and the spatial distribution of invasive species. Lessons could be learnt from successful monitoring systems developed for large protected areas, mainly with the goal for biodiversity conservation but increasingly involving monitoring alien species (Box 9.2).

9.4 Impact and risk metrics

Many attempts have been made to provide a practical scheme for categorizing impacts of invasive species to inform management prioritization and planning. The Environmental Impact Classification for Alien Taxa (EICAT) scheme is probably the most useful scheme for objectively classifying alien species into different impact classes across taxa and environments, based on the magnitude and reversibility of their deleterious environmental impacts (Blackburn *et al.* 2014). EICAT has gained international support through the IUCN and the Global Invasive Alien Species Information Partnership (UNEP 2014), and guidelines have been developed to facilitate consistent and comparable application of the scheme by users (Hawkins *et al.* 2015). The scheme provides a standardized approach for estimating the third EBV for invasion monitoring (see section 9.2). We elaborate on the EICAT scheme in the following.

The EICAT scheme excludes economic or societal impacts, focusing solely on environmental impacts. It identifies species that have deleterious abiotic or biotic impacts, and is not intended to compare deleterious with beneficial impacts in the case of determining the net value of an introduced species. Species are classified on the basis of the evidence of their *most severe documented* impacts across introduced regions. Importantly, the scheme is not a predictive model of impact but a system based on the worst recorded case to flag species with high potential impacts. Unlike risk assessments (see following), it only considers consequences, not likelihoods, of alien species occurrence and invasion. It provides a transparent protocol for translating the broad diversity and range of impact mechanisms into a ranking of environmental impact magnitudes. Species can move up and down impact categories as the quality of evidence improves, as conditions change, or as an invasion proceeds. It can be applied to impacts assessed at a range of spatial scales, from global to national or regional. It can be applied at a range of spatial scales, from global to national or regional. EICAT is based on the Generic Impact Scoring System (GISS) that was developed by Nentwig and colleagues (2010), which they later extended (2016).

Blackburn and colleagues (2014) update the approach to align with the new impact scheme implemented in the Global Invasive Species Database (GISD) by the IUCN Species Survival Commission

(SSC) Invasive Species Specialist Group, including largely mechanisms related to biotic interactions (e.g. competition and parasitism), disturbance (e.g. trampling and flammability) and ecosystem engineering (e.g. bio-fouling).

Furthermore, the EICAT scheme is consistently based on the effects of alien taxa on different organizational levels in the native community and assigns a species to one of five sequential categories of impact according to the magnitude of the impacts on the native communities (Fig. 9.5). The process of categorization involves the collation of all available published evidence on impact for the members of an invasive taxon from all parts of the focal regions to which it has been introduced; it uses the highest level of deleterious impact to inform the classification. Assigning a taxon to any category is an evidence-based proactive way of informing management. Taxa with no recorded evidence of impact are placed in a data-deficient category which flags that research is needed to establish what the magnitude of impact is of the respective alien taxa. Consequently, Blackburn and colleagues (2014) suggest, for practical purposes, a categorization of uncertainty into three levels (high, medium, and low confidence), similar to the approaches used by the Intergovernmental Panel on Climate Change (IPCC) (Mastrandrea et al. 2010) and the European and Mediterranean Plant Protection Organization (EPPO) (Holt et al. 2012; Kenis et al. 2012).

EICAT can feed into both pre- and post-border risk assessments and inform decision making across the board. However, on its own, EICAT is not suitable for use as a management decision tool as impact is rather context-dependent, and information on impacts is often scarce, especially for species at the early stages of invasion. Risk assessments on the other hand incorporate not only known impacts but also the likelihood of invasion and impact. Many risk-assessment models have been developed for alien plant species (Leung et al. 2012; Kumschick and Richardson 2013), several of which were developed specifically for trees and other woody plants (Tucker and Richardson 1995; Reichard and Hamilton 1997; Widrlechner et al. 2004). Any scheme investigating risk should, by definition, consider both likelihoods and consequences. Blackburn and colleagues' (2014) EICAT scheme focuses only on the consequences. By contrast, Wilson and colleagues (2014) described a simple way of allocating tree species to different categories of risk incorporating both likelihoods and consequences where the likelihood of an invasion is measured based on potential status and the likelihood of introduction or extent of planting, with climatic suitability used as an estimate of potential status. To this end, invasiveness elsewhere has also been used to inform the likelihood of a given taxon being invasive in a focal region, especially for regions with similar climates and introduction pathways.

Both invasiveness and impact elsewhere can also be used in approaches for multi-species risk assessment, for instance as training data for using artificial neural networks (such as self-organizing

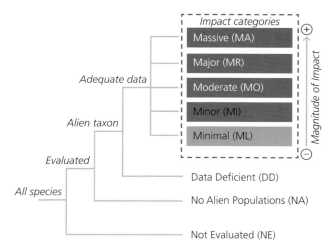

Fig. 9.5 The different categories in the environmental impact classification for alien taxa (EICAT) scheme (Blackburn et al. 2014). Massive (MV): leading to local extinction, as well as irreversible changes in ecosystem functioning. Major (MR): causing changes in community composition such as local extinction and ecosystem changes which are reversible when the alien taxon is removed. Moderate (MO): causing declines in the population size of native species, without changes to ecosystem functioning. Minor (MN): causing reduction of individual fitness for some native species, without declines in native population sizes. Minimal concern (MC): no deleterious impacts on native species. From Hawkins et al. (2015).

maps) to estimate the likelihood of species being established and becoming invasive in a given country (Worner and Gevrey 2006; Paini *et al.* 2010). In the absence of information, the invasiveness of congeners can be used to estimate the *a priori* expectation of an invasion (Diez *et al.* 2012), as certain genera and other are clearly over-represented in terms of invaders (e.g. Rejmánek and Richardson 2013 for trees and shrubs). Moreover, an invader's traits can determine its invasion performance when interacting with native species and potentially its impacts (e.g. Kumschick *et al.* 2013) (see Chapter 6) under particular characteristics of recipient ecosystem (see Chapter 7 and 8). Such invasive traits can be identified using multivariate statistics (including species distribution models) (e.g. Molnar *et al.* 2008; Pyšek *et al.* 2012; van Kleunen *et al.* 2010) and network approaches (e.g. Hui *et al.* 2016; Minoarivelo and Hui 2016; see Chapter 10). Risk assessment can be further visualized through risk mapping that emphasizes pathways of introduction and spread; such spatially explicit risk maps provide a good platform for management prioritization (see following).

9.5 Prioritization

Although less than 20 per cent of alien species cause recognizable problems (Vilà *et al.* 2010; Lockwood *et al.* 2013), the impacts of such species are substantial and persistent on both the environment and embedded socio-economic systems (Jeschke *et al.* 2014) and thus justify management interventions. Because invasion management is often extremely expensive, available resources need to be prioritized and allocated cost-effectively (Krug *et al.* 2009). However, a large number of diverse invasive species are considered for management; this can further complicate the already limited and unbalanced budget for invasion monitoring and controlling (Hulme 2009). Problems and opportunities need to be objectively ranked or prioritized (McGeoch *et al.* 2016) according to the severity of actual and potential impacts on biodiversity and ecosystems (Carrasco *et al.* 2010; Kumschick *et al.* 2012), and management decisions should be based on the impact, budget constraints, and the cost-effectiveness of management actions (Dawson *et al.* 2015; see sections 9.7–9.11).

As recognized by the Convention on Biological Diversity (specifically Aichi Target 9), prioritization of both invasive species and invasion pathways is indispensable for invasion management. Another essential requirement for effective prioritization is the identification of highly valued sites that are most susceptible to invasion, and integrated prioritization needs to reach a balance when allocating limited resources among these three components: species, pathways, and sites (Heikkilä 2011; McGeoch *et al.* 2016). With sufficient information (e.g. Nelson *et al.* 2009), there are classification schemes (e.g. EICAT; see section 9.4) and rules of thumb (Leung *et al.* 2012) that allow managers to identify and prioritize for maximum-impact species, high-risk introduction pathways, and vulnerable sites. To this end, McGeoch and colleagues (2016) merged these three components and defined prioritization as the process of ranking species, pathways, or sites for the purposes of determining their relative environmental (and sometimes also socio-economic) impacts (*sensu* Kumschick *et al.* 2012; Blackburn *et al.* 2014), and for deciding on the relative priority of actions to effectively and efficiently prevent or mitigate the impact of invasive alien species in a balanced and transparent way (Fig. 9.6; see also Sutherland *et al.* 2006; Benke *et al.* 2011).

Firstly, species-based prioritization is the most common and best-developed component (Heikkilä 2011; Kumschick and Richardson 2013; Kumschick *et al.* 2015). Several schemes are available for ranking species according to risk and impact (Pyšek and Richardson 2010) such as the biosecurity frameworks (e.g. ISPM 2004; Baker *et al.* 2008; EPPO 2011) and the EICAT scheme (Blackburn *et al.* 2014; see section 9.4). Secondly, when species-based prioritization fails, especially for unintentional introductions or when little information is available on species (Leung *et al.* 2014), prioritizing critical introduction and dispersal pathways could effectively slow the rate of incoming propagules and further spread (McGeoch *et al.* 2016). Based on 32 different pathways of introduction (e.g. those associated with agriculture, forestry, or the nursery trade; Hulme *et al.* 2008), a detailed categorization of pathways has been developed by the IUCN Invasive Species Specialist Group, and endorsed by the CBD (UNEP 2014; Essl *et al.* 2015). Horticultural, pet, and

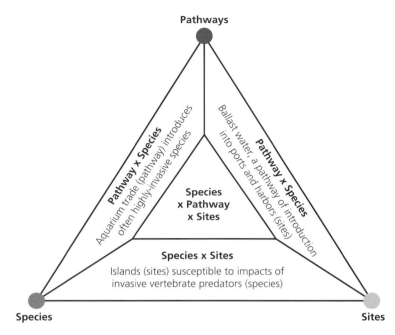

Fig. 9.6 The three components of an integrated prioritization in invasion management. Texts show examples for ornamental species in gardens as escapees (pathway) into adjacent protected areas (sites). From McGeoch *et al.* (2016).

aquarium escapees have been identified as the most frequent pathways for introduction and spread of invasive species (Chang *et al.* 2009; Roy *et al.* 2014). Finally, site-based prioritization recognizes the uneven risk of entry and establishment by invasive species across susceptible landscapes and regions (e.g. water catchments; Yemshanov *et al.* 2013; Bobeldyk *et al.* 2015) and the need for asset protection (e.g. to mitigate threats to sensitive habitats or endangered species with known distributions; Leung *et al.* 2012; McGeoch *et al.* 2016).

Acknowledging the three components, especially considering sites, in invasion management demands spatially explicit prioritization. As an example, Roura-Pascual and colleagues (2010) conducted a multi-criteria decision analysis (de Steiguer *et al.* 2003) aimed at prioritizing sites for the management of the most important invasive alien plants at the landscape scale in South Africa's Cape Floristic Region. Firstly, four management scenarios were developed through a participatory process involving managers and researchers to determine the relative weight of model factors (Fig. 9.7, left). Using a sensitivity analysis, Roura-Pascual and colleagues (2010) assessed the sensitivity of these weights and used estimated median weights to create the

priority maps for these four management scenarios (Fig. 9.7, right). Comparisons of these priority maps revealed substantial variation and also similarities between these four management scenarios. Management scenarios and strategies critically determine the outcome of spatial prioritization. Consequently, particular differences in environmental conditions and the status of the invasions preclude the generalization of prioritization results across regions and call for the adjustment of prioritization strategies to fit the particularities of each region (Roura-Pascual *et al.* 2010).

Besides the multi-criteria decision analysis for priority mapping by arriving at compromises between multiple shareholders and information sources discussed, above risk mapping is another valuable method for prioritizing invasion management actions and coordinating monitoring efforts. The potential distribution of an invasive species in its novel range can be predicted using species distribution models (see Chapter 3), and risk mapping can be considered as an extension of such models for predicting the risk of introduction and establishment of an alien species in a focal region (Jimenez-Valverde *et al.* 2011). For instance, based on freshwater fish occurrences, environmental

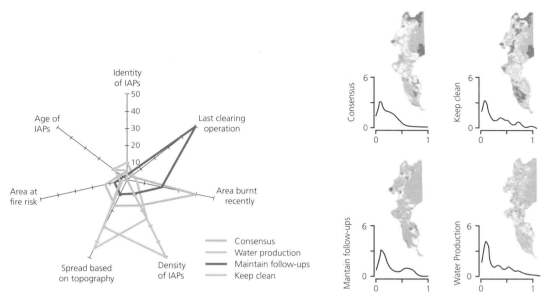

Fig. 9.7 (Left) Visualization of strategies considered for clearing of invasive alien plants (IAPs) in the selected-weights sensitivity analysis. (Right) Priority maps for the management of IAPs in the Cape Peninsula, South Africa, under four different management strategies. Line graphs show the probability distribution of the priority values. The size of the grid cells is 10 × 10 km. From Roura-Pascual *et al.* (2010). Reprinted with permission from John Wiley & Sons.

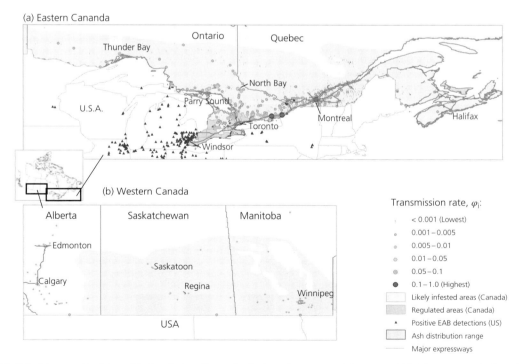

Fig. 9.8 Risk of the emerald ash borer (*Agrilus planipennis*) experiencing long-distance spread via commercial freight transportation for major Canadian municipalities based on the second-degree stochastic dominance criteria. From Yemshanov *et al.* (2012). Reprinted with permission from John Wiley & Sons.

variables, and human activity indicators, Leprieur and colleagues (2009) identified six major global invasion hotspots for freshwater fish invasions, with human activities being the main driver of global fish invasions. Priority was given to developing countries with growing economies which could face high impacts from invasive fish on the native freshwater biota. Risk mapping highlights the role of spread pathways and vectors (e.g., Kaplan *et al.* 2014), in contrast to the classic setting of species distribution models which rely largely on matching species occurrences with the environment. As such, risk mapping provides a tool for merging outcomes from prioritizations for pathways and sites. Uncertainty associated with the introduction and spread of invasive species can be projected (Fig. 9.8; Yemshanov *et al.* 2012). For multiple species, overlaid risk maps or the use of multi-criteria decision analysis could bring together all three components for prioritization and facilitate integrated decision-making in invasion management.

9.6 Invasion debt

The extent and severity of the economic and environmental impacts of biological invasions have increased rapidly over the past few decades (Pimentel *et al.* 2001; Butchart *et al.* 2010; Pereira *et al.* 2013), posing three challenges to management. Firstly, biological invasions are both passengers and drivers of global change, together with climatic, land cover,

economic, ecological, and demographic variables which form complex wicked management problems (Pyšek *et al.* 2010; Essl *et al.* 2011; Richardson and Pyšek 2012). Secondly, the status and extent of alien species are generally poorly monitored (McGeoch *et al.* 2010), resulting in an underestimate of the extent and impact of the invasion problem. Finally, there are often time lags between introduction and spread of alien species (Kowarik 1995; Essl *et al.* 2012). The time between introduction and naturalization can be several decades for some taxa (Caley *et al.* 2008; Larkin 2012), and at a broad-scale, introduced species can take up to several centuries to occupy their full potential distribution (Wilson *et al.* 2007; Williamson *et al.* 2009). A large number of future invasive species have already been introduced and are progressing at different rates along the introduction–naturalization–invasion continuum (Richardson and Pyšek 2012).

Alien species that are already introduced but are not yet invasive can cause delays in management, planning, and policies, posing an *invasion debt* (Seabloom *et al.* 2006; Essl *et al.* 2011; Rouget *et al.* 2016) or the scenario of *waiting for invasions* (Jerde and Lewis 2007). Rouget and colleagues (2016) defined invasion debt as the additional amount of invasion that could take place in the future in a given region. Invasion debt can therefore be divided into four components (Fig. 9.9) that are related to different invasion stages an introduced species has to go through (see also Jerde and Lewis 2007).

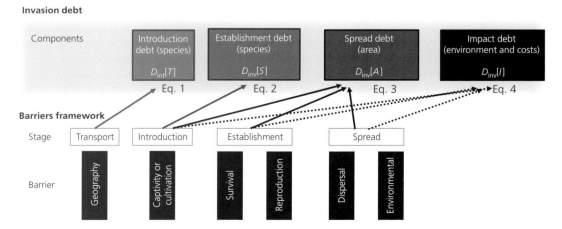

Fig. 9.9 Conceptual framework for quantifying invasion debt in terms of the number of species, potential area invaded, and impact, aligned with the unified invasion framework of Blackburn and colleagues (2011). From Rouget *et al.* (2016). Reprinted with permission from John Wiley & Sons.

Introduction debt is the additional number of species that will be introduced to a given area over a given period. Many of these introductions will fail, only a small fraction will establish, and a subset of those that establish will become invasive (Williamson and Fitter 1996). Introduction debt can be quantified as the sum, across the global not-yet-introduced species pool (S), of the probability for an alien species being introduced to the focal area during a specific period. This probability will vary between species and over time (e.g. new pathways or change in species trade patterns; Seebens *et al.* 2015). Let p_j be the probability of a non-introduced species, j, to be introduced to the focal area during a specific year. The introduction debt (Fig. 9.9) from now (year 1) to the future (year T) can be written as:

$$D_{\text{int}}[T] = \sum_{j \in S} \sum_{t=1}^{T} (1 - p_j)^{t-1} p_j \qquad (9.1)$$

As T approaches infinity, introduction debt will approach S. Let S_I be the current introduced species pool, and the pool of introduced species at year T will be $S_I(T) = S_I + D_{\text{int}}[T]$.

Establishment debt is a species-based component and represents the difference in species richness between the current and potential invasive pool originating from the introduced species pool (Fig. 9.9). Note that establishment debt from species yet to be introduced was not considered but can be easily added. Many factors influence the establishment process (Rouget *et al.* 2016), including: (1) the number of already introduced casual species; (2) the environmental suitability for each species to establish a viable population; (3) species traits that are related to establishment and invasiveness (Pyšek and Richardson 2007); (4) residence time in an area (Wilson *et al.* 2007); and (5) propagule pressure. Establishment debt can be either quantified as the sum of the combined probability across all the factors mentioned earlier for all introduced species, or as the number of species which are likely to establish. The establishment debt (species-based invasion debt; Fig. 9.9) from now (year 1) to year T due to current non-invasive introduced species (i.e. excluding those from introduction debt) is:

$$D_{inv(S)}[T] = \sum_{j \in S_{NI}} \sum_{t=1}^{T} \left((1 - \beta_j)^{t-1} \beta_j \sum_{t'=t}^{T} (1 - \rho_j)^{t'-t} \rho_j \right) \qquad (9.2)$$

where S_{NI} is the current non-invasive introduced species pool, β_j is the annual establishment probability of introduced species j; ρ_j is the annual probability of established species j becoming invasive. Novoa and colleagues (2015) estimated the establishment debt for Cactaceae.

Area-based invasion debt, termed spread debt (Fig. 9.9), is expressed as the additional area that invasive species could occupy in the focal region. Spread debt takes into account both species that are already invasive and introduced species which are not yet invasive but which have a high risk of spreading in the future. Note again that spread debt from species yet to be introduced was not taken into account in Rouget and colleagues (2016) but can be easily added. This component of invasion debt is determined by (1) the probability that a species will become invasive; (2) the environmental suitability of a region for each species; (3) the rate of spread (both natural and human-mediated) of the species; and (4) propagule pressure. The spread debt from now to year T can be partitioned into two components:

$$D_{ins(A)}[T] = D_{inv(A1)}[T] + D_{inv(A2)}[T] \qquad (9.3)$$

The first term on the right is the debt from current invasive species, and the second term is the debt from non-invasive but already introduced species (Rouget *et al.* 2016). Moore and colleagues (2011) calculated the spread debt for *Acacia paradoxa* in South Africa.

Impact-based invasion debt measures the additional impact that invasive species could have in the focal region. Impacts of biological invasions include both negative environmental effects and socio-economic costs (see section 9.4). Impact is directly proportional to the area affected, and per-capita impacts increase with time, often in a non-linear fashion (see section 9.9; Yokomizo *et al.* 2009; Simberloff *et al.* 2013). Consequently, impacts can be estimated from a few factors such as the area invaded, the characteristics of the invaded area (e.g. high biodiversity areas versus anthropogenic habitats), changes in ecosystem functioning, and the economic costs of managing the invasive species. Impact-based invasion debt will be typically expressed as the financial cost of invasive species impact in newly invaded areas (see section 9.9). The impact-based invasion debt from now to year T can also be partitioned into two components (Rouget *et al.* 2016):

$$D_{ins(I)}[T] = D_{inv(I1)}[T] + D_{inv(I2)}[T] \qquad (9.4)$$

The first term on the right, $D_{inv(I1)}$, indicates the impact-based invasion debt for the currently invaded area, and the second term, $D_{inv(I2)}$, indicates the impact-based invasion debt for future invaded area.

Based on potential range, propagule pressure, and invasion status (introduced or invasive), Rouget and colleagues (2016) calculated establishment debt (the additional number of invasive species) across 13 regions of the world for Australian *Acacia* species, a model system for the elucidation of many aspects of invasion ecology (Richardson *et al.* 2011; Kueffer *et al.* 2013). They discovered a considerable increase in the number of invasive *Acacia* species; for example, five out of the seven introduced acacia species in South America are predicted to become established. They were able to quantify three components of invasion debt (establishment, spread, and impacts) in South Africa as better data were available. In this country, 14 *Acacia* species are already invasive and another four species could become established out of the 45 species already introduced but not yet invasive; 17 are unlikely to become established. The spread debt was estimated to almost double the current extent of acacia invasions with a median increase value of 2330 km^2 per species. Over a 20-year period, the spread debt could reach over 10 000 km^2 for *A. dealbata* and *A. mearnsii*. Given the estimated cost associated with managing *Acacia* invasions, this translates to a cost of up to US\$174 million per species. If left unmanaged, the clearing cost of the invasion debt of Australian acacias in South Africa over the next 20 years will exceed US\$500 million. Predicting future invasions is riddled with uncertainty but the concept of invasion debt provides a simple, practical approach for quantifying the extent of impending invasions. Separating invasion debt into components corresponding to invasion phases allows for estimating the contribution of the different invasion stages to the debt (Rouget *et al.* 2016).

9.7 Political will

Global trade has become the main mode of transport for many invasive species (Pimentel *et al.* 2005), which means that biological invasions, especially those resulting from accidentally introduced species, can be considered a market failure of international trade. The burgeoning costs of managing invasions and the damages of invasion to society at large call for policies such as a tariff on relevant trade (McAusland and Costello 2004). Margolis and colleagues (2005) explored the political process likely to govern the formation of tariffs on invasion-related trades by incorporating the cost of an invasive species into the Grossman and Helpman (1994) political economy model. This model considers the interaction of interest groups and a government that cares about general welfare, and it has been successful in predicting trade protection in the United States (Gawande 1997; Goldberg and Maggi 1999) and Turkey (Mitra *et al.* 2002). Margolis and colleagues' (2005) extension of this model made clear how cost and damage of invasions affect the dynamics of political will on taxing relevant trades.

Margolis and colleagues (2005) assumed that contributions to the government and donations to re-election campaigns from lobbyists can interfere with government decisions. In particular, before the government makes its decision, each lobby draws up a contribution schedule which is a binding agreement that commits the lobbyist to contribute a specific amount for every policy choice the government might make on issues of interest. These lobbies are the private owners of certain specialized products, and their profits depend on the price alone. The schedule of contributions is considered a function of domestic prices of the trading items and the damage from associated invasive species. To ensure its political party remains in power, the government needs to concern both general welfare and campaign contributions. A governing party has a better chance of being re-elected if social welfare becomes higher and more money is donated to election campaigns.

To simplify the model, Margolis and colleagues (2005) assumed that the incumbents place a constant weight on social welfare versus campaign contributions, regardless of the origin of contributions or whether the improvement of social welfare means a rise of environmental quality, consumer surplus, or government revenue. The objective of the government can thus be depicted as the following function (Margolis *et al.* 2005):

$$G = \sum_{j \in L} C_j(p) + aW(p, \delta \cdot m(p)) \qquad (9.5)$$

where p is the price vector, δ is a vector of constant marginal expected external damage per unit import (if good i is exported, $\delta_i = 0$ in the case of a consumption-related externality), m is the import vector, C_j is the contribution to incumbents given by the jth lobby, a is the exogenous weighting of consumer welfare (W), and L is the set lobbies. The government objective function differs from the Grossman and Helpman model only by the inclusion of a damage term in the welfare function. The sequence of events is: firstly, lobbies select contribution schedules $C(p)$ and then the government, observing those contribution schedules, sets the tariff vector τ to maximize G subject to $p = p^W + \tau$ where p^W is the world price vector and is assumed exogenously determined.

After considering a few features of the optimal schedules that lobbies are likely to exploit truthful strategies that make the lobby indifferent to whichever tariff policy eventually chosen, the fact that non-truthful strategies can be undermined by coalitions and that the Nash equilibrium in this auction game must maximize the joint pay off of the government and each lobby, Margolis and colleagues (2005) derived the following Nash equilibrium of tariff:

$$\tau_j = \delta_j + \frac{I_j - \alpha_L}{a + \alpha_L} \frac{y_j(p_j)}{\left| m_j'(p_j) \right|} \qquad (9.6)$$

where $I_j = 1$ if the industry has a lobby, and zero otherwise, y_j is the supply and the import, α_L the fraction of the population in the lobby, and a the weight the government places on public welfare. The Nash equilibrium of tariff includes two components (right-hand side of eqn. 9.6): internalization of invasion damage and disguised protectionism. Evidently, the tariff always exceeds the marginal

invasion damage of the import for industry having an organized lobby ($I_j = 1$). Political contributions induce a gap between equilibrium and optimal tariff, disguised as trade protectionism. The two components of the tariff cannot be distinguished without knowledge of either the social damage vector δ or the weight the government places on public welfare a. Without nations giving international bodies the right to decide how much value should be placed on the public goods endangered by invasive species, such disguised protectionism cannot be avoided.

Margolis and colleagues' (2005) model unveiled the hidden linkage of biological invasion to political games and global trade. Although free trade allows countries to specialize in producing goods and thus generates more welfare for trading nations, trade with alien hitch-hikers could lead to negative impacts on the environment and thus public welfare should be restricted. Interest groups can reduce the potential damages of invasion by affecting a policy-maker's choice about the tariff. However, private political donations can lead to trade protectionism by causing the policy-maker to select a tariff that exceeds the level for optimal social welfare. Invasion damages create the gap between optimal and realized tariff, which defines disguised protectionism. Good intentions for reducing invasion risks can be leveraged by lobby groups into policies for trade protectionism. With the continuous negotiations of bilateral and multilateral trade—for example, under the framework of World Trade Organization—political parties and lobby groups could further adjust the government's will regarding mitigating the risks and damages of biological invasions through posing a tariff so to affect not only trade but also public welfare and the chance for the governing party to remain in power.

9.8 Bioeconomics of optimal control

Options for managing biological invasions are determined by the biology of the invasive species, features of the invaded ecosystem and budget constraints, and the management decision on control effort based on the damage of invasive species to the recipient social-ecological system. Management interventions to address biological invasions, such as prevention, monitoring, containment, and eradication, are costly and can often be enforced only for a short period due to limited budget and political will. Efficient modelling of invasion management needs to merge considerations from both ecological dynamics and economic behaviour. This integrated approach is termed bioeconomic modelling (Clark 1990) and can account for the impact of current control choices on future conditions and thus the intertemporal trade-offs to determine optimal control strategies (Epanchin-Niell and Hastings 2010). The typical framework for a bioeconomic model is optimal control theory, a well-developed mathematical theory for modifying the behaviour of dynamical systems through control of system inputs.

To formulate an optimal control problem, we first need to describe the invasion dynamics to be controlled. It is often a differential equation describing the population dynamics or the spatial spread of the invasive species (see chapters in Part I):

$$\dot{n} = f(n, u, t) \tag{9.7}$$

where $n(t)$ is the state variable and in this case represents the population size at time t; $u(t)$ is the control variable at time t, describing the amount of management intervention in managing the focal species. For instance, we can normally specify the function of population dynamics as $f = rn(1 - n/K) - \beta u$, where r and K are intrinsic population growth rate and carrying capacity, as in a logistic model, and β the cost-effectiveness of unit control cost (effort). Secondly, we need to define the constraints to the system. These constraints bound the range of population size n, control effort u, and duration of control (T).

Finally, we need to specify a performance criterion or an objective function J to be maximized (e.g. welfare or positive impact from invasion) or minimized (e.g. damage or negative impact from invasion). In most case, the performance measure is expressed as an integral over the control duration:

$$J = \int_0^T L(n, u, t) dt + h(n_T, T) \tag{9.8}$$

where h is the terminal cost function; if h is independent of n_T, it is called the Mayer problem; if h

is a zero function, it is called the Lagrange problem. Function L depicts the welfare or the damage at time t given the population size $n(t)$ and control effort $u(t)$. When costs or benefits are summed or compared across time, monetary values must be converted into equivalent units (their present value) through discounting (Clark 1990): $L = \exp(-d \cdot t)l(n, u)$, where d is the discount rate. Current benefits are valued more than future benefits because the former could be invested to increase their value through time and can provide more opportunities than if those benefits were delayed. This means that the effect of a control action will be worth more if applied earlier. Similarly, early action and prevention are often favoured in bioeconomic models of biological invasions.

The cost of applying an additional unit of control is an important measure in economics, and is known as the *marginal cost* (defined as the derivative (slope) of the cost function with respect to the amount of control; Clark 1990; Epanchin-Niell and Hastings 2010). When the cost per unit of control is independent of the total amount of control, the total cost will increase in proportion to the amount of control (i.e. a linear cost function with a cost marginal cost). If controlling invasions in a larger area is more expensive per unit area than in a small area, the marginal cost increases with the amount of control. This is often due to the increasing difficulty in accessing a limited resource (Eiswerth and Johnson 2002). This is especially true in invasion management as the last few individuals or propagules (e.g. during the ongoing attempt to eradicate Indian house crows in Cape Town) are difficult to locate and target, incurring much higher costs than when the invasive population size is large. When invasion management requires large upfront costs (e.g. facilities for rearing biological control agents), the marginal control cost will normally decline with the total amount of control due to the economies of scale (Clout and Williams 2009). Similarly, the marginal benefit of control can be defined as the monetary damage avoided by applying an additional unit of control (Epanchin-Niell and Hastings 2010).

In a typical problem of minimizing total management costs and invasion damages, the optimal control strategy is to set the control effort to the level when the marginal cost of control equals the marginal benefit (Fig. 9.10a; Clark 1990; Pindyck and Rubinfeld 2001). In other words, the optimal strategy is to keep increasing the control effort until the cost of applying the next unit of control becomes greater than the damages it would prevent (Epanchin-Niell and Hastings 2010). For dynamic optimization, the effect of current control choices on future costs and damages needs to be considered. In particular, the optimal strategy is to increase control effort until the point when the marginal cost of control equals the sum of current and future damages prevented from the additional unit of control (Fig. 9.10b; Olson 2006; Epanchin-Niell and Hastings 2010).

To formulate a typical bioeconomic model for an invasive species problem we need to clarify four components that are required to frame the problem of optimal invasion management: population dynamics of the focal invasive species; costs of control efforts; cost-effectiveness of controlling; and a monetary measure of invasion damages (impact). Firstly, a good understanding of invasive population demography and spread dynamics is essential for deriving optimal management strategies from bioeconomic models (Saphores and Shogren 2005; Tobin *et al.* 2007; Yokomizo *et al.* 2009). Vital rates of population demography and habitat suitability are minimum requirements for formulating population dynamics (see chapters in Part I). Secondly, the cost of unit control effort is often already recorded in monetary values. For instance, labour costs are often used as the proxy for control costs and normally differ between actions of management interventions (e.g. the cost of containment is often lower than the cost of eradication). However, control costs can be dependent on the amount and intensity of control (Sharov and Liebhold 1998; Burnett *et al.* 2007). Thirdly, the intended actions often cannot have a perfect effect (e.g. containment of grass invasions might face low efficiency; Kettenring and Adams 2011), and the cost-effectiveness of controlling can be assessed from randomized experimental studies (Lindenmayer *et al.* 2015) or using counterfactual estimates in non-experimental contexts (Imbens and Wooldridge 2009; McConnachie *et al.* 2016).

The last component for formulating a bioeconomic model, the monetary measure of invasion damages, is arguably the most important but is also the most difficult to quantify (Olson 2006; McIntosh *et al.* 2009); this is often used together with the control cost as the target function to minimize or maximize. These damages (or oppositely the benefits) can be estimated as invasion impact (sec-

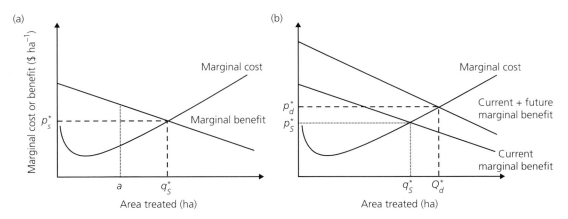

Fig. 9.10 Illustration of marginal costs and benefits (prevented damages) and optimal control levels for (a) static and (b) dynamic optimization management. The figure only illustrates the scenario of decreasing marginal benefits for controlling additional hectares of an invasion. Optimal strategy calls for controlling q_s^* hectares in static management when the cost of controlling the last hectare equals p_s^* dollars per hectare. Choosing to treat only a hectares would be inefficient, as the benefit of treating an additional hectare is higher than the cost of treatment. In dynamic management, future benefits of current control are considered, leading to the optimal control with a greater effort Q_d^* and higher cost per capita P_d^*. It is worth noting that marginal future benefits depend on future control choices and cannot be represented as a simple function. From Epanchin-Niell and Hastings (2010). Reprinted with permission from John Wiley & Sons.

tion 9.4; e.g. Blackburn *et al.* 2014) or opportunity cost such as the loss of cumulative public welfare due to invasion (section 9.9; e.g. Leung *et al.* 2002). Some bioeconomic models of invasion management have attempted to incorporate the monetary estimates of invasion damages explicitly (e.g. Sharov and Liebhold 1998; Eiswerth and Johnson 2002; Potapov *et al.* 2007; Wilen 2007; Olson and Roy 2008; Potapov and Lewis 2008). Other studies have ignored the invasion damages but focus only on minimizing the control costs of invasions (e.g. Menz *et al.* 1980; Higgins *et al.* 2000; Taylor and Hastings 2004) and thus should not be considered proper bioeconomic models. In some cases, the amount of control and the time to achieve management goal have also been used as the target function (Hof 1998; Wadsworth *et al.* 2000; Grevstad 2005). Moreover, all target functions in invasion management are normally optimized, subject to a budget constraint.

As examples of the optimal control theory, Saphores and Shogren (2005) examined two prob-lems associated with the management of alien pests using Bellman's optimality principle. The optimal policy is to first wait and do nothing, and then to apply a control measure as soon as the alien pest density reaches a threshold that was decided by minimizing the present value of expected damages plus control costs. They also assessed the optimal expenditures for monitoring versus controlling, with another threshold identified earlier which the management action should switch from monitoring to controlling. In practice, a time lag exists between detection and management decision (Bar-Ilan and Strange 1996; Mehta *et al.* 2007) and should be added to the formulation. Many studies have also considered multiple invasion management stages, especially between prevention and control stages (e.g., Leung *et al.* 2005; Olson and Roy 2005); these largely favour prevention but there are exceptions. Although the solution to optimal control is highly context- and species-dependent, several rules of thumb do exist (Box 9.3; Epanchin-Niell and Hastings 2010):

Box 9.3 Bioeconomic rules for invasion control

Budget: Budgetary constraints can hinder the control of invasive species and increase the total costs and damages from an invasion by preventing the optimal levels of control from being applied (Taylor and Hastings 2004; Odom *et al.* 2005; Cacho *et al.* 2008).

Connectivity: When the chances of re-invasion are high, the benefits of eradication are reduced (Simberloff 2003). Nonetheless, repeated eradication may still be warranted for highly damaging invasive species (Liebhold and Tobin 2008). With increasing landscape subdivision, individual managers are likely to conduct less control, because the likelihood of re-invasion from uncontrolled neighbouring invasions can be high and because their perceived landscape size is smaller (Wilen 2007; Epanchin-Niell *et al.* 2009).

Cost: Higher control costs lead to lower optimal levels of control. If the control costs per unit do not change with the total area controlled, a once-off eradication (if feasible) should be conducted to prevent further spread. If the per unit control costs increase with the total area controlled, slower eradication may be justified (Sharov and Liebhold 1998). If the control costs per unit decrease with the total area invaded, eradication is less likely to be optimal as the costs of eliminating the last few individuals could outrun the costs and damages from long-term control and damages (Burnett *et al.* 2007). If marginal control costs are high relative to marginal damages when populations are low, eradication is not optimal and delaying control until the invasion has progressed could be optimal (Burnett *et al.* 2007; Olson and Roy 2008).

Damage: Species that cause high levels of damage per unit of invaded area demand higher control efforts (Sharov and Liebhold 1998; Wilen 2007; Carrasco *et al.* 2009; Lewis *et al.* 2009). Under the assumption of constant marginal damages, optimal levels of control effort tend to decrease as the amount of uninvaded landscape declines (Sharov and Liebhold 1998; Wilen 2007).

Delay: Delaying control efforts also increases the total costs and damages from an invasion by reducing control options and increasing eradication costs (Smith *et al.* 1999; Higgins *et al.* 2000; Taylor and Hastings 2004), shifting optimal control goals towards less aggressive control policies (e.g. from eradication to slowing; Sharov and Liebhold 1998; Wilen 2007).

Demography (growth and spread): High rates of spread cause more rapid accumulation of damages, increasing the benefits of control. Consequently, higher rates of spread can increase the likelihood that eradication will be an optimal policy (Sharov and Liebhold 1998; Olson and Roy 2008). If the growth rate increases with invasion size, control has the additional benefit of reducing future growth rate (e.g. Eiswerth and Johnson 2002; Burnett *et al.* 2007). The existence of a density threshold below which population growth is negative can increase the prospects of eradication as an optimal control goal (Liebhold and Bascompte 2003). The optimal control goal can shift from eradication to slowing the spread to abandonment with increasing size of an invasion (Sharov and Liebhold 1998). This is because larger invasions cost more to eradicate, which decreases the returns from eradication and reduces the likelihood of eradication

being the optimal policy (Olson and Roy 2002; Carrasco *et al.* 2009).

Discount: A high discount rate shifts optimal control policies towards lower control effort (e.g. Sharov and Liebhold 1998; Olson and Roy 2002; Wilen 2007; Lewis *et al.* 2009), because high discount rates de-emphasize future damages, thereby reducing the perceived benefits of control. Eradication of small invasions is optimal if the growth rate is larger than the discount rate (Olson and Roy 2002). Shorter time horizons reduce the perceived benefits of control and thus control effort (Blackwood *et al.* 2010; Lewis *et al.* 2009).

Efficiency: High control efficiencies are necessary to eradicate species with compensatory growth after control (Wadsworth *et al.* 2000), whereas less efficient control measures might be adequate for species susceptible to Allee effects (Liebhold and Bascompte 2003; Liebhold and Tobin 2008).

9.9 Economic valuation

As mentioned earlier, the most important component for accurately identifying the optimal control strategy for invasion management is the economic valuation of monetary damages caused by biological invasions (Olson 2006). Damages can be stochastic (Deen *et al.* 1993) and dependent on the size of invasion populations (Cacho *et al.* 2008). Damages are potentially much more difficult to estimate than costs. Damages include direct and indirect economic impacts such as impacts on ecosystem services (e.g. the productivity of the land, water availability, native species diversity, recreation; van Wilgen *et al.* 2008). Some impacts may be relatively straightforward to quantify in monetary terms, such as the damages caused by zebra mussels to power plants due to clogged water pipes (Leung *et al.* 2002), or reduced water production from catchments due to dense stands of invasive trees (Le Maitre *et al.* 1996). For most other damages, such as those caused by altered fire regimes or diverse changes to native communities, determining monetary values is much more challenging (McIntosh *et al.* 2009). Even determining the ecological impact of invasions on natural communities is difficult (Tilman 1999; Stohlgren and Schnase

2006; Kull *et al.* 2011). For instance, valuing damages that have no observable monetary value, such as biodiversity loss, can be particularly challenging (McIntosh *et al.* 2009).

Economic impacts of an invasive population are density-dependent (Parker *et al.* 1999), and the density-impact relationships follow diverse forms in the literature (Standish *et al.* 2001; Alvarez and Cushman 2002; Nava-Camberos *et al.* 2001; Parsons *et al.* 2005; Hester *et al.* 2006; Brown *et al.* 2007) and can differ for the same species in different habitats (Robinson *et al.* 2005). The form of the density-impact relationship can affect the optimal management policy. Yokomizo and colleagues (2009) proposed three basic non-linear shapes (Fig. 9.11), contrasting with the linear curve which shows economic impact as directly proportional to population density. Seed contamination in grain crop trade by wild radish, *Raphanus raphanistrum*, has a density-impact curve with a low threshold (I in Fig. 9.11; Panetta *et al.* 1988), whereas the weed Paterson's curse, *Echium plantagineum*, follows a high-threshold curve (IV in Fig. 9.11; Seaman *et al.* 1989). Trade restrictions can be applied to grain crops once a certain threshold level of weed-seed contaminant is reached (Davis *et al.* 1999). For low-threshold species, the optimal management strategy is to do nothing when facing large environmental fluctuations, rapid density-dependent recruitment, and the low cost of impact (relative to control costs), but the optimal strategy switches to maximal management investment when facing low population densities, slow recruitment, and the high cost of impact. For high-threshold species, large investments in management are not necessary unless they are easy to eradicate and incur large cost of impact. The risk of using the wrong density-impact relationship is from over-investment in unnecessary management (Yokomizo et al. 2009).

Economists have provided insights to evaluate the monetary damages associated with different states of natural systems including caused by invasion-driven loss of ecosystem service (Freeman 1993; Perrings et al. 2000). For instance, damages can be estimated as lost revenues caused by an invasion, such as from crop or forage loss (e.g. Eiswerth and van Kooten 2007). The damage function often assumes that the economic system and the ecological system affect each other in a one-sided way (Daily 1997; Vitousek et al. 1997) but not reciprocally, thus ignoring the possibility of co-evolution between human and natural systems (Crocker and Tschirhart 1992). As changes in an ecosystem (e.g. due to biological invasion) could affect human behaviour and subsequently their actions to the altered ecosystems, the coupling of both systems should be considered in economic valuation and

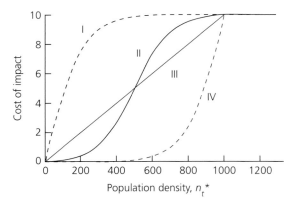

Fig. 9.11 Four potential relationships between the economic cost of impact and population density: I, low-threshold curve; II, S-shaped curve; III, linear curve; IV, high-threshold curve. From Yokomizo et al. (2009). Reprinted with permission from John Wiley & Sons.

decision making (Swallow 1996; Sohngen and Mendelsohn 1998; Brown and Layton 2001).

The coupling of socio-economic and ecological processes in invaded ecosystems can be extremely complicated (e.g. the socio-economic linkages of Australian acacias in South Africa; Kull et al. 2011; Richardson et al. 2011). Attention should be given to those processes that are especially essential for reinforcing feedback loops (Gaertner et al. 2014). Settle and colleagues (2002) explored the invasion of alien lake trout, *Salvelinus namaycush*, in Yellowstone Lake, one of the last inland fisheries for the native Yellowstone cutthroat trout, *Oncorhynchus clarki bouvieri*. Invasion damage to the Yellowstone cutthroat trout population could lead to a trophic cascade that would affect many of its charismatic predators such as grizzly bears, and ultimately affect the satisfaction of fisherman, causing an ecosystem externality (Crocker and Tschihart 1992). Settle and colleagues (2002) found that considering eco-economic coupling can greatly affect the model prediction. Without coupling, fishermen continue to fish as before and pose constant pressure on the cutthroat. With coupling, fishermen will eventually divert their interests to other attractions, affecting the park income. They suggested that the gill-netting policy for controlling lake trout was inefficient.

Finnoff and colleagues (2005) consider two feedback loops for controlling zebra mussels, *Dreissena polymorpha*, in a Midwest lake in the United States: the link between the biological system and firms (primarily power plants and water-treatment facilities), and the link between the manager and the firm, with the beliefs of the decision maker regarding invasions central to both loops. Without the first link, the firm ignores changes in the biological system from the invasion, and consequently operates suboptimally, causing losses of opportunity cost from production shortages or surplus. Without the second link, the manager ignores changes in the firm production from the invasion. Including feedbacks drastically affects the behaviours of the firm and its manager and thus the biological and economic consequences of involved systems. Ultimately, the coupling of economic and ecological processes allow Finnoff and colleagues (2005) to formulate the issue using stochastic dynamic programming with discounted cumulative welfare as target function (also see Leung *et al.* 2002).

All bioeconomic models with integrated ecological and economic processes (see sections 9.7 and 9.10) require economic valuation of the relative damages and benefits of invasive species, as well as management costs, usually based on the market values. This is typically estimated as a sum of replacement costs (price multiplied by quantity) in the ecological literature (e.g. Pimentel *et al.* 2000). Many economists have questioned these estimates because the calculation of replacement costs differs fundamentally from using the standard surplus welfare measures developed in economics for evaluating the opportunity cost of losing ecosystem services (Fig. 9.12; McIntosh *et al.* 2009).

McIntosh and colleagues (2007) considered an economic valuation model to establish values for reducing risks from invasive species. They assume that the invasion is inevitable and the time when invasion damages occur is known. The model considers a rational individual who is facing an intertemporal budget constraint (endowed wealth) and tries to maximize his/her utility over pre- and post-invaded systems. The objective of the model is to determine the individual's willingness to pay to delay the invasion. Before the invasion, the individual's utility is a constant generated from the average environmental quality. After the invasion, utility equals consumption minus market damage from the invasion and the lower level of environmental quality. The individual is assumed to confront a pure consumption loans market at a given interest rate, without debt before death. The value of delaying a harmful invasion by a marginal amount of time is considered to be equal to the willingness to pay to stay in the non-invaded state for the increment of time. As an example, McIntosh and colleagues (2010) explored the economic value of delaying inevitable environmental damage from aquatic invasive species in inland water bodies of the United States. Surveys were divided based on which groups of aquatic invasive species were invading: fishes, molluscs, crustaceans, aquatic plants, or all these simultaneously. Donations were elicited from respondents to delay given impacts and severities of aquatic invasive species in regional lakes and rivers for three scenarios: (1) delaying low impacts for one year; (2) delaying high impacts for one year; and (3) delaying high impacts for ten years.

Assuming that all aquatic species groups invade simultaneously, the average household willingness to pay was a one-time payment, in US dollars, of $34 to delay zero to low impacts for one year, $48 to delay low to high impacts for one year, and $218 to delay low to high impacts for a decade. This means that the willingness to pay of all US households are $3.97 billion, $5.55 billion, and $25.4 billion, respectively, which is much higher than US$0.4 billion invested by the US Federal Government in 2006 for prevention and early detection/rapid response of all invasive species. Two key results emerge from the rational economic valuation tests (McIntosh *et al.* 2007, 2010). Firstly, delaying biological invasion and its associated impacts, even for only a short period, is valuable to the individuals who suffer utility or monetary losses from invasive species. The donation depends on the amount of utility loss

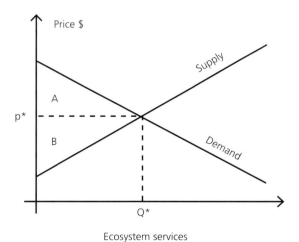

Fig. 9.12 Supply and demand for ecosystem services. Area A is consumer surplus, B producer surplus; p* and Q* are market equilibrium price and quantity. The demand falls but the supply increases with the increase of the price. The demand reflects the buyer's maximum willingness to pay, and the supply reflects the seller's minimum willingness to accept. Areas A (consumer surplus) and B (producer surplus) together represent the total economic benefit of the ecosystem services, which defines the opportunity cost to society should this ecosystem service be disrupted by biological invasion. The area within the dashed lines (p*×Q*) represents the replacement cost of the service. Redrawn from McIntosh *et al.* (2009).

from good to bad environmental states due to invasions. Individuals suffer higher loss donate more, and even those individuals who suffer little from the invasion are willing to donate if invasions can be delayed for a longer period of time. Secondly, individuals, especially those wealthy, well informed, and with properties on waterways, donate more for a longer delay of invasion or for delaying more invasive species. Evidently, proper economic evaluation sets a feasible range for management actions and an important component for bioeconomic modelling.

9.10 Dynamic programming

The optimal control problems in invasion management are often formulated as a continuous system under control. In practice, both the state variable and control variable are usually discrete; for example, annual count of population size and annual budget for invasion management. In this sense, the optimal control problems can be transformed into a dynamic programming problem. Dynamic programming is a formal mathematical approach to decision theory. It implements mathematical programming algorithms for solving complex, stochastic, state-dependent optimization problems that are, in general, impossible to analyse and comprehend using the purely analytic framework of continuous optimization. Dynamic programming has a long history of use in economics and engineering and, with recent advances in computing power, has been increasingly used in addressing population-level optimization problems (Clark 1990; Possingham 1996). In particular, stochastic dynamic programming (SDP) is a powerful tool for comparing alternative management plans, framed often based on both environmental and economic costs and benefits (Walters and Hilborn 1978; Kennedy 1981). It has been widely used in biological conservation (Possingham 1997; Bode *et al.* 2009) and has also been applied in pest and weed control (Shoemaker 1982; Plant and Mangel 1987; Pandey and Medd 1991), sustainable fisheries (Hilborn and Walters 1992), and wildlife management (Reed 1974; Anderson 1975; Milner-Gulland 1997; Possingham and Tuck 1998).

SDP can be applied in invasion management in the following generic framework. In particular, the state variable that describing the invasion dynamics of alien biomass or abundance can be now divided into multiple consecutive steps (e.g. annual counts of animals):

$$n_{t+1} = f\left(n_t, u_t, t, \varepsilon_t\right) \tag{9.9}$$

where ε_t is an exogenous noise to the invasion dynamics. The status n_t can be further categorized into a finite set of s states in SDP ($n_t = 1, 2, \ldots, s$), and the transition between states can be governed by a probability transition matrix, with the probability of the invasion in state n_t transitioning to state n_{t+1} under the management control of u given by

$$p_{ij}(u, t) = prob\left\{n_{t+1} = j \mid n_t = i, u_t = u\right\} \tag{9.10}$$

The stochastic dynamic programming is often solved using backward induction in a recursive fashion:

$$J(n_t, t) = \max_{u_t}\left\{L(n_t, u_t, t) + \sum_{i=1}^{s} p_{n_t, i}(u_t, t) J(i, t+1)\right\}$$
$$\tag{9.11}$$

where the final performance function $J(n_T, T)$ needs to be specified; other symbols have similar meanings as in the optimal control problem.

Related to biological invasions, Shea and Possingham (2000) applied SDP to determine the optimal release strategies for biological control agents to maximize the establishment of agent populations in up to P sites by some terminal time T, with only N insects available for annual release. Each site can only be in one of three possible states: empty, insecure, or established. At an insecure site, the agent is present but the population is small and has a relatively high probability of extinction. At an established site, the agent is well established and is relatively unlikely to become extinct. Individuals from established sites can be used as the secondary source to augment those from the main breeding population for release. Three processes are assumed to affect the transitions between these three states. Firstly, an inoculum with x number of agents released has a probability $p(x)$ of establishment, and becomes an insecure population. Secondly, all the populations can become extinct, with established populations having a lower probability of local extinction, e_s, than insecure populations, e_n. Finally, insecure populations can become established with the probability of g. Shea and Possingham (2000) defined three possible management strategies: many small releases; few large releases; or mixed releases (with the number of 'large' sites randomly chosen first). Extra insects are collected from well-established sites as an augmentation to the routinely released agents from rearing. The weighted value of having s established and n insecure sites is considered the indicator of successful release, $J = (s + n/4)/P$. Finding the optimal solution involves backwards iteration in time (Mangel and Clark 1988), iterated back from T, through $T–1$, $T–2$, to find the optimal strategy set at $t = 0$. Thus for each backward transition, the value function, $J_{t-1}(s,n)$, at time $t–1$ is determined by that at time t as the following:

$$J_{t-1}(s,n) = \max_{K=1,2,3} \sum_{i=0}^{P} \sum_{j=0}^{P-i} \mathbf{T_k}\big((s,n)(i,j)\big)J_t(i,j) \quad (9.12)$$

where T_K is the transition matrix when you use strategy K. The optimal strategy depends on the success of previous releases and the relationship between inoculum size and probability of establishment. The optimal strategy is intuitive: larger releases when there are few established sites (to augment the numbers available for release in future years) or when the probability of establishment of small releases is low; and more, smaller releases at other times.

Leung and colleagues (2002) used SDP as the mathematical basis for their bioeconomic model of invasion risk analysis. In particular, they used a logistic growth model plus uncertainty (ε) for the invasion dynamics, $dN/dt = rN(1 - N/k) + \varepsilon$. The state variable θ is then the discrete level of population abundance. The probability of invasion (I') per time interval was considered as a function of base rate invasion probability (I) and prevention effort (E), $I' = I \cdot \exp(-\lambda E)$. Importantly, Leung and colleagues (2002) introduced welfare as the performance target. Welfare (w) for any given state is a function of production (Q), price per unit of production (p) and the cost (C) per unit of labour (L),

capital (K), prevention (E), and control (X): $w = pQ - CLL - CKK - CEE - CXX$. Production ($Q$) followed an economic Cobb–Douglas functional form (Archer and Shogren 1996), relating labour, capital investment, and damage due to the pest (G). The welfare (w) of each state (θ) was then used in the SDP:

$$\max_{X,E} W_{\theta,t} = w_{\theta,X,E} + \sum P_{\theta,X,E,i} W_{i,t+1} \quad (9.13)$$

where W was the cumulative welfare from the end time horizon (T) to the current time (t), and P the probability of moving from state θ to state i, given strategies X and E (chosen to maximize W). The cumulative welfare W was also calculated by moving backwards from $t = T$ to $t = 1$. Thus, for each state at each time interval, the optimal strategies and future trajectories are known. Leung and colleagues (2002) applied their model to a real-world example with zebra mussels causing damages to facilities in a power plant. They compared a lake that was initially uninvaded with an invaded lake, simulated over a long time frame of 25 years and a shorter one of 5 years. In each of the scenarios (Fig. 9.13), they considered the consequences for cumulative welfare of alternative strategies (Fig. 9.13a–c) and examined the average optimal welfare and the associated optimal expenditures on control (Fig. 9.13d–f), labour, capital, and prevention (Fig. 9.13g–i) over time, as well as the invasion probabilities associated with optimal management strategies for 5 and 25 year durations (Fig. 9.13j–l).

Hastings and colleagues (2006) recently showed that optimal control strategies for the removal of smooth cordgrass, *Spartina alterniflora*, can be simply determined using linear programming. This has three implications. Firstly, because the optimal solution to a linear programming problem lies at a vertex of the constraints, an optimal control strategy in any given year will always be to invest the maximum effort possible into the removal of a single stage class, only investing resources in other classes if the first class has been completely removed. Secondly, linear programming is easier to solve than SDP. This makes linear programming a useful alternative for optimal invasion management. Thirdly, linear programming, albeit simpler than SDP, can still provide valuable insights on why certain con-

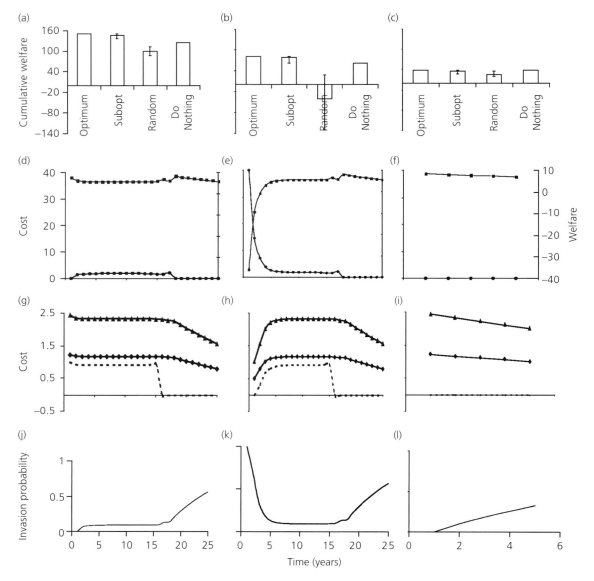

Fig. 9.13 An analysis of a hypothetical invasion. Each datum was a projected value at each time interval, weighted by the probability of being in a state, summed across all states. The panels show the projected cumulative welfares (a–c), optimal welfare (squares), and control expenditure (circles (d–f)), labour (triangles), capital (diamonds), and prevention expenditure (dashed line (g–i)), and the invasion probability that society should accept for optimal welfare (j–l). 'Optimal' used optimal strategies, 'subopt' used random strategies during one time interval, 'random' used random strategies at all intervals, and 'do nothing' spent nothing. The error bars represent one standard deviation. Shown for 25 year uninvaded (a, d, g, and j), 25 year invaded (b, e, h, and k) and 5 year uninvaded (c, f, l, and l) time horizons. From Leung *et al.* (2002). By permission of the Royal Society.

trol strategies are more efficient than others, helping us to comprehend strategies developed from more complex models with greater biological realism and system specificity (Buckley *et al.* 2003).

When considering more complicated scenarios with age structures or multiple interacting species, genetic algorithms could be used. Such algorithms consider a cohort of possible control strategies under

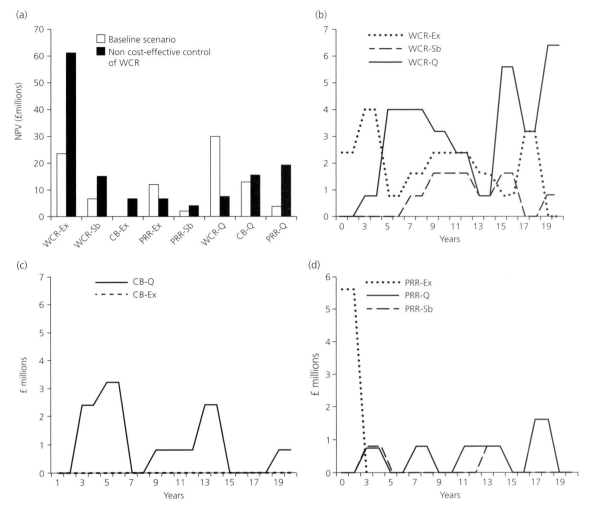

Fig. 9.14 (a) Optimal allocation of management resources for the exclusion, detection, and control of Colorado potato beetle (CB), bacterium for potato ring rot (PRR), and western corn rootworm (WCR) in the UK. In the 'non-cost-effective control of WCR' scenario, the unit cost of control of WCR was increased ten times. Ex: exclusion; Sb: search before discovery; Q: control (removal and surveillance after discovery). (b), (c), and (d) expenditure in the optimal control paths on exclusion, detection, and control of WCR, CB, and PRR respectively. For details see Carrasco *et al.* (2010). With permission from Elsevier.

evolutionary-like processes (e.g. mutation, cross-over, and selection). For instance, Taylor and Hastings (2004) used a genetic algorithm to decide the optimal allocation of effort between removal of high-density and low-density plants of *Spartina alterniflora*. Carrasco and colleagues (2010) used a genetic algorithm to explore the optimal management options under a common budget for controlling multiple invasive species, including the invasive western

corn rootworm (*Diabrotica virgifera*), Colorado potato beetle (*Leptinotarsa decemlineata*), and the bacterium *Clavibacter michiganensis sepedonicus* responsible for the disease potato ring rot in the United Kingdom (Fig. 9.14). Their results show that government agencies should allocate less exclusion and more control resources to introduced species characterized by Allee effects, low rate of satellite colonies generation, and that present low propagule pressure; that is, prioritizing introduced species that have a high invasion debt can increase management efficiency. Importantly, management measures based on the risk analysis of a single invasive species might not be optimal for controlling other invasive species that share a common limited budget. Optimal strategies from single-species, single-control models need to be used with caution. Of course, optimal control and SDP can address similar complex problems equally well (e.g. multiple species with uncertainty; Marten and Moore 2011).

9.11 Spatial optimization

The optimal control strategies derived in previous sections for controlling particular invasive species are spatially implicit (e.g. Parker 2000; Buckley *et al.* 2003; Taylor and Hastings 2004; Hall and Hastings 2007). However, invasion management in practice often needs to consider where the control effort can be best located to maximize its effect. A number of studies using spatially explicit simulation models have clearly shown that where control effort is initially targeted can have a profound effect on the outcome (Moody and Mack 1988; Higgins *et al.* 2000; Wadsworth *et al.* 2000; Hulme 2003). For example, Wadsworth and colleagues (2000) suggest that for species with limited dispersal capacity, the optimal control should target those satellite populations, and such optimal control strategies do not apply for species with the capacity for long-distance dispersal, particularly plant species with seeds dispersed by vectors (e.g. along waterways). Instead, priority should be given to first controlling large core populations so that the primary propagule supply to long-distance dispersal is reduced, thus blocking the further emergence of satellite populations (Hulme 2003). In the following, we consider a few such studies so that readers can grasp the issues and complexity related to the spatial optimization of locating management effort.

Epanchin-Niell and Wilen (2012) develop a spatially explicit model of discrete lattice space that allows for the growth and spread of an invasive species and differential control over both time and space. The initial invasive population is characterized by the size, shape, and location of occupied cells in the lattice space. In particular, Epanchin-Niell and Wilen (2012) assumed that the invasive species causes damages proportional to the area invaded. They consider two control actions: clearing of invaded patches and preventing spread from invaded to uninvaded patches. The cost of preventing establishment of the invasion in a patch depends linearly on the propagule pressure from adjacent invaded cells. Once a cell has been invaded, it remains invaded unless the invasion is removed from the cell at a particular cost. The cost of clearing thus scales linearly with the area cleared. For a cleared cell to remain uninvaded in the following time periods, control must be applied to prevent reinvasion. Once the entire landscape has been cleared, no subsequent control costs are needed. Combinations of these two control actions (clearing and preventing) can be used to eradicate, contain, slow, or redirect the spread of the invasion. They ask, upon discovery, how this spatially explicit invasion should be managed so that the total discounted costs and damages incurred from the invasion can be minimized.

Assuming that the landscape is bounded by unsuitable habitat, Epanchin-Niell and Wilen (2012) found that the initial invasion location of an invasion affects both potential long-term damages and costs of control (Fig. 9.15). Invasions initiated from the centre of the landscape generate higher potential damages as the invasion can spread through the landscape more rapidly, while control costs may be lower for invasions that begin close to landscape boundaries. For optimally controlled invasions with similar characteristics, the net present value of costs and damages is thus higher for central invasions than for invasions that begin distally. The optimal strategy contains the invasion along its frontal edges, directing the spread towards the corner of the landscape, after which the invasion is contained in perpetuity. Using landscape boundaries, this strategy reduces the long-term management costs by 25 per cent.

Besides the size and location of the initial invasion, the shape and contiguity of the invasion can also affect optimal control policies in complex and interesting ways (Epanchin-Niell and Wilen 2012). A critical feature of invasion geometry is the effect of the length of invasion edge on containment costs. For example, containment is more likely to

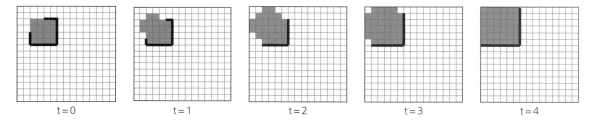

Fig. 9.15 Optimal control of an invasion in a 15 × 15 cell landscape by a 4 × 4 patch of cells near a corner of the landscape. Invaded area is shown in grey. Spread prevention shown by thick black lines. From Epanchin-Niell and Wilen (2012). With permission from Elsevier.

be optimal for a compact invasion than for a similarly located and sized but patchy invasion which obviously has a higher edge to area ratio. Due to the high cost of containing patchy invasions, eradication becomes the optimal strategy for most conditions than it is for compact invasions. Moreover, because landscape shape can affect the likelihood that an invasion will appear close to landscape boundaries, invasions in well-connected landscapes generally result in higher potential damages and warrant more control effort. Irregular shapes of the landscape could create constrictions to spread and thus affect optimal control policies by affecting the costs of containment. Landscape features, such as bottlenecks, can be used strategically to reduce long-term containment costs.

Control costs and damages can vary across the landscape, and this heterogeneity can affect optimal policy (Epanchin-Niell and Wilen 2012). For example, the optimal policy generally requires intense controls to prevent or delay the invasion of high-valued patches. A small initial invasion in a landscape that contains a distant but high-value patch of land could potentially incur large damages from invasion (Fig. 9.16). Intuitively, the optimal policy initially prevents spread in the direction of both high-valued patches and large open spaces and directs spread towards the nearby landscape boundary. As the invasion grows, control is even temporarily abandoned (periods 3–6) until the invasion eventually reaches the edge of the high-value area. Then control is applied to slow the spread and prevent encroachment into high-valued patches. The slowing policy eventually is abandoned, and the containment of invasion eventually turns into the conservation of high-valued patches.

In some cases, spatially explicit and implicit models can produce largely consistent optimal control strategies. For instance, Fedrico and colleagues (2013) considered a simple two-species consumer (rabbits) and resource (grass) system using an individual-based model. The model allows for controlling rabbits using two alternative strategies. For the homogeneous harvesting strategy, rabbits are randomly selected. For the heterogeneous harvesting strategy, different regions can be set to allow or not to allow control. They found that optimal control derived from a simplified non-spatial model is often quite close to the 'best' control option from the much more complex and potentially realistic individual-based model. This means that it is feasible to utilize rule-of-thumb control strategies derived from simple models to manage populations in an optimal manner.

In contrast, other studies have demonstrated a large discrepancy between results of spatially explicit models and management rules of thumb developed using non-spatial models. One such rule of thumb is related to whether priority should be given to the removal of invaders on the range front/edge or the core (Moody and Mack 1988). As mentioned in Epanchin-Niell and Wilen's (2012) work, established invasions can contribute to invasion expansion both through the growth of the main invasion and the creation of new satellite populations. Firstly, greater control may be optimal for smaller, satellite invasions because eradication and containment costs are lower. Moreover, peripheral populations play a more direct role in the spread of a population, particularly if the process of spatial sorting leads to the selection of individuals with strong dispersal ability at the periphery of a popu-

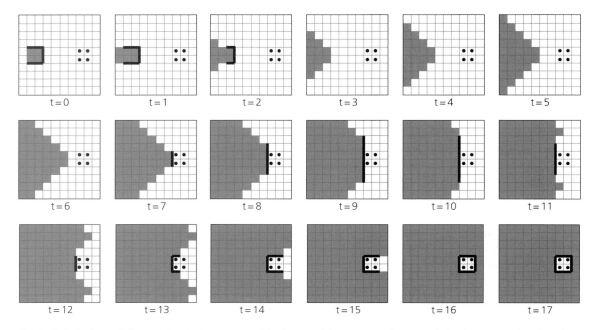

Fig. 9.16 Optimal control of a 2 × 2 invasion in a 15 × 15 cell landscape with heterogeneous damages. A high value 2 × 2 patch, indicated by small black diamonds, lies in front of the initial invasion. Invaded area is shown in grey. Spread prevention shown by thick black lines. From Epanchin-Niell and Wilen (2012). With permission from Elsevier.

lation (Shine *et al.* 2011). Core populations, instead, contribute less to spatial spread despite their major contribution to population growth. Secondly, optimal control for each patch of an invasion depends on the entire invasion and landscape. Optimal solutions for spatially explicit optimization problems generate a far richer set of solution characteristics than work that treats space only implicitly (Epanchin-Niell and Wilen 2012).

This simple rule of thumb of targeting satellite populations before core populations has been especially well supported for slowing the spread of invasive trees (e.g. Watson 1985; Moody and Mack 1988; Doren and Jones 1997; Higgins *et al.* 2000). However, this does not mean that targeting core populations first is always suboptimal. Maxwell and colleagues (2009) found that if core populations are the only source contributing to population growth, targeting such populations can be a better management strategy. Moreover, as core populations typically have the longest residence time and potentially greatest density and per-capita impact, treating core populations first may be the best option for mitigating

impact. For instance, to control the spread of common rhododendron, *Rhododendron ponticum*, in the United Kingdom, the optimal strategy is to target mature reproductive plants first because of the large numbers of seeds they produce (Edwards 2006; Harris *et al.* 2009). Consequently, the location of the source of propagules can affect the optimal control strategy. This suggests that optimal spatial control strategies may differ according to a plant's life history and demographic characteristics, as these will influence the location and densities of seed-producing plants.

On the other hand, budgetary constraints (amount and duration) and management target (e.g. impact mitigation versus slowing spread), as well as management cost and invasion damage, together determine the constraints to the optimal control problem, and such constraints often eventually determine the best optimal strategy. Successful containment is obviously preferable to a failed eradication (Myers *et al.* 2000). Eradication requires a high commitment of management budget and duration. If management effort is not sufficient to remove the core or older populations, targeting edge populations will at least contain the spread and reduce the chance of invasion into high-valued areas (Doren and Jones 1997). Therefore, several competing strategies exist for managing plant invasions, reflecting different constraints made about the processes driving the invasion management. Depending on these constraints and the context of the problem, a management strategy can target or prioritize peripheral sites (Moody and Mack 1988), sites with high density (thus decreasing propagule pressure; Von Holle and Simberloff 2005), and/or populations on highly suitable habitat (Higgins *et al.* 2000).

Habitat heterogeneity and management scales bring more complexity into designing optimal control and prioritization strategies. For instance, Caplat and colleagues (2014) developed a spatial simulation model to test how best to prioritize the control of invasive tree populations spatially when habitat quality varies spatially and how to allocate control effort among two different management units when trees spread from many permanently established source populations (e.g. plantations). Within the local management unit they tested how best to control the spread of an invasive tree from one or two sources. The *selection strategies* reflect different assumptions about the processes driving tree invasions: the importance of peripheral sites ('distance' strategy) (Moody and Mack 1988), propagule pressure ('density' strategy) (Maxwell and Ghersa 1992), and the role of habitat ('habitat' strategy). At the regional scale, Caplat and colleagues (2014) tested how to distribute effort across management units most efficiently. Specifically, they defined three effort *allocation strategies* for allocating effort between units: evenly ('equal' strategy), where each unit gets an equal share; by habitat ('habitat' strategy), where the share is scaled by the amount of suitable habitat in each management unit; and by extent of the invasion (the number of invaded cells in each management unit; 'extent' strategy).

Caplat and colleagues (2014) found that multiple permanent propagule sources (plantations) favour the strategy of targeting peripheral sites (i.e. the 'distance' strategy), and that this strategy largely outperforms other strategies targeting quality habitat or high-density populations. In practical terms, however, this requires that the location of source populations is known and that peripheral sites can be identified and are accessible for treatment. Of the other two strategies, selecting by habitat outperformed selecting by density in most cases (Caplat *et al.* 2014), probably because a tree established in good habitat is likely to disperse offspring into good habitat in the spatially autocorrelated landscape. When spread is fast and habitat is rare, the 'density' strategy performed relatively better. Overall, the best strategy to control tree invasions is to manage sites most distant from the source. If distant trees cannot be accessed, then one should select stands by habitat suitability.

The allocation strategy was not important in the majority of settings, although the equal allocation performed well in most settings (Caplat *et al.* 2014). The fact that the 'equal' strategy rarely outperformed the other strategies confirms the idea that it is a generalist strategy, often outcompeting more sophisticated schemes for allocating control and monitoring effort (see Hui *et al.* 2011). However, with fast spread, some of the effort would be wasted on management units with little population growth when using an equal distribution of effort. In this scenario, the 'habitat' and 'extent' strategies often performed better as large, fast-growing populations are also often located in suitable habitat. Overall, to control tree spread over a region having multiple source populations, effort can be allocated equally to all management units where trees are detected, with two exceptions: (1) where habitat is highly heterogeneous between units, matching management resources to habitat quality can prevent future spread, and (2) for relatively rapid invasions, resources should be allocated to areas with the largest populations.

Evidently, the optimal strategy from spatial optimization of management effort is highly context-dependent. Key issues are assumptions about costs of removal and containment, damages from the invasion, and the description of population dynamics. The population biology approach suggests that, in general, edge populations contribute the most to range expansion and should be removed first to slow down spread, whereas the metapopulation approach suggests prioritizing core populations that supply most of the new propagules to mitigate invasion impact and damages (Taylor and Hastings

2004). The debate on prioritizing for controlling edge populations versus core ones will continue, with increased evidence by exploring more diverse, practical, and efficient spatial control strategies, such as establishing barrier zones (Sharov and Liebhold 1998; Morgan *et al.* 2006; With 2004). More importantly, most spatial optimization models only consider management cost but do not integrate with economic processes (e.g. effect of invasion damages on decisions and discount rate). Consequently, in the next decade, spatial optimization should tie in more closely with related socio-economic processes. This should lead to spatially explicit bioeconomic models and a standard modelling framework for identifying spatially explicit and temporally relevant strategies for invasion management (Moody and Mack 1988; Hof 1998; Wadsworth *et al.* 2000; Grevstad 2005; Whittle *et al.* 2007; Giljohann *et al.* 2011).

References

Allen, R.B., Wright, E.F., MacLeod, C.J., *et al.* (2009) *Designing an Inventory and Monitoring Programme for the Department of Conservation's Natural Heritage Management System*. Landcare Research Contract Report LC0809/153. Prepared for the Department of Conservation, Wellington, New Zealand.

Alvarez, M.E. and Cushman, J.H. (2002) Community-level consequences of a plant invasion: effects on three habitats in coastal California. Ecological Applications, 12, 1434–44.

Anderson, D.R. (1975) Optimal exploitation strategies for an animal population in a Markovian environment: a theory and an example. Ecology, 56, 1281–97.

Archer, D.W. and Shogren, J.F. (1996) Endogenous risk in weed control management. Agricultural Economics, 14, 103–22.

Azaele, S., Cornell, S.J., and Kunin, W.E. (2012) Downscaling species occupancy from coarse spatial scales. Ecological Applications, 22, 1004–14.

Baker, R.H.A., Black, R., Copp, G.H., *et al.* (2008) The UK risk assessment scheme for all non-native species. NeoBiota, 7, 46–57.

Bar-Ilan, A. and Strange, W.C. (1996) Investment lags. American Economic Review, 86, 610–22.

Barwell, L.J., Azaele, S., Kunin, W.E., *et al.* (2014) Can coarse-grain patterns in insect atlas data predict local occupancy? Diversity and Distributions, 20, 895–907.

Bellard, C., Cassey, P., and Blackburn, T.M. (2016) Alien species as a driver of recent extinctions. Biology Letters, 12, 20150623.

Benke, K.K., Steel, J.L., and Weiss, J.E. (2011) Risk assessment models for invasive species: uncertainty in rankings from multi-criteria analysis. Biological Invasions, 13, 239–53.

Blackburn, T.M., Essl, F., Evans, T., *et al.* (2014) A unified classification of alien species based on the magnitude of their environmental impacts. PLoS Biology, 12, e1001850.

Blackburn, T.M., Pyšek, P., Bacher, S., *et al.* (2011) A proposed unified framework for biological invasions. Trends in Ecology and Evolution, 26, 333–9.

Blackwood, J., Hastings, A., and Costello, C. (2010) Cost-effective management of invasive species using linear-quadratic control. Ecological Economics, 69, 519–27.

Bobeldyk, A.M., Rüegg, J., and Lamberti, G.A. (2015) Freshwater hotspots of biological invasion are a function of species-pathway interactions. Hydrobiologia, 746, 363–73.

Bode, M., Rout, T., Hawkins, C., and Wintle, B. (2009) Efficiently locating conservation boundaries: searching for the Tasmanian devil facial tumour disease front. Biological Conservation, 142, 1333–9.

Bottrill, M.C., Joseph, L.N., Carwardine, J., *et al.* (2009) Finite conservation funds mean triage is unavoidable. Trends in Ecology and Evolution, 24, 183–4.

Brotons, L., Thuiller, W., Araújo, M.B., *et al.* (2004) Presence-absence versus presence-only modelling methods for predicting bird habitat suitability. Ecography, 27, 437–48.

Brown, G. and Layton, D. (2001) A market solution for preserving biodiversity: the black rhino. In: Shogren, J.F. and Tschirhart, J. (eds) *Protecting Endangered Species in the United States: Biological Needs, Political Realities, Economic Choices*. New York, NY: Cambridge University Press.

Brown, P.R., Huth, N.I., Banks, P.B., *et al.* (2007) Relationship between abundance of rodents and damage to agricultural crops. Agriculture, Ecosystems and Environment, 120, 405–15.

Buckley, Y.M., Briese, D.T., and Rees, M. (2003) Demography and management of the invasive plant species Hypericum perforatum. I. Using multi-level mixed-effects models for characterizing growth, survival and fecundity in a long-term data set. Journal of Applied Ecology, 40, 481–93.

Buckley, Y.M., Brockerhoff, E., Langer, L., *et al.* (2005) Slowing down a pine invasion despite uncertainty in demography and dispersal. Journal of Applied Ecology, 42, 1020–30.

Burnett, K., Kaiser, B., and Roumasset, J. (2007) Economic lessons from control efforts for an invasive species: *Miconia calvescens* in Hawaii. Journal of Forest Economics, 13, 151–67.

Butchart, S.H., Walpole, M., Collen, B., *et al.* (2010) Global biodiversity: indicators of recent declines. Science, 328, 1164–8.

Cacho, O.J., Wise, R.M., Hester, S.M., *et al.* (2008) Bioeconomic modeling for control of weeds in natural environments. Ecological Economics, 65, 559–68.

Caley, P., Groves, R.H., and Barker, R. (2008) Estimating the invasion success of introduced plants. Diversity and Distributions, 14, 196–203.

Canhos, D.A., Sousa-Baena, M.S., de Souza, S., *et al.* (2015) The importance of biodiversity E-infrastructures for megadiverse countries. PLoS Biology, 13, e1002204.

Caplat, P., Hui, C., Maxwell, B.D., *et al.* (2014) Cross-scale management strategies for optimal control of trees invading from source plantations. Biological Invasions, 16, 677–90.

Carlson, M. and Schmiegelow, F. (2002) Cost-effective sampling design applied to large-scale monitoring of boreal birds. Conservation Ecology, 6, 11.

Carrasco, M. (2009) Cross-modal attention enhances perceived contrast. Proceedings of the National Academy of Sciences of the USA, 106, 22039–40.

Carrasco, L.R., Baker, R., MacLeod, A., *et al.* (2010) Optimal and robust control of invasive alien species spreading in homogeneous landscapes. Journal of the Royal Society Interface, 7, 529–40.

Caswell, H. (2001) *Matrix Population Models.* Chichester: John Wiley & Sons.

Catford, J.A., Vesk, P.A., Richardson, D.M., *et al.* (2012) Quantifying levels of biological invasion: towards the objective classification of invaded and invasible ecosystems. Global Change Biology, 18, 44–62.

Chang, A.L., Grossman, J.D., Spezio, T.S., *et al.* (2009) Tackling aquatic invasions: risks and opportunities for the aquarium fish industry. Biological Invasions, 11, 773–85.

Clark, C.W. (1990) *Mathematical Bioeconomics: The Optimal Management of Renewable Resources*, 2nd edn. New York, NY: Wiley-Interscience.

Clout, M.N. and Williams, P.A. (2009) *Invasive Species Management: A Handbook of Principles and Techniques.* New York, NY: Oxford University Press.

Crocker, T.D. and Tschirhart, J. (1992) Ecosystems, externalities, and economies. Environmental and Resource Economics, 2, 551–67.

Daily, G. (ed.) (1997) *Nature's Services: Societal Dependence on Natural Ecosystems.* Washington, DC: Island Press.

Davis, M.A., Wrage, K.J., Reich, P.B., *et al.* (1999) Survival, growth, and photosynthesis of tree seedlings competing with herbaceous vegetation along a water-light-nitrogen gradient. Plant Ecology, 145, 341–50.

Dawson, J., Oppel, S., Cuthbert, R.J., *et al.* (2015) Prioritizing islands for the eradication of invasive vertebrates in the United Kingdom overseas territories. Conservation Biology, 29, 143–53.

Deen, W., Weersink, A., Turvey, C.G., *et al.* (1993) Weed control decision rules under uncertainty. Review of Agricultural Economics, 15, 39–50.

Dejean, T., Valentini, A., Miquel, C., *et al.* (2012) Improved detection of an alien invasive species through environmental DNA barcoding: the example of the American bullfrog *Lithobates catesbeianus.* Journal of Applied Ecology, 49, 953–9.

de Steiguer, J.E., Liberti, L., Schuler, A., *et al.* (2003) *Multi-Criteria Decision Models for Forestry and Natural Resources Management: An Annotated Bibliography.* Newtown Square, PA: United States Department of Agriculture, Forest Service, Northeastern Research Station. General Technical Report NE-307.

Diez, J.M., Hulme, P.E., and Duncan, R.P. (2012) Using prior information to build probabilistic invasive species risk assessments. Biological Invasions, 14, 681–91.

Doren, R.F. and Jones, D.T. (1997) Plant management in Everglades National Park. In: Simberloff, D., Schmitz, D.C., and Brown, T.C. (eds) *Strangers in Paradise.* Washington, DC: Island Press.

Dryver, A.L. and Thompson, S.K. (2005) Improved unbiased estimators in adaptive cluster sampling. Journal of the Royal Statistical Society: Series B (Statistical Methodology), 67, 157–66.

Dungan, J.L., Perry, J.N., Dale, M.R.T., *et al.* (2002) A balanced view of scale in spatial statistical analysis. Ecography, 25, 626–40.

Edwards, C. (2006) *Managing and Controlling Invasive Rhododendron. Forestry Commission Practice Guide.* Edinburgh: Forest Commission.

Eiswerth, M.E. and Johnson, W.S. (2002) Managing nonindigenous invasive species: insights from dynamic analysis. Environmental and Resource Economics, 23, 319–42.

Eiswerth, M.E. and Van Kooten, G.C. (2007) Dynamic programming and learning models for management of a nonnative species. Canadian Journal of Agricultural Economics/Revue canadienne d'agroeconomie, 55, 485–98.

Elith, J. and Leathwick, J.R. (2009) Species distribution models: ecological explanation and prediction across space and time. Annual Review of Ecology, Evolution, and Systematics, 40, 677.

Epanchin-Niell, R.S. and Hastings, A. (2010) Controlling established invaders: integrating economics and spread dynamics to determine optimal management. Ecology Letters, 13, 528–41.

Epanchin-Niell, R.S., Hufford, M.B., Aslan, C.E., *et al.* (2009) Controlling invasive species in complex social landscapes. Frontiers in Ecology and the Environment, 8, 210–16.

Epanchin-Niell, R.S. and Wilen, J.E. (2012) Optimal spatial control of biological invasions. Journal of Environmental Economics and Management, 63, 260–70.

EPPO (2011) EPPO Reporting Service n. 4, 11. Available at: <http://www.eppo.org/ PUBLICATIONS/reporting/reporting_service.htm>.

Essl, F., Bacher, S., Blackburn, T.M., *et al.* (2015) Crossing frontiers in tackling pathways of biological invasions. Bioscience, 65, 769–82.

Essl, F., Dullinger, S., Rabitsch, W., *et al.* (2011) Socioeconomic legacy yields an invasion debt. Proceedings of the National Academy of Sciences of the USA, 108, 203–7.

Essl, F., Mang, T., and Moser, D. (2012) Ancient and recent alien species in temperate forests: steady state and time lags. Biological Invasions, 14, 1331–42.

Federico, P., Gross, L.J., Lenhart, S., *et al.* (2013) Optimal control in individual-based models: implications from aggregated methods. American Naturalist, 181, 64–77.

Finnoff, D., Shogren, J.F., Leung, B., *et al.* (2005) The importance of bioeconomic feedback in invasive species management. Ecological Economics, 52, 367–81.

Fortin, M.-J. and Dale, M.R.T. (2005) *Spatial Analysis: A Guide for Ecologists*. Cambridge: Cambridge University Press.

Foxcroft, L.C. and Freitag-Ronaldson, S. (2007) Seven decades of institutional learning: managing alien plant invasions in the Kruger National Park, South Africa. Oryx, 41, 160–7.

Foxcroft, L.C. and Richardson, D.M. (2003) Managing alien plant invasions in the Kruger National Park, South Africa. In: Child, L.E., Brock, J.H., Brundu, G., *et al.* (eds) *Plant Invasions: Ecological Threats and Management Solutions*. Leiden: Backhuys Publishers, pp. 385–403.

Foxcroft, L.C., Richardson, D.M., Rouget, M., *et al.* (2009) Patterns of alien plant distribution at multiple spatial scales in a large national park: implications for ecology, management and monitoring. Diversity and Distributions, 15, 367–378.

Foxcroft, L.C., Richardson, D.M., and Wilson, J.R. (2008) Ornamental plants as invasive aliens: problems and solutions in Kruger National Park, South Africa. Environmental Management, 41, 32–51.

Freeman III, A.M. (1993) *The Measurement of Environmental and Resource Values: Theory and Methods*. Washington, DC: Resources for the Future.

Gaertner, M., Biggs, R., Te Beest, M., *et al.* (2014) Invasive plants as drivers of regime shifts: identifying high priority invaders that alter feedback relationships. Diversity and Distributions, 20, 733–44.

Gaston, K.J. (2003) *The Structure and Dynamics of Geographic Ranges*. Oxford: Oxford University Press.

Gaston, K.J. and Fuller, R.A. (2009) The sizes of species' geographic ranges. Journal of Applied Ecology, 46, 1–9.

Gawande, K. (1997) US non-tariff barriers as privately provided public goods. Journal of Public Economics, 64, 61–81.

Geijzendorffer, I.R., Regan, E.C., Pereira, H.M., *et al.* (2015) Bridging the gap between biodiversity data and policy reporting needs: an Essential Biodiversity Variables perspective. Journal of Applied Ecology, 53, 1341–50.

Giljohann, K.M., Hauser, C.E., Williams, N.S., *et al.* (2011) Optimizing invasive species control across space: willow invasion management in the Australian Alps. Journal of Applied Ecology, 48, 1286–94.

Goldberg, P.K. and Maggi, G. (1999) Protection for sale: an empirical investigation. American Economic Review, 89, 1135–55.

Goodman, P.S. (2003) Assessing management effectiveness and setting priorities in protected areas in KwaZulu-Natal. BioScience, 53, 843–50.

Gotelli, N.J. and Graves, G.R. (1996) *Null Models in Ecology*. Washington, DC: Smithsonian Institution Press.

Grevstad, F.S. (2005) Simulating control strategies for a spatially structured weed invasion: Spartina alterniflora (Loisel) in Pacific Coast estuaries. Biological Invasions, 7, 665–77.

Grossman, G.M. and Helpman, E. (1994) Protection for Sale. American Economic Review, 84, 833–50.

Guo, Q. (2011) Counting 'exotics'. NeoBiota, 9, 71.

Hall, R.J. and Hastings, A. (2007) Minimizing invader impacts: striking the right balance between removal and restoration. Journal of Theoretical Biology, 249, 437–44.

Harris, C.M., Park, K.J., Atkinson, R., *et al.* (2009) Invasive species control: incorporating demographic data and seed dispersal into a management model for Rhododendron ponticum. Ecological Informatics, 4, 226–33.

Hartley, S. and Kunin, W.E. (2003) Scale dependency of rarity, extinction risk, and conservation priority. Conservation Biology, 17, 1559–70.

Hastings, A., Hall, R.J., and Taylor, C.M. (2006) A simple approach to optimal control of invasive species. Theoretical Population Biology, 70, 431–5.

Hauser, C.E. and McCarthy, M.A. (2009) Streamlining 'search and destroy': cost-effective surveillance for invasive species management. Ecology Letters, 12, 683–92.

Hawkins, C.L., Bacher, S., Essl, F., *et al.* (2015) Framework and guidelines for implementing the proposed IUCN Environmental Impact Classification for Alien Taxa (EICAT). Diversity and Distributions, 21, 1360–3.

Heikkilä, J. (2011) A review of risk prioritisation schemes of pathogens, pests and weeds: principles and practices. Agricultural and Food Science, 20, 15–28.

Hester, S.M., Sinden, J.A., and Cacho, O.J. (2006) *Weed Invasions in Natural Environments: Toward a Framework for Estimating the Cost of Changes in Output of Ecosystem Services*. Armidale: University of New England. Agricultural and Resource Economics Working Paper No. 2006–2007.

Higgins, S.I., Richardson, D.M., and Cowling, R.M. (2000) Using a dynamic landscape model for planning the management of alien plant invasions. Ecological Applications, 10, 1833–48.

Hilborn, R. and Walters, C.J. (1992) *Quantitative Fisheries Stochastic Assessment: Choice, Dynamics and Uncertainty.* New York, NY: Chapman and Hall.

Hof, J. (1998) Optimizing spatial and dynamic population-based control strategies for invading forest pests. Natural Resource Modelling, 11, 197–216.

Holt, J., Leach, A.W., Knight, J.D., *et al.* (2012) Tools for visualizing and integrating pest risk assessment ratings and uncertainties. EPPO Bulletin, 42, 35–41.

Hui, C. (2009) A Bayesian solution to the modifiable areal unit problem. In: Hassanien, A.-E., Abraham, A., and Herrera, F. (eds) *Foundations of Computational Intelligence*, Vol. 2. Berlin: Springer, pp. 175–96.

Hui, C., Foxcroft, L.C., Richardson, D.M., *et al.* (2011) Defining optimal sampling effort for large-scale monitoring of invasive alien plants: a Bayesian method for estimating abundance and distribution. Journal of Applied Ecology, 48, 768–76.

Hui, C. and McGeoch, M.A. (2014) Zeta diversity as a concept and metric that unifies incidence-based biodiversity patterns. American Naturalist, 184, 684–94.

Hui, C., McGeoch, M.A., Reyers, B., *et al.* (2009) Extrapolating population size from the occupancy–abundance relationship and the scaling pattern of occupancy. Ecological Applications, 19, 2038–48.

Hui, C., McGeoch, M.A., and Warren, M. (2006) A spatially explicit approach to estimating species occupancy and spatial correlation. Journal of Animal Ecology, 75, 140–7.

Hui, C., Richardson, D.M., Visser, V., *et al.* (2014) Macroecology meets invasion ecology: performance of Australian acacias and eucalypts around the world foretold by features of their native ranges. Biological Invasions, 16, 565–76.

Hui, C., Richardson, D.M., Landi, P., *et al.* (2016) Defining invasiveness and invasibility in ecological networks. Biological Invasions, 18, 971–83.

Hui, C., Veldtman, R., and McGeoch, M.A. (2010) Measures, perceptions and scaling patterns of aggregated species distributions. Ecography, 33, 95–102.

Hulme, P.E. (2003) Biological invasions: winning the science battles but losing the conservation war? Oryx, 37, 178–93.

Hulme, P.E. (2009) Trade, transport and trouble: managing invasive species pathways in an era of globalization. Journal of Applied Ecology, 46, 10–18.

Hulme, P.E. (2012) Weed risk assessment: a way forward or a waste of time? Journal of Applied Ecology, 49, 10–19.

Hulme, P.E., Bacher, S., Kenis, M., *et al.* (2008) Grasping at the routes of biological invasions: a framework for integrating pathways into policy. Journal of Applied Ecology, 45, 403–14.

Imbens, G. and Wooldridge, J. (2009) Recent Developments in the Econometrics of Program Evaluation. Journal of Economic Literature, 47, 5–86.

FAO. ISPM No. 11 Pest risk analysis for quarantine pests including analysis of environmental risks and living modified organism. Secretary of the International Plant Protection Convention. Rome, Italy, 2004.

IUCN, I., 2012. Red List of Threatened Species: Version 2011.2.

Jerde, C.L. and Lewis, M.A. (2007) Waiting for invasions: a framework for the arrival of nonindigenous species. American Naturalist, 170, 1–9.

Jeschke, J.M., Bacher, S., Blackburn, T.M., *et al.* (2014) Defining the Impact of Non-Native Species. Conservation Biology, 28, 1188–94.

Jetz, W., McPherson, J.M., and Guralnick, R.P. (2012) Integrating biodiversity distribution knowledge: toward a global map of life. Trends in Ecology and Evolution, 27, 151–9.

Jiménez-Valverde, A., Peterson, A.T., Soberón, J., *et al.* (2011) Use of niche models in invasive species risk assessments. Biological Invasions, 13, 2785–97.

Johnson, S.P. (1993) *The Earth Summit: The United Nations Conference on Environment and Development.* London: Graham and Trotman/Martinus Nijhoff.

Joseph, L.N., Field, S.A., Wilcox, C., *et al.* (2006) Presence-absence versus abundance data for monitoring threatened species. Conservation Biology, 20, 1679–87.

Kaplan, H., van Niekerk, A., Le Roux, J.J., *et al.* (2014) Incorporating risk mapping at multiple spatial scales into eradication management plans. Biological Invasions, 16, 691–703.

Katsanevakis, S. and Roy, H.E. (2015) Alien species related information systems and information management. Management of Biological Invasions, 6, 115–17.

Kenis, M., Bacher, S., Baker, R.H.A., *et al.* (2012) New protocols to assess the environmental impact of pests in the EPPO decision-support scheme for pest risk analysis. EPPO Bulletin, 42, 21–7.

Kennedy, J.O.S. (1981) Applications of dynamic programming to agriculture, forestry and fisheries: review and prognosis. Review of Marketing and Agricultural Economics, 49, 141–73.

Kettenring, K.M. and Adams, C.R. (2011) Lessons learned from invasive plant control experiments: a systematic review and meta-analysis. Journal of Applied Ecology, 48, 970–9.

Koop, A.L. and Horvitz, C.C. (2005) Projection matrix analysis of the demography of an invasive, nonnative shrub (*Ardisia elliptica*). Ecology, 86, 2661–72.

Kowarik, I., Pyšek, P., Prach, K., *et al.* (1995) Time lags in biological invasions with regard to the success and failure of alien species. In: *Plant Invasions: General Aspects and Special Problems*. Workshop held at Kostelec nad Černými lesy, Czech Republic, 16–19 September 1993. SPB Academic Publishing, pp. 15–38.

Krug, R.M., Roura-Pascual, N., and Richardson, D.M. (2009) Prioritising areas for the management of invasive alien plants in the CFR: different strategies, different priorities? South African Journal of Botany, 75, 408–9.

Kueffer, C., Pyšek, P., and Richardson, D.M. (2013) Integrative invasion science: model systems, multi-site studies, focused meta-analysis and invasion syndromes. New Phytologist, 200, 615–33.

Kull, C.A., Shackleton, C.M., Cunningham, P.J., *et al.* (2011) Adoption, use and perception of Australian acacias around the world. Diversity and Distributions, 17, 822–36.

Kumschick S., Bacher, S., and Blackburn, T.M. (2013) What determines the impact of alien birds and mammals in Europe? Biological Invasions, 15, 785–97.

Kumschick, S., Bacher, S., Dawson, W., *et al.* (2012) A conceptual framework for prioritization of invasive alien species for management according to their impact. Neo-Biota, 15, 69.

Kumschick, S., Bacher, S., Evans, T., *et al.* (2015) Comparing impacts of alien plants and animals in Europe using a standard scoring system. Journal of Applied Ecology, 52, 552–61.

Kumschick, S. and Richardson, D.M. (2013) Species-based risk assessments for biological invasions: advances and challenges. Diversity and Distributions, 19, 1095–105.

Larkin, D.J. (2012) Lengths and correlates of lag phases in upper-Midwest plant invasions. Biological Invasions, 14, 827–38.

Latombe, G., Pyšek, P., Jeschke, J.M., *et al.* (2016) A vision for global monitoring of biological invasions. Biological Conservation, doi:10.1016/j.biocon.2016.06.013.

Le Maitre, D.C., van Wilgen, B.W., Chapman, R.A., *et al.* (1996) Invasive plants and water resources in the western Cape Province, South Africa: modelling the consequences of a lack of management. Journal of Applied Ecology, 33, 161–72.

Leprieur, F., Olden, J.D., Lek, S., *et al.* (2009) Contrasting patterns and mechanisms of spatial turnover for native and exotic freshwater fishes in Europe. Journal of Biogeography, 36, 1899–912.

Le Roux, J. and Wieczorek, A.M. (2009) Molecular systematics and population genetics of biological invasions: towards a better understanding of invasive species management. Annals of Applied Biology, 154, 1–17.

Leung, B., Finnoff, D., Shogren, J.F., *et al.* (2005) Managing invasive species: rules of thumb for rapid assessment. Ecological Economics, 55, 24–36.

Leung, B., Lodge, D.M., Finnoff, D., *et al.* (2002) An ounce of prevention or a pound of cure: bioeconomic risk analysis of invasive species. Proceedings of the Royal Society B: Biological Sciences, 269, 2407–13.

Leung, B., Roura-Pascual, N., Bacher, S., *et al.* (2012) TEASIng apart alien species risk assessments: a framework for best practices. Ecology Letters, 15, 1475–93.

Leung, B., Springborn, M.R., Turner, J.A., *et al.* (2014) Pathway-level risk analysis: the net present value of an invasive species policy in the US. Frontiers in Ecology and the Environment, 12, 273–9.

Lewis, M., Potapov, A.B., and Finnoff, D. (2009) Modeling integrated decision-making responses to invasive species. In: Keller, R., Lodge, D., Lewis, M., *et al.* (eds) *Bioeconomics of Invasive Species: Integrating Ecology, Economics, Policy, and Management*. New York, NY: Oxford University Press, pp. 180–204.

Liebhold, A. and Bascompte, J. (2003) The Allee effect, stochastic dynamics and the eradication of alien species. Ecology Letters, 6, 133–40.

Liebhold, A.M. and Tobin, P.C. (2008) Population ecology of insect invasions and their management. Annual Review of Entomology, 53, 387–408.

Lindenmayer, D.B., Wood, J., MacGregor, C., *et al.* (2015) A long-term experimental case study of the ecological effectiveness and cost effectiveness of invasive plant management in achieving conservation goals: Bitou Bush control in Booderee National Park in Eastern Australia. PLoS One, 10, e0128482.

Lockwood, J.L., Hoopes, M.F., and Marchetti, M.P. (2013) *Invasion Ecology*. Chichester: John Wiley & Sons.

Mace, G.M., Collar, N.J., Gaston, K.J., *et al.* (2008) Quantification of extinction risk: IUCN's system for classifying threatened species. Conservation Biology, 22, 1424–42.

MacKenzie, D.I., Nichols, J.D., Hines, J.E., *et al.* (2003) Estimating site occupancy, colonization, and local extinction when a species is detected imperfectly. Ecology, 84, 2200–7.

MacKenzie, D.I., Nichols, J.D., Lachman, G.B., *et al.* (2002) Estimating site occupancy rates when detection probabilities are less than one. Ecology, 83, 2248–55.

Mangel, M. and Clark, C. (1988) *Dynamic Modeling in Behavioral Ecology*. Princeton, NJ: Princeton University Press.

Marco, D.E. and Páez, S.A. (2000) Invasion of *Gleditsia triacanthos* in *Lithraea ternifolia* montane forests of central Argentina. Environmental Management, 26, 409–19.

Marescaux, J. and Van Doninck, K. (2013) Using DNA barcoding to differentiate invasive *Dreissena* species (Mollusca, Bivalvia). ZooKeys, 365, 235.

Margolis, M., Shogren, J.F., and Fischer, C. (2005) How trade politics affect invasive species control. Ecological Economics, 52, 305–13.

Marten, A.L. and Moore, C.C. (2011) An options based bioeconomic model for biological and chemical control of invasive species. Ecological Economics, 70, 2050–61.

Mastrandrea, M.D., Field, C.B., Stocker, T.F., *et al.* (2010) Guidance note for lead authors of the IPCC Fifth Assessment Report on consistent treatment of uncertainties. Geneva: Intergovernmental Panel on Climate Change (IPCC). Available at: <http://www.ipcc.ch/pdf/supporting-material/uncertainty-guidance-note.pdf>.

Maxwell, B.D. and Ghersa, C. (1992) The influence of weed seed dispersion versus the effect of competition on crop yield. Weed Technology, 6, 196–204.

Maxwell, B.D., Lehnhoff, E., and Rew, L.J. (2009) The rationale for monitoring invasive plant populations as a crucial step for management. Invasive Plant Science and Management, 2, 1–9.

McAusland, C. and Costello, C. (2004) Avoiding invasives: trade-related policies for controlling unintentional exotic species introductions. Journal of Environmental Economics and Management, 48, 954–77.

McConnachie, M.M., van Wilgen, B.W., Ferraro P.J., *et al.* (2016) Using counterfactuals to evaluate the cost-effectiveness of controlling biological invasions. Ecological Applications, 26, 475–483.

McGeoch, M.A., Butchart, S.H., Spear, D., *et al.* (2010) Global indicators of biological invasion: species numbers, biodiversity impact and policy responses. Diversity and Distributions, 16, 95–108.

McGeoch, M.A., Genovesi, P., Bellingham, P.J., *et al.* (2016) Prioritizing species, pathways, and sites to achieve conservation targets for biological invasion. Biological Invasions, 18, 299–314.

McGeoch, M.A. and Latombe, G. (2016) Characterizing common and range expanding species. Journal of Biogeography, 43, 217–28.

McGeoch, M.A. and Squires, Z. (2015) An Essential Biodiversity Variable approach to monitoring biological invasions: Guide for Countries. GEO BON Technical Series, 2, 13. Available at: <http://www.geobon.org/Downloads/reports/GEOBON/2015/MonitoringBiologicalInvasions.pdf>.

McGeoch, M.A., Shaw, J.D., Terauds, A., *et al.* (2015) Monitoring biological invasion across the broader Antarctic: a baseline and indicator framework. Global Environmental Change, 32, 108–25.

McIntosh, C., Finnoff, D., Settle, C., *et al.* (2009). Economic valuation and invasive species. In: Keller, R., Lodge, D., Lewis, M., *et al.* (eds) *Bioeconomics of Invasive Species: Integrating Ecology, Economics, Policy, and Management.* New York, NY: Oxford University Press, pp. 151–79.

McIntosh, C.R., Shogren, J.F., and Finnoff, D.C. (2007) Invasive species and delaying the inevitable: results from a pilot valuation experiment. Journal of Agricultural and Applied Economics, 39, 83–95.

McIntosh, C.R., Shogren, J.F., and Finnoff, D.C. (2010) Invasive species and delaying the inevitable: valuation evidence from a national survey. Ecological Economics, 69, 632–40.

Mehta, S.V., Haight, R.G., Homans, F.R., *et al.* (2007) Optimal detection and control strategies for invasive species management. Ecological Economics, 61, 237–45.

Menz, K.M., Coote, B.G., and Auld, B.A. (1980) Spatial aspects of weed control. Agricultural Systems, 6, 67–75.

Milner-Gulland, E.J. (1997) A stochastic dynamic programming model for the management of the saiga antelope. Ecological Applications, 7, 130–42.

Minoarivelo, H.O. and Hui, C. (2016) Trait-mediated interaction leads to structural emergence in mutualistic networks. Evolutionary Ecology, 30, 105–21.

Mitra, D., Thomakos, D.D., and Ulubaşoğlu, M.A. (2002) 'Protection for Sale' in a developing country: democracy vs. dictatorship. Review of Economics and Statistics, 84, 497–508.

Molnar, J.L., Gamboa, R.L., Revenga, C., *et al.* (2008) Assessing the global threat of invasive species to marine biodiversity. Frontiers in Ecology and the Environment, 6, 485–92.

Moody, M.E. and Mack, R.N. (1988) Controlling the spread of plant invasions: the importance of nascent foci. Journal of Applied Ecology, 25, 1009–21.

Moore, J.L., Runge, M.C., Webber, B.L., *et al.* (2011) Contain or eradicate? Optimizing the management goal for Australian acacia invasions in the face of uncertainty. Diversity and Distributions, 17, 1047–59.

Morgan, D.R., Nugent, G., and Warburton, B. (2006) Benefits and feasibility of local elimination of possum populations. Wildlife Research, 33, 605–14.

Münzbergová, Z., Hadincová, V., Wild, J., *et al.* (2013) Variability in the contribution of different life stages to population growth as a key factor in the invasion success of Pinus strobus. PLoS One, 8, e56953.

Myers, J.H., Simberloff, D., Kuris, A.M., *et al.* (2000) Eradication revisited: dealing with exotic species. TREE, 15, 316–20.

Nativi, S., Mazzetti, P., Santoro, M., *et al.* (2015) Big Data challenges in building the global earth observation system of systems. Environmental Modelling and Software, 68, 1–26.

Nava-Camberos, U., Riley, D.G., and Harris, M.K. (2001) Density-yield relationships and economic injury levels for *Bemisia argentifolii* (Homoptera: Aleyrodidae) in cantaloupe in Texas. Journal of Economic Entomology, 94, 180–9.

Nelson, E., Mendoza, G., Regetz, J., *et al.* (2009) Modeling multiple ecosystem services, biodiversity conservation, commodity production, and tradeoffs at landscape scales. Frontiers in Ecology and the Environment, 7, 4–11.

Nentwig, W., Bacher, S., Pyšek, P., *et al.* (2016) The Generic Impact Scoring System (GISS): a standardized tool to quantify the impacts of alien species. Environmental Monitoring and Assessment, 188, 315.

Nentwig, W., Kühnel, E., and Bacher, S. (2010) A generic impact-scoring system applied to alien mammals in Europe. Conservation Biology, 24, 302–11.

Novoa, A., Le Roux, J.J., Robertson, M.P., *et al.* (2015) Introduced and invasive cactus species: a global review. AoB Plants, 7, plu078.

Nunez, M.A. and Pauchard, A. (2010) Biological invasions in developing and developed countries: does one model fit all? Biological Invasions, 12, 707–14.

Odom, D., Sinden, J.A., Cacho, O., *et al.* (2005) Economic issues in the management of plants invading natural environments: Scotch broom in Barrington Tops National Park. Biological Invasions, 7, 445–57.

Olson, L.J. (2006) The economics of terrestrial invasive species: a review of the literature. Agricultural and Resource Economics Review, 35, 178.

Olson, L.J. and Roy, S. (2002) The economics of controlling a stochastic biological invasion. American Journal of Agricultural Economics, 84, 1311–16.

Olson, L.J. and Roy, S. (2005) On prevention and control of an uncertain biological invasion. Applied Economic Perspectives and Policy, 27, 491–7.

Olson, L.J. and Roy, S. (2008) Controlling a biological invasion: a non-classical dynamic economic model. Economic Theory, 36, 453–69.

Paini, D.R., Worner, S.P., Cook, D.C., *et al.* (2010) Using a self-organizing map to predict invasive species: sensitivity to data errors and a comparison with expert opinion. Journal of Applied Ecology, 47, 290–8.

Pandey, S. and Medd, R.W. (1991) A stochastic dynamic programming framework for weed control decision making: an application to *Avena fatua* L. Agricultural Economics, 6, 115–28.

Panetta, F., Csurhes, S., Markula, A., *et al.* (2011) Predicting the cost of eradication for 41 Class 1 declared weeds in Queensland. Plant Protection Quarterly, 26, 42.

Panetta, F.D., Gilbey, D.J., and D'Antuono, M.F. (1988) Survival and fecundity of wild radish (*Raphanus raphanistrum* L.) plants in relation to cropping, time of emergence and chemical control. Crop and Pasture Science, 39, 385–97.

Parker, I.M. (2000) Invasion dynamics of *Cytisus scoparius*: a matrix model approach. Ecological Applications, 10, 726–43.

Parker, I.M., Simberloff, D., Lonsdale, W.M., *et al.* (1999) Impact: toward a framework for understanding the ecological effects of invaders. Biological Invasions, 1, 3–19.

Parsons, K., Quiring, D., Piene, H., *et al.* (2005) Relationship between balsam fir sawfly density and defoliation in balsam fir. Forest Ecology and Management, 205, 325–31.

Pereira, H.M., Ferrier, S., Walters, M., *et al.* (2013) Essential biodiversity variables. Science, 339, 277–8.

Perrings, C., Williamson, M., and Dalmazzone, S. (eds) (2000) *The Economics of Biological Invasions*. Cheltenham: Edward Elgar.

Petit, R.J. (2004) Biological invasions at the gene level. Diversity and Distributions, 10, 159–65.

Pichancourt, J.B., Chadès, I., Firn, J., *et al.* (2012) Simple rules to contain an invasive species with a complex life cycle and high dispersal capacity. Journal of Applied Ecology, 49, 52–62.

Pimentel, D., Lach, L., Zuniga, R., *et al.* (2000) Environmental and economic costs of nonindigenous species in the United States. BioScience, 50, 53–65.

Pimentel, D., McNair, S., Janecka, J., *et al.* (2001) Economic and environmental threats of alien plant, animal, and microbe invasions. Agriculture, Ecosystems and Environment, 84, 1–20.

Pimentel, D., Zuniga, R., and Morrison, D. (2005) Update on the environmental and economic costs associated with alien-invasive species in the United States. Ecological Economics, 52, 273–88.

Pindyck, R. and Rubinfeld, D. (2001). *Microeconomics*, 5th edn. Upper Saddle River, NJ: Prentice Hall.

Plant, R.E. and Mangel, M. (1987) Modeling and simulation in agricultural pest management. SIAM Review, 29, 235–62.

Possingham, H.P. (1996) Decision theory and biodiversity management: how to manage a metapopulation. In: Floyd, R.B., Sheppard, A.W., and De Barro, P.J. (eds) *Frontiers of Population Ecology*. Melbourne: CSIRO Publishing, pp. 391–8.

Possingham, H.P. (1997) State-dependent decision analysis for conservation biology. In: Pickett, S.T.A., Ostfeld, R.S., Shachak, M., *et al.* (eds) *The Ecological Basis of Conservation: Heterogeneity, Ecosystems and Biodiversity*. New York, NY: Chapman and Hall, pp. 298–304.

Possingham, H.P. and Tuck, G. (1998) Fire management strategies that minimise the probability of population extinction for early and mid-successional species. In: Fletcher, D., Kavaliers, L., and Manly, B.J.F.(eds) *Statistics in Ecology and Environmental Monitoring*. Dunedin: University of Otago Press, pp. 157–67.

Potapov, A.B. and Lewis, M.A. (2008) Allee effect and control of lake system invasion. Bulletin of Mathematical Biology, 70, 1371–97.

Potapov, A.B., Lewis, M.A., and Finnoff, D.C. (2007) Optimal control of biological invasions in lake networks. Natural Resource Modeling, 20, 351–79.

Pyšek, P. and Richardson, D.M. (2007) Traits associated with invasiveness in alien plants: where do we stand? In: Nentwig, W. (ed.) *Biological Invasions*. Berlin: Springer, pp. 97–125.

Pyšek, P., Jarošík, V., Hulme, P.E., *et al.* (2012) A global assessment of invasive plant impacts on resident species, communities and ecosystems: the interaction of impact measures, invading species' traits and environment. Global Change Biology, 18, 1725–37.

Pyšek, P., Jarošík, V., Hulme, P.E., *et al.* (2010) Disentangling the role of environmental and human pressures on biological invasions across Europe. Proceedings of the National Academy of Sciences of the USA, 107, 12157–62.

Pyšek, P. and Richardson, D.M. (2010) Invasive species, environmental change and management, and health. Annual Review of Environment and Resources, 35, 25–55.

Pyšek, P., Richardson, D.M., Pergl, J., *et al.* (2008) Geographical and taxonomic biases in invasion ecology. Trends in Ecology and Evolution, 23, 237–44.

Pyšek, P., Richardson, D.M., Rejmánek, M., *et al.* (2004). Alien plants in checklists and floras: towards better communication between taxonomists and ecologists. Taxon, 53, 131–43.

Randall, R.P. (2007) *The Introduced Flora of Australia and its Weed Status*. Adelaide: CRC for Australian Weed Management.

Reed, W.J. (1974) A stochastic model for the economic management of a renewable animal resource. Mathematical Biosciences, 22, 313–37.

Reichard, S.H. and Hamilton, C.W. (1997) Predicting invasions of woody plants introduced into North America. Conservation Biology, 11, 193–203.

Rejmánek, M. and Richardson, D.M. (2013) Trees and shrubs as invasive alien species-2013 update of the global database. Diversity and Distributions, 19, 1093–4.

Rejmánek, M., Richardson, D.M., and Pyšek, P. (2013) Plant invasions and invasibility of plant communities. In: van der Maarel, E. and Franklin, J. (eds) *Vegetation Ecology*, 2. New York, NY: Wiley, pp. 387–424.

Richardson, D.M. (1988). Age structure and regeneration after fire in a self-sown *Pinus halepensis* forest on the Cape Peninsula, South Africa. South African Journal of Botany, 54, 140–44.

Richardson, D.M., Carruthers, J., Hui, C., *et al.* (2011) Human-mediated introductions of Australian acacias–a global experiment in biogeography. Diversity and Distributions, 17, 771–87.

Richardson, D.M., Hui, C., Nuñez, M.A., *et al.* (2014) Tree invasions: patterns, processes, challenges and opportunities. Biological invasions, 16, 473–81.

Richardson, D.M. and Kluge, R.L. (2008) Seed banks of invasive Australian *Acacia* species in South Africa: role in invasiveness and options for management. Perspectives in Plant Ecology Evolution and Systematics, 10, 161–77.

Richardson, D.M. and Pyšek, P. (2012) Naturalization of introduced plants: ecological drivers of biogeographical patterns. New Phytologist, 196, 383–96.

Richardson, D.M., Pyšek, P., Rejmánek, M., *et al.* (2000) Naturalization and invasion of alien plants: concepts and definitions. Diversity and Distributions, 6, 93–107

Richardson, D.M. and Rejmánek, M. (2011) Trees and shrubs as invasive alien species: a global review. Diversity and Distributions, 17, 788–809.

Richardson, D.M. and Ricciardi, A. (2013) Misleading criticisms of invasion science: a field guide. Diversity and Distributions, 19, 1461–7.

Richardson, D.M. and Whittaker, R.J. (2010) Conservation biogeography-foundations, concepts and challenges. Diversity and Distributions, 16, 313–20.

Robinson, C.J., Smyth, D., and Whitehead, P.J. (2005) Bush tucker, bush pets, and bush threats: cooperative management of feral animals in Australia's Kakadu National Park. Conservation Biology, 19, 1385–91.

Rouget, M., Robertson, M.P., Wilson, J.R.U., *et al.* (2016) Invasion debt—quantifying future biological invasions. Diversity and Distributions, 22, 445–56.

Roura-Pascual, N., Krug, R.M., Richardson, D.M., *et al.* (2010) Spatially-explicit sensitivity analysis for conservation management: exploring the influence of decisions in invasive alien plant management. Diversity and Distributions, 16, 426–38.

Rout, T.M., Thompson, C.J., and McCarthy, M.A. (2009) Robust decisions for declaring eradication of invasive species. Journal of Applied Ecology, 46, 782–6.

Roy, H.E., Brown, P.M., Adriaens, T., *et al.* (2016) The harlequin ladybird, *Harmonia axyridis*: global perspectives on invasion history and ecology. Biological Invasions, 18, 997–1044.

Roy, H.E., Peyton, J., Aldridge, D.C., *et al.* (2014) Horizon scanning for invasive alien species with the potential to threaten biodiversity in Great Britain. Global Change Biology, 20, 3859–71.

Royle, J.A., Kéry, M., Gautier, R., *et al.* (2007) Hierarchical spatial models of abundance and occurrence from imperfect survey data. Ecological Monographs, 77, 465–81.

Royle, J.A., Nichols, J.D., and Kéry, M. (2005) Modelling occurrence and abundance of species when detection is imperfect. Oikos, 110, 353–9.

Saphores, J.D.M. and Shogren, J.F. (2005) Managing exotic pests under uncertainty: optimal control actions and bioeconomic investigations. Ecological Economics, 52, 327–39.

Scoble, J. and Lowe, A.J. (2010) A case for incorporating phylogeography and landscape genetics into species distribution modelling approaches to improve climate adaptation and conservation planning. Diversity and Distributions, 16, 343–53.

Seabloom, E.W., Williams, J.W., Slayback, D., et al. (2006) Human impacts, plant invasion, and imperiled plant species in California. Ecological Applications, 16, 1338–50.

Seaman, J.T., Turvey, W.S., Ottaway, S.J., et al. (1989) Investigations into the toxicity of Echium plantagineum in sheep: 1. Field grazing experiments. Australian Veterinary Journal, 66, 279–85.

Sebert-Cuvillier, E., Paccaut, F., Chabrerie, O., et al. (2007) Local population dynamics of an invasive tree species with a complex life-history cycle: a stochastic matrix model. Ecological Modelling, 201, 127–43.

Seebens, H., Essl, F., Dawson, W., et al. (2015) Global trade will accelerate plant invasions in emerging economies under climate change. Global Change Biology, 21, 4128–40.

Settle, C., Crocker, T.D., and Shogren, J.F. (2002) On the joint determination of biological and economic systems. Ecological Economics, 42, 301–11.

Sharov, A.A. and Liebhold, A.M. (1998) Bioeconomics of managing the spread of exotic pest species with barrier zones. Ecological Applications, 8, 833–45.

Shea, K. and Possingham, H.P. (2000) Optimal release strategies for biological control agents: an application of stochastic dynamic programming to population management. Journal of Applied Ecology, 37, 77–86.

Shine, R., Brown, G.P., and Phillips, B.L. (2011). An evolutionary process that assembles phenotypes through space rather than through time. Proceedings of the National Academy of Sciences of the USA, 108, 5708–11.

Shoemaker, C.A. (1982) Optimal integrated control of univoltine pest populations with age structure. Operations Research, 30, 4–61.

Simberloff, D. (2003) Eradication-preventing invasions at the outset. Weed Science, 51, 247–53.

Simberloff, D. (2011) How common are invasion-induced ecosystem impacts? Biological Invasions, 13, 1255–68.

Simberloff, D. and Gibbons, L. (2004) Now you see them, now you don't!—population crashes of established introduced species. Biological Invasions, 6, 161–72.

Simberloff, D., Martin, J.L., Genovesi, P., et al. (2013) Impacts of biological invasions: what's what and the way forward. Trends in Ecology and Evolution, 28, 58–66.

Skidmore, A.K., Pettorelli, N., Coops, N.C., et al. (2015) Environmental science: agree on biodiversity metrics to track from space. Nature, 523, 403–5.

Smith, H.A., Johnson, W.S., Shonkwiler, J.S., et al. (1999) The implications of variable or constant expansion rates in invasive weed infestations. Weed Science, 47, 62–6.

Sohngen, B. and Mendelsohn, R. (1998) Valuing the impact of large-scale ecological change in a market: the effect of climate change on US timber. American Economic Review, 88, 686–710.

Standish, R.J., Robertson, A.W., and Williams, P.A. (2001) The impact of an invasive weed Tradescantia fluminensis on native forest regeneration. Journal of Applied Ecology, 38, 1253–63.

Stohlgren, T.J. and Schnase, J.L. (2006) Risk analysis for biological hazards: what we need to know about invasive species. Risk Analysis, 26, 163–73.

Stokes, K.E., O'neill, K.P., Montgomery, W.I., et al. (2006) The importance of stakeholder engagement in invasive species management: a cross-jurisdictional perspective in Ireland. Biodiversity and Conservation, 15, 2829–52.

Sutherland, W.J., Armstrong-Brown, S., Armsworth, P.R., et al. (2006) The identification of 100 ecological questions of high policy relevance in the UK. Journal of Applied Ecology, 43, 617–27.

Swallow, S.K. (1996) Resource capital theory and ecosystem economics: developing non-renewable habitats with heterogeneous quality. Southern Economic Journal, 63, 106–23.

Taylor, C.M. and Hastings, A. (2004) Finding optimal control strategies for invasive species: a density-structured model for Spartina alterniflora. Journal of Applied Ecology, 41, 1049–57.

Thompson, S.K., 1990. Adaptive cluster sampling. Journal of the American Statistical Association, 85, 1050–9.

Thompson, S.K. (2002) Sampling, 2nd edn. New York, NY: John Wiley & Sons.

Thompson, G.D., Bellstedt, D.U., Richardson, D.M., et al. (2015) A tree well travelled: global genetic structure of the invasive tree Acacia saligna. Journal of Biogeography, 42, 305–14.

Thompson, G.D., Robertson, M.P., Webber, B.L., et al. (2011) Predicting the sub-specific identity of invasive species using distribution models: Acacia saligna as an example. Diversity and Distributions, 17, 1001–14.

Thuiller, W., Richardson, D.M., Pyšek, P., et al. (2005) Niche-based modelling as a tool for predicting the risk of alien plant invasions at a global scale. Global Change Biology, 11, 2234–50.

Tilman, D. (1999) The ecological consequences of changes in biodiversity: a search for general principles 101. Ecology, 80, 1455–74.

Tittensor, D.P., Walpole, M., Hill, S.L., et al. (2014) A mid-term analysis of progress toward international biodiversity targets. Science, 346, 241–4.

Tobin, P.C., Whitmire, S.L., Johnson, D.M., et al. (2007) Invasion speed is affected by geographical variation in the strength of Allee effects. Ecology Letters, 10, 36–43.

Tucker, K.C. and Richardson, D.M. (1995) An expert system for screening potentially invasive alien plants in South African fynbos. Journal of Environmental Management, 44, 309–38.

UNEP (2011) COP 6 Decision VI/23. Strategic Plan for Biodiversity 2011–2020. Further information related to technical rationale for Aichi Biodiversity Targets, including potential indicators and targets. UNEP/CBD/COP/10/INF/12/Rev 1. Nahoya: UNEP.

United Nations Environment Programme (UNEP). Available at: <http://www.pops.int>.

Urquhart, N.S., Paulsen, S.G., and Larsen, D.P. (1998) Monitoring for policy-relevant regional trends over time. Ecological Applications, 8, 246–57.

Van Kleunen, M., Weber, E., and Fischer, M. (2010) A meta-analysis of trait differences between invasive and non-invasive plant species. Ecology Letters, 13, 235–45.

Van Kleunen, M., Dawson, W., Essl, F., et al. (2015) Global exchange and accumulation of non-native plants. Nature, 525, 100–3.

Van Wilgen, B.W., Reyers, B., Le Maitre, D.C., et al. (2008) A biome-scale assessment of the impact of invasive alien plants on ecosystem services in South Africa. Journal of Environmental Management, 89, 336–49.

Van Wilgen, N.J., Roura-Pascual, N., and Richardson, D.M. (2009) A quantitative climate-match score for risk-assessment screening of reptile and amphibian introductions. Environmental Management 44, 590–607.

Vayssier-Taussat, M., Moutailler, S., Michelet, L., et al. (2013) Next generation sequencing uncovers unexpected bacterial pathogens in ticks in western Europe. PLoS One, 8, e81439.

Vilà, M., Basnou, C., Pyšek, P., et al. (2010) How well do we understand the impacts of alien species on ecosystem services? A pan-European cross-species assessment. Frontiers in Ecology and the Environment, 8, 135–44.

Visser, V., Langdon, B., Pauchard, A., et al. (2014) Unlocking the potential of Google Earth as a tool in invasion science. Biological Invasions, 16, 513–34.

Vitousek, P.M., Mooney, H.A., Lubchenco, J., et al. (1997) Human domination of Earth's ecosystems. Science, 277, 494–9.

Von Holle, B.V. and Simberloff, D. (2005) Ecological resistance to biological invasion overwhelmed by propagule pressure. Ecology, 86, 3212–18.

Wadsworth, R.A., Collingham, Y.C., Willis, S.G., et al. (2000) Simulating the spread and management of alien riparian weeds: are they out of control? Journal of Applied Ecology, 37, 28–38.

Walters, C.J. and Hilborn, R. (1978) Ecological optimization and adaptive management. Annual Review of Ecology and Systematics, 9, 157–88.

Watson, A.K. (ed.) (1985) Leafy Spurge. Monograph 3. Champaign, IL: Weed Science Society of America.

Whittle, A.J., Lenhart, S., and Gross, L.J. (2007) Optimal control for management of an invasive plant species. Mathematical Biosciences and Engineering, 4, 101–12.

Widrlechner, M.P., Thompson, J.R., Iles, J.K., et al. (2004). Models for predicting the risk of naturalization of non-native woody plants in Iowa. Journal of Environmental Horticulture, 22, 23–31.

Wilen, J.E. (2007) Economics of spatial-dynamic processes. American Journal of Agricultural Economics, 89, 1134–44.

Williamson, M., Dehnen-Schmutz, K., Kühn, I., et al. (2009) The distribution of range sizes of native and alien plants in four European countries and the effects of residence time. Diversity and Distributions, 15, 158–66.

Williamson, M.H. and Fitter, A. (1996) The characters of successful invaders. Biological Conservation, 78, 163–70.

Wilson, R.J., Thomas, C.D., Fox, R., et al. (2004) Spatial patterns in species distributions reveal biodiversity change. Nature, 432, 393–6.

Wilson, J.R.U., Caplat, P., Dickie, I.A., et al. (2014) A standardized set of metrics to assess and monitor tree invasions. Biological Invasions, 16, 535–51.

Wilson, J.R.U., Gairifo, C., Gibson, M.R., et al. (2011) Risk assessment, eradication, and biological control: global efforts to limit Australian acacia invasions. Diversity and Distributions, 17, 1030–46.

Wilson, J.R.U., Richardson, D.M., Rouget, M., et al. (2007) Residence time and potential range: crucial considerations in modelling plant invasions. Diversity and Distributions, 13, 11–22.

With, K.A. (2004) Assessing the risk of invasive spread in fragmented landscapes. Risk Analysis, 24, 803–15.

Worner, S.P. and Gevrey, M. (2006) Modelling global insect pest species assemblages to determine risk of invasion. Journal of Applied Ecology, 43, 858–67.

Yemshanov, D., Koch, F.H., Ducey, M., et al. (2013) Mapping ecological risks with a portfolio-based technique: incorporating uncertainty and decision-making preferences. Diversity and Distributions, 19, 567–79.

Yemshanov, D., McKenney, D.W., and Pedlar, J.H. (2012) Mapping forest composition from the Canadian National Forest Inventory and land cover classification maps. Environmental Monitoring and Assessment, 184, 4655–69.

Yokomizo, H., Possingham, H.P., Thomas, M.B., et al. (2009) Managing the impact of invasive species: the value of knowing the density-impact curve. Ecological Applications, 19, 376–86.

PART III

Synthesis

CHAPTER 10

Complex adaptive networks

10.1 Ecological networks under invasion

Previous sections of the book have explored separately the eco-evolutionary experience of invasive species (chapters in Part 1) and the features of invaded ecosystems (chapters in Part 2). In the final two chapters, we do not aim to reiterate all the key knowledge components of invasion ecology derived from traditional experimental and observational methods of population and community ecology; rather, we explore the invasion syndrome afresh from the perspective of complexity science. Complexity science, a burgeoning research field that began in the late 1980s, explores the self-organization and functioning of complex adaptive systems (CAS) which comprise a large number of interacting, co-evolving components that can adaptively respond to changes and perturbations (see Box 10.1).

An ecosystem under biological invasion precisely fits the definition of a CAS (Levin 1999; Anand *et al.* 2010; Solé and Goodwin 2000; Parrott and Meyer 2012), whereas the ecological network of biotic interactions among species, native or alien, embedded in the ecosystem has become a powerful modelling proxy for capturing ecosystem complexity (Bascompte and Jordano 2007). In particular, the dynamics of invading species and the response of ecological networks are largely self-organized through a multitude of interactions and feedback loops, where human beings and their agency also considered part of the network. In this chapter, we use ecological networks as a model of the invaded ecosystem and examine the emergence of network structure and function.

The search for generality in invasion ecology has progressed largely through the quantification of the drivers behind two concepts: the invasiveness of alien species and the invasibility of recipient ecosystems. Generalities regarding the two concepts were initially sought separately, but more recently much effort has been directed towards merging the concepts (e.g. Richardson and Pyšek 2006). 'Invasibility' defines the properties of a community, habitat, or ecosystem that determine its vulnerability to invasion (Lonsdale 1999). Early studies tended to use the concept deterministically (particular systems were deemed either susceptible or resistant to invasion), whereas more recent work recognizes the probabilistic nature of invasibility. To further elucidate the concept of invasibility, we first need to determine how an ecosystem is structured and how it responds to perturbations such as those imposed by biological invasions. This relates to the diversity–invasibility hypothesis that was first discussed by Elton (1958) with reference to the potential ecological resistance of species-rich communities to invasion.

The development of the diversity–invasibility hypothesis followed three threads (Fridley 2011). Firstly, evolutionary ecologists have followed Darwin's (1859) view that more diverse regional assemblages have greater evolutionary innovation; that is, species-rich ecosystems are evolutionarily unstable and thus innovative (e.g. Dobzhansky 1950; Doebeli 2011). Secondly, community ecologists have followed MacArthur's (1972) limiting similarity theory of competitive exclusion through niche resemblance in seeking to determine whether susceptibility to invasion is correlated with the species

Invasion Dynamics. Cang Hui and David M. Richardson, Oxford University Press (2017).
© Cang Hui and David M. Richardson. DOI 10.1093/acprof:oso/9780198745334.001.0001

Box 10.1 Complex adaptive systems

A complex adaptive system (CAS) is a dynamic system comprising multiple interacting parts that can adaptively and collectively respond to perturbations. The boundary between the system and its environment is often hard to define, and thus poses no separate and distinct environment as the target of system adaptation. Rather, the system and its environment co-adapt and co-evolve, making a CAS capable of coping with a changing environment. CASs are therefore often used for exploring the complex behaviours across a wide spatio-temporal scale emerging from non-linear abiotic and biotic interactions among the multitude of elements within a loosely bounded open system. Complexity in system patterns and structures arises from the interrelationship, interaction, and interconnectivity of elements within a system, and also between the system and its environment (Gell-Mann 1995).

Complexity theory has emerged as a vibrant transdisciplinary field. It grew from work in cybernetics and systems theory at the Santa Fe Institute in the 1980s (Kauffman 1993). From its beginnings, complexity theory has been cross-disciplinary and has relied heavily on computer simulations as the main research instrument. Natural systems, such as brains, immune systems, ecosystems and societies, and artificial systems, such as parallel and distributed computing systems, artificial intelligence, neural networks, evolutionary programs, have become the typical models for studying the behaviour, structure, and function of CAS. Numerical and computational approaches, either handling large quantities of data or building simulations (e.g. using agent-based models), have been used to explore these systems.

Seven features separate CASs from traditional closed and rigid systems (Chan 2001): (i) distributed control; (ii) connectivity; (iii) co-evolution; (iv) initial condition sensitivity; (v) emergent order; (vi) being far from equilibrium; and (vii) the state of paradox. We elaborate these seven features using scenarios of biological invasions. (i) No single centralized mechanism governs the invasibility of recipient ecosystems, and no single trait or syndrome of traits determines the invasiveness of alien species. The resistance from resident species to invasion is not the mere sum of the effect of each species; this means that species richness is often not a good indicator of biotic resistance. (ii) Impact from biological invasion is not only felt by the species that interact directly with the invasive species but also by other species through mediated interactions. (iii) The invasiveness and performance of an invader not only depends on its own traits and propagule pressure but also on the context of recipient ecosystems and traits/densities of resident species. (iv) The community assemblage of an invaded ecosystem might not simply converge to a modified climax, but could potentially lead to alternative system regimes due to priority effects and invasional meltdown. (v) Invasion patterns are retrospectively observable, and characteristics of successful invasions are identifiable and invasion impacts can be explained. (vi) Besides characteristics of invaders and recipient ecosystems, introduction rate and propagule pressure, as well as disturbance, also critically determine invasion performance and the features of novel ecosystems. (vii) Although characteristics of high invasion performance can be clearly identified and strongly supported by evidence from the past or other areas, predictions based on such strong knowledge is rather poor, a paradoxical phenomenon known as spontaneous self-organization in complexity science (Prigogine 1977; Kauffman 1995).

---

richness of communities (e.g. Stohlgren *et al.* 2003; Tilman 2004). Finally, network ecologists have followed Elton's (1958) formal argument and have explored in more detail, following May's (1973) proposition, whether complexity begets instability (McCann 2000; Allesina and Tang 2012); that is, whether the architecture of a network is indicative of its stability against perturbation and its invasibility. There is a long history of debate around May's (1973) proposition on the complexity–stability relationship; this has centred on whether system complexity, for example as measured by network size

and connectance, promotes or inhibits network stability. Consensus has yet to be reached on this issue (Justus 2008).

Resolution of the debates around Elton's (1958) diversity–invasibility hypothesis and May's (1973) proposition of the complexity–stability relationship lies at the heart of understanding the structure and function emergence of CAS. Current discussions in this regard focus on three aspects. Firstly, system complexity can be elaborated by considering system architecture. In particular, network architecture can be depicted by a number of metrics (e.g. nestedness and modularity) that reflect different facets of system complexity. Theoretical models are being contested to bring a cohesive view on the emergence of network architectures. Secondly, system stability is multifaceted and can be quantified in many ways; for example, according to resilience, persistence, and robustness. For instance, ecological stability has traditionally been defined according to *resilience* which denotes how systems respond to small perturbations. Invasibility, on the other hand, depicts how systems respond to species introductions and could be considered a mirror concept to *robustness* which is defined on the basis of system responses to species removal. As will be illustrated in this chapter, both invasibility and robustness are related more to evolutionary stability, especially when considering trait-mediated interactions in ecological networks. Finally, system stability needs to be connected with particular features of system architecture. Progress has been made in documenting the architecture of empirical ecological networks, although documented network architectures remain to be clearly connected with ecosystem function and stability, including invasibility.

Following these three aspects in resolving the debate on the complexity–stability relationship, we first introduce typical metrics and features of ecological networks (section 10.2), and then discuss selected models to explain the emergence of network architectures (section 10.3). Next, we define network invasibility (section 10.4) and explore its relationship with other network architectural and stability metrics (section 10.5). Finally, we explore the invasibility of an adaptive network and examine how specialized mutualistic interaction and disturbance can create opportunity niches for invasion (section 10.6). Knowledge on the emergence of network architecture and stability further allows us to portray the profile of a successful invader in adaptive ecological networks (section 10.7). We end by exploring the emergence of network stability and the direction of system evolution (section 10.8). As these sections are the first attempts to merge concepts in complexity science and invasion science, they are meant to be prognostic rather than conclusive.

10.2 Ecological network architecture

As the recipient ecosystem often comprises many interacting species, an ecological network of biotic interactions among resident species provides an excellent model for exploring the inherent complexity regarding ecosystem stability and function. Two types of ecological networks are typically examined in the ecological literature: mutualistic and antagonistic. The role of competition is not discussed here as it is normally considered to only affect within-trophic species coexistence and community structure (see Chapter 8). The architecture of an ecological network can be measured as particular features of the interaction matrix, depicting whether and how strongly two nodes (species) interact. Typical features include *connectance* (the proportion of realized interactions among all possible ones; Olesen and Jordano 2002), *nestedness* (the degree to which specialists only interact with a subset of species with which generalists interact; Bascompte *et al.* 2003), and *compartmentalization* (the level of modularity at which a network can be grouped into delimited modules; Newman 2006). Network complexity normally refers to a combined factor of network size (the number of nodes) and connectance (May 1973; Allesina and Tang 2012).

Mutualistic networks are mostly formed due to the reciprocal dependence of plants on their pollinators, nodule-forming bacteria, mycorrhizal fungi, frugivores or seed dispersers. There are some animal-to-animal mutualistic interactions (e.g. between aphids and ants) but they are not diverse enough to be meaningfully considered as networks. Mutualistic dependence can lead to complicated co-evolutionary dynamics (e.g. Rodríguez-Gironés and Llandres 2008; Zhang *et al.* 2013), which further contributes to the weaving of complex webs and affects how ecosystems function and the extent to which stability is retained under perturbations (Kaiser-Bunbury *et al.* 2010).

Mutualistic networks exhibit several important patterns. Firstly, early studies on the level of species specialization (number of interactions) and specificity (matching traits between interacting species) in food webs (Waser *et al.* 1996; Vázquez and Aizen 2003) were motivated by the examination of the distribution of node degrees of interaction between species (i.e. the number of other species in the network with which a given species interacts). It was found that most species are poorly connected in mutualistic networks and that only a small number are well connected, resulting in a right-skewed node-degree distribution that usually follows a truncated power law (Jordano *et al.* 2003). Secondly, mutualistic networks often show highly asymmetrically nested structures (Bascompte and Jordano 2007), with plant species depending strongly on animal species, but not necessarily the reverse (Richardson *et al.* 2000; Bascompte *et al.* 2006). Finally, an important feature of mutualistic networks is the existence of modules (Dicks *et al.* 2002). These three patterns of mutualistic networks are not independent of each other, and the emergence of these patterns should be explained simultaneously by a single model integrating essential eco-evolutionary processes. Consensus on the architecture of mutualistic networks has largely been reached in the literature (e.g. Bascompte *et al.* 2003; Olesen *et al.* 2007; Guimarães *et al.* 2007; Thébault and Fontaine 2010; Mello *et al.* 2011).

Antagonistic interactions, such as herbivory, parasitism, and predation, represent the process of resource exploitation in ecological networks (van Alphen *et al.* 2003), and are important to the provision of ecosystem function and services (Dobson *et al.* 2008). The process of resource exploitation can also create a nested structure, fostering high species richness (Bastola *et al.* 2009) and enhancing network resilience against perturbations (Fortuna and Bascompte 2006; Burgos *et al.* 2007; Memmott *et al.* 2004). However, highly nested antagonistic networks are also often associated with reduced species persistence (James *et al.* 2012) and unstable community dynamics (Allesina and Tang 2012).

The chase-and-run feature of 'Red Queen' dynamics that are inherent in antagonistic interactions can also potentially divide species into modules where consumers within a module are likely to share the same function and exploit similar resources (Guimera *et al.* 2010; Krasnov *et al.* 2012; Thebault *et al.* 2010), thereby forming a compartmentalized network. Compartmentalization then leads to a high level of modularity and stabilizes ecological networks by confining the effect of perturbations within modules (Guimera *et al.* 2010; Stouffer and Bascompte 2011; Fortuna *et al.* 2010). However, there is still considerable debate around issues relating to antagonistic networks, for instance on whether they should be more compartmentalized than would be expected under random conditions (e.g. Poisot 2013).

10.3 Network emergence via interaction switching

The establishment of biotic interactions is affected by both phylogenetic constraints of co-evolution (Ehrlich and Raven 1964) and ecological fitting (Janzen 1985) (see Chapter 6 for details). Consequently, to explain pattern emergence in ecological networks, a group of ontogenetic models have incorporated the evolutionary history (phylogeny) of the interactive taxa, whereas other models have attempted to capture species packing in the network through ecological fitting and species rewiring. The number of studies on the dynamics of complex ecological networks is increasing rapidly, and it is not our intention to review the field of ecological networks (see instead, Pascual and Dunne 2005; Bascompte and Jordano 2013). Here we only consider selected models to facilitate our understanding of biological invasions into ecological networks from the perspective of CASs.

Several studies have explored how the evolution of individual species affects the dynamics of interaction networks. In particular, Rezende and colleagues (2007a) examined the evolution of traits along phylogenetic trees, which partially explained the nested architecture in mutualistic networks. Rezende and colleagues (2007b) further found that in 50 per cent of the empirical networks they examined, phylogenetic proximity between paired species was positively correlated with their interaction similarities, suggesting a tendency for closely related species to interact with the same partners. This means that observed network architecture could partially reflect the distinct phylogenetic structure and history of involved taxa (e.g. phylogenetic asymmetry and accelerated evolution). Some studies have further hypothesized the effect of reciprocal selection between involved taxa. For example, Guimarães and colleagues (2011) modelled the co-evolution of traits in mutualistic networks and found a higher convergence of traits in super-generalist species, which in turn plays an important role in the maintenance of the network organization and stability. Nuismer and colleagues (2013) showed that co-evolutionary selection may increase network connectance while altering the pattern of nestedness. This suggests the potential of network emergence through adaptive diversification (Doebeli and Ispolatov 2010; Doebeli 2011), adding support to Ehrlich and Raven's (1964) co-evolutionary origin of biotic interactions.

If interaction patterns persist over evolutionary time scales, as can be expected particularly when interactions are closely linked to morphology and thus to underlying genotypes (Rezende *et al.* 2007a), species can inherit the interaction patterns of their ancestors. For instance, Minoarivelo and colleagues (2014) designed a model describing the evolution of pairwise interactions as Markov processes along joint phylogenies (top row in Fig. 10.1), drawing on phylogenetic models of molecular evolution (Felsenstein 2004). Phylogenetic signal was detected in 21 per cent of the 53 empirical networks compiled by Rezende and team (2007a). The model was selected as the best fit in 34 per cent networks for predicting the node degree distribution (bottom row in Fig. 10.1). However, for these 53 mutualistic networks the gain/loss events of ecological interaction were only estimated to have occurred on average 4.12 times per phylogenetic path from the root to a leaf node. Evidently, the rate of gain/loss events needs to be extremely low to conserve the ancestral interaction pattern. This extremely low rate of gain/loss events thus cannot reflect rapid fluctuations in interactions (Petanidou *et al.* 2008), but only major phylogenetic or phylogeographic events (e.g. the gain/loss of wings in insects; Whiting *et al.* 2003).

On the other hand, ecological fitting requires species to switch their partners in response to changes of context such as the temporal presence/absence of previous and new mutualists from disturbance and biological invasion (Valdovinos *et al.* 2010). This adaptive nature of selecting and switching interaction partners (i.e. tuning target resources) in a CAS is essential for the survival of consumers that are competing for available resources (van Baalen *et al.* 2001; Staniczenko *et al.* 2010; Kimbrell and Holt 2005). Adaptive interaction switches occur when the quantity and quality of available resources change. Consumers prefer to select highly profitable resources rather than consuming all resources available to them, as specified in optimal foraging theory (Stephens and Krebs 1986). Concurrently, they will exploit abundant resources rather than rare ones to minimize risks of hunger from food scarcity (Fossette *et al.* 2012; Zhang and Hui 2014). This adaptive behaviour of interaction switching could well be the predominant process that pushes an ecological network towards becoming compartmentalized or nested, thereby forming a potentially stable and resilient complex community (Valdovinos *et al.* 2010; Kondoh 2003).

Evolution is largely driven by mutation, natural selection, and random drift. Ecological fitting in a CAS should also reflect these different evolutionary forces. Several models have been proposed to capture adaptive interaction switching into network emergence. There is, however, much scope for improvement and refinement in this area, especially regarding the predictive power of such models and the need to incorporate all essential forces of evolution (Saavedra *et al.* 2009; Suweis *et al.* 2013). Adaptive interaction switching implemented in most of these models only emphasizes adaptation from one or two particular processes, specifically by using the classical optimality methodology in evolutionary ecology (Suweis *et al.* 2013), or optimization-based analytical treatment for adaptive behaviour (Valdovinos *et al.* 2010; Kondoh 2003). To capture the essence of network evolution, the roles of both optimization and random drift need to be appreciated, as together they represent a combination of both adaptive and non-adaptive forces that potentially counter-balance each other in structuring network architecture.

A hybrid model that emphasizes both adaptation and drift offers a more complete picture of network evolution through adaptive interaction switching (Zhang *et al.* 2011; Nuwagaba *et al.* 2015). Consumers in the model are allowed not only to eliminate selectively the unfit resources from their diets based on the benefits and abundance of these resources (i.e. an adaptive process) but also to try new

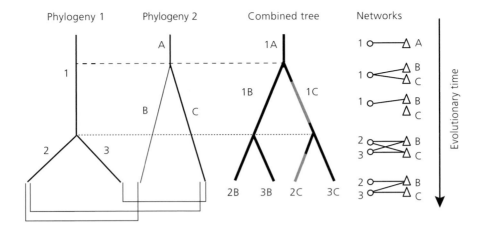

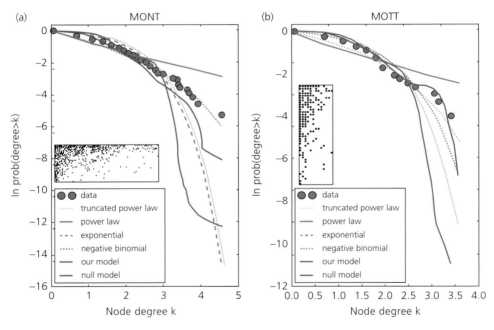

Fig. 10.1 (Top) Example of the combination of two phylogenetic trees and the evolution of biotic interactions as a binary trait (on/off) along the combined phylogeny. Every combination of branches of the two phylogenetic trees is represented by a corresponding branch in the combined tree. The branch in the combined tree is coloured grey when the interaction between the corresponding species pair is lost. The top right panel shows resulting interaction networks. (Bottom) Log-log plot of the complementary cumulative distribution of the connectivity of a frugivory network with 206 species at Monteverde (MONT) in Costa Rica and a pollination network with 55 species in North Carolina (MOTT) (Rezende *et al.* 2007a). Inset: adjacency matrix representation. The degree distribution in MONT is the best fit by a truncated power law while it is best fit by the Markov process model in MOTT. From Minoarivelo *et al.* (2014). Reproduced with permission from John Wiley & Sons.

resources randomly (i.e. random drift). This hybrid behavioural rule depicts adaptation in line with Alfred Russell Wallace's view of natural selection—via the elimination of the unfit, and random drift as the innovation in foraging behaviours–separating it from other rules of adaptive interaction switching. This hybrid behavioural rule not only emphasizes the adaptive process by which the consumer gradually improves its resource utilization efficiency by retaining highly beneficial and abundant resources and eliminating less beneficial and rare ones, but also allows for behavioural innovation whereby new resources can be exploited by consumers via the random drift of interactions. Several hypotheses on the effect of biotic interaction on invasion per-

formance (e.g. host jump, new associations, enemy release, and EICA; see Chapter 6) support this behaviour of adaptive interaction switching. Through the interaction switch, the modularity of a random network with similar size (number of species) and complexity (connectance) to observation but randomly assigned interaction strengths, gradually converges to a stable equilibrium similar to the observed modularity (top left, Fig. 10.2), as do other network architecture measures, such as nestedness and node degree distribution, to list a few, that can also simultaneously emerge (Fig. 10.2; Zhang *et al.* 2011; Nuwagaba *et al.* 2015).

Key drivers behind the emergence of network architecture can now be elucidated (Fig. 10.3).

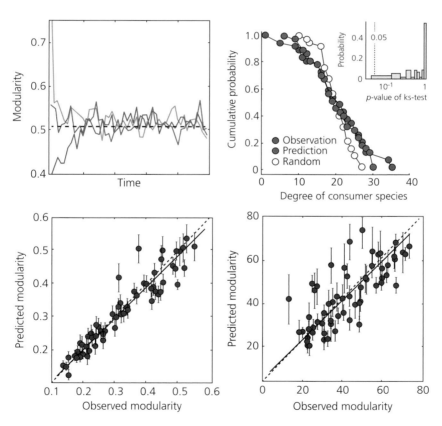

Fig. 10.2 (Top left): The effect of initial network structure and model parameters on the dynamics of modularity over time for a stream food web (Thompson and Townsend 1999) (Top right): Comparison of the predicted and observed node-degree distributions of consumer species for the stream food web (Townsend *et al.* 1998). (Bottom): Observed versus predicted modularity and nestedness from the adaptive interaction switching model for 28 plant–herbivore and 33 host–parasite networks. From Nuwagaba *et al.* (2015).

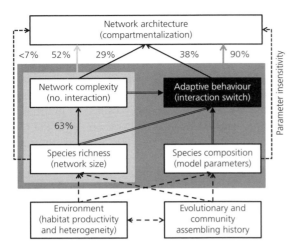

Fig. 10.3 A conceptual framework of the emergence of network architecture. Network architectures (here, modularity) of 61 real ecological networks are explained by network size, complexity (the number of interactions), and interaction switch using a generalized linear model. Dashed lines indicate factors not included in the model; double-lined arrows indicate model inputs; dotted lines indicate either parameter insensitivity or insignificant variation explained. Percentage indicates variation explained (adjusted R²). From Nuwagaba *et al.* (2015).

Firstly, evolutionary history, community assembly processes, biological invasions, and environmental characteristics (e.g. productivity and heterogeneity) could largely determine the number of species (i.e. network size) and the composition/turnover of species and traits (i.e. model parameters) that a community can accommodate (Krasnov *et al.* 2012; Rezende *et al.* 2007a; Dunne *et al.* 2002, 2013; Loreau 2010; see also Chapter 8). Secondly, extracting the contribution of network size and complexity, the hybrid behavioural rule of adaptive interaction switching (adaptation plus drift) alone could explain nearly 40 per cent of variation in the architecture (here, modularity) of 61 observed networks (Nuwagaba *et al.* 2015). Adaptive interaction switching (ecological fitting) and the context of co-evolutionary history, local community composition, and environmental features could together explain 90 per cent variation in the architecture of these 61 networks (Fig. 10.3). This conceptual framework of network emergence also highlights priorities for future research. Further work is needed to elucidate (i) the factors that determine the number of species that a community can hold (i.e. network size), and (ii) the mechanisms of how network size indirectly affects network architectures by directly dictating network complexity. Studies on species packing in local and regional communities (e.g. on Darwin's naturalization hypothesis and the integration of alien species in native species assemblages; see Chapters 6 and 8) could shed light on these issues.

The framework of CAS thus provides a solid foundation for explaining the architecture and function emergence of adaptive ecological networks. Adaptive interaction switching can occur at three different levels. At the individual level, it is important to choose adaptively with whom to interact (habitat and diet selection) or to avoid (anti-predation strategies). Such preferential interactions could simply arise from optimal or adaptive foraging where individuals aim to maximize their energy intake rate while minimize risks for starvation and predation (Zhang and Hui 2014). Consequently, interaction switching can occur rapidly for individuals at a pace even faster than the typical ecological time scale (e.g. host switch in parasites; van Baalen *et al.* 2001). At the population level, successful interactions often require involved species to possess a certain level of matching or complementary traits, such as matching phenology for flowering time in plants and the foraging activity of insect pollinators (Waser 2015). Through adaptively changing the level of trait complementarity and plasticity, populations can adjust interaction strength to their own advantage (Santamaria and Rodriguez-Girones 2007). At the species level, species can completely lose or regain the possibility of interactions through developing incompatible traits of forbidden links and setting up barriers to exploitation (Jordano *et al.* 2003), potentially through adaptive diversification and evolutionary branching (Dieckmann and Doebeli 1999; Zhang *et al.* 2013) and geographic barriers. Therefore, processes of interaction switching (e.g. adaptive behaviour, assortative interaction, and random drift), together with constraints on network size and complexity (e.g. assemblage history and the co-evolution of trait complementarity) are dominant forces that give rise to the realized network architecture.

10.4 Invasibility of ecological networks

Characters of the recipient ecosystem that influence its susceptibility to the establishment and spread of invasive species describe its invasibility (Lonsdale 1999; Alpert *et al.* 2000). These characters that define invasibility include not only physical factors such as broad climatic and environmental suitability for invaders but also the architecture of networks of biotic interactions. As such, the concept of invasibility can be incorporated in the framework of CASs and perhaps defined as the instability of adaptive ecological networks (see following), so to improve the elucidation of characters of ecosystems that make them vulnerable to particular invaders. Full comprehension of the proposed framework of invasibility in adaptive ecological networks requires us first to establish two important concepts in community and evolutionary ecology: *community saturation* and *evolutionary stability*—both inspired by Elton's (1958) argument that more diverse ecosystems should be more resistant to invasion (see also Chapter 6).

The concept of *community saturation* emerged from theoretical principles of limiting similarity between coexisting species due to niche competition (MacArthur 1972; Abrams 1983), where strong interspecific interactions can preclude the establishment of introduced species which has an intermediate optimal niche position in-between the positions of two resident species (along the resource/niche axis). Without disturbance, a saturated community can be defined as a particular community assemblage that, once achieved though not necessarily so, can withstand any invasions from a feasible species pool, provided that all species are bounded by life-history trade-offs (Diaz *et al.* 2016; Kunstler *et al.* 2016). Specifically, the deviation from a saturated community can be measured by the cumulative niche space that permits invasion of a given recipient ecosystem. Very few, if any, ecological communities are truly saturated, as local communities can be strongly affected by regional species dynamics and stochasticity (Abrams 1998; Loreau 2000; see Chapter 8).

Evolutionary stability was initially proposed as a concept of an unbeatable strategy in games, known as the evolutionarily stable strategy (ESS; Maynard Smith and Price 1973). An ESS can be typically identified by maximizing the pay off or fitness function of a strategy, and has become the target function of many optimality models in behavioural and evolutionary ecology and economics (Charnov 1979; Parker and Maynard Smith 1990). The ESS of an ecological network can be defined as a suite of species with their particular traits and population densities (for the case of density-dependent fitness) that, once assembled, cannot be invaded by an alien species given low propagule pressure, irrespective of the life-history or other traits of the species. If all the resident species of a community are inferior competitors to an invasive species, the community assembly is clearly not an ESS. Note that a community not at an ESS does not necessarily imply the lack of an ESS. However, in theory, the ESS of ecological networks does not always exist; rather, the concepts of community saturation and ESS serve as a theoretical benchmark of a system that is un-invasible against which invasibility can be measured. Importantly, these two concepts are only relevant when the rate of propagule influx is low. When propagule pressure is high, the behaviour of the system will be overridden by the influx of propagules.

Invasibility and invasiveness are not isolated concepts but are strongly interwoven. As an example, Darwin (1859) posited, in what has become known as his 'naturalization hypothesis', that introduced species should be more successful (i.e. more likely to invade) when the recipient community lacks congeneric or ecologically similar species (Duncan and Williams 2002). The underlying logic of this statement relies on the assumption that closely related species show greater overlap of life-histories, traits, and therefore niche requirements such that an intact community would be less invasible to congeners and other closely related species than to more distantly related species. A counterargument which has received some empirical support is that trait similarity among related species might predict habitat suitability and result in higher invasibility when congeners are considered (Duncan and Williams 2002; see also Chapter 6). Either way, it is clear that the two core concepts in invasion ecology—invasibility and invasiveness—are context-dependent and are closely related (e.g. Richardson and Pyšek 2006; Thuiller *et al.* 2010).

Assembly-level models have a long history in community ecology (e.g. Drake 1990; Morton and Law 1997). They normally assume infrequent colonization of new species from a predetermined regional species pool. Some recent assembly-level models further allow limited evolutionary processes (e.g. Drossel *et al.* 2001; McKane 2004; Münkemüller and Gallien 2015) and adaptive responses to disturbance (Kondoh 2003; Loeuille and Loreau 2005; Zhang *et al.* 2011; Suweis *et al.* 2013; Nuwagaba *et al.* 2015; Minoarivelo and Hui 2016a). Generating a resident network of multiple heterotrophic species can be accomplished in two ways. Firstly, many studies have followed the simple procedure of randomly assigning trait values and parameters (e.g. trait-dependent growth rates and carrying capacity) for all initial species, running the model until equilibrium is reached, and then removing those species with population sizes below a certain threshold (Holland and DeAngelis 2010). At this stage, the network is considered to be at its *ecological equilibrium*. Once the recipient community has reached this equilibrium, we could consider the invasiveness of a potential introduced species, following evolutionary invasion analysis (Box 10.2), as its *invasion fitness*, defined as its per-capita population growth rate when propagule pressure is trivial (close to zero) and when population sizes of resident species in the community are at their equilibria (Fig. 10.4a). Invasion fitness is a good proxy of invasiveness for an introduced species: if the trait of an introduced species lies within the positive-fitness intervals along the zero invasion fitness line (Fig. 10.4a), the introduced species will experience positive invasion fitness and thus be able to establish and invade the resident community. If trait values land within the negative-fitness intervals, the species will be repelled by the resident community (Fig. 10.4a).

For a given introduced species with a particular trait, if there is a native resident species with an identical or similar trait (i.e. the trait of introduced species is close to any traits of resident species—dots along the horizontal axis in Fig. 10.4a), the invasion fitness will then be close to zero. Because of the zero population growth, such species are less likely to establish and immediately become invasive simply due to demographic stochasticity (the case of neutral coexistence). However, for exactly the same reason, an introduced species is also less likely to be immediately expelled. However, this scenario when an invader has the same trait/niche as a resident species has met with mixed evidence. An update of the classic limiting similarity theory is clearly needed to incorporate further the scenario of neutral coexistence that complete niche overlap does not lead to immediate competitive exclusion but rather coexistence, especially when the propagule influx is not stopped (e.g. Abrams and Rueffler 2009; Mayfield and Levine 2010; Hui *et al.* 2016; see Chapter 6).

If the trait of an alien species is distinct from those of any resident species (i.e. located between dots along the horizontal axis in Fig. 10.4a), it is then likely to become either highly invasive (peaks in positive fitness curves in Fig. 10.4a) or be quickly expelled from the network (valleys in negative fitness curves in Fig. 10.4a), with a 50/50 chance for successful invasion in a species-rich network due to the constraints on any dynamic systems given the continuity of the invasion fitness function (from the fundamental theorem of algebra and the central limit theorem). In this regard, the invasibility of the recipient ecological network can be defined as the total width of opportunity niche in the trait space (i.e. the summation of all the positive-fitness intervals).

As an alternative to the aforementioned static trait approach of building a community at its ecological equilibrium, we could also generate a model community as an adaptive ecological network, where species within the network can co-evolve according to adaptive dynamics (Box 10.2), or where species with different traits can be continuously introduced into the community from a large species pool (i.e. a meta-community; Gilpin and Hanski 1991; Vellend 2010). This approach will potentially lead to a saturated ecological network or an ESS if that exists (Fig. 10.4b). By definition, no alien species can invade a saturated network or an evolutionarily stable network as the invasion fitness of any introduced species is equal to or less than zero (Fig. 10.4b). However, with the community hosting an increasing number of resident species, the fitness landscape will be flattened, making it increasingly easy for new invaders to establish—a case of invasional meltdown (Fig. 10.4b). Even if the saturated assembly does exist, a community under constant

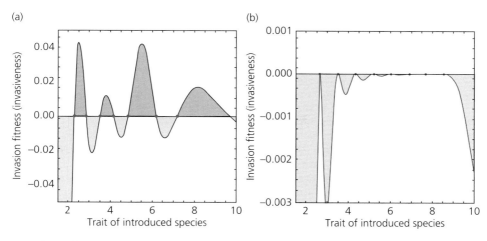

Fig. 10.4 Invasion fitness of an introduced species as a function of its trait value relative to the trait values of the resident species in the ecological networks. Dots along the horizontal axis indicate the trait values of native resident species. (a) A randomly generated network at its ecological equilibrium. Only introduced species with traits that lie in the positive-fitness intervals can establish and invade the recipient ecological network; introduced species with traits that lie in the negative-fitness intervals will be repelled by the network (i.e. eliminated from the resident species/ecosystem). Invasibility is thus defined as the total width of the positive-fitness intervals along the zero-fitness line. (b) A saturated assembly is defined as the recipient network with zero invasibility. Curves produced from an adaptive dynamics model based on the adaptive dynamics of trait evolution in food webs (Loeuille and Loreau 2005; Brännström *et al.* 2011). From Hui *et al.* (2016). With permission from Springer.

Box 10.2 Evolutionary invasion analysis

The evolutionary trajectory is traditionally regarded as the path of organisms attempting to maximize their fitness by optimizing their life-history strategies (which can be loosely defined as traits). Such a perspective involves two assumptions. Firstly, there is a fitness measure that can be maximized in the attainable trait set. Secondly, this optimal trait can be reached, from the current condition, through incremental evolutionary changes. The first assumption leads to the definition of the evolutionarily stable strategy (ESS): there exists a trait or a suite of traits that have a competitive advantage over all other attainable traits; in other words, it can resist the invasion of all other traits. The second assumption refers to the convergence (asymptotic) stability of this optimal trait; that is, a trait close to the optimal trait can be invaded/replaced by a trait even closer to the optimal trait through directional selection. A convergence-stable ESS is called a continuously stable strategy (CSS).

Evolutionary invasion analysis comprises a set of quantitative techniques designed to address these two assumptions: conditions for the existence of an ESS and for a rare trait to invade a resident population (Otta and Day 2007; Hurford *et al.* 2010). The classic setup considers an invading trait not far from the resident trait solely through mutation; that is, we are looking for a *local* CSS, strategies that are convergence-stable and cannot be invaded by local traits. Specifically, we first allow the population dynamics of a given trait to settle at its equilibrium. If this trait is a local CSS, an adjacent trait will not be able to invade the system; if not, some adjacent traits could invade the system and replace the resident trait in the population. However, given the pervasiveness of global environmental changes, many non-native species with distinct traits or genotypes are constantly being introduced to ecosystems, suggesting the increasing relevance of searching for the *global* CSS in a complex adaptive network. That is, not only do we allow the system to be invaded by traits adjacent to the resident ones, but also by traits that are distinct from the resident traits.

Adaptive dynamics, a powerful analytical tool for studying the evolution and succession of quantitative traits or phenotypic characters, was developed in 1990s by game theorists (e.g. Nowak and Sigmund 1990), population geneticists (e.g. Abrams *et al.* 1993), and theoretical ecologists (e.g. Metz *et al.* 1992; Dieckmann and Law 1996). It studies evolutionary changes induced by rare and small mutations when

(Continued)

Box 10.2 *Continued*

fitness is density- or frequency-dependent (Waxman and Gavrilets 2005). As individuals interact within a community, their fitness not only depends on their own traits but also on the frequency or density of other traits of different species. The evolution of traits can be evaluated by examining the survival of rare invaders in a community dominated by resident populations at their stable equilibria.

The canonical equation of adaptive dynamics describes trait dynamics under directional selection through the continuous invasion of rare mutants or invaders into resident populations (Geritz *et al.* 1998; Doebeli and Dieckmann 2000; Vukics *et al.* 2003). Consider n interacting species. Let $\boldsymbol{x} = (x_1, x_2, \ldots, x_n)^{\mathsf{T}}$ be the vector of all traits for the n species, where x_k is a vector of all traits of species k. A mutation of species k with the trait vector y_k, or an introduced species with its traits close to those of species k, has a chance to invade when its *invasion fitness*, $f_k(y_k, \boldsymbol{x})$ (i.e. its per-capita growth rate; Metz *et al.* 1992), exceeds that of the resident population of the species, $f_k(x_k, \boldsymbol{x})$. As the resident populations are assumed to be at their ecological equilibria when the rare mutation or species introduction occurs, the fitness of the resident populations equals zero, $f_k(x_k, \boldsymbol{x}) = 0$. That is, the condition for invasion success is: $f_k(y_k, \boldsymbol{x}) > 0$.

The evolutionary dynamics of the trait vector x_k of the kth species is controlled by the mutation matrix (M_k) and selection gradient $(\nabla_{y_k} f_k(y_k, \boldsymbol{x})|_{y_k = x_k})$ as outlined in the following canonical equation of adaptive dynamics (Dieckmann and Law 1996; Leimar 2009):

$$\dot{x}_k = M_k \cdot \nabla_{y_k} f_k(y_k, \boldsymbol{x})|_{y_k = x_k} .$$

The mutation matrix represents genetic variation, $M_k = \mu_k \hat{n}_k \sigma_k / 2$, with μ_k and $\hat{n}_k$ the mutation rate (propagule pressure), and equilibrium population size of species k, and σ_k the covariance matrix of trait increment from correlated mutation (i.e. trade-offs between traits) (Dieckmann and Law 1996; Leimar 2009). The selection gradient $\nabla_{y_k} f_k(y_k, \boldsymbol{x})|_{y_k = x_k}$ indicates the direction of natural selection, of which the ith component $(\partial f_k(y_k, \boldsymbol{x}) / \partial y_k)|_{y_k = x_k}$ represents the selection direction of the ith trait of species k. A trait coalition of all species in the community (i.e. species identity is disguised by its traits), $\hat{x}$, becomes an *evolutionary singularity* if the selection gradients for all n species vanish, $\nabla_{y_k} f_k(y_k, \boldsymbol{x})|_{y_k = x_k, x = \hat{x}} = 0$. According to the Lyapunov stability criterion, a singularity becomes *convergence stable* (i.e. an evolutionary attractor) if all eigenvalues of the Jacobian matrix of the canonical equations in the community have negative real parts at the singularity; otherwise it

becomes unstable (Dercole and Rinaldi 2008). A singularity becomes *evolutionarily stable* (i.e. an ESS) if it cannot be invaded by any trait mutants in the vicinity of the trait space, $f_k(y_k, \hat{x}) > 0$. Note that evolutionarily stable and convergence stable are different concepts. When a singularity is both evolutionarily and convergence-stable, it is then named a continuously stable strategy (CSS; Eshel *et al.* 1997). Intuitively, convergence-stable determines whether a singularity is reachable through directional selection, whereas evolutionarily stable depicts whether a singularity is located at a fitness minimum or maximum.

When a singularity is convergence-stable but evolutionarily unstable, some species in the community will experience disruptive selection which could possibly lead to evolutionary branching (Doebeli and Dieckmann 2000; Dercole and Rinaldi 2008). Successful evolutionary branching relies on the competitive coexistence of the two newly emerged sister lineages, a condition known as *mutual invadability* (Geritz *et al.* 1998). Hybrids from the two sister lineages suffer from low fitness due to the shape of fitness landscape near the evolutionary branching point, and opposing directional selection will separate the traits of these two lineages even further. Once reproductive isolation and incompatibility is reached due to the hybridization barrier of low fitness, this evolutionary branching can be considered a speciation event.

The field of evolutionary invasion analysis is developing rapidly due to its potential to handle a wide range of questions (e.g. the emergence of social norms and cooperative behaviours; the evolution of complex life cycles). However, we here also point out a few aspects that require further refinement. Firstly, for simplicity, approaches of evolutionary invasion analysis, such as adaptive dynamics, often ignore the variation of traits and only consider the evolution of average traits (Dieckmann and Law 1996; Champagnat *et al.* 2001), although individual-based models do support the evolutionary dynamics portrayed by these approaches. However, studying only the evolution of average traits may overlook many important ecological and evolutionary features (Bolnick *et al.* 2011). For instance, the mean phenotypic trait could be misleading when the distribution of trait values is bi- or multi-mode (Cohen 2009). Following the initial proposition of Levin and Segel (1985), Cohen (2003) coined the term 'evolutionary distribution' that considers the evolution of trait distribution in a continuous space using reaction-diffusion-type models, with ecological processes captured by the reaction term and trait dynamics by the diffusion term (Doebeli and Ispolatov 2010).

Secondly, trait is typically used as a direct proxy of genotype in evolutionary invasion analysis, and the issue of phenotypic plasticity is ignored. The plasticity of functional traits, depicted as reaction norms (Angilletta *et al.* 2003), could profoundly affect invasion success and evolutionary trajectories. A new line of theories on the evolution of plasticity, potentially based on epigenetics, are needed (e.g. Duncan *et al.* 2014). Thirdly, numerical analyses of mean traits often rely on the separation of ecological and evolutionary time scales. This assumption is challenged by recent observations that show that evolutionary processes can occur rapidly even over time scales similar to ecological processes, and that rapid adaptation and evolution fre-quently occur during invasion (Yoshida *et al.* 2003; Prentis *et al.* 2008; Jones *et al.* 2009). Lastly, evolutionary invasion analysis concerns only convergence-stable singularities as only these singularities can be reached from incremental directional selection. However, invasion ecology needs to deal not only with introductions of invaders with similar or adjacent traits to those of the resident species but also with introductions of species with distinct traits; singularities that are convergence unstable are therefore also of concern (Minoarivelo and Hui 2016b). In any case, evolutionary invasion analysis provides us with a platform and starting point for defining the invasibility of a complex ecological network.

bombardment of invasive intrusion is unlikely to be either at equilibrium or remain saturated, but somewhere between the two extremes.

This demonstration with trait-based knowledge suggests that to be capable of directly assessing the invasibility of a recipient ecosystem, we need comprehensive information about the functional traits that affect biotic interaction strengths and thus population dynamics, as well as overall densities. In practice, however, sufficient data are rarely available for estimating invasibility directly, even though global monitoring and web-based data sharing initiatives are seeking to make full records of species densities and traits in prioritized ecosystems available (see Chapter 9). For a rapid assessment of invasibility, we need an indirect or alternative way of capturing the interaction matrix. An effective and conceptually and logistically tractable approach is to construct the interaction matrix of the recipient ecosystem (Hui *et al.* 2016). Ideally, the matrix should include the full list of species in the recipient ecosystem. This can also be relaxed by only considering the potential networks that an introduced species is likely to impact or interact with; this is referred to as an invader-centric ecological network (Morales-Castilla *et al.* 2015; Hui *et al.* 2016). Although accurate interaction strengths may enhance predictive power, qualitative matrices are usually adequate to understand network behaviours (Quirk and Ruppert 1965; May 1973; Jeffries 1974). By correlating network structure with invasibility, we could potentially use particular network structures to indicate invasibility (see section 10.5).

10.5 Network architecture and stability

Many metrics of network architecture have been proposed to describe diverse features of network structure based on a matrix of interaction strength. These metrics can be divided into two categories: those that portray the role of particular species in the network (e.g. centrality), and those that portray the structure of entire networks (e.g. nestedness and modularity). Network invasibility as proposed in the previous section is related to the latter. In particular, three most important features of a matrix—its sparsity, symmetry, and hierarchy—can be measured using three widely used metrics for quantitative networks: connectance, modularity, and nestedness (Tylianakis *et al.* 2007; Clauset *et al.* 2008; Dormann and Strauß 2014). Another well-known measure of network structure is the level of specialization which can be quantified according to the index H_2' proposed by Blüthgen and colleagues (2006) which measures the overall deviation of species' realized degrees from their expected ones, ranging from 0 (no specialization) to 1 (perfect specialization).

Network stability, in contrast to network architecture, deals with the ways in which networks respond to perturbations (Yodzis 1981). Different approaches for measuring network stability have been proposed, the most widely used of which are Lyapunov asymptotic stability, resilience, persistence, and robustness (May 1974; Pimm and Lawton 1978; Dunne *et al.* 2002; Donohue *et al.* 2013; see also Chapter 7), or, more generally, structural stability (Rohr *et al.* 2014) and evolutionary stability (Brännström *et al.* 2011; Minoarivelo and Hui 2016b). Network *resilience* can be measured as the logarithm of the absolute value of the dominant eigenvalue of the Jacobian matrix at equilibrium (DeAngelis 1980; Okuyama and Holland 2008; Encinas-Viso *et al.* 2012), depicting how quickly the system returns to a steady state after being perturbed (DeAngelis 1980). Network *robustness* depicts how a system responds to species removal (Dunne *et al.* 2002) and is measured as the fraction of species that had to be removed to result in a total loss of $\geq$ 50 per cent of all species. *Disruptiveness*, a measure of evolutionary instability, can be computed as the average over all species of the strength of disruptive selection in the system (Brännström *et al.* 2011), with the strength of disruptive selection for a particular species measured as the curvature of its invasion fitness. As emphasized earlier in this chapter, invasibility is a new aspect of system stability that can be measured as the fraction of traits of successful invaders among all possible traits invading a community and that is closely related to evolutionary stability.

Recent progress in resolving the debate on the diversity–invasibility hypothesis and the complexity–stability relationship has moved further, from expanding the metrics and knowledge of network architecture and stability separately, to exploring the causal relationship between the architecture and stability of many mutualistic and antagonistic networks (e.g. Memmott *et al.* 2004; Eklöf and Ebenman 2006; Bascompte *et al.* 2006; Burgos *et al.* 2007; Estrada 2007; Bastola *et al.* 2009; Kiers *et al.* 2010; Thébault and Fontaine 2010; Brose 2011; Stouffer and Bascompte 2011; James *et al.* 2012). A number of dynamic network models with adaptive and random species rewiring have been proposed as candidates for explaining emergent network structures and associated network stability (Kondoh 2003; Vacher *et al.* 2008; Zhang *et al.* 2011; Suweis *et al.* 2013; Minoarivelo *et al.* 2014; Nuwagaba *et al.* 2015). Following this line of research, the role of particular network architectures in stabilizing networks has been hotly debated.

Although the relationships between network architecture and invasibility are not necessarily causal correlations, we could still contemplate forms of the relationships. Firstly, May (1973) devised a prerequisite which was further generalized by Allesina and Tang (2012) to ensure stability in a complex network, $\alpha(SC)^{1/2} < \beta$, where α stands for the standard deviation of interaction strength, S species richness, C connectance, and β self-regulating force (e.g. negative density-dependence). A highly connected network (large C) could violate this condition and thus cause network instability, creating opportunity niches for invasion and thus augmenting invasibility. Secondly, a highly nested network suggests a strong hierarchical architecture, potentially from sorting species through multiple ecological filters, creating unbalanced energy/material flow from specialists to generalists and opportunity niches for introduced specialists that can more efficiently exploit resources than resident generalists. In other words, network asymmetry creates opportunity niches for specialists and thus enhances invasibility. Finally, a highly compartmentalized network is formed by clearly bounded modules, and these modules could have spatially or temporally partitioned available niches and habitats. This nevertheless provides opportunities for the generalist invaders that can explore two or more modules. Species possessing traits with high plasticity or tolerance, and those with complex life cycles (through ontogenetic niche shift), could thus invade highly compartmentalized networks.

A high level of network connectance has been predicted to enhance the resistance of food webs to invasion (Romanuk *et al.* 2009), although this view is contested (Baiser *et al.* 2010; Lurgi *et al.* 2014). Modularity (i.e. the structure of compartmentalization in networks) is observed to be lower in invaded pollination networks and food webs than in uninvaded ones (Albrecht *et al.* 2014; Lurgi *et al.* 2014). Invaded pollination networks are more nested and normally contain more species than uninvaded networks (Padrón *et al.* 2009; Traveset *et al.* 2013; Stouffer *et al.* 2014). However, these empirical observations do not allow us to determine whether these network features have triggered or facilitated the invasion, or are simply a consequence of the invasion (Traveset and Richardson 2006).

The relationship between network architecture and stability is particularly complex. On one hand, patterns of connectance and nestedness observed in mutualistic networks can facilitate the coexistence of species, thereby contributing positively to network stability (Bastolla *et al.* 2009; Thébault and Fontaine 2010; Rohr *et al.* 2014). Network complexity, measured as network size and connectivity (number of interactions), can also enhance network resilience (Okuyama and Holland 2008). On the other hand, theoretical studies have shown that these features specific to mutualistic networks can also be detrimental to network stability. For instance, extremely low levels of nestedness are associated with high levels of modularity and thus destabilize the community due to the detrimental effect of modularity on the stability of mutualistic networks (Campbell *et al.* 2012; Thébault and Fontaine 2010). In particular, when interaction strength is further considered, stability becomes negatively affected by both connectance and nestedness (Allesina and Tang 2012; Vieira and Almeida-Neto 2015). Inconsistency of the correlation between network architecture and stability seems to be attributable to the confusion in choosing appropriate measures of network stability, as each one only measures one specific facet of stability and thus often leads to contradictions when interpreted as general stability for comparison (Vallina and Quéré 2011).

Importantly, the relationship between network architecture and invasibility could be weak. Trivial changes in the overall network architecture follow-ing the introduction of an alien species have been documented empirically (Vilà *et al.* 2009; Padrón *et al.* 2009) and were tentatively explained by the peripheral role of the invader in the network. In particular, Albrecht and colleagues (2014) analysed 20 independent pairs of invaded and uninvaded pollination networks and found that the overall number of modules was not changed by invasion but that modules were more interconnected by the super-generalist invaders. This suggested the potential yet inconclusive role of super-generalist invaders in dismantling the compartmentalized structure in native pollination networks.

Testing simulated mutualistic networks provides a systematic way of exploring the relationship between network architecture and invasibility (Campbell *et al.* 2015; Minoarivelo and Hui 2016b). By comparing the pre- and post-invasion architecture of simulated pollination networks, Campbell and colleagues (2015) found that while connectance decreased, nestedness increased as a result of invasions. By examining the performance of an invader in simulated adaptive mutualistic networks, Minoarivel and Hui (2016b) found that although network architectural metrics and network stability/instability metrics are significantly related, the relationship is quite weak (Fig. 10.5). Apart from the two clusters formed by most network architectural metrics and most network stability/instability met-

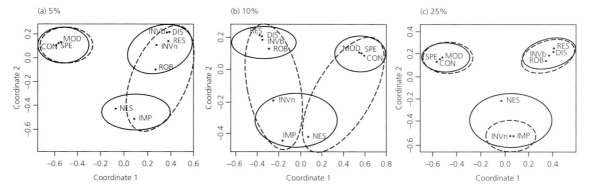

Fig. 10.5 Visualization of the relationship between all network metrics in a two-dimensional space using a multi-dimensional scaling analysis, for different initial propagule size. Invader propagule size is respectively (a) 5 per cent, (b) 10 per cent, and (c) 25 per cent of the average native population density. Clusters formed by a k-mean clustering analysis are shown by the solid circles and those formed by a hierarchical clustering analysis are shown by the broken circles. CON: connectance; SPE: specialization; MOD: modularity; NES: nestedness; IMP: impact; ROB: robustness; INVb: invasibility; INVn: average invasiveness; RES: resilience; DIS: disruptiveness. From Minoarivelo and Hui (2016b). With permission from Minoarivelo & Stellenbosch University.

rics respectively, the third cluster contains nestedness, invasion impact, and the average invasiveness of invaders in a network (Fig. 10.5). Members of the third cluster are either divided into the other two main clusters or left in isolation when using hierarchical clustering analysis, with nestedness weakly related to both the other network architectural metrics and the group of network stability/instability metrics. There is a very strong positive relationship between resilience, invasibility, and disruptiveness. Measurement of invasion impact (the reduction of total abundance of all resident species due to invasion) is the least related to the other network stability/instability metrics. Although invasibility and disruptiveness measure network instability rather than network stability, they are still positively correlated with robustness and resilience. For example, the most robust and resilient community is also the most disruptive and invasible one. This may indicate that the two groups of stability/invasibility metrics describe two different but closely related concepts.

Due to its weak relationship with network stability, network architecture is only one component of network characteristics and does not fully determine the overall functioning of the interactions. However, this does not mean that we should completely discard the role of network architecture in stabilizing/destabilizing mutualistic networks (Bastolla *et al.* 2009; Thébault and Fontaine 2010; Allesina and Tang 2012; Rohr *et al.* 2014; Vieira and Almeida-Neto 2015). In particular, the finding that nestedness is negatively correlated with network resilience and robustness agrees with previous studies (Allesina and Tang 2012; Campbell *et al.* 2012). Importantly, the more robust and resilient the network, the more invasible it is (Minoarivelo and Hui 2016b). This means that mutualistic networks which are not susceptible to species removal (i.e. robust) and are able to return quickly to a steady state after perturbation (i.e. resilient) also tend to be susceptible to invasions. This is understandable because the ability of a community to adapt quickly in the face of perturbations also confers on it the ability to accommodate the invader and make it more invasive. Moreover, the positive relationships between network stability metrics (resilience, robustness) and network instability metrics (invasibility,

invasiveness, disruptiveness, impact) highlight the importance of an appropriate definition of stability for each study. This highlights the view of Ives and Carpenter (2007) that stability metrics cannot be interpreted outside the context of defining environmental drivers of the change.

10.6 Co-evolving networks under invasion

What could be the reason for apparent existence of opportunity niches left untouched by co-evolution, but which are open to invasion? We discuss here two potential reasons: high level of network/interaction specialization, and disturbance. Considering the addition of a species in an adaptive ecological network could lead to more diverse evolutionary scenarios, as both the new arrival and the resident species will embark on the course of co-evolution (see Chapter 6). In particular, in a mutualistic network, when the interactions between mutualistic partners are highly specialized, the enforced positive frequency-dependent selection on traits could give an advantage to trait alignment (Fig. 10.6a), leading to trait convergence in the community (Elias *et al.* 2008; Guimarães *et al.* 2011; Nuismer *et al.* 2012). With the support from strong mutualism, these aligned traits can possess much higher population densities than other traits with no mutualistic partners in the community, posing strong asymmetric competition and a barrier preventing other species from further exploring mutualistic benefits—a manifestation of negative density-dependent selection through competitive exclusion (Hardin 1960). Consequently, an empty niche is often left unexploited in such a highly specialized mutualistic network on each side of the animal and plant trait space (Fig. 10.6a), as demonstrated by intentionally introducing a species to an adaptive mutualistic network (Minorivelo and Hui 2016a). The presence of both positive and negative frequency-dependent selection results in an unpredictable, volatile evolutionary trajectory, making the system susceptible to invasion (Richardson *et al.* 2000; Wilder *et al.* 2011; Traveset and Richardson 2014).

Exploitation of empty niches by alien species could cause a drastic change in the evolutionary

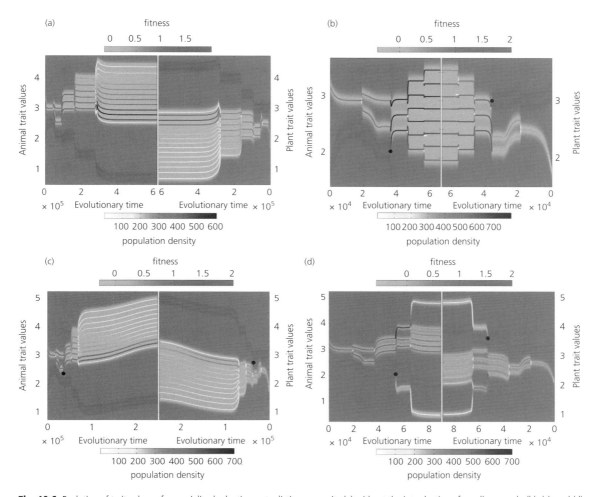

Fig. 10.6 Evolution of trait values of a specialized adaptive mutualistic community (a) without the introduction of an alien morph; (b), (c), and (d) with introductions of alien morphs at different locations of the niche space. The fitness landscape is presented as the background described in the greylevel bars at the top. Population densities are described in the greylevel bars at the bottom. Black dots show where aliens are introduced. From Minoarivelo (2016). With permission from Minoarivelo & Stellenbosch University.

trajectory of the entire network. The system often undergoes a strong directional selection, leading to either an ESS where all empty niches are exploited and all traits are highly aligned (a rare case, Fig. 10.6b), or the emergence of alternative empty niches after the invasion (a likely case, Fig. 10.6c and d), thus opening the community for additional invasion. The emerged new fitness peak of empty niches often locates at the boundary value of a trait cluster. That is, when the mutualistic interaction is highly specialized, new and distinct empty niches are likely to be created after the invasion when the

system often also undergoes strongly directional selection. The fate of an introduced species depends not only on its trait value but also on the level of specialization of the community. Although alien species with distinct traits could be released from strong competition, they suffer from low mutualistic support and have a low chance of establishment, particularly in specialized mutualistic networks (Traveset and Richardson 2014). The terrain of invasion fitness is evidently ragged and dynamic, and is shaped by many frequency-dependent selection forces. This can create opportunity niches for invasions (Fig. 10.6). Hence, the debate on whether or not Darwin's naturalization hypothesis holds is context-dependent in mutualistic networks and must consider the eco-evolutionary experience of the native and introduced species (Saul *et al.* 2013).

Many factors can disturb the resource supply and drive the change and cyclic fluctuation of resources including changes in land use, alteration of fire regimes, seasonality, and weather conditions. Fluctuation in resource availability is a key factor that mediates the susceptibility of an ecosystem to invasion (Davis *et al.* 2000) and can create an opportunity niche with positive invasion fitness that is temporally available for invasion, especially for alien species with traits pre-adapted for optimal resource consumption (Fig. 10.7; Hui *et al.* 2016). Although perturbation might not create a persistent opportunity niche for invasion, environmental stochasticity can markedly increase the network invasibility in otherwise resistant assemblages even in a saturated community. The paradox of enrichment (Rosenzweig 1971) states that enhanced resource levels can be accompanied by instability in a food chain. Such instability from temporal resource enrichment provides opportunities for invasions to establish in the community. Evidently, the enrich-

ment and fluctuation of resources can both affect invasibility, both playing similar roles in influencing network invasibility.

Overall, opportunity niches for invasion could exist in specialized and disturbed networks. Biological invasion could drastically alter the evolutionary path of resident species which often experience assembly-wide directional selection, pushing rapid evolution in resident species (see Chapters 5 and 6), and even disruptive selection, triggering sympatric diversification. An adaptive ecological network, being a multi-agent CAS, is optimized through evolution especially for the coexistence of these agents (resident species), thus emphasizing mechanisms of species coexistence but de-emphasizing the low chance (one versus many) of outcompeting other resident species (see Chapter 8). This is a typical strategy in multi-player games—to survive/tolerate rather than to outcompete, especially given the typical flat fitness landscape that has resulted from co-evolution. It creates a system phenomenon of 'easy in, difficult out' of new or rare species to absorb the impact of perturbations and incursions (see section 10.8), allowing the system to harbour many 'sleeper weeds' (introduced species in the lag phase; see Chapter 2) and a massive invasion debt (see Chapter 9) that can be temporally released from the available of opportunity niches in specialized networks or from disturbance.

10.7 Profiling an invader

The characteristics of successful invaders were discussed in several chapters in Part I. Here, we discuss invasion performance only from the perspective of an adaptive ecological network which relies on two features that affect its role in the network: interaction strength (i.e. linkage weight) and the level of generalization (i.e. node degree), relative to those of the resident species in the network. Features of both the invader and the native community have been shown to play critical roles in predicting the success and impact of an invasion (Shea and Chesson 2002; Gurevitch *et al.* 2011). Consequently, invasion success is better estimated by comparing sets of invasive and native resident species on their foraging, reproduction, and/or dispersal traits (Pyšek and Richardson 2007), especially the suite of traits

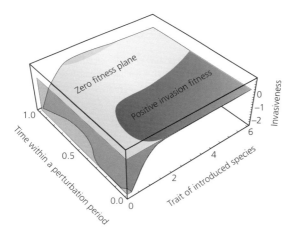

Fig. 10.7 Invasion fitness of an introduced species as a function of its trait value in an ecological network. The resource of the saturated assembly in Fig. 10.4b is disturbed by adding a periodic perturbation. An alien species with the trait and timing of introduction in the positive invasion fitness zone can invade the ecological network; otherwise it will be repelled from the network. From Hui *et al.* (2016). With permission from Springer.

that separate invasive species from native species (e.g. see Chapter 4).

Biotic interactions are directly mediated by involved traits, and the difference and matching between traits of resident species and invaders thus determine the interaction strength (Chapter 6); that is, whether invaders possess distinct traits or similar traits to resident species. Specifically, the strength of interactions is often mediated by matching between functional traits of interacting species (Jousselin *et al.* 2003; Santamaría and Rodríguez-Gironés 2007; Stang *et al.* 2009). A certain degree of similarity between phenotypic traits of the invader and the resident species was predicted to give a good indication of strong mutualistic interactions (Gibson *et al.* 2012), although species with high invasiveness and impact in pollination networks also often acquire traits atypical of native species (Fig. 10.8; Aizen *et al.* 2008; Campbell *et al.* 2015; Minoarivelo and Hui 2016b, but see Morales and Traveset 2009). Therefore, the relative position of introduced traits to those of resident species in a 3D trait space, potentially separating traits of mutualists, competitors, and antagonists into different dimensions, could provide a trait-based tool for assessing invasion performance (Pyšek and Richardson 2007; van Kleunen *et al.* 2010).

Node degree, on the other hand, reflects the level of generalization and specialization, often associated with interaction promiscuity and phenotypic plasticity (Chapter 6). A high level of interaction generalization has been shown to be important in determining invasion success in food webs (Romanuk *et al.* 2009; Lurgi *et al.* 2014) and in mutualistic networks (Traveset and Richardson 2014). Indeed, there is strong evidence that invasive plants in plant-pollinator networks have higher levels of interaction generalization than natives (Albrecht *et al.* 2014). The overall interaction can even be monopolized by super-generalist invaders in pollination networks (Aizen *et al.* 2008; Bartomeus *et al.* 2008; Vilà *et al.* 2009). Introduced species become highly invasive when they have the similar level of generalization to the average trait of resident species or when they are more generalist species but with extreme traits (Fig. 10.8; Minoarivelo and Hui 2016b). The importance of high interaction generalization to invasiveness requires concern when the invader trait is distinctive from native traits (Aizen *et al.* 2008; Bartomeus *et al.* 2008; Vilà *et al.* 2009; Albrecht *et al.* 2014), with successful invaders often interacting with either the most specialist natives (Stouffer *et al.* 2014) or the most generalist natives (Padrón *et al.* 2009).

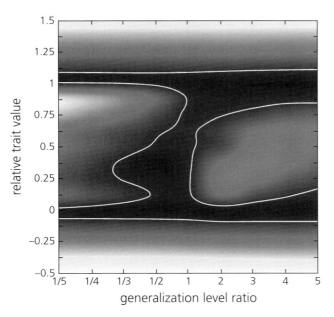

Fig. 10.8 Invasion performance (population growth rate) of an introduced species in adaptive mutualistic networks (considering both trait-mediated competition and mutualism), as a function of trait value and level of mutualism generalization relative to resident species. Darker areas indicate better performance, with white lines representing zero growth rate contours. Introduced species has a trait similar to the minimum or maximum resident trait will have a relative trait value equal to 0 or 1, respectively; the level of generalization is measured as the tolerance of the invader to trait difference for feasible mutualistic interactions, with the generalization level ratio = 1 indicating an equal level of generalization to those of the resident species, < 1 for specialist invaders and > 1 generalist invaders.
From Minoarivelo and Hui (2016b). With permission from Minoarivelo & Stellenbosch University.

As far as the impact of invasion is concerned, the alien species needs to be highly general in terms of interaction (Aizen *et al.* 2008; Albrecht *et al.* 2014). If the invader is a generalist, many other species will depend on it. Because of the cascading effect of interactions, the overall impact of this invader on the network can be high (Dunne and Williams 2009; Bascompte and Stouffer 2009; Traveset and Richardson 2014). Minoarivelo and Hui (2016b) found that invasive species with similar traits and level of generalization to those of resident species have the highest impact in mutualistic networks. Invaders with traits atypical of the native mutualistic community only pose little impact on resident species due to weak competition from mismatched traits (Morales and Traveset 2009, but see Campbell *et al.* 2015). It is worth to highlight especially for risk assessment that traits for fast alien population growth (invasiveness) are related to but different from traits for high impact (reducing the total abundance of all resident species). Lessons could be learned from CAS works on the role of both linkage weight (strength of interaction) and node degree (specialist versus generalist) in affecting network functioning. The horizontal H-shaped result for profiling successful invaders in simulated networks (e.g. Fig. 10.8) warrants further investigation for other types of ecological networks (e.g. antagonistic networks).

10.8 Evolving towards marginal stability

The long-standing debate on the complexity–stability relationship in complex ecological systems can be systematically addressed in various ways (McCann 2000). We can compare matrices generated from random null models with those based on observed interactions to examine the role of network architecture (e.g. Prado and Lewinsohn 2004). Binary and weighted interaction matrices can be contrasted to assess the role of skewed distribution of interaction strength. Alternatively, we can compare antagonistic and mutualistic interactions to determine the roles of different types of biotic interactions. However, the philosophical underpinnings of such comparisons treat different features of ecological networks as separate entities, thus ignoring that these features are multifaceted descriptions

of network functioning. To appreciate the inherent multifacetedness in depicting network functioning, a potential solution to this debate could be to agree that both architecture and stability features emerge through spontaneous self-organization of adaptive ecological networks, which is a critical feature of CAS. Therefore, the complexity–stability relationship of an ecological network could simply be a by-product of system evolution and functioning.

As discussed in section 10.3, the interaction matrix in recent models of adaptive interaction switching changes constantly, reflecting an adaptive network responding to changes (Staniczenko *et al.* 2010; Kondoh 2003). By allowing consumers to readjust their exploited resources by updating the interaction matrix, adaptive network models portray a multiplayer game and can successfully capture the spontaneous emergence of all features of network architecture. What then are the direction of system evolution and associated dynamics of stability in such adaptive ecological networks? Running the same adaptive interaction switching models (Zhang *et al.* 2011; Nuwagaba *et al.* 2015) suggests that a complex adaptive network evolves towards a clear target: *marginal stability* (Fig. 10.9),

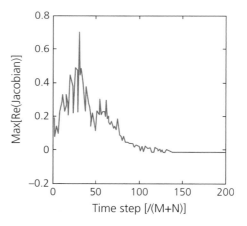

Fig. 10.9 Effect of the adaptive process of interaction switching on the maximum real part of the eigenvalues of the Jacobian matrix (a metric of system resilience). The simulation is based on the binary interaction matrix of the stream food web in a pine plantation in New Zealand (Thompson and Townsend 2003). Positive values lead to instability and the system evolves towards marginal stability with the lead eigenvalue just below (or around) zero, with no intention to further strengthen the stability or resilience of the system. From Nuwagaba *et al.* (2015).

which is consistent with the seven features of a CAS (Box 10.1). The adaptive interaction switch allows the abundance of species to fluctuate without necessarily leading to extinction when facing perturbations (van Baalen *et al.* 2001); this is vital for rebalancing the network back to its equilibrium, thereby buffering ecosystems against perturbations (Staniczenko *et al.* 2010). It is, thus, a strong structuralizing force (for the emergence of network architecture) and a marginal stabilizing force that largely stabilizes ecological networks but still allows alien species to invade and rare species be expelled from the network (Suweis *et al.* 2013; Zhang *et al.* 2011). This makes an adaptive network a better proxy than a rigid network for depicting real ecological communities, and the adaptive nature is important for forecasting the response of ecosystems to environmental changes and perturbations (Valdovinos *et al.* 2010; Kaiser-Bunbury *et al.* 2010).

This is a grand multiplayer game. To survive in such a game, species often have to have multiple contingency plans with which to handle ecological or evolutionary selection pressures. Ecologically, species can adjust the extent and structure of their geographical range, or simply shift their range, forming different aggregation patterns of biodiversity (e.g. Hui and McGeoch 2014). They can also invoke different strategies of population dynamics to release the pressure, for example, population fluctuation. Evolutionary pressure, by contrast, can affect the traits of both consumers and resources, and consequently often determines the availability and accessibility of resources at a physiological level (Spottiswoode and Stevens 2012; Zhang *et al.* 2013). Species can modify their functional traits convergently or divergently by changing their morphology, phenology, tolerance, performance, and plasticity; such changes are reinforced by heritable genotypes, leading to diverse evolutionary trajectories. Consequently, the architecture of an ecological network is also shaped by the imprint of its evolutionary history. Both evolutionary history and the introduction of alien species bring species with different traits into play and act to craft the architecture of an ecological network (Minoarivelo *et al.* 2014; Rezende *et al.* 2007a; Nuismer *et al.* 2013), whereas adaptive interaction switching or species rewiring or rapid trait evolution allows the system to be refined through ecological fitting

(Valdovinos *et al.* 2010; Zhang *et al.* 2011; Nuwagaba *et al.* 2015). At the system level, adaptive ecological networks are optimizing (yet not optimized) for biomass productivity and energy dissipation while maintaining marginal stability. Network architectures and invasibility of recipient ecosystems are thus by-products of both co-evolution and ecological fitting, through spontaneous self-organization, in response to anthropogenic challenges facing embedded social-ecological systems.

References

Abrams, P. (1983) The theory of limiting similarity. Annual Review of Ecology and Systematics, 14, 359–76.

Abrams, P.A. (1998) High competition with low similarity and low competition with high similarity: exploitative and apparent competition in consumer-resource systems. American Naturalist, 152, 114–28.

Abrams, P.A., Matsuda, H., and Harada, Y. (1993) Evolutionarily unstable fitness maxima and stable fitness minima of continuous traits. Evolutionary Ecology, 7, 465–47.

Abrams, P.A. and Rueffler, C. (2009) Coexistence and limiting similarity of consumer species competing for a linear array of resources. Ecology, 90, 812–22.

Aizen, M., Morales, C., and Morales, J.M. (2008) Invasive mutualists erode native pollination webs. PLoS Biology, 6, 396–403.

Albrecht, M., Padrón, B., Bartomeus, I., *et al.* (2014) Consequences of plant invasions on compartmentalization and species' roles in plant-pollinator networks. Proceedings of the Royal Society B: Biological Sciences, 281, 20140773.

Allesina, S. and Tang, S. (2012) Stability criteria for complex ecosystems. Nature, 483, 205–8.

Alpert, P., Bone, E., and Holzapfel, B.E. (2000) Invasiveness, invasibility and the role of environmental stress in the spread of non-native plants. Perspectives in Plant Ecology, Evolution and Systematics, 3, 52–66.

Anand, M., Gonzalez, A., Guichard, F., *et al.* (2010) Ecological systems as complex systems: challenges for an emerging science. Diversity, 2, 395–410.

Angilletta, M.J., Wilson, R.S., Navas, C.A., *et al.* (2003) Tradeoffs and the evolution of thermal reaction norms. Trends in Ecology and Evolution, 18, 234–40.

Baiser, B., Russel, G.J., and Lockwood, J.L. (2010) Connectance determines invasion success via trophic interactions in model food webs. Oikos, 119, 1970–6.

Bartomeus, I., Vilà, M., and Santamaria, L. (2008) Contrasting effects of invasive plants in plant-pollinator networks. Oecologia, 155, 761–70.

Bascompte, J. and Jordano, P. (2007) Plant-animal mutualistic networks: the architecture of biodiversity. Annual Review of Ecology, Evolution and Systematics, 38, 567–93.

Bascompte, J. and Jordano, P. (2013) *Mutualistic Networks*. Princeton, NJ: Princeton University Press.

Bascompte, J., Jordano, P., Melian, C.J., *et al.* (2003) The nested assembly of plant-animal mutualistic networks. Proceedings of the National Academy of Sciences of the USA, 100, 9383–7.

Bascompte, J., Jordano, P., and Olesen, J.M. (2006) Asymmetric coevolutionary networks facilitate biodiversity maintenance. Science, 312, 431–3.

Bascompte, J. and Stouffer, D.B. (2009) The assembly and disassembly of ecological networks. Philosophical Transactions of the Royal Society B: Biological Sciences, 364, 1781–7.

Bastola, U., Fortuna, M.A., Pascual-Garcia, A., *et al.* (2009) The architecture of mutualistic networks minimizes competition and increase biodiversity. Nature, 458, 1018–21.

Blüthgen, N., Menzel, F., and Blüthgen, N. (2006) Measuring specialization in species interaction networks. BMC Ecology, 6, 9.

Bolnick, D.I., Amarasekare, P., Araújo, M.S., *et al.* (2011) Why intraspecific trait variation matters in community ecology. Trends in Ecology and Evolution, 26, 183–92.

Brose, U. (2011) Extinctions in complex, size-structured communities. Basic and Applied Ecology, 12, 557–61.

Brännströms, Å.,Loeuille, N., and Loreau, M. (2011) Emergence and maintenance of biodiversity in an evolutionary food-web model. Theoretical Ecology, 4, 467–78.

Burgos, E., Ceva, H., Perazzo, R.P.J., *et al.* (2007) Why nestedness in mutualistic networks? Journal of Theoretical Biology, 249, 307–13.

Campbell, C., Yang, S., Albert, R., *et al.* (2015) Plant-pollinator community network response to species invasion depends on both invader and community characteristics. Oikos, 124, 406–13.

Campbell, C., Yang, S., Shea, K., *et al.* (2012) Topology of plant-pollinator networks that are vulnerable to collapse from species extinction. Physical Review E, 86, 021924.

Champagnat, R., Valette, R., Hochon, J.C., *et al.* (2001) Modeling, simulation and analysis of batch production systems. Discrete Event Dynamic Systems, 11, 119–36.

Chan, S. (2001) Complex adaptive systems. ESD.83 Research Seminar in Engineering Systems, Massachusetts Institute of Technology.

Charnov, E.L. (1979) Simultaneous hermaphroditism and sexual selection. Proceedings of the National Academy of Sciences of the USA, 76, 2480–4.

Clauset, A., Moore, C., and Newman, M.E. (2008) Hierarchical structure and the prediction of missing links in networks. Nature, 453, 98–101.

Cohen, Y. (2003) Distributed predator-prey co-evolution. Evolutionary Ecology Research, 5, 819–34.

Cohen, Y. (2009) Evolutionary distributions. Evolutionary Ecology Research, 11, 611–35.

Darwin, C. (1859) *The Origin of Species*. London: John Murray.

Davis, M.A., Grime, J.P., and Thompson, K. (2000) Fluctuating resources in plant communities: a general theory of invasibility. Journal of Ecology, 88, 528–34.

DeAngelis, D.L. (1980) Energy flow, nutrient cycling, and ecosystem resilience. Ecology, 61, 764–71.

Dercole, F. and Rinaldi, S. (2008) *Analysis of Evolutionary Processes: The Adaptive Dynamics Approach and its Applications*. Princeton, NJ: Princeton University Press.

Dicks, L.V., Corbet, S.A., and Pywell, R.F. (2002) Compartmentalization in plant-insect flower visitor webs. Journal of Animal Ecology, 71, 32–43.

Dieckmann, U. and Doebeli, M. (1999) On the origin of species by sympatric speciation. Nature, 400, 354–7.

Dieckmann, U. and Law, R. (1996) The dynamical theory of coevolution: a derivation from stochastic ecological processes. Journal of Mathematical Biology, 34, 579–612.

Dobson, A., Lafferty, K.D., Kuris, A.M., *et al.* (2008) Homage to Linnaeus: how many parasites? How many hosts? Proceedings of the National Academy of Sciences of the USA, 105, 11482–9.

Dobzhansky, T. (1950) Evolution in the tropics. American Scientist, 38, 209–21.

Doebeli, M. (2011) *Adaptive Diversification (MPB-48)*. Princeton, NJ: Princeton University Press.

Doebeli, M. and Dieckmann, U. (2000) Evolutionary branching and sympatric speciation caused by different types of ecological interactions. American Naturalist, 156, S77–101.

Doebeli, M. and Ispolatov, I. (2010) Complexity and diversity. Science, 328, 494–7.

Donohue, I., Petchey, O.L., Montoya, J.M., *et al.* (2013) On the dimensionality of ecological stability. Ecology Letters, 16, 421–9.

Dormann, C.F. and Strauß, R. (2014) A method for detecting modules in quantitative bipartite networks. Methods in Ecology and Evolution, 5, 90–98.

Drake, J.A. (1990) The mechanics of community assembly and succession. Journal of Theoretical Biology, 147, 213–33.

Drossel, B., Higgs, P.G., and McKane, A.J. (2001) The influence of predator-prey population dynamics on the long-term evolution of food web structure. Journal of Theoretical Biology, 208, 91–107.

Duncan, E.J., Gluckman, P.D., and Dearden, P.K. (2014) Epigenetics, plasticity, and evolution: how do we link epigenetic change to phenotype? Journal of Experimental Zoology Part B: Molecular and Developmental Evolution, 322, 208–20.

Duncan, R.P. and Williams, P.A. (2002) Ecology: Darwin's naturalization hypothesis challenged. Nature, 417, 608–9.

Dunne, J.A., Lafferty, K.D., Dobson, A.P., *et al.* (2013) Parasites affect food web structure primarily through Increased diversity and complexity. PLoS Biology, 11, e1001579.

Dunne, J. and Williams, R.J. (2009) Cascading extinctions and community collapse in model food webs. Philosophical Transactions of the Royal Society B: Biological Sciences, 364, 1711–23.

Dunne, J.A., Williams, R.J., and Martinez, N.D. (2002) Network structure and biodiversity loss in food webs: robustness increases with connectance. Ecology Letters, 5, 558–67.

Díaz, S., Kattge, J., Cornelissen, J.H., *et al.* (2016) The global spectrum of plant form and function. Nature, 529, 167–71.

Ehrlich, P.R. and Raven, P.H. (1964) Butterflies and plants: a study in coevolution. Evolution, 18, 586–608.

Eklöf, A. and Ebenman, B.O. (2006) Species loss and secondary extinctions in simple and complex model communities. Journal of Animal Ecology, 75, 239–46.

Elias, M., Gompert, Z., Jiggins, C., *et al.* (2008) Mutualistic interactions drive ecological niche convergence in a diverse butterfly community. PLoS One, 6, e300.

Elton, C.S. (1958) *The Ecology of Invasions by Animals and Plants*. London: Methuen.

Encinas-Viso, F., Revilla, T.A., and Etienne, R.S. (2012) Phenology drives mutualistic network structure and diversity. Ecology Letters, 15, 198–208.

Eshel, I., Motro, U., and Sansone, E. (1997) Continuous stability and evolutionary convergence. Journal of Theoretical Biology, 185, 333–43.

Estrada, E. (2007) Topological structural classes of complex networks. Physical Review E, 75, 016103.

Felsenstein, J. (2004) *Inferring Phylogenies*, Vol. 2. Sunderland, MA: Sinauer Associates.

Fortuna, M.A. and Bascompte, J. (2006) Habitat loss and the structure of plant—animal mutualistic networks. Ecology Letters, 9, 278–83.

Fortuna, M.A., Stouffer, D.B., Olesen, J.M., *et al.* (2010) Nestedness versus modularity in ecological networks: two sides of the same coin? Journal of Animal Ecology, 79, 811–17.

Fossette, S., Glleiss, A.C., Casey, J.P., *et al.* (2012) Does prey size matter? Novel observations of feeding in the leatherback turtle (Dermochelyscoriacea) allow a test of predator-prey size relationships. Biology Letters, 8, 351–4.

Fridley, J.D. (2011) Biodiversity as a bulwark against invasion: conceptual threads since Elton. In: Richardson, D.M. (ed.) *Fifty Years of Invasion Ecology: The Legacy of Charles Elton*. Oxford: Wiley-Blackwell, pp.121–30.

Gell-Mann, M. (1995) What is complexity? Complexity, 1, 16–19.

Geritz, S.A., Mesze, G., and Metz, J.A. (1998) Evolutionarily singular strategies and the adaptive growth and branching of the evolutionary tree. Evolutionary Ecology, 12, 35–57.

Gibson, M.R., Richardson, D.M., and Pauw, A. (2012) Can floral traits predict an invasive plant's impact on native plant-pollinator communities? Journal of Ecology, 100, 1216–23.

Gilpin, M.E. and Hanski, I. (1991) *Metapopulation Dynamics*. London: Academic Press.

Guimarães, P.R., Jordano, P., and Thompson, J.N. (2011) Evolution and coevolution in mutualistic networks. Ecology Letters, 14, 877–885.

Guimarães, P.R., Rico-Gray, V., Oliveira, P.S., *et al.* (2007) Interaction intimacy affects structure and coevolutionary dynamics in mutualistic networks. Current Biology, 17, 1797–1803.

Guimera, R, Stouffer, D.B., Sales-Pardo, M., *et al.* (2010) Origin of compartmentalisation in food webs. Ecology,91, 2941–51.

Gurevitch, J., Wardle, G.M., Inderjit, *et al.* (2011) Emergent insights from the synthesis of conceptual frameworks for biological invasions. Ecology and Evolution, 14, 407–18.

Hardin, G. (1960) The competitive exclusion principle. Science, 131, 1292–7.

Holland, J.N. and DeAngelis, D.L. (2010) A consumer-resource approach to the density-dependent population dynamics of mutualism. Ecology, 91, 1286–95.

Hui, C. and McGeoch, M.A. (2014) Zeta diversity as a concept and metric that unifies incidence-based biodiversity patterns. American Naturalist, 184, 684–94.

Hui, C., Richardson, D.M., Landi, P., *et al.* (2016) Defining invasiveness and invasibility in ecological networks. Biological Invasions, 18, 971–83.

Hurford, A., Cownden, D., and Day, T. (2010) Next-generation tools for evolutionary invasion analysis. Journal of the Royal Society Interface, 7, 561–71.

Ives, A. and Carpenter, S.R. (2007) Stability and diversity of ecosystems. Science, 317, 58–62.

James, A., Pitchford, J.W., and Plank, M.J. (2012) Disentangling nestedness from models of ecological complexity. Nature, 487, 227–30.

Janzen, D.H. (1985) On ecological fitting. Oikos, 45, 308–10.

Jeffries, C. (1974) Qualitative stability and digraphs in model ecosystems. Ecology, 55, 1415–19.

Jones, R.A.,Ferrière, R., and Bronstein, J.L. (2009) Eco-evolutionary dynamics of mutualists and exploiters. American Naturalist, 174, 780–94.

Jordano, P., Bascompte, J., and Olesen, J.M. (2003) Invariant properties in coevolutionary networks of plant-animal interactions. Ecology Letters, 6, 69–81.

Jousselin, E. Rasplus, J.Y., and Kjellberg, F. (2003) Convergence and coevolution in a mutualism: evidence from a molecular phylogeny of *Ficus*. Evolution, 57, 1255–69.

Justus, J. (2008) Complexity, diversity, stability. In: Sarkar, S. and Plutynski, A. (eds) *A Companion to the Philosophy of Biology*. Oxford: Blackwell, pp. 321–50..

Kaiser-Bunbury, C.N., Muff, S., Memmott, J., *et al.* (2010) The robustness of pollination networks to the loss of species and interactions: a quantitative approach incorporating pollinator behaviour. Ecology Letters, 13, 442–52.

Kauffman, S. (1995) *At Home in the Universe*. Oxford: Oxford University Press.

Kauffman, S.A (1993) *The Origins of Order: Self Organization and Selection in Evolution*. Oxford: Oxford University Press.

Kiers, T.E., Palmer, T.M., Ives, A.R., *et al.* (2010) Mutualisms in a changing world: an evolutionary perspective. Ecology Letters, 13, 1459–74.

Kimbrell, T. and Holt, R.D. (2005) Individual behaviour, space and predator evolution promote persistence in a two-patch system with predator switching. Evolutionary Ecology Research, 7, 53–71.

Kondoh, M. (2003) Foraging adaptation and the relationship between food-web complexity and stability. Science, 299, 1388–91.

Krasnov, B.R., Fortuna, M.A., Mouillot, D., *et al.* (2012) Phylogenetic signal in module composition and species connectivity in compartmentalized host-parasite networks. American Naturalist, 179, 501–11.

Kunstler, G., Falster, D., Coomes, D.A., *et al.* (2016) Plant functional traits have globally consistent effects on competition. Nature, 529, 204–7.

Leimar, O. (2009) Multidimensional convergence stability. Evolutionary Ecology Research, 11, 191–208.

Levin, S.A. (1999) Towards a science of ecological management. Conservation Ecology, 3, 6.

Levin, S.A. and Segel, L.A. (1985) Pattern generation in space and aspect. SIAM Review, 27, 45–67.

Loeuille, N. and Loreau, M. (2005) Evolutionary emergence of size-structured food webs. Proceedings of the National Academy of Sciences of the USA, 102, 5761–6.

Lonsdale, W.M. (1999) Global patterns of plant invasions and the concept of invasibility. Ecology, 80, 1522–36.

Loreau, M. (2000) Are communities saturated? On the relationship between α, β and γ diversity. Ecology Letters, 3, 73–6.

Loreau, M. (2010) *From Populations to Ecosystems: Theoretical Foundations for a New Ecological Synthesis*. Princeton, NJ: Princeton University Press.

Lurgi, M., Galiana, N., López, B.C., *et al.* (2014) Network complexity and species traits mediate the effects of biological invasions on dynamic food webs. Frontiers in Ecology and Evolution, 2, 00036.

MacArthur, R.H. (1972) *Geographical Ecology*. New York, NY: Harper and Row.

May, R.M. (1973) *Stability and Complexity in Model Ecosystems*. Princeton, NJ: Princeton University Press.

Mayfield, M.M. and Levine, J.M. (2010) Opposing effects of competitive exclusion on the phylogenetic structure of communities. Ecology Letters, 13, 1085–93.

Maynard Smith, J. and Price, G.R. (1973) Thelogic of animal conflict. Nature, 246, 15–18.

McCann, K.S. (2000) The diversity–stability debate. Nature, 405, 228–33.

McKane, A.J. (2004) Evolving complex food webs. European Physical Journal B, 38, 287–95.

Mello, M.A.R., Marquitti, F.M.D., Guimarães Jr, P.R., *et al.* (2011) The missing part of seed dispersal networks: structure and robustness of bat-fruit interactions. PLoS One, 6, e17395.

Memmott, J., Waser, N.M., and Price, M.V. (2004) Tolerance of pollination networks to species extinctions. Proceedings of the Royal Society B: Biological Sciences, 271, 2605–11.

Metz, J.A., Nisbet, R.M., and Geritz, S.A. (1992) How should we define 'fitness' for general ecological scenarios? Trends in Ecology and Evolution, 7, 198–202.

Minoarivelo, H.O. (2016). The eco-evolutionary dynamics of mutualistic networks: from pattern of emergence to stability. PhD thesis, Stellenbosch University, Stellenbosch.

Minoarivelo, H.O. and Hui, C. (2016a) Trait-mediated interaction leads to structural emergence in mutualistic networks. Evolutionary Ecology, 30, 105–21.

Minoarivelo, H.O. and Hui, C. (2016b) Invading a mutualistic network: to be or not to be similar. Ecology and Evolution, 6, 4981–96.

Minoarivelo, H.O., Hui, C., Terblanche, J.S., *et al.* (2014) Detecting phylogenetic signal in mutualistic interaction networks using a Markov process model. Oikos, 123, 1250–60.

Morales, C.L. and Traveset, A. (2009) A meta-analysis of impacts of alien vs. native plants on pollinator visitation and reproductive success of co-flowering native plants. Ecology Letters, 12, 716–28.

Morales-Castilla, I., Matias, M.G., *et al.* (2015) Inferring biotic interactions from proxies. Trends in Ecology and Evolution, 30, 347–56.

Morton, R.D. and Law, R. (1997) Regional species pools and the assembly of local ecological communities. Journal of Theoretical Biology, 187, 321–31.

Münkemüller, T. and Gallien, L. (2015) VirtualCom: a simulation model for eco-evolutionary community assembly and invasion. Methods in Ecology and Evolution, 6, 735–43.

Newman, M.E. (2006) Modularity and community structure in networks. Proceedings of the National Academy of Sciences of the USA, 103, 8577–82.

Nowak, M. and Sigmund, K. (1990) The evolution of stochastic strategies in the prisoner's dilemma. Acta Applicandae Mathematicae, 20, 247–65.

Nuismer, S.L., Jordano, P., and Bascompte, J. (2013) Co-evolution and the architecture of mutualistic networks. Evolution, 67, 338–54.

Nuismer, S.L., MacPherson, A., and Rosenblum, E.B. (2012) Crossing the threshold: gene flow, dominance and the critical level of standing genetic variation required for adaptation to novel environments. Journal of Evolutionary Biology, 25, 2665–71.

Nuwagaba, S., Zhang, F., and Hui, C. (2015) A hybrid behavioural rule of adaptation and drift explains the emergent architecture of antagonistic networks. Proceedings of the Royal Society B: Biological Sciences, 282, 20150320.

Okuyama, T. and Holland, J.N. (2008) Network structural properties mediate the stability of mutualistic communities. Ecology Letters, 11, 208–16.

Olesen, J.M., Bascompte, J., Dupont, Y.L., et al. (2007) The modularity of pollination networks. Proceedings of the National Academy of Sciences of the USA, 104, 19891–6.

Olesen, J.M. and Jordano, P. (2002) Geographic patterns in plant-pollinator mutualistic networks. Ecology, 83, 2416–24.

Otto, S.P. and Day, T. (2007) A Biologist's Guide to Mathematical Modeling in Ecology and Evolution. Princeton, NJ: Princeton University Press.

Padrón, B., Traveset, A., Biedenweg, T., Díaz, D., et al. (2009) Impact of alien plant invaders on pollination networks in two archipelagos. PLoS One, 4, e6275.

Parker, G.A. and Maynard Smith, J. (1990) Optimality theory in evolutionary biology. Nature, 348, 27–33.

Parrott, L. and Meyer, W.S. (2012) Future landscapes: managing within complexity. Frontiers in Ecology and the Environment, 10, 382–9.

Pascual, M. and Dunne, J.A. (eds) (2005) Ecological Networks: Linking Structure to Dynamics in Food Webs. Oxford: Oxford University Press.

Petanidou, T., Kallimanis, A.S., Tzanopoulos, J., et al. (2008) Long-term observation of a pollination network: fluctuation in species and interactions, relative invariance of network structure and implications for estimates of specialization. Ecology Letters, 11, 564–75.

Pimm, S.L. and Lawton, J.H. (1978) On feeding on more than one trophic level. Nature, 275, 542–4.

Poisot, T. (2013) An a posteriori measure of network modularity. F1000 Research, 2, 130.

Prado, P.I. and Lewinsohn, T.M. (2004) Compartments in insect-plant and their consequences for community structure. Journal of Animal Ecology, 73, 1168–78.

Prentis, P.J., Wilson, J.R.U., Dormontt, E.E., et al. (2008) Adaptive evolution in invasive species. Trends in Plant Science, 13, 288–94.

Prigogine, I. and Nicolis, G. (1977) Self-Organization in Nonequilibrium Systems: From Dissipative Structures to Order through Fluctuations. New York, NY: Wiley.

Pyšek, P. and Richardson, D.M. (2007). Traits associated with invasiveness in alien plants: where do we stand? In: Nentwig, W. (ed.) Biological Invasions. Berlin: Springer, pp. 97–125.

Quirk, J. and Ruppert, R. (1965) Qualitative economics and the stability of equilibrium. Review of Economic Studies, 32, 311–26.

Rezende, E.L., Jordano, P., and Bascompte, J. (2007b) Effects of phenotypic complementarity and phylogeny on the nested structure of mutualistic networks. Oikos, 116, 1919–29.

Rezende, E.L., Lavabre, J.E., Guimarães, P.R., et al. (2007a) Non-random coextinctions in phylogenetically structured mutualistic networks. Nature, 448, 925–8.

Richardson, D.M., Allsopp, N., D'Antonio, C.M., et al. (2000) Plant invasions: the role of mutalism. Biological Reviews, 75, 65–93.

Richardson, D.M. and Pyšek, P. (2006) Plant invasions: merging the concepts of species invasiveness and community invasibility. Progress in Physical Geography, 30, 409–31.

Rodríguez-Gironés, M.A. and Llandres, A.L. (2008) Resource competition triggers the co-evolution of long tongues and deep corolla tubes. PLoS One, 3, e2992.

Rohr, R.P., Saavedra, S., and Bascompte, J. (2014) On the structural stability of mutualistic systems. Science, 345, 1253497.

Romanuk, T.N., Zhou, Y., Brose, U., et al. (2009) Predicting invasion success in complex ecological networks. Philosophical Transactions of the Royal Society B: Biological Sciences, 364, 1743–54.

Rosenzweig, M.L. (1971) Paradox of enrichment: destabilization of exploitation ecosystems in ecological time. Science, 171, 385–7.

Saavedra, S., Reed-Tsochas, F., and Uzzi, B. (2009) A simple model of bipartite cooperation for ecological and organizational networks. Nature, 457, 463–6.

Santamaría, L. and Rodríguez-Gironés, M.A. (2007) Linkage rules for plant-pollinator networks: trait complementarity or exploitation barriers? PLoS Biology, 5, 354–62.

Saul, W.C., Jeschke, J.M., and Heger, T. (2013) The role of eco-evolutionary experience in invasion success. NeoBiota, 17, 57–74.

Shea, K. and Chesson, P. (2002) Community ecology theory as a framework for biological invasions. Trends in Ecology and Evolution, 17, 170–6.

Solé, R. and Goodwin, B. (2000) *Signs of Life: How Complexity Pervades Biology*. New York, NY: Basic Books.

Spottiswoode, C.N. and Stevens, M. (2012) Host-parasite arms races and rapid changes in bird egg appearance. American Naturalist, 179, 633–48.

Stang, M., Klinkhamer, P.G., Waser N.M., *et al.* (2009) Size specific interaction patterns and size matching in a plant-pollinator interaction web. Annals of Botany, 103, 1459–69.

Staniczenko, P., Lewis, O.T., Jones, N.S., *et al.* (2010) Structural dynamics and robustness of food webs. Ecology Letters, 13, 891–9.

Stephens, D.W. and Krebs, J.R. (1986) *Foraging Theory*. Princeton, NJ: Princeton University Press.

Stohlgren, T.J., Barnett, D.T., and Kartesz, J.T. (2003) The rich get richer: patterns of plant invasions in the United States. Frontiers in Ecology and the Environment, 1, 11–14.

Stouffer, D.B. and Bascompte, J. (2011) Compartmentalisation increases food-web persistence. Proceedings of the National Academy of Sciences of the USA, 108, 3648–52.

Stouffer, D.B., Cirtwill, A.R., and Bascompte, J. (2014) How exotic plants integrate into pollination networks. Journal of Ecology, 102, 1442–50.

Suweis, S., Simini, F., Banavar, J.R., *et al.* (2013) Emergence of structural and dynamical properties of ecological mutualistic networks. Nature, 500, 449–52.

Thompson, R.M. and Townsend, C.R. (1999) The effect of seasonal variation on the community structure and food-web attributes of two streams: implications for food-web science. Oikos, 87, 75–88.

Thuiller, W., Gallien, L., Boulangeat, I., *et al.* (2010) Resolving Darwin's naturalization conundrum: a quest for evidence. Diversity and Distributions, 16, 461–75.

Thébault, E. and Fontaine, C. (2010) Stability of ecological communities and the architecture of mutualistic and trophic networks. Science, 329, 853–6.

Tilman, D. (2004) Niche tradeoffs, neutrality, and community structure: a stochastic theory of resource competition, invasion, and community assembly. Proceedings of the National Academy of Sciences of the USA, 101, 10854–61.

Townsend, C.R., Thompson, R.M., McIntosh, A.R., *et al.* (1998) Disturbance, resource supply, and food-web architecture in streams. Ecology Letters, 1, 200–9.

Traveset, A., Heleno, R., Chamorro, S., *et al.* (2013) Invaders of pollination networks in the Galápagos Islands:

emergence of novel communities. Proceedings of the Royal Society B: Biological Sciences, 280, 20123040.

Traveset, A. and Richardson, D.M., 2006. Biological invasions as disruptors of plant reproductive mutualisms. Trends in Ecology and Evolution, 21, 208–16.

Traveset, A. and Richardson, D.M. (2014) Mutualistic interactions and biological invasions. Annual Reviews in Ecology, Evolution and Systematics, 45, 89–113.

Tylianakis, J.M., Tscharntke, T., and Lewis, O.T. (2007) Habitat modification alters the structure of tropical host-parasitoid food webs. Nature, 445, 202–5.

Vacher, C., Piou, D., and Desprez-Loustau, M.L. (2008) Architecture of an antagonistic tree/fungus network: the asymmetric influence of past evolutionary history. PLoS One, 3, e1740.

Valdovinos, F.S., Ramos-Jiliberto, R., Garay-Narváez, L., *et al.* (2010) Consequences of adaptive behaviour for the structure and dynamics of food webs. Ecology Letters, 13, 1546–59.

Vallina, S.M. and Quéré, C.L. (2011) Stability of complex food webs: resilience, resistance and the average interaction strength. Journal of Theoretical Biology, 272, 160–73.

Van Alphen, J.J.M., Bernstein, C., and Driessen, G. (2003) Information acquisition and time allocation in insect parasitoids. Trends in Ecology and Evolution, 18, 81–7.

Van Baalen, M., Křivan, V., van Rijn, P.C., *et al.* (2001) Alternative food, switching predators, and the persistence of predator-prey systems. American Naturalist, 157(5), 512–24.

Van Kleunen, M., Weber, E., and Fischer, M. (2010) A meta-analysis of trait differences between invasive and non-invasive plant species. Ecology Letters, 13, 235–45.

Vázquez, D.P. and Aizen, M.A. (2003) Null model analyses of specialization in plant-pollinator interactions. Ecology, 84, 2493–501.

Velland, M. (2010) Conceptual synthesis in community ecology. Quarterly Review of Biology, 85, 183–206.

Vieira, M.C. and Almeida-Neto, M. (2015) A simple stochastic model for complex coextinctions in mutualistic networks: robustness decreases with connectance. Ecology Letters, 18, 144–52.

Vilà, M., Bartomeus, I., Dietzsch, A.C., *et al.* (2009) Invasive plant integration into native plant-pollinator networks across Europe. Proceedings of the Royal Society B: Biological Sciences, 276, 3887–93.

Vukics, A., Asbóth, J., and Meszéna, G. (2003) Speciation in multidimensional evolutionary space. Physical Review E, 68, 041903.

Waser, N.M. (2015) Competition for pollination and the evolution of flowering time. American Naturalist, 185, iii–v.

Waser, N.M., Chittka, L., Price, M.V., *et al.* (1996) Generalization in pollination systems, and why it matters. Ecology, 77, 1043–60.

Waxman, D. and Gavrilets, S. (2005) Twenty questions on adaptive dynamics. Journal of Evolutionary Biology, 18, 1139–54.

Whiting, M.F., Bradler, S., and Maxwell, T. (2003) Loss and recovery of wings in stick insects. Nature, 421, 264–7.

Wilder, S.M., Holway, D.A., Suarez, A.V., *et al.* (2011) Intercontinental differences in resource use reveal the importance of mutualisms in fire ant invasions. Proceedings of the National Academy of Sciences of the USA, 108, 20639–44.

Yodzis, P. (1981) The stability of real ecosystems. Nature, 289, 674–6.

Yoshida, T., Jones, L.E., Ellner, S.P., *et al.* (2003) Rapid evolution drives ecological dynamics in a predator-prey system. Nature, 424, 303–6.

Zhang, F. and Hui, C. (2014) Recent experience-driven behaviour optimizes foraging. Animal Behaviour, 88, 13–19.

Zhang, F., Hui, C., and Pauw, A. (2013) Adaptive divergence in Darwin's race: how coevolution can generate trait diversity in a pollination system. Evolution, 67, 548–60.

Zhang, F., Hui, C., and Terblanche, J.S. (2011) An interaction switch predicts the nested architecture of mutualistic networks. Ecology Letters, 14, 797–803.

CHAPTER 11

Managing biological invasions in the Anthropocene

11.1 Background

Previous chapters have described how organisms introduced by humans infiltrate recipient ecosystems, how some of them proliferate and spread, and the many ways that invasive species interact with resident organisms and communities, often leading to profound alterations to ecosystems. We have reviewed the rapid progress in recent decades towards an understanding of invasion dynamics. This research has followed many approaches, using a wide range of methods, and has borrowed insights and tools from many diverse disciplines. Important insights are emerging from increasingly sophisticated manipulative experiments, from the many natural experiments created through introductions, many forms of intra-regional dissemination, invasions, and interactions with other facets of global change, and from diverse theoretical and modelling studies (Table 11.1). We have touched briefly on the many ways in which biological invasions affect how people view and value various components of ecosystems. Almost every issue of journals in the fields of biogeography, ecology, and environmental management contains something relevant to invasion dynamics. There is increasing engagement from the humanities in issues pertaining to invasive species, and there has been a rapid increase in the level of interdisciplinarity in invasion research, both within biological sub-disciplines and, increasingly, between biology and other fields.

The study of biological invasions is a vibrant field. Previous chapters have shown that invasions provide endless opportunities for gaining insights into the determinants of biological diversity and distributions. Despite important advances, the more we learn about any aspect of invasions the more questions arise, and the more hypotheses are proposed (Catford et al. 2009; Davis 2011). Our understanding of the 'nuts and bolts' of invasions is based largely on insights from a fairly small sample of taxa and recipient ecosystems, and invasion dynamics in many geographical areas and ecosystem types remain poorly studied (Pyšek et al. 2008). We need to address these biases, but not every invasive species and invaded ecosystem can be studied. Given the huge and growing extent of the phenomenon and the escalating scale and complexity of problems associated with managing invasive species, what approaches and directions hold the most promise for generating new insights that will be particularly useful for management? Table 11.1 provides some suggestions on the key unknowns in the field: research on these topics is needed to advance our capacity to understand, and therefore to manage, invasions.

Despite the progress towards understanding the determinants of invasiveness, it is rather disconcerting that the most robust predictor of whether a species will invade a particular habitat or region remains whether the species has invaded elsewhere (e.g. Rejmánek et al. 2005). Our understanding of the impacts of invasive species is also sketchy,

Invasion Dynamics. Cang Hui and David M. Richardson, Oxford University Press (2017).
© Cang Hui and David M. Richardson. DOI 10.1093/acprof:oso/9780198745334.001.0001

Table 11.1 A summary of key aspects of invasion dynamics addressed in Chapters 2–10, the approaches/disciplines applied, and priorities for future research.

Chapter	Key aspects	Approach/discipline	Priorities for future research
2: The dynamics of spread	• Four patterns of invasive spread (exponential/sigmoidal; linear, biphasic, and acceleration) • Exponential expansion as the null mode of spread • Stratified dispersal versus habitat transition in explaining biphasic expansion • Multiple reasons for lag phases • Shocks and adaptive biotic interactions behind boom-and-bust patterns • Context dependence and discordance in spread	• Data collection • Non-linear and segment regression • Historical reconstruction • Population ecology	• Standard reporting • Testing against four parametric forms of spread • Elucidating context-dependent spread • Forecasting of lag phases
3: Modelling spatial dynamics	• A general modelling framework: core, context, and methods • Lagrangian models of random walks and step-selection functions • Eulerian models of reaction-diffusion models and Skellam's formula of wave speed • Integrodifference equations for implementing different dispersal kernels • Implementing environmental heterogeneity, stochasticity, and biotic interactions as modelling context • Species distribution models • Agent-based modelling 101: individual-based, lattice, and network models	• Standard modular framework for model building • Spatial ecology • Computer science • Numerical ecology	• Transparent models • Parameterization of complex models • Demography-based species distribution models • Model validation
4: From dispersal to boosted range expansion	• Boosted range expansion and Reid's paradox • Dispersal syndromes and spatial sorting behind the acceleration of range expansion • Core–edge comparison of dispersal syndromes • The classic case of cane toads in Australia and other examples • Three stages of dispersal (departure, transience, and settlement) and factors behind each • Long-distance dispersal leads to fat-tailed dispersal kernels • Human-mediated dispersal as the main booster • Human movement follows Levy flight, but with critical difference • Skellam's formula revised to incorporate the variety of dispersal kernels, spatial sorting, and propagule pressure	• Dispersal syndromes • Trait comparisons • Population genetics • Physiology • Movement ecology • Dispersal ecology • Data science	• Technology, devices, and approaches for better tracking of seed dispersal by wind • Technology, devices, and approaches for better tracking of human-mediated dispersal
5: Non-equilibrium dynamics	• The debates around the *balance of nature* • Time-series analysis of population variability • Positive density dependence: Allee effects lead to range pinning • High dispersal rates not good for populations suffering from Allee effects • Identify the role of density dependence and other environmental factors in population dynamics by autoregressive models • Niche shift in the novel range: who shifts and is it good for invasion? • Spatial synchrony of multiple populations: metapopulations and portfolio theory of multi-site population viability • Foretell temporal dynamics and invasiveness from spatial patterns • Spatial autoregressive models for exploring invasion dynamics far from equilibrium	• Time-series analysis • Population viability analysis • Autoregressive models • Statistical ecology • Species distribution models • Population ecology • Spatial ecology	• Invasion management considering density dependence • Explaining and/or forecasting niche shift • Space-for-time substitution revisited

(continued)

Table 11.1 (*Continued*)

Chapter	Key aspects	Approach/discipline	Priorities for future research
6: Biotic interactions	• Direct pairwise interaction explained from a simple model • 19 hypothesis related to direct biotic interactions • Antagonism-related hypotheses (e.g. enemy release, EICA, and new associations) • Mutualism-related hypotheses (e.g. mutualist facilitation, host jumping, and co-introduction) • Competition-related hypotheses (e.g. biotic resistance, novel weapons, empty niche, limiting similarity, and niche resemblance) • Mediated interactions through shared predators/parasites and resources • Biological invasions as passengers or drivers of environmental changes • The many ways of measuring and quantifying interaction strength and related hypotheses (e.g. intermediate disturbance, resource–enemy release, fluctuating resource, and opportunity windows) • Spatial interactions incorporated, modelled, and tested • Interaction promiscuity and switching facilitate invasion (e.g. enhanced mutualism) • Novel eco-evolutionary experiences: pre- and rapid adaptation, drift, founder effects, and Darwin's naturalisation hypothesis	• Greenhouse/laboratory experiments • Manipulative experiments • Species distribution models • Contest experiments • Population ecology • Physiology • Population genetics	• Standard interaction strength measures • Documenting interaction strength • Documenting rapid evolution • Clarifying confusion between promiscuity and the generalist/specialist dichotomy
7: Regime shifts	• Regime shifts, bifurcation, and reinforcing feedbacks • The multifaceted concept of system stability • Criterion for system reshuffling and losing its current regime • Evolutionary regimes • Early warning signals of regime shift: critical slowing down, flickering, and others • Detecting the existence of system bistability and examples • Using effect size ratio to identify reinforcing feedbacks during invasion • Facilitating invasion through ecosystem engineering of resources and disturbance regimes	• Systems science • Compartment models • Flow charts • Stability analysis • Dynamical systems • Ecosystem engineering • Ecosystem science	• Early warning signals for invasions with major impacts • Detecting bistability and tipping points • The role of ecosystem engineering and niche construction • Standardized reporting of regimes and reinforcing feedbacks
8: Community assembly and succession	• Disturbance and secondary succession facilitate invasion • Invasion dissembles local communities by dismantling aggregation that reduces competition • Regional succession from invasion reduces compartmentalization • Eight mechanisms of species coexistence • Niche versus neutral, deterministic versus stochastic in community assembly • Fitness difference and niche overlap determine invasion performance • Invasion paradox explained from bottom-up mechanisms of coexistence through scale and heterogeneity • A framework of assessing the impact of invasion on community assembly through species turnover and diversity signature • Island biogeography of alien species • Alien biomes and hypotheses of their geographic boundaries	• Community ecology • Disturbance • Succession • Species turnover • Biogeography • Spatial ecology	• Elucidating the role of disturbance and succession in facilitating invasions • Testing the unified framework of invasion performance based on fitness difference and niche overlap • Methods identifying key assembly processes • Alien biomes and their boundaries explained and tested

Chapter	Topics	Disciplines	Research needs
9: Monitoring and management	• Occurrence, status, and impact as essential variables for alien monitoring • Essential and necessary variables for monitoring plant invasions • Modular strategy for invasion management • Lessons learned from monitoring plant invasions in protected areas • Metrics of impact and risks (e.g. EICAT) • Integrated prioritization of species, pathways, and sites • Spatial prioritization and risk mapping • Invasion debt and the risk of doing nothing • Political will influenced by lobby groups • Four components of a bioeconomic model: dynamics, cost, cost-effectiveness, and impact • Rules of thumb for invasion control • Economic valuation: utility versus opportunity costs • Solving optimal management of invasion control using dynamic programming • Context-dependent results from spatial optimization: controlling edge versus core populations	• Invasion management • Global change biology • Political ecology • Data sharing • Data science • Optimal control • Mathematical ecology • Bioeconomics • Multi-criteria decision analysis • Economic behaviour • Sampling and monitoring • Computer science	• Global co-ordination of data collection, reporting, and sharing • Implementation of efficient data monitoring schemes • Standardized documenting and reporting • Accurate economic valuation of the impacts of invasive species • Measuring control efficiency • Multi-species prioritization and optimal control • Spatial optimization and prioritization
10: Complex adaptive networks	• From Elton's diversity–invasibility hypothesis to the debate on May's complexity–stability relationship • What is a complex adaptive system? • Architecture of ecological networks, typically and under invasion • Co-evolution versus ecological fitting in explaining network emergence • Adaptive interaction switching of Wallace's elimination of the unfit explains network emergence • Invasibility of ecological networks defined and rooted in the concepts of community saturation and evolutionarily stable strategy (ESS) • Building a community at ecological versus evolutionary equilibrium • What is evolutionary invasion analysis and adaptive dynamics? • Clustering of metrics of network architecture (e.g. nestedness and modularity) and stability (e.g. resilience, robustness, and invasibility) • System evolution towards marginal stability • Opportunity niche and biological invasion of a mutualistic network • Profiling a network invader: trait matching and node degrees	• Systems science • Network science • Ecological network • Simulations • Trait-mediated interactions • Trait evolution • System emergence • Complex adaptive systems • Evolutionary invasion analysis • Adaptive dynamics • Mathematical ecology • Phylogeny	• Trait-based frameworks for invasiveness and invasibility • The importance of co-evolution versus ecological fitting in invaded ecosystems clarified • Improved frameworks for assessing effects of invasions on system functioning and ecosystem services • Testing of the hypothesis of system evolution towards marginal stability • Invasion management considering linkage weight and node degree. • Documentation of invader-centric network matrices • Identification of opportunity niches in ecological networks

and frameworks that have been proposed to-date are derived from insights from studies of a small sample of cases, mostly those invasions that cause the most rapid and dramatic impacts. Most invasive species, however, have slower and more subtle impacts; such impacts may nevertheless have profound implications for ecosystems over longer time scales (e.g. Traveset and Richardson 2006) and much more work is needed to better understand the drivers of such impacts (Strayer 2012; Kumschick *et al.* 2015). Exciting work is underway on numerous fronts to address the determinants of invasiveness and invasibility, and advances in these areas should go a long way towards countering the criticisms that have been levelled against invasion science (Table 1.1 in Chapter 1). However, given the huge complexity of ecosystems and the many additional layers created by the myriad drivers of biotic shuffling and other facets of global change that are constantly adjusting everything that affects alien species along the introduction–naturalization–invasion continuum (Kueffer *et al.* 2013), there are clearly inherent limitations on predictability in invasion science. In our view, expectations for what should be expected in terms of a 'predictive understanding of invasions' (which some would argue should be the ultimate aim of invasion science) need to be tempered to a degree. 'Silver bullet' or 'one-size-fits-all' explanations and generalizations for all types of invasions are futile aims. The same applies to management: the best prescriptions for effective intervention involve reviewing and adapting methods that have been applied elsewhere for particular situations. Despite the accumulation of 'best practice' procedures for dealing with particular invasive species using various approaches (e.g. Code of Best Practices for Classical Biological Control of Weeds; Balciunas 2000), every management intervention is an experiment. Unfortunately, far too often managers repeat failed experiments, and more effective ways of sharing management experiences for particular groups of species and categories of ecosystems and habitats are needed (Wilson *et al.* 2011).

Kueffer and colleagues (2013) proposed that a profitable and pragmatic way forward to improve our understanding of the full suite of interacting factors that influence invasions is to give more attention to intensive studies of particular model organisms and ecosystems, to comparative studies across study systems, and to undertake quantitative syntheses constrained to clearly defined theoretical domains. Following this approach does not deny that every introduction/invasion event is, to a degree, unique. Rather, it seeks to construct a series of model examples for which the full range of complex interactions has (or can be) thoroughly elucidated and which can serve as baselines or benchmarks against which other invasions can be contrasted. To this end, Kueffer and colleagues (2013) proposed ways of improving the means by which information about species biology and ecosystem characteristics can be integrated with particular invasion contexts such as an invasion stage, invaded habitat, and/or socio-economic context, thereby defining particular 'invasion syndromes' (Fig. 11.1). As an example of an invasion syndrome, and an application of this approach (termed 'integrative invasion science'), Kueffer and colleagues (2013) discussed plant invasions in mountains. Application of the approach outlined Figure 11.1, involving an understanding of species traits, ecological, evolutionary, and anthropogenic factors that facilitate such invasions, provides the foundation for a transparent framing of the phenomenon. Elucidation of the drivers of such invasions, which have historically mainly involved the self-propagated spread of alien species from lowlands with high levels of human influence to more natural high-elevation sites, requires an understanding of multiple interacting factors (including metapopulation dynamics, dispersal, phenotypic plasticity, evolutionary change, and land-use gradients). Clarification of such an invasion syndrome facilitates new insights and a level of prediction across regions, assuming that similar drivers that shaped past invasions remain the main drivers of invasion into the future. Importantly, the characterization of such a syndrome also identifies new areas for research under changing or predicted future scenarios—for example, the increased importance of direct introductions of mountain plants to new high-elevation sites. Implicit in the formulation of guidelines for this model of 'integrative invasion science' is the recognition that a theory of invasions that will be useful for management will

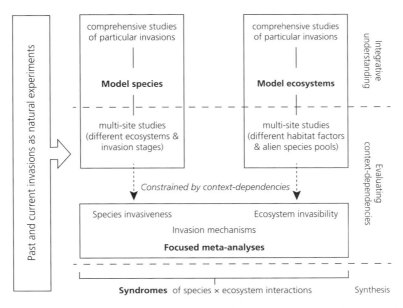

Fig. 11.1 Schematic diagram of the interrelationships among emerging approaches in integrative invasion science. From Kueffer *et al.* (2013). Reproduced with permission from John Wiley & Sons.

be 'a somewhat heterogeneous conglomerate of elements of varying generality and predictive power' (Kueffer *et al.* 2013).

Human actions are, by definition, the proximate driver of invasions as they facilitate the dispersal of species to regions beyond their normal dispersal distance. Humans also affect every aspect of progression along the introduction–naturalization–invasion continuum, for example by aiding (directly and indirectly) the dissemination of the alien species within its new range, and altering the susceptibility of ecosystems to invasion and disruption. Invasions are often 'drivers' of change, but are also frequently 'passengers' of change (MacDougall and Turkington 2005). Whether drivers or passengers, invasive species are firmly embedded in social-ecological systems worldwide. Over the last two decades, there has been a rapid increase in interest in the 'human dimensions' of biological invasions (e.g. McNeely 2001; Le Maitre *et al.* 2004). The need for new insights in this area has become increasingly obvious and urgent as the complexity of managing invasions as part of a 'lethal cocktail' of drivers of global change and environmental degradation has grown (Richardson 2011). Areas dominated by human influences have replaced natural ecosystems in many of the world's most biodiverse and productive regions, and now cover nearly twice the global area of wildlands (Ellis 2013). Non-native species are common, and often dominant, in 'anthromes' (Fig. 11.2). Problems with invasive species are increasing in all ecosystems, even in large protected areas—until fairly recently the bastions of natural ecosystems (Foxcroft *et al.* 2013). Very large parts of the planet thus comprise systems with abiotic, biotic, and social elements (and the interactions between these components) that, because of human influence, differ from those that prevailed historically, and have a tendency to self-organize and manifest novel qualities without intensive human management (Hobbs *et al.* 2014). Across most of the planet, no aspect of ecosystem functioning, nor any determinant of ecosystem services, can be understood or managed without considering the current or potential effects of invasive species. This chapter investigates the growing complexity of dealing with biological invasions in a world that is increasingly dominated by human activities (Ellis 2015) and where biological invasions are a key component of global change (Franklin *et al.* 2016).

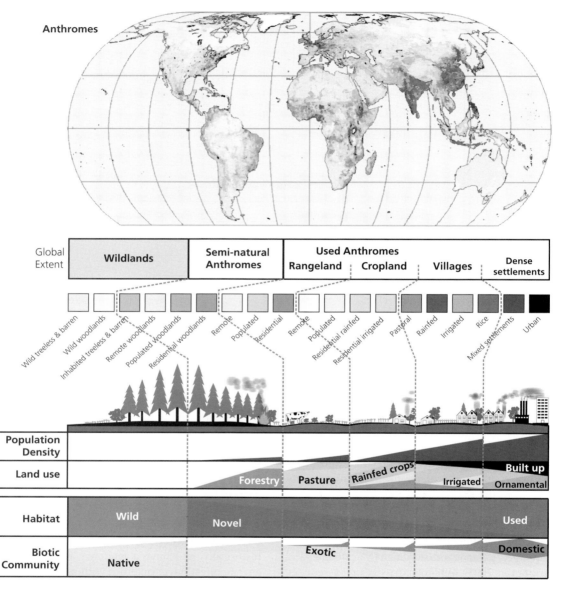

Fig. 11.2 The global extent of contemporary anthromes (top; for the year 2000) and associated socio-economic and ecological characteristics, including human population density, land use, and integrity of habitat and biotic communities (bottom). From Martin *et al.* (2014). Reproduced with permission from John Wiley & Sons. (see plate 4)

11.2 Biological invasions and social-ecological systems

Social-ecological systems challenge most management strategies and policies because of the high levels of complexity (Levin *et al.* 2013). Chapter 10 introduced the topic of complex adaptive systems (CASs) and showed that an ecosystem invaded by alien species neatly fits the definition of a CAS. As the functioning of CAS emerges from self-organized interactions of lower-level agents, it is not necessarily optimized for performance (productivity, biodiversity, and stability). Due to the alliance and rivalry and the trade-offs and feedbacks between independent decision-making agents, management interventions that aim to safeguard or maximize certain socio-ecological services could trigger conflicts of interest and jeopardize standing system functioning (Anderies *et al.* 2007), often leading to unintended consequences (e.g. the *hydra effect* of enhanced invasion performance following management intended to curb the invasion; Abrams 2009). For efficient prevention of invasions and management of already established invaders, key challenges are to elucidate the essence of ecological networks (see Chapter 10) and to understand the socio-economic behaviours and drivers that can reciprocally affect invasion performance and ecosystem function. In particular, since anthropogenic factors are the main drivers of invasions in many socio-ecological systems, invasion-related anthropogenic factors need special attention in invasion management.

There is great scope for improving the efficiency of the management of biological invasions by drawing on advances in the theory of complex adaptive systems. Rammel and colleagues (2007), in discussing natural resource management in general, argue that management issues exist at a range of hierarchical levels with different spatial and temporal scales and are 'shaped by a mutual yet non-deterministic co-evolutionary dialogue'. This dialogue refers to environmental changes tied to changes and adaptations that appear within social-ecological systems. In the case of invasions, such changes and adaptations could refer to changes in the extent and magnitude of invasions, levels of impacts, changes in perceptions among different sectors of society, conflicts of interest, and changing infrastructural and legislative landscapes. The following lessons from CASs on steering system trajectories (Chapter 10) could help invasion managers in addressing these challenges.

- Invasion management should not be a simple reaction to the action of biological invasion but should be directed at managing the co-evolving socio-ecological systems. Invasion management thus often faces hysteresis, resistance, and tipping points. Due to the strong priority effects that often exist in such non-linear systems, management needs to be implemented not only promptly to take advantage of the priority effect but also

cautiously to deal with inertia in the system, with both adaptation and innovation in mind.
- Policies for invasion management need to take account of the system hierarchy, and to recognize that the system is not only embedded within larger systems but also has many sub-systems that potentially create conflicts within management units, between local management units, and between local and regional management units.
- System complexity (diversity and connectivity) and functioning (productivity, efficiency, and resilience) are related. The diversity of agent types within a social-ecological system, whether a generalist or specialist agent, with unique or shared/common agendas, could lead to alliances that create compartmentalization or nested hierarchical structure so as to achieve maximal policy acceptance and effect for integrated invasion management.

In the rest of this chapter we consider three frameworks that have been used to analyse social-ecological systems to assist in framing and thereby identifying potential solutions to problems relating to biological invasions, namely (i) the driver, pressure, state, impact, response (DPSIR) framework (Eurostat 1999); (ii) the concept of 'wicked problems' (Rittel and Webber 1973); and (iii) 'novel ecosystems' (Hobbs *et al.* 2006).

11.3 The driver, pressure, state, impact, response (DPSIR) framework

DPSIR is a causal framework for describing interactions between society and the environment. The components of the framework are: **D**riving forces, **P**ressures, **S**tates, **I**mpacts, and **R**esponses. First presented in its current form by the European Environmental Agency (1995), DPSIR facilitates the understanding of complex systems by elucidating the relationships between large-scale socio-economic trends exerting pressures on the environment, the condition of the environment, and the response of the society to such conditions. The framework has been widely used for guiding decision making in natural resource management (Bowen and Riley 2003; Borja *et al.* 2006), and for analysing factors that facilitate the spread of invasive species (Elliott 2003). Figure 11.3 illustrates a typical parameterization the DPSIR framework for a biological invasion

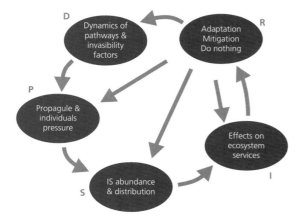

Fig. 11.3 Driver, Pressure, State, Impact, Response (DPSIR) categories for biological invasions. From Rodríguez-Labajos *et al.* (2009). With permission from Elsevier.

scenario in a social-ecological system. Some applications of DPSIR add additional components. For example, Mangi and colleagues (2007), addressing reef fisheries management complexities in Kenya, included a sixth component called barriers to effective management to highlight explicitly limitations that constrain the efficiency of the responses in solving the problem.

There have been several detailed applications of the DPSIR framework to assist in addressing key complexities in objective decision making relating to invasive species issues. Roura-Pascual and colleagues (2009) used DPSIR, together with another decision-making tool (analytic hierarchy process), to design a strategy for prioritizing management of the most important invasive alien plant species in South Africa's Cape Floristic Region (CFR). The two methods were applied to explore the complexity of problems and to identify appropriate response strategies. Efficient management of invasive plants in this region is constrained by multiple interacting environmental and socio-economic factors (Fig. 11.4). Factors related to the fire-prone nature of the ecosystem and the characteristics of

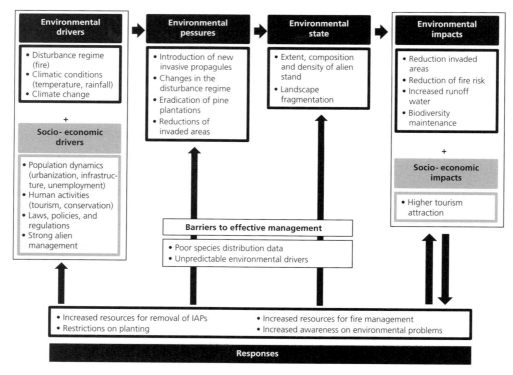

Fig. 11.4 The driving forces, pressure, state, impacts, responses (DPSIR) framework for understanding the spread of invasive alien plants (IAPs) on the Cape Peninsula, South Africa. From Roura-Pascual *et al.* (2009). With permission from Elsevier.

the invasive stands emerged as pivotal features for setting spatially explicit priorities for management. The analyses provided an objective and quantifiable perspective for improving the management efficiency. It suggested that considerable improvements in the management of invasive plants in the region could be achieved by: (1) coordinating and synchronizing control operations at both regional and local scales; (2) establishing clear mandates on how to proceed with control operations; and (3) improving the quality of distribution data for invasive alien plants. Conceptualization of the social-ecological system in the CFR affected by massive plant invasions using DPSIR proved useful for developing scenarios to identify broad-scale management options to steer invaded landscapes in different parts of the region to desirable futures (Roura-Pascual *et al.* 2011).

11.4 Biological invasions as a wicked problem

Biological invasions and other drivers of global change, and interactions between them, are creating novel socio-ecological conditions that are both difficult to fully understand and solve because of the inherent socio-ecological complexities and constant change. The resulting multitude of conflicting perspectives, objectives, and management goals can make the problem almost impossible to characterize, let alone solve, to the satisfaction of all stakeholders. Such problems have been termed 'wicked problems' (Rittel and Webber 1973).

A wicked problem is defined as one with the following properties (Woodford *et al.* 2016):

- The problem is not understood until a solution has been developed. Different stakeholders might disagree on some or all aspects of another stakeholder's definition to the problem, if they are personally invested in pursuing a particular solution.
- There is no stopping rule. Because neither the problem nor its potential solutions are definitive, there can be no definition of complete success—the stage at which problem-solving activities can cease.
- Solutions to the problem are not right or wrong. Rather, solutions can be viewed as 'better' or 'worse' by consensus of the stakeholders.

- Each problem is essentially unique. The source of wickedness lies in the social complexity of the stakeholders, and this varies from case to case.
- Every solution to the problem is a 'one-shot' operation. An enacted solution results in the emergence of new aspects of the problem which need to be dealt with in turn, using follow-up solutions.
- Wicked problems have no given alternative solutions. Many potential solutions could be identified, but only some will be appropriate to pursue, depending on the problem's individual nature and social context.

Problems associated with the management of invasive species have previously been termed wicked problems. The term was used by Evans and colleagues (2008) in discussing problems encountered when managing aquatic pests in Florida, by McNeely (2013) with reference to the management of plant introductions in protected areas, and by Seastedt (2014) in describing issues surrounding biological control. The management of invasions is particularly susceptible to wickedness because of the conflicting social pressures. Differing values and risks assigned to particular taxa by affected parties can lead to social conflicts around their management. The wickedness of a problem varies from case to case. Not all criteria of 'wickedness' might be satisfied. Some criteria may outweigh others in making a particular problem more or less wicked, and the wickedness of a problem can vary by region or country according to the perspectives of the different stakeholders involved. In each case, however, understanding the fundamental nature of the problem determines how it can be managed.

Woodford and colleagues (2016) reviewed four case studies that reflect the main goals in managers of biological invasions: prevention, eradication, and impact reduction. In assessing the drivers and extent of wickedness in each case study, they found that a disconnect between the perception and reality of tameness or wickedness can profoundly influence the likelihood of successful management. For example, managing species introductions can be wicked, but shifting from species-focused to vector-focused risk management can greatly reduce the complexity, making it a tame problem (Fig. 11.5A).

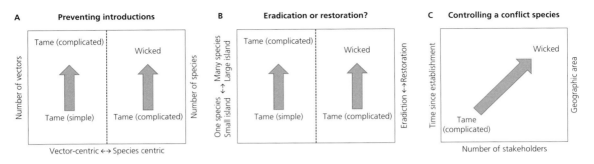

Fig. 11.5 Conceptual diagrams for four case studies selected to elucidate the wickedness inherent in the management of invasive alien species and how the wickedness might be counteracted. A: managing ballast water in the Laurentian Great Lakes; B: the eradication of invasive species from islands; C: (two studies): controlling invasive alien pine trees in South Africa, and managing invasive rainbow trout around the world. From Woodford *et al.* (2016).

The scope and scale of the overall management goal will also determine the wickedness of the problem and the achievability of management solutions (*cf.* eradication and ecosystem restoration as aims in Fig. 11.B). Finally, managing species that have both positive and negative impacts requires engagement with all stakeholders as well as scenario-based planning (Fig. 11.5C). Effective management of invasions requires either recognizing unavoidable wickedness, or avoiding it through an altered management perspective.

11.5 Biological invasions and 'novel ecosystems'

The growing scale of biological invasions, and the complex synergies between invasions and other facets of global change that generate multifaceted influences on ecosystems are causing daunting problems for managers. In many ecosystems, especially those with high levels of human influence (Fig. 11.2), defining goals for management is problematic. Management of habitats degraded by invasive species to restore the system to some 'pristine' or some other condition approximating the pre-invasion situation is frequently impractical, undesirable (to at least some stakeholders), or practically impossible. One reason for this is that the invasive species, and often other drivers acting synergistically with them, alter ecosystems so as to preclude many native species for long periods after the invasive species and other stressors have been

removed. In such cases, narrowly focused management interventions that aim simply to remove invasive species exacerbate rather than reduce problems.

There is increasing support for revising strategies for management of *some* invaded ecosystems to embrace the notion of 'novel ecosystems' (Fig. 11.6). A novel ecosystem is 'a system of abiotic, biotic, and social components (and their interactions) that, by virtue of human influence, differ from those that prevailed historically, having a tendency to self-organize and manifest novel qualities without intensive human management' (Hobbs *et al.* 2013; see also Hobbs *et al.* 2006). Biological invasions are important 'builders and shapers' of novel ecosystems thus defined (Richardson and Gaertner 2013). The novel ecosystem concept raises many fundamental questions, including those relating to factors that enable native species to persist with invaders and the role of elements of biodiversity in regulating ecosystem functioning. Since 'novel ecosystems' are the result of deliberate or inadvertent human action and their key novel feature is the potential for changes in ecosystem functioning, consideration needs to be given to developing appropriate management goals and approaches under new conditions. This may involve viewing the role of alien species, and even invasive species, more pragmatically in the context of shifting species' ranges and changing communities in the wider context of global change biology, and even considering some 'new' species as desirable elements for maintaining certain ecosystem services in particular contexts.

Removing alien species from such, often human-dominated, systems is often impractical, and management is in some cases more effectively directed at managing such novel ecosystems to provide sustainable delivery of certain functions or services.

A range of options is available for the management of ecosystems identified as historical, hybrid, and novel as shown in Figure 11.6. These depend on the goals that are selected for particular situations, which may be biodiversity protection, conservation of ecosystem functioning and services, maintenance of historical continuity, and provision

of natural resources for local human livelihoods. One of the main criticisms levelled at the 'novel ecosystem' approach is that levels of anthropogenic disturbance and biotic novelty required before an ecosystem should be classified as hybrid or novel (as in Fig. 11.6) are very seldom quantified which makes it difficult or impossible to objectively place a given system into one of the system states shown in Figure 11.6. Another concern is that embracing such a model automatically devalues efforts at maintaining ecosystems free of invasive species and those directed at eradicating, containing, and managing

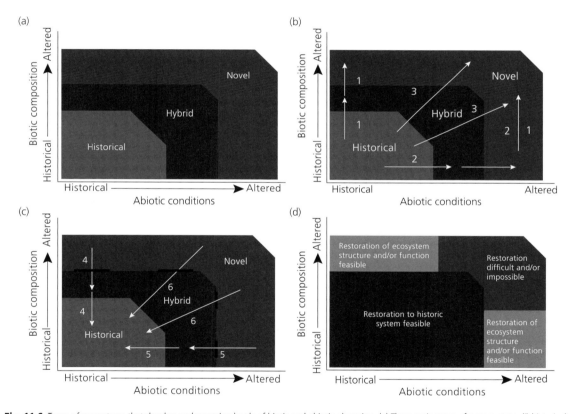

Fig. 11.6 Types of ecosystems that develop under varying levels of biotic and abiotic alteration. (a) Three main types of system state: (i) historical, (ii) hybrid, and (iii) novel, within which systems have been potentially irreversibly changed by large modifications to abiotic conditions or biotic composition. (b) Potential pathways of development of ecosystems in the face of changing biotic composition (loss or addition of species) and abiotic change (land use or climate). (c) Reversing the pathways of development in (a) requires removal of invasive species (Pathway 4) and/ or amelioration of altered environmental conditions (Pathways 5 and 6). Black lines indicate the presence of potential restoration thresholds that prevent the system moving back to a less altered state without significant management input. (d) The state space can be divided into an area within which restoration to a system within the historic range of variability remains feasible (which includes some or most hybrid systems), areas within which restoration of ecosystem structure and/or function can be achieved without a return to historic system characteristics, and an area within which restoration is likely to be difficult or impossible and hence alternative management objectives are required. From Hobbs *et al.* (2009). With permission from Elsevier.

invasions and other drivers of degradation in other ways (e.g. Murcia *et al.* 2014). These authors argue that 'novel ecosystems' 'are a Trojan horse for conservation', and that 'they provide a license to trash nature if they provide ecosystem services' (Simberloff *et al.* 2015). These are valid concerns and more discussion is required in this regard, but there seems little doubt that pragmatic approaches such as the novel ecosystem concept are needed in an increasingly human-dominated world (Hobbs *et al.* 2014).

11.6 Invasion science 2050

Good progress has been made in recent years in addressing many issues relating to invasions, but substantial challenges remain. Table 11.1 summarizes some of the key requirements for improving our understanding of 'invasion dynamics' as defined in this book. Improved insights are also urgently needed on aspects that were not addressed in this book, notably the global and regional drivers of pathways of introduction of alien species (Essl *et al.* 2015). The agenda for research on invasions is changing rapidly in response to the emergence of new challenges. Richardson (2011) chronicled the evolution of 'invasion science'—the collective term that is now widely applied to the study of the full range of issues pertaining to the phenomenon of biological invasions, including emerging fields that may be lumped in the domain of 'biosecurity'. Issues relating to the management of risks posed by invasive species to the economy, environment, and human health through exclusion (the prevention of initial introduction of a species), mitigation, adaptation, control, and eradication are rapidly moving to centre stage in the theatre of invasion science. These issues are likely to feature increasingly and prominently in new versions of invasion science, with human dimensions coming to the fore in shaping and weaving the mosaic of 'invasion science 2050' (a vision of the field in the year 2050).

Managing invasions as part of increasingly complex social-ecological systems in the face of rapid global change demands innovative new approaches to elucidate invasion-driven reinforcing feedback loops (Gaertner *et al.* 2014) and to reduce impacts associated with invasive species. As one important feature of CAS, these socio-ecological systems exhibit a *state of paradox*—although patterns derived from observations are highly non-random and thus ordered (e.g. invasive traits can be identified by comparing invasive and non-invasive species), predictions based on such observed knowledge are, however, usually poor, with high levels of uncertainty (Melbourne and Hastings 2009). A high deviation from random expectation in observed invasion dynamics supposedly indicates a high certainty for predicting and forecasting invasion performance. The low predictability or high uncertainty associated with invasion performance and embedded complex social-ecological systems could be related to the lack of appreciation and methodology of the role of feedback loops driven by biological invasions in invaded ecosystems. Feedback/coupling loops are central to the functioning of all living systems (Hoagland and Dodson 1995). Invasive species not only negotiate realized niches in recipient ecosystems, but also imprint their impact onto the resident species, interactions, and processes in the ever-evolving ecological system. Holling (2001) outlined the dynamics of a social-ecological system undergoing an adaptive cycle (Fig. 11.7). This suggests that an invaded ecosystem could pass through four phases—from early exploitation by invasive species (r) to the transformation of ecosystem resources (K), then from the release (Ω) of current ecosystem regime to the reorganization (α) into a novel ecosystem. Clearly, current statistical approaches (e.g. linear models and categorical trees) are clumsy for quantifying such loops. To achieve

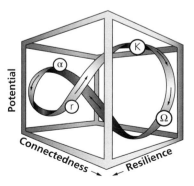

Fig. 11.7 Panarchy, the adaptive cycle of complex social-ecological systems. Four letters r, K, α, Ω indicate the ecosystem function of exploitation, conservation, release, and reorganization. From Holling (2001).

better predictability of invasion performance and ecosystem invasibility, we need better data (especially time series) and novel approaches (e.g. further development of approaches for adaptive networks) that are capable of quantifying and examining such cyclical features in social-ecological systems that are being transformed by biological invasions. This is not just a call for next-generation quantitative methodologies to improve detection and measurement of feedback loops in ecology. It is also an appeal for a paradigm shift in ecology and invasion science to embrace adaptive cycles and network thinking.

References

Abrams, P.A. (2009) When does greater mortality increase population size? The long history and diverse mechanisms underlying the hydra effect. Ecology Letters, 12, 462–74.

Anderies, J.M., Rodriguez, A.A., Janssen, M.A., et al. (2007) Panaceas, uncertainty, and the robust control framework in sustainability science. Proceedings of the National Academy of Sciences of the USA, 104, 15194–99.

Balciunas, J.K. (2000) Code of Best Practices for Classical Biological Control of Weeds (2000). In: Spencer, N.R. (ed.) Proceedings of the X International Symposium on Biological Control of Weeds. Bozeman, MT: Montana State University, p. 435.

Borja, A., Galparsoro, I., Solaun, O., et al. (2006) The European Water Framework Directive and the DPSIR, a methodological approach to assess the risk of failing to achieve good ecological status. Estuarine Coastal and Shelf Science, 66, 84–96.

Bowen, R.E. and Riley, C. (2003) Socio-economic indicators and integrated coastal management. Ocean and Coastal Management, 46, 299–312.

Catford, J.A., Jansson, R., and Nilsson, C. (2009) Reducing redundancy in invasion ecology by integrating hypotheses into a single theoretical framework. Diversity and Distributions, 15, 22–40.

Davis, M.A. (2011) Researching invasive species 50 years after Elton: a cautionary tale. In: Richardson, D.M. (ed.) Fifty Years of Invasion Ecology. The Legacy of Charles Elton. Oxford: Wiley-Blackwell, pp. 269–76.

Elliott, M. (2003) Biological pollutants and biological pollution—an increasing cause for concern. Marine Pollution Bulletin, 46, 275–80.

Ellis, E.C. (2013) Sustaining biodiversity and people in the world's anthropogenic biomes. Current Opinion in Environmental Sustainability, 5, 368–72.

Ellis, E.C. (2015) Ecology in an anthropogenic biosphere. Ecological Monographs, 85, 287–331.

Essl, F., Bacher, S., Blackburn, T.M., et al. (2015) Crossing frontiers in tackling pathways of biological invasions. BioScience, 65, 769–82.

European Environmental Agency (1995) Europe's Environment: The Dobris Assessment. Copenhagen: European Environmental Agency.

Eurostat (1999). Towards Environmental Pressure Indicators for the EU. Luxembourg: Eurostat.

Evans, J.M., Wilkie, A.C., and Burkhardt, J. (2008) Adaptive management of nonnative species: moving beyond the 'either-or' through experimental pluralism. Journal of Agricultural and Environmental Ethics, 21, 521–39.

Foxcroft, L.C., Pyšek, P., Richardson, D. M., et al. (eds) (2013) Plant Invasions in Protected Areas: Patterns, Problems and Challenges. Invading Nature Series Vol 7. Berlin: Springer.

Franklin, J., Serra-Diaz, J.M., Syphard, A.D., et al. (2016) Global change and terrestrial plant community dynamics. Proceedings of the National Academy of Sciences of the USA, 113, 3725–34.

Gaertner, M., Biggs, R., te Beest, M., et al. (2014) Invasive plants as drivers of regime shifts: identifying high priority invaders that alter feedback relationships. Diversity and Distributions, 20, 733–44.

Hoagland, M.B. and Dodson, B. (1995) The Way Life Works. London: Times Books.

Hobbs, R.J., Arico, S., Aronson, J., et al. (2006) Novel ecosystems: theoretical and management aspects of the new ecological world order. Global Ecology and Biogeography, 15, 1–7.

Hobbs, R.J., Higgs, E.S., and Hall, C.A. (eds) (2013) Defining novel ecosystems. In: Hobbs, R.J., Higgs, E.S., and Hall, C.A. (eds) Novel Ecosystems: Intervening in the New Ecological World Order. Oxford: Wiley-Blackwell, pp. 58–60.

Hobbs, R.J., Higgs, E.S., Hall, C.M., et al. (2014) Managing the whole landscape: historical, hybrid and novel ecosystems. Frontiers in Ecology and the Environment, 12, 557–64.

Hobbs, R.J., Higgs, E.S., and Harris, J.A. (2009) Novel ecosystems: implications for conservation and restoration. Trends in Ecology and Evolution, 24, 599–605.

Holling, C.S. (2001) Understanding the complexity of economic, ecological and social systems. Ecosystems, 4, 390–405.

Kueffer, C., Pyšek, P., and Richardson, D.M. (2013) Integrative invasion science: model systems, multi-site studies, focused meta-analysis, and invasion syndromes. New Phytologist, 200, 615–33.

Kumschick, S., Gaertner, M., Vilà, M., et al. (2015) Ecological impacts of alien species: quantification, scope, caveats and recommendations. BioScience, 65, 55–63.

Le Maitre, D.C., Richardson, D.M., and Chapman, R.A. (2004) Alien plant invasions in South Africa: driving forces and the human dimension. South African Journal of Science, 100, 103–12.

Levin, S., Xepapadeas, T., Crépin, A.-S., *et al.* (2013) Social-ecological systems as complex adaptive systems: modelling and policy implications. Environment and Development Economics, 18, 111–32.

MacDougall, A.S. and Turkington, R. (2005) Are invasive species the drivers or passengers of change in degraded ecosystems? Ecology, 86, 42–55.

Mangi, S.C., Roberts, C.M., and Rodwell, L.D. (2007) Reef fisheries management in Kenya: preliminary approach using the driver-pressure-state-impacts-response (DP-SIR) scheme of indicators. Ocean and Coastal Management, 50, 463–80.

Martin, L.J., Quinn, J.E., Ellis, E.C., *et al.* (2014) Conservation opportunities across the world's anthromes. Diversity and Distributions, 20, 745–55.

McNeely, J.A. (ed.) (2001) *The Great Reshuffling: Human Dimensions of Invasive Alien Species.* Gland: IUCN, The World Conservation Union.

McNeely, J.A. (2013) Global efforts to address the wicked problem of invasive alien species. In: Foxcroft, L.C., Pyšek, P., Richardson, D.M., *et al.* (eds) *Plant Invasions in Protected Areas: Patterns, Problems and Challenges.* Springer: Dordrecht, pp. 61–88.

Melbourne, B.A. and Hastings, A. (2009) Highly variable spread rates in replicated biological invasions: fundamental limits to predictability. Science, 325, 1536–9

Murcia, C., Aronson, J., Kattan, G.H., *et al.* (2014) A critique of the 'novel ecosystem' concept. Trends in Ecology and Evolution, 29, 548–53.

Pyšek, P., Richardson, D.M., Pergl, J., *et al.* (2008) Geographical and taxonomical biases in invasion ecology. Trends in Ecology and Evolution, 23, 237–44.

Rammel, C., Stagl, S., and Wilfing, H. (2007) Managing complex adaptive systems—A co-evolutionary perspective on natural resource management. Ecological Economics, 63, 9–21.

Rejmánek, M., Richardson, D.M., Higgins, S.I., *et al.* (2005) Ecology of invasive plants—state of the art. In: Mooney, H.A., Mack, R.N., McNeely, J.A., *et al.* (eds) *Invasive Alien Species: A New Synthesis.* Washington, DC: Island Press, pp. 104–61.

Richardson, D.M. (2011) Invasion science: the roads travelled and the roads ahead. In: Richardson, D.M. (ed.) *Fifty Years of Invasion Ecology. The Legacy of Charles Elton.* Oxford: Wiley-Blackwell, pp. 397–407.

Richardson, D.M. and Gaertner, M. (2013) Plant invasions as builders and shapers of novel ecosystems. In: Hobbs, R.J., Higgs, E.C., and Hall, C.M. (eds) *Novel Ecosystems: Intervening in the New Ecological World Order.* Oxford: Wiley-Blackwell, pp. 102–14.

Rittel, H. and Webber, M.M. (1973) Dilemmas in a general theory of planning. Policy Science, 4, 155–69.

Roura-Pascual, N., Richardson, D.M., Chapman, R.A., *et al.* (2011) Managing biological invasions: charting courses to desirable futures in the Cape Floristic Region, South Africa. Regional Environmental Change, 11, 311–20.

Roura-Pascual, N., Richardson, D.M., Krug, R.M., *et al.* (2009) Ecology and management of alien plant invasions in South African fynbos: accommodating key complexities in objective decision making. Biological Conservation, 142, 1595–604.

Seastedt, T.R. (2014) Biological control of invasive plant species: a reassessment for the Anthropocene. New Phytologist, 205, 490–502.

Simberloff, D., Murcia, C., and Aronson, J. (2015) *Ensia.* Available at: <http://ensia.com/voices/novel-ecosystems-are-a-trojan-horse-for-conservation/>.

Strayer, D.L. (2012) Eight questions about invasions and ecosystem functioning. Ecology Letters, 15, 1199–210.

Traveset, A. and Richardson, D.M. (2006) Biological invasions as disruptors of plant reproductive mutualisms. Trends in Ecology and Evolution, 21, 208–16.

Wilson, J.R.U., Gairifo, C., Gibson, M.R., *et al.* (2011) Risk assessment, eradication, and biological control: global efforts to limit Australian acacia invasions. Diversity and Distributions, 17, 1030–46.

Woodford, D.J., Richardson, D.M., MacIsaac, H.J., *et al.* (2016) Confronting the wicked problem of managing biological invasions. NeoBiota, 31, 63–86.

Glossary

Adaptive diversification—Diversification (polymorphism and speciation) from frequency-dependent selection, often through **trait-mediated interactions**.

Agent-based models—Models that explore system-level behaviour and patterns by simulating the actions and interactions of embedded agents (components or compartments) of the system that are largely autonomous and adaptive.

Alien biome—A biogeographical cluster of co-occurring **alien species**.

Alien species (synonyms: adventive, exotic, foreign, introduced, non-indigenous, non-native species)—Species whose presence in a region is attributable to human actions that enabled them to overcome fundamental biogeographical barriers (i.e. human-mediated extra-range dispersal).

Allee effects—The positive density dependence that many species suffer when the population size or density is low, primarily due to difficulty in finding mates. It is relevant to **biological invasion** as the initial population size and/or genetic diversity (known as the *founder effect*) of many introduced species is low. Allee effects favour **alien species** with self-compatibility and often cause the phenomenon of **range pinning** during range expansion.

Anthrome (synonym: anthropogenic biome)—Global ecological and biogeographical patterns created through sustained human-mediated interactions with ecosystems.

Anthropocene—The geological era during which human activity has become the dominant influence on climate and the environment.

Bioeconomic models—Optimal-control-type models that are designed for identifying optimal strategies for managing **invasive species**. Such models normally require the formulation of population dynamics of the **alien species**, management cost, control efficiency, and a target function for optimization (normally to minimize monetary impact of **invasion** or the loss of public welfare); the solution is often sought within the budget limit with further consideration on discount rate of the budget.

Biological invasions (synonyms: bioinvasions, biotic invasions, species invasions)—The phenomenon of, and suite of processes involved in determining: (1) the transport of organisms, through human activity (intentionally or accidentally, via **introduction pathways**) to areas outside the potential range of those organisms as defined by their natural dispersal mechanisms and biogeographical barriers; and (2) the fate of such organisms in their new ranges, including their ability to survive, establish, reproduce, disperse, spread, proliferate, and interact with **resident** biota, and exert influence in many ways on and in invaded ecosystems.

Biosecurity—The management of risks posed by organisms to the economy, environment and human health through exclusion (the prevention of initial **introduction** of a species), mitigation, adaptation, control, and **eradication**.

Biotic homogenization—The addition to, and often the partial if not extensive replacement of, local biotas by **alien species**, which can result in decreased compositional turnover (β diversity) of species between distant areas, both in terms of taxonomic and phylogenetic similarity.

Biotic resistance—Resistance by **resident species** to the establishment (or post-establishment survival, proliferation, and spread) of **alien species**.

Biotic resistance hypothesis—A notion, derived from **limiting similarity theory**, which argues that species-rich communities can withstand and even resist **biological invasions** (but see **invasion paradox**).

Biphasic expansion—The pattern of range expansion of many **alien species** which shows an initial phase of slow expansion, followed by a phase with much faster spreading rate.

Casual species—Those **alien species** that do not form self-replacing populations in the invaded region and whose persistence depends on repeated **introductions** of propagules. The term is generally used for plants.

Cellular automaton (also see **Lattice model**)—An early form of **agent-based models** that describe only discrete states of agents (e.g. on/off). In spatial models,

309

cellular automata divide the space into grids (cells) and consider only discrete states of these spatial units (e.g. occupied or not). These models normally consider only dispersal between adjacent cells, although more generalized versions can implement different dispersal strategies.

Co-evolution—Trait evolution through two (or more) species engaging in biotic interactions that reciprocally affect each other's fitness, posing frequency-dependent selection on each other.

Co-introduction hypothesis—The view that the simultaneous introduction of **alien species** and their mutualists is required to ensure the establishment or enhanced invasive performance in the new ranges.

Colonization pressure—A variant of the concept of **propagule pressure**; defined as the number of species introduced or released to a single location, some of which will go on to establish a self-sustaining population and some of which will not. Lockwood and colleagues (2009, Diversity and Distributions, 15, 904–10) argue that colonization pressure should serve as a null hypothesis for understanding temporal or spatial differences in **alien species** richness, as the more species are introduced, the more we should expect to establish.

Community assembly (synonyms: community succession; related: community assembly rules)—The formation and succession of species assemblages in a community, often driven by disturbance, and increasingly mediated by and **biological invasions** (see also **community disassembly**).

Community disassembly—The dismantling of existing co-existence mechanisms, resulting in altered (and often depauperate) species assemblages due to **biological invasions** and other stressors.

Complex adaptive system (CAS)—A decentralized system consisting of many diverse and autonomous components or parts (agents) which are interrelated, interdependent, and linked through many interconnections, and which behave as a unified whole in learning from experience and in adjusting (not just reacting) to changes in the environment.

Complexity theory—A set of concepts that attempts to explain complex phenomena not explainable by traditional (mechanistic) theories. It integrates ideas derived from chaos theory, cognitive psychology, computer science, evolutionary biology, general systems theory, fuzzy logic, information theory, and other fields to deal with the natural and artificial systems as they are, and not by simplifying them (breaking them down into their constituent parts). It recognizes that complex behaviour emerges from a few simple rules, and that all complex systems are networks of many interdependent parts which interact according to those rules.

Complexity–stability relationship—The relationship between system complexity (as measured by both the number of and connections between components) and stability (measured primarily by system **resilience** and persistence). The debate on this relationship is rooted in the need for inference of system function from system structure, or inferring processes from observed patterns. The relationship that complexity leads to the loss of system stability has, arguably, led to the modular design of many complex systems (e.g. airplanes; to reduce system complexity so to ensure stability).

Critical slowing down (CSD)—see **early warning signal**.

Darwin's naturalization hypothesis—The notion that **alien species** with close **native** relatives in their introduced range may have reduced chances of establishment and **invasion**; based on ideas formulated by Charles Darwin in Chapter 3 of *The Origin of Species*, borrowing ideas from Alphonse de Candolle.

Density dependence—Demographic mechanisms that regulate population dynamics, notably when the population size is at extremes, including positive density dependence (see **Allee effects**) that operates when the population size is low, and negative density dependence that operates when the population size is close to the carrying capacity. Density dependence often regulates population dynamics through curbed or hampered individual fitness.

Dispersal kernel—A probability density function characterizing the spatial distribution of propagules dispersed from a common source. The 'dispersal distance kernel' more precisely describes the distribution of seed dispersal distances (and hence the distribution of the product of displacement velocity, V, and retention time, P).

Dispersal pathway—The combination of processes and opportunities resulting in the movement of propagules from one area to another, including aspects of the vectors involved, features of the original and recipient environments, and the nature and timing of what exactly is moved. The definition thus combines phenomenological and mechanistic aspects (Wilson *et al.* (2009) Trends in Ecology and Evolution, 24, 136–44) (see also **Introduction pathway**).

Dispersal vector—An agent transporting seeds or other dispersal units. Dispersal vectors can be biotic (e.g. birds) or abiotic (e.g. wind).

Disturbance—A temporal change, either regular or irregular (uncertain), in the environmental conditions that can trigger population fluctuations and secondary succession. Disturbance is an important driver of **biological invasions**.

Diversity–invasibility hypothesis—The proposition that more biologically diverse communities are less susceptible to **invasion** by **alien species** or novel genotypes (related terms and concepts: **biotic-resistance hypothesis; diversity-resistance hypothesis**). See also **biotic resistance, invasional meltdown**.

Early warning signal (EWS)—Detectable system behaviours that appears when the system approaches the tipping point of its current regime. An important but still not universally accepted EWS is the **critical slowing down** (CSD) which occurs when the system experiences zero **resilience** at the tipping point, leading to prolonged system responses and large variations.

Ecological fitting—The emergence and formation of biotic interaction without the **co-evolution** of involved species, but through matching or compatible traits, often after rapid trials and learning. **Biological invasions** provide strong evidence for the formation of biotic interactions through ecological fitting (also see **new associations hypothesis**).

Ecosystem engineering (synonym: niche construction)—The ability and process of modifying nutrient cycles and disturbance regimes through the actions and metabolism of organisms, often leading to altered selection direction and force. Many **invasive species** are ecosystem engineers and trigger **regime shifts** in recipient ecosystems (see also **invasional meltdown**).

Ecosystem functioning—The cycles of nutrients, biomass, and energy that safeguard the provision of multifaceted ecosystem service and the stability and persistence of embedded species assemblage.

EICA—see **evolution of increased competitive ability hypothesis**.

Empty-niche hypothesis—The view that ecological networks with specialized interactions could hamper the effect of **co-evolution,** leaving unexploited niches from incremental evolution, thus creating opportunities for **alien species** to establish and exploit such empty niches through **ecological fitting**.

Enemy release hypothesis (ERH)—The notion that **alien species** have a better chance of establishing and becoming dominant when released from the negative effects of natural enemies that, in their native range, lead to high mortality rates and reduced productivity. Colautti and colleagues (2004, Ecology Letters, 7, 721–33) argue that the ERH is often accepted without recognizing that all **alien species** will lose at least some natural enemies due to bottlenecks during transport.

Eradication—The extirpation of an entire population of an **alien species** within a designated management unit.

Evolution of increased competitive ability hypothesis (EICA)—A concept that posits that plants introduced to an environment that lack their usual herbivores or disease agents will experience selection favouring individuals that allocate less energy to defence and more to growth and reproduction (Blossey and Nötzold (1995) Journal of Ecology, 83, 887–9).

Evolutionarily stable strategy (ESS)—A strategy which, if adopted by a population in a given environment, cannot be invaded by any alternative strategy that is initially rare.

Evolutionary invasion analysis—A set of tools, including adaptive dynamics, for exploring the evolution of traits or life-history strategies under directional and disruptive frequency dependent selection. A key goal of this analysis is to identify the **evolutionarily stable strategy (ESS)** in the system.

Extralimital expansion—The spread of populations into new habitats in areas outside their natural distribution, typically within geographic or geopolitical boundaries in which the species is **native**.

Fat-tailed dispersal—A **dispersal kernel** with a tail that drops off more slowly than that of any negative exponential kernel.

Fluctuating resources theory of invasibility—A concept, formulated for **alien** plants by Davis and colleagues (2000, Journal of Ecology, 88, 528–34), that predicts that pronounced fluctuations in resource availability enhances community **invasibility** if coinciding with the availability of sufficient propagules to initiate an **invasion**.

Gravity model—Any model, especially those simulating spatial spread, that uses gravity-like connectivity or interaction intensity between two sites/populations. For instance, the rate of migration between two cities can be assumed to be proportional to the production of the population sizes of the two cities divided by square or a specific power of the geographical distance between the two cities.

Host-jumping hypothesis—The notion that **alien plants** without co-introduced mutualists (e.g. co-evolved mycorrhizal fungi) could still perform well in novel ranges by forming new associations with resident generalist or promiscuous mutualists. See also **new associations hypothesis**.

Impact—The description or quantification of how an **alien species** affects the physical, chemical and biological environment.

Individual-based model—see **agent-based models**

Integrodifference equation—A relaxation of the classic **reaction-diffusion models**, from differential equations to difference equations, so that different **dispersal kernels** and complex life cycles can be implemented (Kot *et al.* (1996) Ecology, 77, 2027–42).

Interaction strength—A multifaceted concept. In general, it is assessed when two species under consideration are at equilibrium; the interaction strength of species A on B is the change in A's population growth rate as a function of changes in B's population size.

Interaction switching (synonym: species rewiring)—The phenomenon that species can switch their interaction partners in an ecological network, either randomly or adaptively (Zhang *et al.* (2011) Ecology Letters, 14, 797–803). Interaction switching can happen at different temporal scales, from rapid changes in behaviours to slow changes that coincide with phylogeographical events.

Intermediate disturbance hypothesis—The view that high levels of disturbance will ultimately reduce the biodiversity, whereas low levels of disturbance will enhance the effect of competition and thus reduce the biodiversity through competitive exclusion. Highest biodiversity is thus achieved at intermediate levels of disturbance.

Introduced—see **alien**.

Introduction—Movement of a species, intentionally or accidentally, due to human activity, from an area where it is **native** to a region outside that range ('introduced' is synonymous with **alien**). The act of an introduction (inoculation of propagules) may or may not lead to **invasion**.

Introduction pathway—The processes that result in the **introduction** of **alien** species from one geographical location to another. Hulme and colleagues (2008, Journal of Animal Ecology, 45, 403–14) proposed a universal framework applicable to a wide range of taxonomic groups in terrestrial and aquatic ecosystems. Introduction pathways are a subset of **dispersal pathways**—those that are mediated by human activities.

Introduction–naturalization–invasion (INI) continuum—A conceptualization of the progression of stages and phases in the status of an **alien** organism in a new environment which posits that the organism must negotiate a series of barriers. The extent to which a species is able to negotiate sequential barriers (mediated by **propagule pressure** and **residence time**, and frequently involving a **lag phase**) determines the organism's status as **alien: casual, naturalized**, or **invasive** (Richardson *et al.* (2000) Diversity and Distributions, 6, 93–107).

Invasibility—The properties of a community, habitat, or ecosystem that determine its vulnerability to **invasion**. Early studies tended to use the concept deterministically (particular systems were deemed either invasible or not), but invasibility is more appropriately considered probabilistically, and the degree of invasibility may change markedly over time due, for instance, to changes in biotic or abiotic features of the ecosystem (see also b**iotic resistance, colonization pressure, fluctuating resources theory of invasibility, invasiveness, lag phase**).

Invasion—The multi-stage process whereby an organism negotiates a series of potential barriers in the naturalization–invasion continuum (Richardson *et al.* (2000) Diversity and Distributions, 6, 93–107) (*cf.* **range expansion**).

Invasion debt—A concept that posits that even if **introductions** cease (and/or other drivers of **invasion** are relaxed, for example, **propagule pressure** is reduced), new **invasions** will continue to emerge and already-invasive species will continue to spread and cause potentially greater **impacts**, since large numbers of **alien species** are already present, many of them in a **lag phase** (Essl *et al.* (2010) Proceedings of the National Academy of Sciences of the USA, 108, 203–7; Rouget *et al.* (2016) Diversity and Distributions, 22, 445–56).

Invasion dynamics—Changes in population demography, niches, spatiotemporal patterns, and genetic/phenotypic makeup during the introduction and spread of **alien species**, and the ecological, evolutionary and socio-economic responses of recipient **social-ecological systems**.

Invasion ecology—The study of the causes and consequences of the **introduction** of organisms to areas outside their native range as governed by their dispersal mechanisms and biogeographical barriers. The field deals with all aspects relating to the **introduction** of organisms, their ability to establish, **naturalize**, and **invade** in the target region, their interactions with **resident organisms** in their new location, and the consideration of costs and benefits of their presence and abundance with reference to human value systems. This term is often used interchangeably with 'invasion biology' in the literature; see also **invasion science**.

Invasion paradox—The positive correlation between native and **alien species** at landscape and regional scales that seemingly defies predictions from the **biotic resistance hypothesis**.

Invasion science (synonym: invasion research)—A term used to describe the full spectrum of fields of enquiry that address issues pertaining to **alien species** and **biological invasions**. The field embraces **invasion ecology**, but increasingly involves non-biological lines of enquiry, including economics, ethics, sociology, and inter- and transdisciplinary studies.

Invasional meltdown—A phenomenon whereby **alien species** facilitate one another's establishment, spread, and **impacts** (Simberloff and von Holle (1999) Biological Invasions, 1, 21–32).

Invasive species—**Alien species** that sustain self-replacing populations over several life cycles, produce reproductive offspring, often in very large numbers at considerable distances from the parent and/or site of **introduction**, and have the potential to spread over long distances. Invasive species are a subset of **naturalized** species; not all naturalized species become invasive. This definition explicitly excludes any connotation of **impact**, and is based exclusively on ecological and biogeographical criteria (Wilson *et al.* (2009) Trends in Ecology and Evolution, 24, 586).

Invasiveness—The features of **alien** organisms, such as their life-history traits and modes of reproduction that define their capacity to invade; that is, to overcome various barriers to **invasion**. The level of invasiveness of a species can change over time due to, for example, changes in genetic diversity through hybridization, introgression, or the continued arrival of new propagules of the same species that is already established in a region, but from new and different (meta)populations, such that genetic diversity may increase. This latter concept is important in management strategies, which sometimes assume that less concern needs to be paid to the continued **introduction** of species (the continued arrival of propagules, whether accidental or intentional) that are *already* well-established in a region, overlooking the critical potential for elevated **invasiveness** over time.

Island biogeography—The study of the species composition and species richness on islands (or insular habitats) aimed at establishing and explaining the factors that affect species diversity of a specific community.

Jump dispersal—A category of **long-distance dispersal**, over sometimes substantial scales, whereby connection (gene flow) between the new and original ranges is maintained.

Lag phase—The time between when an **alien species** arrives in a new area and the onset of the phase of rapid, or exponential, increase. Multiple factors are frequently implicated in the persistence or dissolution of the lag phase in **invasions**, including an initial shortage of invasible sites, the absence or shortage of essential mutualists, inadequate genetic diversity, and the relaxation of competition or predation (due to other alterations in the resident biota). Lag phases may also be the result of statistical or sampling artefacts commonly found in time series of **alien species** records.

Lag time—The broad set of lag (the period of time from one event to another) phenomena across the entire **invasion** sequence, which may include, (1) the apparent long-term failure of species to successfully **invade** from potential donor regions to potential recipient regions (until they do, due, for example, to changes in the environments of donor and/or recipient regions, to changes in vectors or to changes in other phenomena), (2) lags in population increase (see **lag phase**), and (3) lags in geographic expansion, whereby a species may appear to remain resident in one relatively small and restricted region for a long period of time, but then begin to

suddenly expand (due, in part, to the fact that spread increases exponentially once multiple foci have had time to establish).

Lattice model—A generalized **cellular automaton**, often with continuous state variables for the agents, often used for modelling the spatial dynamics of species over geographical lattices.

Limiting similarity theory—A concept in theoretical ecology and community ecology that proposes the existence of a maximum level of niche overlap between two given species that will allow continued coexistence.

Long-distance dispersal (LDD)—Dispersal of propagules over a long distance, defined either by the absolute distance travelled or by a set proportion of all propagules that disperse the farthest.

Managed relocation (synonyms: assisted colonization, assisted migration, assisted translocation)—A form of management intervention aimed at reducing the negative effects of global change (especially rapid climate change) on defined biological units such as populations, species, or ecosystems (Richardson *et al.* (2009) Proceedings of the National Academy of Sciences of the USA, 24, 9721–4).

Marginal stability—System stability, when measured by the maximum eigenvalue of the system Jacobin, that is close to the zero maximum eigenvalue which is a tipping point for the system to lose resilience. **Complex adaptive systems** do not often evolve towards maximum stability, but only marginal stability; system stability is not the target for system evolution (Nuwagaba *et al.* (2015) Proceedings of the Royal Society B: Biological Sciences, 282, 20150320).

Metacommunity—Multiple communities maintained at regional level, with migrations of propagules between local communities and from regional species pool constantly invading local communities.

Monitoring—Rapid and accurate assessment of occurrence, status and impact of invasions across a region, often incurring much lower management effort/cost than implementing other control strategies (Latombe *et al.* (2016) Biological Conservation, doi:10.1016/j.biocon.2016.06.013).

Native species (synonym: indigenous species)—Species that have evolved in a given area or that arrived there by natural means (via **range expansion**), without the intentional or accidental intervention of humans from an area where they are native.

Natural experiment—Studies that interpret observations at different sites (or other observational units) in nature as the outcome of past processes that were influenced by specific differences between observational units.

Naturalized species (synonym: established species)—Those **alien species** that sustain self-replacing populations for several life cycles or a given period of time (ten years is advocated for plants) without direct intervention by people, or despite human intervention. The term is currently mainly used with reference to terrestrial plant **invasions,** although it was previously widely used for mammals.

Network architecture—The structure of the adjacency matrix of a network, depicting the interaction strength between any two nodes in a network. Metrics, such as nestedness, modularity, and node-degree distribution, are used to depict the matrix/network structure.

Network emergence—Processes and models that could give rise to the formation of particular network architecture and functions from an initial random network.

Neutral-niche continuum—The notion that community assembly can be located along a continuum from a purely neutral to a completely niche-based assembly.

New associations hypothesis—See **host-jump hypothesis.**

Niche resemblance hypothesis—The notion that resident species with similar niches to that of an invader could ensure the availability of suitable habitat for the invader, thus reducing the chance of the invader being expelled. Invaders with niches that resemble those of resident species often do not spread immediately, but could become 'sleeper weeds' waiting for opportunities created by disturbance (Hui *et al.* (2016) Biological Invasions, 18, 971–83).

Niche shift (synonym: niche adaptation)—**Alien species** often face different biotic and abiotic environments compared to those in their native ranges; in some cases **alien species** rapidly shift their realized niches.

Niche-mosaic hypothesis—The notion that the succession of regional metacommunities could lead to distinct mosaics of species assemblages, with each mosaic having distinct niche-based functions but with species within each mosaic having species of resembling functions (neutral clusters) (Hui *et al.* (2013) Nature Communications, 4, 2454).

Novel ecosystems—Those comprising species that occur in combinations and relative abundances that have not occurred previously at a given location or biome. Such ecosystems result from either the degradation or **invasion** of natural ecosystems (those dominated by **native species**) or the abandonment of intensively managed systems (Hobbs *et al.* (2006) Global Ecology and Biogeography, 15, 1–7).

Novel weapons hypothesis—The idea that some **alien plant species** may become invasive because they produce biologically active secondary metabolites that are not produced by species in invaded communities, and that such novelty provides the **alien species** with advantages against native competitors, consumers, or microbes that are not adapted to tolerate the chemical.

Optimal control theory—See **bioeconomic models.**

Optimal foraging theory—A theory stating that natural selection favours animals whose behavioural strategies maximize their net energy intake per unit time spent foraging.

Percolation theory—A theory that describes the phase transition of connectivity in graphs. In its ecological application, the connectivity of occupied cells increases non-linearly with the increase of occupancy, drastically increasing when the occupancy is greater than the percolation point.

Phenotypic plasticity—The ability of one genotype to produce a range of phenotypes in different environments. If the phenotypes produced in a novel environment are also associated with better performance, it is then called adaptive phenotypic plasticity (Via *et al.* (1995) Trends in Ecology and Evolution, 10, 212–17).

Portfolio theory—A economic theory that posits that risk and uncertainty can be reduced by diversifying investments; in population viability analysis, the total population variation can be reduced if the dynamics of two local populations have negative covariance (Schindler *et al.* (2015) Frontiers in Ecology and the Environment, 13, 257–63).

Prioritization—An approach to ensure the cost-effective allocation of management resources among multiple **invasive species,** pathways and sites, and often also to reach consensus among multiple agents at play (McGeoch *et al.* (2016) Biological Invasions, 18, 299–314).

Priority effect—The order and timing of species colonization during community assembly can affect community structures across scales, with early pioneer species playing important roles in deciding the trajectory of community succession (Chase (2003) Oecologia, 136, 489–98).

Propagule pressure—A concept that encompasses variation in the quantity, quality, composition, and rate of supply of **alien** organisms resulting from the transport conditions and pathways between source and recipient regions (see also **colonization pressure**) (Lockwood *et al.* (2005) Trends in Ecology and Evolution, 5, 223–8).

Prospect theory—A theory in behavioural economics that replaces the classic utility theory that explains decision making based on the utility of different assets, with a probability theory that the value of an asset is assigned to the gains and losses from the customers' standing asset portfolio (Kahneman and Tversky (1974) Econometrica, 47, 263–91). It has been used for developing adaptive foraging theory with the foraging behaviours/decisions based on recent experiences of the forager (Zhang and Hui (2014) Animal Behaviour, 88, 13–19).

Random walk—An individual-based approach for animal movement, with the path portrayed as consecutive steps with randomly assigned turning angles.

Range expansion—The process whereby a species spreads into new areas (usually new regions, rather than local-scale movements) due to natural dispersal, though the expansion may be assisted by human-mediated changes to the environment. Differs from **invasion** in that human-mediated extra-range dispersal (i.e. across a biogeographical barrier) is not implicated; the concept can be applied to both **native** and **alien species**.

Range pinning—The confinement of a species to one environment due to **Allee effects** that reduce the chance for a limited number of colonizers to successfully establish in another environment.

Rapid evolution—**Invasive species** in their novel ranges are often under strong direction selection due to altered selection force, combined with strong founder effect, that could lead to rapid changes in functional traits in the **invasive species** or closely interacting **native species**, often within less than a hundred generations (e.g. Stuart *et al.* (2014) Science, 346, 463–6).

Reaction-diffusion model—Partial differential equations designed for modelling chemical reactions but which have been widely used in ecological modelling of spatial dynamics, with reaction representing population growth and diffusion dispersal.

Regime shift—Large, abrupt, and persistent changes in the structure and function of an ecosystem, due to either bifurcation from changing ecosystem processes or jumping basins of attraction due to a large disturbance.

Reid's paradox (of rapid plant migration)—The failure of ordinary reaction–diffusion equations, which assume a Gaussian tail to the dispersal curve, to reconstruct certain range expansions, notably the post-glacial rate of spread of trees. Only by invoking **long-distance dispersal** can such expansions be accurately reconstructed.

Residence time—The time since the **introduction** of a species to a region; since the introduction date is usually derived from post-hoc records and is likely inaccurate, the term *minimum residence time* is often used. The extent of invasion of **alien species** generally increases with increasing residence time as species have more time to fill their potential ranges.

Resident biota/organisms—Species that are present in a community, habitat or region at the time of **introduction** of an **alien species**. The pool of resident species includes both **native species** and **alien species** introduced previously. (See also **biotic resistance, novel ecosystems**).

Resilience—Capacity of a **social-ecological system** to maintain similar structure, functioning, and feedbacks despite shocks and perturbations.

Resource-enemy release hypothesis—The notion that fast-growing plant species adapted to high resource availability have weaker constitutive defences against enemies, and therefore incur relatively large costs when enemies are present. It is argued that these fast-growing species benefit most from **enemy release**, and that the two mechanisms can act in concert to cause **invasion**; this could explain both the strong effects of resource availability on **invasion** and the extraordinary success of some **alien species** (Blumenthal (2005) Science, 310, 243–4).

Risk assessment (RA)—The estimation of the quantitative or qualitative value of risk (the likelihood of an event occurring within a specified time frame and the consequences if it occurs). In the context of **invasion science**, RA is usually undertaken to evaluate the likelihood of the entry, establishment and spread of a species (intentionally or accidentally) in a given region, negotiating given barriers in the **naturalization–invasion continuum**, and the extent and severity of ecological, social, and economic **impacts**.

Robustness—The ability of an ecosystem to withstand species loss, often measured as the number of species removed so that the number of secondary extinctions could reach 50 per cent of the original species richness.

Settling-Down Hypothesis—Diminishing effect of stochasticity with residence time in structuring community assemblages. That is, species in older assemblages are expected to belong to distinct functional clusters, with species matching the habitat within the cluster. In contrast, more recent introductions should have a poorer match as many species are initially randomly introduced to sub-optimal clusters of sites.

Space-for-time substitution—Effect of community succession can be mirrored by examining communities at different stages of succession. Instead of comparing community before and after invasion, we could compare pristine and invaded communities next to each other. Also, the spatial structure of species could reveal its temporal trends and invasiveness.

Social-ecological system—An ecological system intricately linked with and affected by one or more social systems.

Spatial sorting—Strong dispersers will accumulate at the advancing range front and potentially also through assortative mating at the range front, pushing for the acceleration of range expansion.

Spatial synchrony—The covariance of time series of two geographically distinct populations could be explained by shared environmental forcing (Moran's effect), common natural enemies, and migration (metapopulations).

Species distribution models (SDM)—SDMs are numerical tools that combine observations of species occurrence or abundance with environmental estimates. They are used to gain ecological and evolutionary insights and to predict distributions across landscapes, sometimes requiring extrapolation in space and time.

Species turnover—Compositional changes of species assemblages across sites (or time), pairwise turnover is typically measured as beta diversity and multi-site turnover as zeta diversity.

Storage effect—A coexistence mechanism which tries to explain how such a wide variety of similar species are able to coexist within the same ecological community or guild through temporal and spatial variation in the environment.

Trait-mediated interaction—Using trait matching or trait difference to infer interaction strength.

Unified framework for biological invasions—A framework that reconciles and integrates key features of the most commonly used invasion frameworks into a single conceptual model that is applicable to all **biological invasions**. It combines previous stage-based and barrier models, and provides a terminology and categorization for populations at different points in the invasion process (Blackburn *et al.* (2011) Trends in Ecology and Evolution, 26, 333–9).

Wicked problems—Management problems where the cause-and-effect relationships between components, be they logistical components or stakeholders involved in management, are unordered and thus have solutions that are not obvious and require collaboration among stakeholders to determine appropriate actions.

Taxonomic index

General index

locators in **bold** refer to definitions